MicroRNA Advances and Application in Plant Biology

Plant Biology, Sustainability and Climate Change

MicroRNA Advances and Application in Plant Biology

Series editors

Azamal Husen

Muhammad Iqbal

Edited by

Rahul Datta

Department of Geology and Pedology, Faculty of Forestry and Wood Technology, Mendel University in Brno, Czech Republic

Meenakshi Sharma

Department of Botany and Life Sciences, Daulat Ram College, University of Delhi, Delhi, India

Sachidanand Singh

Department of Biotechnology, School of Energy Technology, Pandit Deendayal Energy University, Gandhinagar, Gujarat, India

ELSEVIER

ACADEMIC PRESS

An imprint of Elsevier

Academic Press is an imprint of Elsevier
125 London Wall, London EC2Y 5AS, United Kingdom
525 B Street, Suite 1650, San Diego, CA 92101, United States
50 Hampshire Street, 5th Floor, Cambridge, MA 02139, United States

Notices

Knowledge and best practice in this field are constantly changing. As new research and experience broaden our understanding, changes in research methods, professional practices, or medical treatment may become necessary.

Practitioners and researchers must always rely on their own experience and knowledge in evaluating and using any information, methods, compounds, or experiments described herein. In using such information or methods they should be mindful of their own safety and the safety of others, including parties for whom they have a professional responsibility.

To the fullest extent of the law, neither the Publisher nor the authors, contributors, or editors, assume any liability for any injury and/or damage to persons or property as a matter of products liability, negligence or otherwise, or from any use or operation of any methods, products, instructions, or ideas contained in the material herein.

ISBN: 978-0-443-21682-4

For information on all Academic Press publications visit our website at https://www.elsevier.com/books-and-journals

Publisher: Mara Conner
Acquisitions Editor: Nancy Maragioglio
Editorial Project Manager: Kyle Gravel
Production Project Manager: Erragounta Saibabu Rao
Cover Designer: Vicky Pearson Esser

Typeset by TNQ Technologies

Contents

Contributors

E. Abhishek
ICAR-Indian Agricultural Research Institute, New Delhi, India

Ankita Acharya
Department of Pharmaceutical Technology, JIS University, Agarpara, Kolkata, India

Kasturi Banerjee
Department of Bio-Sciences, School of Bio-Sciences & Technology (SBST), Vellore Institute of Technology (VIT), Vellore, Tamil Nadu, India

Megha Barot
Research and Development Cell, Parul University, Vadodara, Gujarat, India; Department of Environmental Science, Parul Institute of Applied Sciences, Parul University, Vadodara, Gujarat, India

Kotte Bhargava
ICAR-Indian Agricultural Research Institute, New Delhi, India

P. Bhavana
ICAR-Research Complex for Eastern Region, Farming Systems Research Centre Hill and Plateau Region, Ranchi, Jharkhand, India

Deblina Bhowmik
Institute of Chemical Technology Mumbai, Marathwada Campus, Jalna, Maharashtra, India

Sai Anisha Bhupathiraju
Department of Bio-Sciences, School of Bio-Sciences & Technology (SBST), Vellore Institute of Technology (VIT), Vellore, Tamil Nadu, India

Simona Čerteková
Slovak University of Agriculture in Nitra, Faculty of Agrobiology and Food Resources, Institute of Plant and Environmental Sciences, Nitra, Slovakia

Rituparna Chatterjee
Department of Pharmaceutical Technology, JIS University, Agarpara, Kolkata, India

Ranjanee Chattopadhayay
Amity Institute of Biotechnology, Amity University, Noida, Uttar Pradesh, India

Sanhita Chowdhury
Department of Entomology, Post Graduate College of Agriculture, Dr. Rajendra Prasad Central Agricultural University, Samastipur, Bihar, India

Rahul Datta
Department of Agrochemistry, Soil Science, Microbiology and Plant Nutrition, Faculty of AgriSciences, Mendel University in Brno, Brno, Czech Republic

Vinod Kumar Dubey
Department of Entomology, Post Graduate College of Agriculture, Dr. Rajendra Prasad Central Agricultural University, Samastipur, Bihar, India

Ritam Dutta
Department of Pharmaceutical Technology, JIS University, Agarpara, Kolkata, India

Rohan Dutta
Department of Bio-Sciences, School of Bio-Sciences & Technology (SBST), Vellore Institute of Technology (VIT), Vellore, Tamil Nadu, India

Md Faiyazuddin
School of Pharmacy, Al-Karim University, Katihar, Bihar, India

Amol D. Gholap
Department of Pharmaceutics, St. John Institute of Pharmacy and Research, Palghar, Maharashtra, India

Sonali A. Gholap
Department of Data Analytics, CRISIL House, Mumbai, Maharashtra, India

Shubhrima Ghosh
Trinity Translational Medicine Institute, School of Medicine, Trinity College Dublin, Dublin, Ireland; School of Biological, Health and Sports Sciences, Technological University Dublin, Dublin, Ireland

Neha Gochhwal
DPG Degree College, Gurugram, Haryana, India

Ľubomír Harenčár
Department of Molecular Biology and Biotechnology, Institute of Plant Genetics and Biotechnology, Plant Science and Biodiversity Centre, Slovak Academy of Sciences, Nitra, Slovakia

Navnath T. Hatvate
Institute of Chemical Technology Mumbai, Marathwada Campus, Jalna, Maharashtra, India

Tanvi Jain
Faculty of Biotechnology, Institute of Biosciences and Technology, Shri Ramswaroop Memorial University, Barabanki, Uttar Pradesh, India

Srabani Karmakar
Kingston College of Science, Barasat, Kolkata, India

S. Keshavkant
School of Studies in Biotechnology, Pt. Ravishankar Shukla University, Raipur, Chhattisgarh, India

Noopur Khare
Faculty of Biotechnology, Institute of Biosciences and Technology, Shri Ramswaroop Memorial University, Barabanki, Uttar Pradesh, India; Bhai Gurdas Institute of Engineering and Technology, Sangrur, Punjab, India

Pragati Khare
Department of Pharmacy, Bhagwant University, Ajmer, Rajasthan, India

Saloni Koli
Department of Bio-Sciences, School of Bio-Sciences & Technology (SBST), Vellore Institute of Technology (VIT), Vellore, Tamil Nadu, India

Amardeep Kour
Punjab Agricultural University, Regional Research Station, Faridkot, Punjab, India

G. Vamsi Krishna
Department of Plant Pathology, Agricultural College, ANGRAU, Guntur, Andhra Pradesh, India

Matúš Kučka
Slovak University of Agriculture in Nitra, Faculty of Agrobiology and Food Resources, Institute of Plant and Environmental Sciences, Nitra, Slovakia

Pawan Kumar
Department of Botany, Shivaji College, University of Delhi, New Delhi, India

Chanchal Kumari
Amity Institute of Biotechnology, Amity University, Noida, Uttar Pradesh, India

V. Prasanna Kumari
Department of Plant Pathology, Agricultural College, ANGRAU, Guntur, Andhra Pradesh, India

Kiran M. Londhe
Department of Botany, Dr. Babasaheb Ambedkar Marathwada University, Chhatrapati Sambhajinagar, Maharashtra, India

Indu Malik
DPG Degree College, Gurugram, Haryana, India

Mehak Manchanda
Amity Institute of Biotechnology, Amity University, Noida, Uttar Pradesh, India

Anita Garg Mangla
Department of Biochemistry, Daulat Ram College, University of Delhi, New Delhi, India

Anamika Mishra
Department of Bio-Sciences, School of Bio-Sciences & Technology (SBST), Vellore Institute of Technology (VIT), Vellore, Tamil Nadu, India

Arpan Modi
Axenic Agritech Pvt Ltd, Ahmedabad, Gujarat, India

Sagnik Nag
Jeffrey Cheah School of Medicine and Health Sciences, Monash University Malaysia, Selangor, Malaysia

Singh Chauhan Nagendra
Drug Testing Laboratory Avam Anusandhan Kendra, Raipur, Chhattisgarh, India

Anupam K. Pal
Institute of Pharmacy, Shri Ramswaroop Memorial University, Barabanki, Uttar Pradesh, India

Manju Pandey
Institute of Pharmacy, Shri Ramswaroop Memorial University, Barabanki, Uttar Pradesh, India

Shraddha Pandey
Institute of Pharmacy, Shri Ramswaroop Memorial University, Barabanki, Uttar Pradesh, India

Maravi Preeti
Department of Biotechnology, Bharti Vishwavidyalaya, Durg, Chhattisgarh, India

Arathi Radhakrishnan
Amity Institute of Biotechnology, Amity University, Noida, Uttar Pradesh, India

Katarína Ražná
Slovak University of Agriculture in Nitra, Faculty of Agrobiology and Food Resources, Institute of Plant and Environmental Sciences, Nitra, Slovakia

Lellapalli Rithesh
Department of Plant Pathology, Kerala Agricultural University, Thiruvanathapuram, Kerala, India

Satish Rojekar
Department of Pharmacological Sciences, Icahn School of Medicine at Mount Sinai, New York, NY, United States

Sanjay Kumar Sahoo
Department of Entomology, Post Graduate College of Agriculture, Dr. Rajendra Prasad Central Agricultural University, Samastipur, Bihar, India

Sampada D. Sawant
Department of Pharmaceutics, St. John Institute of Pharmacy and Research, Palghar, Maharashtra, India

Kumar Shantanu
Department of Botany, Deshbandhu College, University of Delhi, New Delhi, India

Meenakshi Sharma
Department of Botany, Daulat Ram College, University of Delhi, New Delhi, India; Daulat Ram College, University of Delhi, Delhi, India

Neha Shekhawat
DPG Degree College, Gurugram, Haryana, India

Nistala Shweta
Department of Biotechnology, Bharti Vishwavidyalaya, Durg, Chhattisgarh, India

Soumya Singh
Faculty of Biotechnology, Institute of Biosciences and Technology, Shri Ramswaroop Memorial University, Barabanki, Uttar Pradesh, India

Sachidanand Singh
Department of Biotechnology, School of Energy and Technology, Pandit Deendayal Energy University, Gandhinagar, Gujarat, India

Apurwa Singh
Institute of Pharmacy, Shri Ramswaroop Memorial University, Barabanki, Uttar Pradesh, India

Riddhima Singh
Amity Institute of Biotechnology, Amity University, Noida, Uttar Pradesh, India

Satnam Singh
Punjab Agricultural University, Regional Research Station, Faridkot, Punjab, India

Sanavar Soham
Department of Botany, Kalindi College, University of Delhi, New Delhi, India

Rajpal Srivastav
Department of Science & Technology, Ministry of Science & Technology, New Delhi, India; Amity Institute of Biotechnology, Amity University, Noida, Uttar Pradesh, India

Shikha Srivastava
Institute of Pharmacy, Shri Ramswaroop Memorial University, Barabanki, Uttar Pradesh, India

Garima Tripathi
Department of Bio-Sciences, School of Bio-Sciences & Technology (SBST), Vellore Institute of Technology (VIT), Vellore, Tamil Nadu, India

P. Kishore Varma
Department of Plant Pathology, RARS, ANGRAU, Guntur, Andhra Pradesh, India

CHAPTER

1 Plant microRNAs: An overview

Meenakshi Sharma[1], Rahul Datta[2], Sachidanand Singh[3], Anita Garg Mangla[4], Rajpal Srivastav[5]

[1]*Department of Botany, Daulat Ram College, University of Delhi, New Delhi, India;* [2]*Department of Agrochemistry, Soil Science, Microbiology and Plant Nutrition, Faculty of AgriSciences, Mendel University in Brno, Brno, Czech Republic;* [3]*Department of Biotechnology, School of Energy Technology, Pandit Deendayal Energy University, Gandhinagar, Gujarat, India;* [4]*Department of Biochemistry, Daulat Ram College, University of Delhi, New Delhi, India;* [5]*Amity Institute of Biotechnology, Amity University, Noida, Uttar Pradesh, India*

1. Introduction to MicroRNAs (miRNAs)

The cell, as the fundamental unit in all living organisms, exhibits diverse types in plants, each characterized by specific structures such as leaves, stems, flowers, and seeds. While cells share common components such as the nucleus, chloroplast, and mitochondria, variations in their formation and function arise. Genomic DNA used in the nucleus encodes instructions for cell differentiation and development. Despite identical genomic DNA, gene silencing enables diverse cellular functions with MicroRNAs (miRNAs). miRNAs are small noncoding RNAs, playing a crucial role in ribo-regulation and influencing plant growth, and divers function at cellular levels.

In the early 1990s the first miRNA, lin-4, reported in *Caenorhabditis elegans* by Lee, Wightman, and Ambros. In 2002 Reinhart et al. conducted studies on miRNAs in the important model plant *Arabidopsis thaliana*, uncovering the presence of these small RNA molecules. Thousands of miRNAs have been identified across various organisms, including mammals, plants, animals, fungi, viruses and registered in public miRNA database (Kozomara and Griffiths-Jones, 2014). In addition to messenger RNA, various small noncoding RNAs, including miRNA, small interfering RNA (siRNA), PIWI-interacting RNAs (piRNAs), small interfering RNAs and transfer RNA-derived small RNAs (tsRNAs), contribute to the regulation of gene expression. Although all of these share short lengths and gene-silencing capabilities, their functions and production mechanisms differ. Eukaryotic small RNAs such as MicroRNAs (miRNAs) (Lagos-Quintana et al., 2001; Lau et al., 2001) short interfering RNAs (siRNAs) (Mello and Conte Jr, 2004), PIWI-interacting RNAs (piRNAs) (Aravin et al., 2007) and transfer RNA-derived small RNA represent four major categories of functional small RNAs. According to the current release on miRbase Expert database till January 2024, thousands of noncoding RNA sequences are

MicroRNA Advances and Application in Plant Biology. https://doi.org/10.1016/B978-0-443-21682-4.00007-5

available. ncRNA (50,767), sncRNA (50,767), small regulatory ncRNA (50,767), miRNA (50,439) piRNA (328), transcript (40,922) and pre-miRNA (40,921) are recorded. Previously noncoding DNA regarded as "junk DNA,", particularly introns, constituting over 98.5% of total DNA, has been found to play novel roles in post-transcriptional gene regulation through RNA interference in animals and plants. MiRNAs, approximately 20–35 nucleotides long, are integral to the endogenous RNA pathway, capable of activating gene expression under specific conditions and shuttling between subcellular compartments to regulate transcription and translation rates (Bartel, 2009; Dang et al., 2011).

MiRNA synthesis occurs in two stages: first, the protein DROSHA ("Double-stranded RNA-specific endoribonuclease") facilitates synthesis in the nucleus, and subsequently, miRNAs are exported to the cytoplasm. The protein DICER (RNase III family of endoribonucleases or helicase with RNase motif) further processes them, generating mature, single-stranded forms spanning 20–35 nucleotides. The Argonaut protein (AGO), forming the RNA-induced silencing complex (RISC) with mature miRNA, engages in gene regulation. AGO assists in aligning the miRNA and creating a roadblock in mRNA, inhibiting the translation process.

In recent years, various types of small RNAs exhibiting distinct characteristics have consistently emerged, leading to their classification based on origins, binding proteins, or secondary structures. Examples include trans-acting siRNAs (tasiRNAs) (Vazquez et al., 2004), repeat-associated siRNAs (rasiRNAs) (Reinhart et al., 2002), natural antisense-siRNAs (natsiRNAs) (Borsani et al., 2005), tiny noncoding RNAs (tncRNAs) (Ambros et al., 2003), heterochromatic small RNAs (hcRNAs) (Zilberman et al., 2003), and small scan RNAs (scnRNAs) (Taverna et al., 2002). The identification and diversification of conserved miRNA families across diverse plant species underscores the evolutionary significance of these small RNAs and multifaceted roles in cellular processes.

1.1 MicroRNAs (miRNAs)

miRNAs are short (20–25 nucleotides) RNA molecules and derived from the double-stranded region of a 60–70 nt RNA hairpin precursor, that regulate gene expression post-transcriptionally.

1.2 Small interfering RNAs (siRNAs)

siRNAs originating from exogenous sources such as viral infections or endogenous elements such as transposons and repetitive DNA undergo biogenesis primarily from long double-stranded RNAs (dsRNAs) through the action of the enzyme Dicer. Endogenous duplex small interfering RNAs (siRNAs) are incorporated into a protein complex, culminating in the assembly of the RNA-induced silencing complex (RISC). The unwinding of the siRNA duplex is facilitated by ATP, leading to the activation of the RISC complex. Guided by the antisense strand of the siRNA, the active RISC complex selectively targets and initiates the degradation of complementary mRNA molecules. This coordinated process highlights the regulatory function of native siRNAs in post-transcriptional gene silencing, adding complexity to the RNA interference machinery in cellular pathways. SiRNAs regulate gene expression at both Transcriptional Gene Silencing (TGS) and Post-transcriptional Gene Silencing (PTGS) levels. Studies indicate that small-RNA silencing pathways influence DNA and chromatin levels in their effects (Moazed, 2009).

1.3 PIWI-interacting RNAs (piRNAs)

PIWI-interacting RNAs (piRNAs) originate from discrete genomic loci known as piRNA clusters, predominantly present in germline cells. Precursors to piRNAs are long single-stranded RNAs, undergoing processing facilitated by PIWI proteins and associated factors. This processing results in the generation of mature piRNAs. Functionally, piRNAs are integral in the suppression of transposable elements, contributing significantly to the preservation of proper germline development and the maintenance of genome stability (Wahba et al., 2021).

1.4 Transfer RNA-derived small RNAs (tsRNAs)

Transfer RNA (tRNA) serves as the precursor molecule for tsRNA. The biogenesis of tsRNA involves multiple pathways, including the cleavage of mature tRNAs or the processing of pre-tRNAs. tsRNA is implicated in various cellular mechanisms, such as translation regulation, mRNA stability, rRNA synthesis, stress responses, RNA reverse transcription, migration, cell cycle, apoptosis etc (Zong et al., 2021). Plant miRNAs play diverse and crucial roles in regulating various aspects of plant biological processes such as root and shoot development, leaf morphology, control of flowering time, and modulation of hormone signaling pathways. Beyond their contributions to normal plant physiology, miRNAs actively involved in responding to stress conditions both abiotic stresses (drought, salinity, and extreme temperatures) and biotic stresses such as pathogens attack.

Several investigation of microRNAs (miRNAs) in both monocotyledonous (monocot) and dicotyledonous (dicot) plants has yielded valuable insights in regulatory networks that govern plant biology. Specifically, in monocot plants such as rice (*Oryza sativa*), wheat (*Triticum aestivum* L.), maize (*Zea mays*), barley (*Hordeum vulgare*), sugarcane (*Saccharum officinarum* L.) and others (Chen et al., 2021). The recent studies on miRNA enhances our understanding of the molecular mechanisms underlying key aspects of plant growth and development in monocotyledonous species (Song et al., 2019).

In the context of monocot plants such as rice, miR156 has emerged as a key regulator influencing the transition from the vegetative phase to the reproductive phase. This is achieved by targeting genes associated with the control of flowering time. Another miRNA, miR167, contributes to the regulation of root development in rice by targeting auxin response factors. In maize, miR156 and miR172 play crucial roles in controlling flowering time and the development of reproductive structures, including tassels and ears. Additionally, miR396 has been implicated in the regulation of leaf development and nutrient homeostasis in maize. This miRNA targets growth-regulating factors (*GRFs*) and phosphatase 2C (*PP2C*) genes, influencing leaf morphogenesis and nutrient balance, respectively. The exploration of miRNA functions in monocot plants unveils sophisticated regulatory mechanisms governing essential processes such as seed germination, root development, leaf morphogenesis, flower development, and grain filling.

The biogenesis of microRNAs (miRNAs) in dicotyledonous (dicot) plants adheres to a conserved pathway, extensively investigated across various species, including rockcress (*Arabidopsis thaliana*), soybean (*Glycine max*), tomato (*Lycopersicon esculentum*), cotton (*Gossypium herbaceum*), Medicago (*Medicago sativa* L.), Miyakogusa (*Lotus japonicus*) (Narjala et al., 2020; Chithung et al., 2023). In Arabidopsis, the miRNA miR156 plays a regulatory role in developmental transitions, influencing flowering time control, leaf development, and phase change. Another miRNA, miR172, influences floral organ identity and the transition from vegetative to reproductive phases. miR398 responds to oxidative stress, modulating the expression of the copper/zinc superoxide dismutase (*CSD1*) gene, crucial for scavenging reactive oxygen species. In soybean, miR159 governs flowering time and seed development, while miR164 and miR166 contribute to leaf development. Tomato exhibits miRNAs such as miR156, miR172, and miR319, which regulate fruit ripening and size. Cotton miRNAs, including miR156, miR164, and miR395, respond to abiotic stresses such as drought and salinity. The study of dicot miRNAs has significantly enriched our understanding of the molecular mechanisms that underlie plant development and responses to environmental conditions (Jones-Rhoades and Bartel, 2004; Song et al., 2019).

Algal microRNAs (miRNAs) have been identified and extensively studied across diverse species within algal groups, including *Chlamydomonas reinhardtii*. Various miRNAs have been acknowledged for regulating different aspects of the alga's physiology, contributing to our understanding of molecular control within algal systems (Voshall et al., 2017). Algal miRNAs participate in various processes, including growth, development, stress responses, and nutrient metabolism. miR414 is involved in responding to nutrient availability, miR912 is linked to flagellar motility, miR169 is implicated in nitrogen deprivation responses, and miR408 is associated with copper homeostasis. Similarly, miRNAs have been identified in other algae, encompassing red algae such as *Porphyra yezoensis* and diatoms such as *Thalassiosira pseudonana* (Gao et al., 2021). Although the exploration of algae miRNAs is in its early stages, further research is necessary to comprehensively understand their functions and regulatory networks across different algal species. Exploration of algal miRNAs offers valuable insights with broad applications, spanning biofuel production, bioremediation, and comprehension of environmental impacts on algal ecosystems.

In 2010, miRNA-like RNA (milRNA) also reported in fungi such as *Neurospora crassa* and later reported in yeast *Saccharomyces cerevisiae*, *Fusarium oxysporum*,

Cryptococcus neoformans, *Metarhizium anisopliae*, *Trichoderma reesei*, *Penicillium marneffei*, and *Sclerotinia sclerotiorum*, although the miRNA pathway and aspects of gene expression control in filamentous species poorly understood across the majority of fungal species (Chen et al., 2014).

MicroRNAs (miRNAs) in mushrooms have gained attention for their regulatory roles in various biological processes. Po-MilR-1 is a novel microRNA with a length of 22 bp in oyster mushroom (*Pleurotus ostreatus*) Examples include miR159, involved in fungal development, and miR166, influencing hyphal growth. Understanding mushroom miRNAs and its regulatory role will help in medicine and food technology (Xu et al., 2021; Marin et al., 2022; Qi et al., 2023).

Manipulation of microRNA (miRNA) expression stands as a compelling strategy for enhancing crucial traits in crops such as stress tolerance, yield, nutrient content, quality, and adaptation to environmental challenges (Raza et al., 2023). Illustratively, in rice, the overexpression of miR393 has been demonstrated to enhance stress tolerance by regulating auxin signaling. In maize, the manipulation of miR164 influences grain yield by affecting meristem activity. Furthermore, targeted alteration of miR399 in Arabidopsis has been shown to enhance phosphate uptake, leading to improved nutrient efficiency. In tomatoes, the manipulation of miR156 has implications for fruit quality attributes such as size and ripening. These examples underscore the potential of miRNA manipulation in shaping diverse crop traits, establishing a promising frontier for precision agriculture and sustainable food production amid changing environmental conditions. It is crucial to note that further research in this field holds the key to unlocking the full potential of miRNA-based crop improvement strategies. A deeper understanding of the molecular mechanisms and broader applicability of miRNA manipulation in diverse crops will contribute to the development of more targeted and effective approaches for crop enhancement (Raza et al., 2023).

2. miRNA biogenesis

MicroRNA (miRNA) biogenesis in plants involves a series of steps that lead to the maturation of functional miRNAs (Fig. 1.1). The process generally comprises three main steps: transcription of miRNA genes, processing of pri-miRNAs and maturation (Song et al., 2019; Wang et al., 2019; Xu and Chen, 2023).

2.1 Transcription of plant miRNA genes

MiRNAs, originating from specific genes called miRNA genes and processed from longer RNA precursors known as primary miRNAs (pri-miRNAs). Plant miRNA genes can be dispersed throughout the genome or clustered together in specific regions. In some cases, miRNA clusters contain multiple miRNA genes that are transcribed together and produce a group of related miRNAs. Clustering of miRNA genes can provide coordinated regulation of gene expression and functional

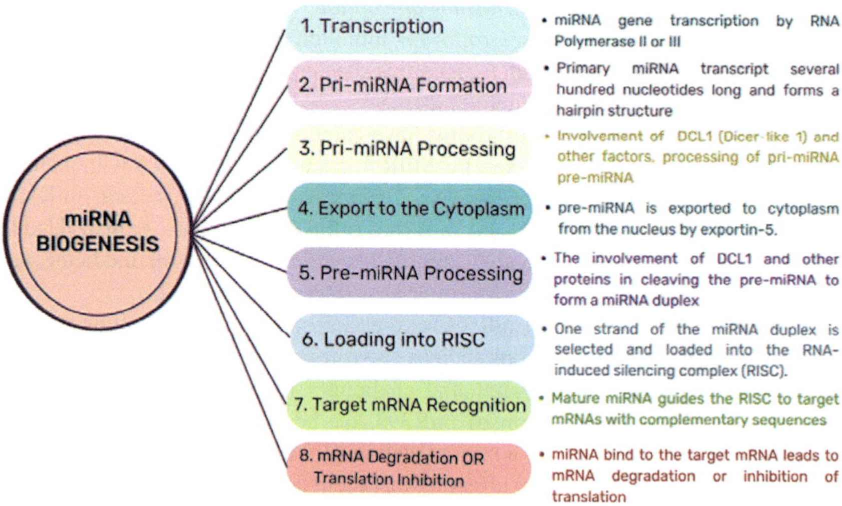

FIGURE 1.1

A general representation of the MicroRNA biogenesis.

interactions among miRNAs. Different plant genomes contain a large number of miRNA genes, contributing to the diversity of small miRNAs. several hundred miRNA genes have been reported in *Arabidopsis thaliana,* rice (*Oryza sativa*) and maize (*Zea mays*). Plant miRNA genes have evolved through gene duplication, genomic rearrangements, and sequence divergence, expanding miRNA gene families. These families consist of multiple genes encoding related miRNAs with similar functions. Despite conserved features, sequence variations within miRNA families across plant species can lead to functional diversification. This allows different miRNA family members to target specific mRNA transcripts or regulate distinct biological processes. Most of the MIR genes reside in the noncoding regions and transcribed independently. Some of the MIR such as MIR400 and MIR402 are found within introns of protein-coding or noncoding genes. Such genes are co-transcribed with their host genes by RNA polymerase II and regulated by RNA splicing. Both cis- and trans-regulatory modules can regulate the transcription of MIR genes in plants. Example: The TATA box located at upstream of the transcription start sites of MIR genes and show cis-regulatory modules. The nomenclature of miRNAs in plants follows a standardized naming convention that provides information about their origin, family, and evolutionary relationships (Yehu et al., 2017). The nomenclature is based on the organism's scientific name, followed by a unique numerical identifier and a letter suffix for members of the same family. The miRNA name begins with the three-letter prefix indicating the scientific name of the organism. For example: "osa" for *Oryza sativa* (rice): osa-miR156, the 156th identified miRNA in Oryza sativa, osa-miR156a and osa-miR156b represent two different members of the miR156 family in Oryza sativa.

The molecular structure of plant miRNA genes usually consist of a transcriptional unit that includes a promoter region, a transcribed region, and a termination sequence. The transcribed region contains the sequence for the pri-miRNA, which is the primary transcript of the miRNA gene. The promoter region of plant miRNA genes controls the expression of the miRNA gene. It contains regulatory elements that determine when and where the miRNA gene is transcribed. Promoters can be specific to individual miRNA genes or shared among a cluster of miRNA genes (Liu et al., 2017).

During MicroRNA (miRNA) biogenesis in plants, Firstly a long primary miRNA is produce through the intergenic miRNA genes transcription by pol II or in some cases pol III resulting in the production of primary miRNA transcripts (pri-miRNAs). During miRNA gene transcription through RNA polymerase II (Pol II)-and transcription factors (TFs such as MYC2, ARF, SORLREP3, and LFY) play a crucial role in the transcription of MicroRNA (miRNA) genes, in-regulatory network governing gene expression (Xu and Chen, 2023). In the context of miRNA transcription, TFs bind to specific DNA sequences in the promoter regions of miRNA genes, either enhancing or repressing their transcription. This interaction influences the recruitment of RNA polymerase II and associated transcriptional machinery, modulating the production of primary miRNA transcripts. Some of the factors such as Mediators, Elongators, ILP1, NTR1, NOT2, CDC5, PP4R3A, TREX-2, SMA1, STV1, CHR2/BRM, PIF4, RBV regulate the MIR gene transcription positively (Li and Yu, 2021). The negative factors are PIFs, SNC1, TPR1 regulate the MIR gene transcription negatively. The combinatorial action of TFs establishes a finely tuned regulatory environment, allowing for precise control over miRNA expression levels. Understanding the interplay between transcription factors and miRNA gene transcription is essential for deciphering the complexities of gene regulatory networks (Samad et al., 2017).

After the initiation, DNA used as template strand and complementary synthesized by adding nucleotides through RNA polymerase II to synthesizes the pri-miRNA. The transcription of miRNA genes in plants is a complex and tightly regulated process and extensively controlled by RNA Pol II, associated factors and epigenetic regulators. The C-terminal domain (CTD) of largest subunit of Pol II is phosphorylated with the help of CYCLIN-DEPENDENT KINASE F; 1 (CDKF; 1) and CYCLIN-DEPENDENT KINASE Ds (CDKDs), which promotes pri-miRNA 5′ capping and 3′ polyadenylation. The pri-miRNAs cotranscriptionally processed into miRNA duplexes due to imperfect folding by DICER-LIKE (DCL) proteins. These pri-miRNAs typically contain one or more double-stranded hairpin structures due to the presence of stem-loop regions with single-stranded RNA extension at both the ends and it serve as precursors for mature miRNAs (Lee et al., 2004). Research has shown that certain human miRNAs are transcribed by RNA polymerase III (Borchert et al., 2006). p53 and miR-34 cluster, MYC and miR-17 cluster, MYOD1 and miR-1 cluster. *(p53 is a tumor suppressor protein that plays a crucial role in regulating the cell cycle, DNA repair, and apoptosis. p53 activates the expression of the miR-34 family, which includes miR-34a, miR-34b, and miR-34c.*

Furthermore, MYC is a proto-oncogene that regulates cell proliferation, growth, and differentiation. MYC is known to regulate the expression of the miR-17-92 cluster, which includes miR-17, miR-18a, miR-19a, miR-20a, miR-19b-1, and miR-92a-1. MYOD1 is a transcription factor involved in the regulation of muscle development and differentiation) (Shen et al., 2019).

One characteristic feature of pri-miRNAs is that they often form hairpin structures. This hairpin structure is a result of the secondary structure formed by the self-complementary base pairing within the pri-miRNA molecule. The pri-miRNAs, which consist appropriate stem length, $\geq$10 bp long terminal loop and single-stranded RNA extension at both the ends are most efficient to underwent in processing to form mature and functional miRNA. During transcription, the pri-miRNA transcripts undergo modifications similar to protein-coding mRNA transcripts, including the addition of a 5′ cap and a polyadenylated (poly-A) tail at the 3′ end. The 5′ cap is a modified guanosine nucleotide that is added to the 5′ end of the RNA transcript. The 5′ cap structure protects the RNA from degradation and is important for efficient translation of protein-coding mRNAs. The poly-A tail is a sequence of adenine nucleotides added to the 3′ end of the RNA transcript. The poly-A tail and 5′ capping contributes to mRNA stability, processing and facilitates nuclear export, and is involved in the initiation of translation for protein-coding mRNAs (Song et al., 2019; Xu and Chen, 2023).

2.1.1 Polycistronic miRNAs

MiRNAs can contribute to polyploidization through processes such as whole genome duplication, segmental duplication of subchromosomal regions, or tandem duplications involving a limited number of genes. In rice, various members of miRNA families are situated in duplicated genomic regions, as documented in studies (Barik et al., 2015). Most of the miRNAs in plants are monocistronic (miRNAs transcribed as single units and give rise to a single mature miRNA) whereas, some miRNAs are polycistronic (miRNAs are transcribed as a single pri-miRNA that can give rise to multiple distinct mature miRNAs). Polycistronic arrangement allows for the coordinated regulation of multiple target genes from a single primary transcript. The presence of polycistronic miRNAs has been identified in various plant species such as *Arabidopsis thaliana*, *Oryza sativa*, and *Populus trichocarpa*. The One very well-known example is the miR-169 family in *Arabidopsis thaliana*. The pri-miR-169 transcript contains three distinct miRNA sequences, each of which targets a different member of the NF-YA transcription factor family. The polycistronic nature of certain miRNAs adds an additional layer of complexity to the regulatory networks in plants. It enables coordinated regulation of multiple genes involved in specific biological processes or pathways. Understanding the functions and regulatory mechanisms of polycistronic miRNAs contributes to our knowledge of plant gene expression and development.

Polycistronic miRNAs are classified into two categories: Homologous and Nonhomologous polycistronic, as delineated by Baldrich et al. (2016). Homologous polycistronic microRNAs (miRNAs Transcribed as single units, yield multiple mature miRNAs that share sequence similarities). This unique arrangement enables synchronized control of homologous target genes within a single transcriptional unit. For instance, in *Arabidopsis thaliana*, miR-169 comprises a polycistronic family targeting distinct NF-YA transcription factors. The functional significance of homologous polycistronic miRNAs contributes to complex plant gene expression regulatory networks. This multifaceted relationship between miRNAs and genomic duplications sheds light on the diverse mechanisms influencing miRNA evolution and function in plants. Nonhomologous polycistronic microRNAs (miRNAs originating from a single primary transcript, these miRNAs generate diverse mature sequences with distinct target specificities). Unlike homologous, nonhomologous polycistronic miRNAs regulate unrelated genes and simultaneously control multiple cellular processes. The complex nonhomologous polycistronic miRNAs arrangement exemplifies the post-transcriptional gene regulation in plants. The well reported examples are miR-172/miR-165 family in Arabidopsis, which target the APETALA2 and Class III homeodomain-leucine zipper genes, respectively. The miR-156/miR-157 family in *Arabidopsis thaliana*, producing miRNAs that regulate SPL transcription factors involved in diverse developmental processes. This nonhomologous polycistronic arrangement underscores the adaptability of plants to modulate multiple, functionally unrelated pathways simultaneously. Deciphering the functional implications of nonhomologous polycistronic miRNAs enhances the understanding of plant gene networks and their adaptive responses.

2.2 Stabilization of pri-miRNAs in plants

The stabilization of primary microRNA transcripts (pri-miRNAs) is a crucial step microRNA (miRNA) biogenesis and for ensuring the efficient production of mature miRNAs. Various factors contribute to pri miRNA stabilization, including RNA-binding proteins, secondary structure elements, and specific sequence motifs within the transcript. Once the pri-miRNAs are transcribed, they undergo modifications such as the addition of a 5′ cap and a 3′ poly(A) tail, which is crucial for enhancing the stability of primary microRNA transcripts (pri-miRNAs). Such modifications give protection to pri-miRNAs from nuclear RNA exosome-mediated degradation during transcription. The 5′ cap structure prevents exonucleolytic degradation from the 5′ end, while the polyadenylation at the 3′ end provides resistance against 3′ to 5′ exonucleolytic degradation.

This cooperative process ensures the stability of pri-miRNAs, facilitating their seamless transition through processing stages to generate mature and functional microRNAs. The coordination of these protective mechanisms underscores the sophistication of the pri-miRNA biogenesis pathway in preserving the accuracy of gene regulatory processes in plants.

The methyltransferase HEN1, which adds a methyl group to the 3′ end of miRNAs, also participates in pri-miRNA stabilization. This methylation protects the miRNA duplex from 3′ to 5′ exonucleolytic degradation, ensuring the stability of the pri-miRNA precursor. RNA-binding proteins, exemplified by HYL1

(HYPONASTIC LEAVES 1) and DDL (DUAL FUNCTION LIGAND), play a pivotal role in safeguarding primary microRNA transcripts (pri-miRNAs) from degradation during biogenesis. These proteins bind to pri-miRNAs, stabilizing them and preventing their premature degradation. In Arabidopsis, HYL1 interacts with pri-miRNAs, aiding in their structural stabilization and ensuring efficient processing into precursor miRNAs. Similarly, DDL contributes to pri-miRNA stability, further protecting these transcripts from degradation (Wang et al., 2018). The collaborative efforts of RNA-binding proteins such as HYL1 and DDL emphasize their pivotal role in preserving the integrity of pri-miRNAs, thereby aiding the subsequent stages of miRNA maturation and regulatory functions. In a separate investigation, the depletion of SUPPRESSORS OF THE PAS2 1 (SOP1), another component of the nuclear exosome that colocalizes with HEN2, was observed to have no impact on pri-miRNA accumulation (Xu and Chen, 2023). HEN2, an RNA-binding protein, adds a methyl group to the 3′ end of pri-miRNAs, preventing exonucleolytic degradation. The RNA-binding protein DAWDLE (DDL) and MAC (MORPHOGENESIS, ARRESTED CELL DIVISION, AND DEATH) subunits, including PLEIOTROPIC REGULATORY LOCUS1 (PRL1), MAC3, and MAC5A, collaboratively contribute to pri-miRNA stability in plants. Meanwhile, the MAC subunits, especially PRL1, MAC3, and MAC5A, are involved in the formation of the MAC complex, which further aids in protecting pri-miRNAs from premature degradation (Bajczyk et al., 2023). This cooperative action highlights the complex involvement of RNA-binding proteins and MAC subunits in ensuring the robust stability of primary microRNA transcripts in plants.

2.3 Pr-miRNA processing or DROSA-DGCR8 complex processing

Processing of pri-miRNAs, also known as DROSHA-DGCR8 complex processing, occurs in the nucleus. The Microprocessor complex, comprising the RNase III enzyme DROSHA and its cofactor DGCR8 (DiGeorge syndrome critical region gene 8), is instrumental in this process. Within the nucleus, various splicing factors and lariat RNAs tightly regulate pri-miRNA processing. Specific sequences within the stem of pri-miRNA are released by endoribonucleases and cleaved by the enzyme DROSHA. This cleavage takes place near the base of the stem-loop structure, resulting in the formation of a hairpin-shaped structure called precursor miRNA (pre-miRNA). The pri-miRNA, approximately 60−70 nucleotides in length after cleavage, undergoes processing by two main protein complexes: DROSHA and DGCR8 in animals, and Dicer-like 1 (DCL1) (Fig. 1.2A) and HYL1 in plants. DGCR8 plays a crucial role in recognizing the stem-loop structure within the pri-miRNA transcript. This interaction is essential for recruiting DROSHA to the pri-miRNA, ensuring accurate positioning for cleavage. The DROSHA nuclease, in conjunction with the RNA-binding protein DGCR8 in the microprocessor complex, carries out endoribonucleolytic cleavage activity, cleaving the long terminal regions of the 5′ and 3′ ends of pri-miRNA.

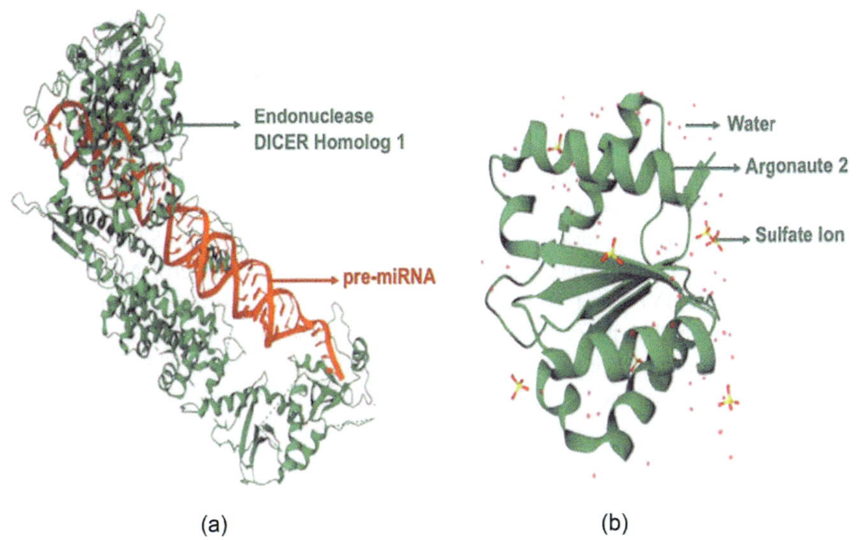

FIGURE 1.2

(A) Cryo-EM structure of *Arabidopsis thaliana* DCL1 in complex with pre-miRNA; (B) crystal structure of *Arabidopsis thaliana* AGO2 mid domain.

Adopted from pdb database (A) https://www.rcsb.org/3d-view/7ELE/1; (B) https://doi.org/10.2210/pdb4G0M/pdb.

The result of this process is the generation of pre-miRNA, a hairpin-shaped structure with approximately 2-nt overhangs at the 3′ ends. The pre-miRNA is then exported from the nucleus to the cytoplasm by the exportin-5/Ran-GTP complex. This intricate series of events highlights the sequential and highly regulated nature of pri-miRNA processing, ultimately leading to the production of mature miRNAs with important regulatory functions. The pri-miRNA is further recognized and processed by the dicing complex into a miRNA/miRNA* duplex with 2-nt overhangs at the 3′ ends in the nucleus to generate precursor miRNAs (pre-miRNAs). This pre-miRNA is then exported from the nucleus to the cytoplasm by the exportin-5/Ran-GTP complex (Bajczyk et al., 2023; Xu and Chen, 2023).

2.3.1 Dicing complex assembly

Following the initial cleavage of pri-miRNAs by the DROSHA-DGCR8 complex in the nucleus, pre-miRNAs are transported to the cytoplasm. Once in the cytoplasm, the dicing complex, composed of proteins such as Dicer and associated cofactors, takes center stage in the miRNA maturation pathway. Dicer, an RNase III endonuclease, is the central component of the dicing complex responsible for cleaving pre-miRNAs into short double-stranded RNA duplexes. These duplexes consist of the mature miRNA strand (guide strand) and its complementary strand (passenger strand). These dsRNA molecules serve as substrates for Dicer-like (DCL) proteins.

Additionally, accessory proteins, such as TRBP (TAR RNA-binding protein), assist in stabilizing the interaction between Dicer and the pre-miRNA, facilitating accurate and efficient processing.

The integral components of the Dicing complex are DCL proteins collaborate with accessory proteins, including Hyponastic Leaves 1 (HYL1) and Serrate (SE), to recognize and cleave precursor miRNAs into miRNA/miRNA* duplexes. This duplex formation is crucial for generating the mature miRNA that guides post-transcriptional gene regulation. The absence of HYL1 (HYPONASTIC LEAVES 1) or SE (SERRATE) in plants leads to a significant impact on microRNA (miRNA) biogenesis. Both HYL1 and SE are key components of the microprocessor complex involved in pri-miRNA processing. When either of these factors is lacking, there is a reduction in the global accumulation of mature miRNAs. Additionally, the levels of pri-miRNAs increase, indicating a crucial role for HYL1 and SE in promoting efficient processing of primary microRNA transcripts and ensuring the production of functional, mature miRNAs in plants.

2.3.2 Regulation of pri-miRNA processing

The regulation of pri-miRNA processing involves multiple factors in a highly organized manner. MAC (MORPHOGENESIS, ARRESTED CELL DIVISION, AND DEATH) subunits, including MAC7 and MAC3, actively participate in assembling the Dicing complex, collaborating with essential components such as HYL1 (HYPONASTIC LEAVES 1). MAC7 and MAC3's cooperative action emphasizes their crucial role in ensuring precise assembly of the Dicing complex for accurate processing of primary microRNA transcripts in plants. RNA helicases such as TOUGH (TGH) aid in unwinding pri-miRNA secondary structures, facilitating subsequent cleavage (Li and Yu, 2021). One example is the MOS4 (MODIFIER OF SNC1,4) and associated complex (MAC) in Arabidopsis, binds with the pri-miRNA precursor, influencing its secondary structure. This complex recruits additional factors such as MAC3 and MAC7, ensuring proper Dicing complex assembly. Additionally, the THO/TREX complex participates in pri-miRNA regulation. It aids in nuclear export of pri-miRNAs, facilitating their processing in the cytoplasm. THO/TREX subunits, such as MOS11, contribute to the coordination of nuclear events during miRNA biogenesis. This network of regulatory steps exemplified by MOS, THO/TREX, and associated complexes underscores the complexity of pri-miRNA processing, highlighting the sophisticated molecular machinery involved in plant gene regulation (Jia et al., 2017). These examples underscore the collaborative action of various factors in the regulation of pri-miRNA processing, ensuring the accurate production of mature and functional microRNAs in plants during miRNA biogenesis. The assembly is highly regulated, involving recognition of pre-miRNA secondary structures and proper loading onto Argonaute proteins. In case of Arabidopsis, DCL1 interacts with HYL1 and SE and plays a central role to cleave precursor miRNAs into miRNA/miRNA* duplexes. Similarly, in rice, OsDCL1 is essential for miRNA biogenesis (Bajczyk et al., 2023; Xu and Chen, 2023).

2.3.3 Secondary structures of pri-miRNAs

The biogenesis of microRNAs (miRNAs) in cells is significantly influenced by the secondary structures of primary microRNA transcripts (pri-miRNAs). The secondary structures of pri-miRNAs increase their stability and accessibility, thus influencing the efficiency of subsequent processing steps and ensuring the production of functional mature miRNAs in plants. Pri-miRNAs commonly adopt stem-loop structures that contain the mature miRNA sequence. This structural arrangement is crucial for the recognition and processing of pri-miRNAs by the microprocessor complex. During microRNA (miRNA) biogenesis, the Microprocessor complex (MC) consists of at least two essential components: DROSHA, an RNaseIII enzyme, and DGCR8, a double-stranded RNA-binding protein. This complex functions by cleaving the primary-miRNA (pri-miRNA), resulting in the liberation of the pre-miRNA stem-loop structure (Connerty et al., 2015). For example, in Arabidopsis, pri-miR156 exhibits a characteristic stem-loop structure, facilitating precise cleavage by Dicer-like 1 (DCL1), during miRNA maturation. The GC bias and specific signatures within pri-miRNAs significantly impact the precision of primary microRNA (pri-miRNA) processing and the subsequent abundance of miRNAs in plants. The GC content in the genome affects the stability of pri-miRNA secondary structures, influencing the accuracy of microprocessor complex recognition and cleavage sites during miRNA maturation. Signature motifs within pri-miRNAs contribute to recognition specificity. The presence of alternative secondary structures in miRNA precursors, characterized by a terminal branched loop, plays a crucial role in pri-miRNA processing and influences their biogenesis. For example, pri-miR165 and pri-miR166 in plant development exhibit a distinctive terminal branched loop structure, recognized by the dicing complex, leading to bi-directional processing. This alteration in recognition pattern influences the precision of cleavage, generating two mature miRNAs from opposite arms of the hairpin. Numerous pri-miRNAs contain introns, and specific introns within the stem-loop region can disrupt the proper secondary structure of pri-miRNAs. To preserve structural integrity, these introns must be removed through splicing. in Arabidopsis, pri-miR163 has an intron within the stem-loop structure that affects miRNA processing. Splicing of this intron is crucial for the correct folding of the pri-miRNA, ensuring precise recognition and processing by the microprocessor complex. Protein factors, particularly RNA-binding proteins such as DAWDLE (DDL) in Arabidopsis, play a crucial role in regulating pri-miRNA processing. DDL interacts with pri-miRNAs, stabilizing them and influencing their secondary structure, which impacts processing mechanism. Several Protein factor play a role in modifying the stability and conformation of pri-miRNAs, influencing the microprocessor complex recognition sites and, consequently, impacting the effectiveness of pri-miRNA cleavage. The configurations of these RNA molecules, folding into stem-loop structures, serve as recognition sites for the microprocessor complex, initiating cleavage and subsequent generation of precursor miRNAs.

3. miRNA maturation

3.1 Pre-miRNAs export to the cytosol

Following nuclear processing by DROSHA, HASTY (HST) protein is responsible for the recognition and transportation of pre-miRNA/miRNA* duplexes from the nucleus to the cytoplasm. HASTY (HST) protein serves as the orthologous counterpart to exportin-5 (EXP5/XPO5) found in animals. In plants, during miRNA maturation export of miRNA/miRNA* duplexes takes place predominantly in the nucleus. Exportin- 5 structure is like a baseball mitt, which hold pre-miRNA using its basic internal surface. it also has a basic Tunnel-like structure at bottom, which specifically interacts with the 3′ overhang of the RNA. It has features that can specifically interact with the structure of pre-miRMA generated by DROSHA (Okada et al., 2009). RAN-GTP functions as a cofactor for exportin 5, facilitating the formation of the transport complex (Okada et al., 2009). CRM1/exportin 1(XPO1) is a general exportin that mediates the nuclear export of proteins in various cellular processes and reported these are involved in the export of AGO1-miRNA complexes from the nucleus to the cytoplasm. Specific miRNAs in *Arabidopsis thaliana* are exported to cytoplasm by Exportin-4 (XPO4) protein. The field of small RNA transport and processing is an active area of research, and new information may emerge over time (Xu and Chen, 2023). In the cytoplasm, the pre-miRNA is further processed by another enzyme called DICER, along with its associated proteins.

3.2 miRNA stabilization and RISC assembly

The miRNA/miRNA* duplex is generated through the cleavage of the primary microRNA transcript (pri-miRNA) by Dicer-like (DCL) proteins. This duplex, spanning approximately 21–23 nucleotides, exhibits protruding single-stranded ends of 2–3 nucleotides at the 3′ termini. Comprising the mature miRNA strand and its complementary counterpart, referred to as miRNA* or the passenger strand, the duplex undergoes further modifications such as 2′-O-methylation and 3′ ends polymerization facilitated by the nuclear HUA ENHANCER1 (HEN1) protein. This process accelerates miRNA turnover, preventing recognition and degradation by exonucleases. The interaction between HEN1, DCL1 (DICER-LIKE 1), and HYL1 (HYPONASTIC LEAVES 1) is pivotal in plant microRNA (miRNA) biogenesis. HEN1's 2′-O-methylation contributes to duplex stability, replacing SE (SERRATE) in the dicing complex after pri-miRNA processing by the microprocessor complex (DCL1, HYL1, and SE). This suggests a potential role for HEN1 in the later stages of miRNA biogenesis. In plant miRNA-mediated gene regulation, AGO1 plays a central role, with loaded miRNA guiding the RNA-induced silencing complex (RISC) (Fig. 1.3). The RISC loading complex, including DICER, Argonaute, and additional proteins such as TARBP2 (TAR RNA-binding protein 2) or PRKRA (Protein Kinase, Interferon-Inducible Double-Stranded RNA-Dependent Activator), participates in miRNA loading onto AGO. TARBP2 serves as a bridge

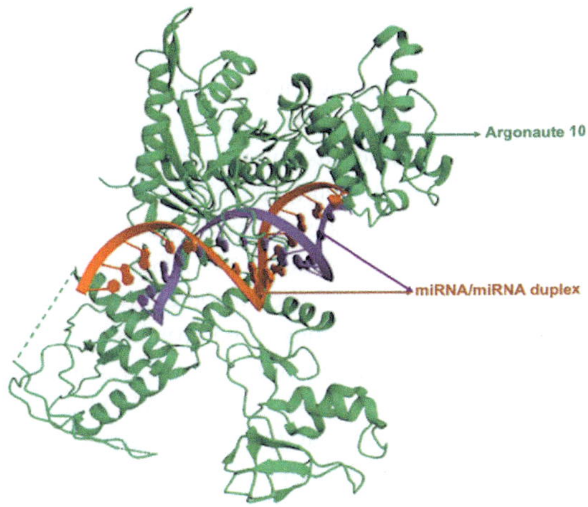

FIGURE 1.3

Cryo-EM structure of Arabidopsis Ago10-guide-target RNA complex in a central duplex conformation.

Modified from pdb database https://doi.org/10.2210/pdb7SWF/pdb.

between DICER and AGO, facilitating miRNA transfer, while PRKRA potentially enhances miRNA stability during the loading process (Xu and Chen, 2023).

3.3 miRNA strand selection and loading of miRNA onto AGO

MiRNA biogenesis comprises a sequence of steps, encompassing strand selection and incorporation into Argonaute (AGO) proteins. The selection process involves designating one strand of the miRNA duplex as the mature, functional miRNA, while the other, referred to as miRNA* (miRNA star), is typically degraded. This determination is crucial, shaping the role of the AGO protein in post-transcriptional gene regulation by guiding it to the target mRNA. The guide strand selection hinges on thermodynamic stability, particularly at the 5′ ends of the miRNA strands. The 5′ end of the guide strand often exhibits a thermodynamically more stable base pairing compared with the 5′ end of the passenger strand. The less thermodynamically stable strand at the 5′ end is preferentially retained as the guide strand, directing AGO-mediated silencing of target mRNAs. Asymmetry in base-pairing at the ends, characterized by mismatches or bulges, also influences the selection process. Complementary to the preference for stability at the 5′ end, the guide strand is also characterized by weaker base pairing at the 3′ end. The 3′ end of the guide strand tends to have less stable base pairing, facilitating efficient unwinding of the duplex during RISC loading.

Loading factors, exemplified by TRBP (TAR RNA-binding protein), and Argonaute proteins (e.g., AGO2) within the RNA-Induced Silencing Complex (RISC), actively contribute to the strand selection process. During miRNA biogenesis, within

RISC, helicase activity facilitates unwinding of the miRNA duplex. TRBP interacts with Dicer, aiding in presenting the miRNA duplex to AGO2 for integration into RISC. Recognition of structural and thermodynamic characteristics of the miRNA duplex informs unwinding, facilitating the transfer of the preferred strand (guide strand) into RISC, where it assumes its mature and functional miRNA state. In the well-studied Arabidopsis model plant, miRNA guide strands typically initiate with a 5′ uridine. This preference associates with the recognition and loading of miRNAs into AGO1, a key effector protein in the AGO family (AGO1 to AGO10 in *Arabidopsis thaliana*). This specificity contributes to gene silencing mediated by miRNAs. Loading miRNAs into various AGO proteins introduces functional diversity and regulatory specificity, adding complexity to plant regulatory networks. Other members of the AGO protein family, such as AGO2 to AGO10, play roles in distinct biological processes, with AGO10 involved in antiviral defense and AGO4 participating in RNA-directed DNA methylation. Protein cofactors such as HYL1, CPL1, HOS5, RS40, RS41, and regulatory checkpoints significantly contribute to the precise selection of miRNA strands in plants (Medley et al., 2021).

HYL1 (HYPONASTIC LEAVES 1): HYL1 is a double-stranded RNA-binding protein that associates with Dicer-Like 1 (DCL1) during miRNA processing. Its role in facilitating accurate strand selection has been demonstrated in *Arabidopsis thaliana*. HYL1 recognizes the unique structural features of the miRNA precursor such as pri-miR159 precursor in Arabidopsis and assists in the proper loading of the guide strand into the miRISC.

CPL1 (CAP-BINDING PROTEIN 80-LIKE 1): CPL1 is involved in 5′ cap formation during miRNA biogenesis. Studies have shown that CPL1 interacts with HYL1 and DCL1, influencing the processing of miRNA precursors. This suggests a regulatory role for CPL1 in the early stages of miRNA biogenesis and strand selection.

HOS5 (HIGH OSMOTIC STRESS GENE EXPRESSION 5): HOS5 is a protein involved in the cold stress response in Arabidopsis. Recent research has revealed its additional role in miRNA strand selection (Xu and Chen, 2023). HOS5 interacts with DCL1 and affects the abundance of specific miRNA isoforms, suggesting a role in miRNA-mediated gene regulation under stress conditions. RS40 and RS41 are components of the THO/TREX complex, known for their involvement in mRNA export. RS40 and RS41 associate with DCL1 and modulate the efficiency of miRNA strand selection, potentially linking miRNA processing to the broader mRNA processing machinery.

Several molecular chaperone such as HEAT SHOCK PROTEIN 90 (HSP90), play crucial role in the assembly of the RNA-induced silencing complex (RISC), specifically the AGO1-containing RISC. HSP90 is a molecular chaperone that assists in the folding, stabilization, and activation of various client proteins. It plays a crucial role in cellular homeostasis, especially under stress conditions. HSP90 binds to AGO1 in an ATP-dependent manner. This implies that the binding of HSP90 to AGO1 is facilitated by the presence of ATP, and the energy released from ATP hydrolysis involved in conformational change of AGO1 to allow sRNA duplex loading for stability of this interaction (Makhnevych and Houry, 2012). The recent studies reported the multifaceted role of the nuclear WD40 domain protein RBV in the process of microRNA (miRNA) biogenesis (Xu and Chen, 2023). RBV is a nuclear protein characterized by the presence of WD40 repeats. WD40 repeats are structural motifs often involved in protein-protein interactions and can have diverse functions. RBV is involved in promoting the loading of miRNAs into AGO1. This implies that RBV contributes to the proper assembly of the miRISC complex by facilitating the incorporation of miRNAs into AGO1 and promoting the transcription of MIR genes.

4. Regulation of microRNA (miRNA) genes transcription

4.1 Developmental stages regulate the transcription of MIR genes

Throughout the journey of plant development, unique transcriptional patterns emerge for microRNA genes, influencing a diverse array of biological processes such as germination, seedling establishment, vegetative growth, flowering or fruit development, reproductive development and senescence (Xu and Chen, 2023). During embryogenesis, specific MIR genes such as MIR156 and MIR172 are activated to guide crucial developmental processes including organ formation and tissue differentiation (Dong et al., 2022). In dynamic plant models such as *Arabidopsis thaliana*, where the MIR159 family is essential in regulating the transition from vegetative to reproductive phases. MIR159 expression is high in vegetative tissues, but as the plant undergoes flowering, MIR159 levels decline. This downregulation is coordinated with the upregulation of genes targeted by MIR159, facilitating the transition to reproductive growth. Similarly, the regulation of MIR172 is associated with flowering time control. In the vegetative phase, high MIR172 expression represses the expression of floral repressor genes, promoting the maintenance of the vegetative state. However, during the transition to flowering, MIR172 levels decrease, allowing the expression of target genes that facilitate the floral transition. Therefore, transcription of MIR172 show multiple layers of regulation. In case of Arabidopsis, miR156 regulates the transition from the vegetative to the reproductive phase by targeting SPL transcription factors such as SPL9 and SPL10. During transition process of embryo-to seedling and early stages of seed development in Arabidopsis, two different transcription factors FUSCA3 (FUS3) and ABSCISIC ACID INSENSETIVE3 (ABI3) directly promote MIR156 transcription respectively. While, both the factors repressing the expression of MIR156 during late development stage (Chiu et al., 2012).

During Seed maturation following regulatory components works:

ABI3-Mediated Regulation: ABI3 (ABSCISIC ACID-INSENSITIVE 3) is a transcription factor involved in abscisic acid (ABA) signaling during seed maturation.

ABI3-Suppression of MIR156D: ABI3 suppresses the MIR156D, which is likely a member of the miR156 family and is known to regulate key developmental processes, including flowering and vegetative phase transitions.

Suppression of MIR156D: ABI3 suppresses the expression of MIR156D during seed maturation. MIR156D is likely a member of the miR156 family, which is known to regulate key developmental processes, including flowering and vegetative phase transitions.

ABI3-downregulation of MIR160B: ABI3 also downregulates MIR160B, a member of the miR160 family. MIR160B regulates AUXIN RESPONSE FACTOR (ARF10 and ARF16), contributing to seed dormancy. Simultaneously, APETALA2 (AP2), a plant transcription factor, exerts multifaceted control over the transcription of MIR172. Binding to specific DNA sequences in the promoter region of MIR172. Interestingly, MIR156 positively modulates the transcription of MIR172 by binding to specific DNA sequences in its promoter region during developmental or environmental challenges. The regulatory process involves MIR172, which generates miRNAs targeting AP2 transcripts, establishing a feedback loop fine-tuning AP2 levels. The collaborative repression of MIR172 by AP2, along with transcriptional corepressors LEUNIG (LUG) and SEUSS (SEU), plays a crucial role in regulating floral organ identity and flowering time in plants (Grigorova et al., 2011). Understanding this developmental stage-specific transcriptional control of MIR genes provides crucial insights about relationship between miRNA and plant development. Such kind of interplay between AP2 and MIR172 highlights the complexity of gene regulatory networks in plants. The AP2 protein is primarily confined to the outer floral whorls, a localization governed by miR172-mediated mRNA translation repression in the inner whorls. This spatial regulation ensures precise control over floral organ development, underscoring the pivotal role of miR172 in modulating AP2 expression for proper flower formation.

Leaf polarity development showcases a complex interplay of transcription factors, notably involving class II homeodomain leucine zipper (HD-ZIPs) HAT3 and ATHB4. These transcription factors collaborate with the class III HD-ZIP protein REVOLUTA (REV) (Wang et al., 2021). Together, they form a complex that directly represses the transcription of MIR165/166 genes through a conserved cis-element in their promoters. Interestingly, REV itself is a target of miR165/166. This regulatory loop establishes a finely tuned mechanism where HAT3, ATHB4, and REV contribute to the precise control of miRNA expression, influencing leaf polarity and ensuring proper development through intricate molecular interactions.

4.2 Epigenetic regulation of MIR gene transcription

Epigenetic regulation is fundamental to controlling MIR gene transcription in plants. DNA methylation, histone modifications, and small RNA-directed pathways collectively shape the epigenetic landscape of MIR loci. Methylation patterns in promoter regions influence transcriptional activity, while histone modifications, such as acetylation and methylation, impact chromatin accessibility. Small RNAs, such as siRNAs and miRNAs, can guide DNA methylation and histone modification, establishing a regulatory feedback loop. In Arabidopsis, hypermethylation of the MIR156 locus leads to decreased miR156 expression, influencing developmental transitions such as flowering time. In *Arabidopsis thaliana*, during regulation of developmental transitions particularly flowering time, is show epigenetic modifications, specially the DNA methylation. One crucial player in this regulatory network is microRNA156 (miR156), a key regulator of the transition from the vegetative to the reproductive phase. Epigenetic modification is associated with condensed chromatin structure and reduced accessibility of the transcriptional machinery to the MIR156 promoter. As a consequence, hypermethylation of the MIR156 locus leads to decreased expression of miR156. MiR156, in its active form, targets and downregulates several transcription factors, including SQUAMOSA PROMOTER BINDING PROTEIN-LIKE (SPL) genes. These SPL genes, when expressed, promote the expression of genes involved in floral development and flowering time control. Therefore, decreased expression of miR156, resulting from hypermethylation of the MIR156 locus, allows the accumulation of SPL transcripts, promoting the transition to flowering. GENERAL CONTROL NON-REPRESSED PROTEIN 5 (GCN5) facilitates the expression of specific MIR genes by adding acetylation marks to H3K14 at the MIR loci. This acetylation modification contributes to chromatin accessibility, promoting transcription of the targeted MIR genes and further highlighting the diverse mechanisms involved in regulating microRNA expression. POLYCOMB REPRESSIVE COMPLEX 2 (PRC2) subunits CURLY LEAF (CLF) and SWINGER (SWN) play a redundant role in vegetative phase change by elevating H3K27me3 marks at MIR156A/MIR156C loci. This histone modification leads to transcriptional repression, highlighting the cooperative function of PRC2 subunits in regulating microRNA expression during plant developmental transitions. In rice, (*Oryza sativa*) histone modifications such as acetylation and methylation, dynamically shape chromatin structure at the MIR398a promoter dynamically adjust in response to oxidative stress, impacting miR398 expression and subsequently regulating copper homeostasis. Furthermore, small RNA-directed DNA methylation (RdDM) mechanisms link epigenetic and miRNA pathways such as MIR156 in maize (*Zea mays*). RdDM is a process where small RNAs guide DNA methylation at specific genomic loci, influencing gene expression. These siRNAs guide effector proteins to these genomic regions, once guided to the target loci by small RNAs, the effector proteins recruit DNA methyltransferases, enzymes responsible for adding methyl groups to cytosine residues in DNA. This DNA methylation leads to changes in chromatin structure, typically resulting in gene silencing. In the case of MIR156, RdDM-induced DNA methylation at its promoter region affects the accessibility of the transcriptional machinery, influencing the expression of the miR156 gene. CHROMOSOME REMODELING2 (CHR2), a chromatin re-modeler, modulates the structure of DNA-protein complexes, influencing the accessibility of MIR gene loci. The SWR1 chromatin remodeling complex (SWR1-C) exemplifies how nucleosome occupancy at MIR gene promoters is dynamically altered, positively or negatively impacting gene expression (Patty and Hainer, 2020). Another instance in plants involves the SWI/SNF complex, highlighting diverse chromatin remodels roles in MIR gene transcription for precise control over developmental and stress-responsive processes. These epigenetic modifications, responsive to environmental challenges and developmental stages, shaping plant phenotypes, dictate the precise temporal and spatial regulation of MIR transcription (Liu and Chang, 2021).

4.3 Hormonal regulation and MIR gene expression

Hormonal regulation also governs MIR gene expression in plants, influencing key developmental processes and stress responses. Abscisic acid (ABA) induces MIR159 in Arabidopsis, targeting MYB transcription factors involved in stress response also targets GAMYB transcription factors involved in gibberellin signaling. Conversely, auxin-responsive elements control MIR167 and MIR393, impacting auxin homeostasis and transport by targeting auxin response factors (ARFs). Gibberellins modulate MIR156/157, impacting developmental transitions by targeting SPL transcription factors. Ethylene influences MIR398 in response to oxidative stress and positively modulates MIR172, influencing flowering time. Jasmonic acid triggers MIR159, affecting MYC transcription factors involved in defense. Cytokinin regulate MIR156 and MIR164, crucial for developmental transitions. The intricate crosstalk between hormones and MIR genes highlights a sophisticated network governing plant growth, development, and stress responses, underscoring the pivotal role of hormonal signaling in shaping the dynamic landscape of microRNA-mediated gene regulation in plants (Dong et al., 2022).

4.4 Feedback regulation and MIR gene expression

miRNAs can regulate their own expression through feedback loops. Some miRNAs target and regulate the expression of TFs or other proteins involved in miRNA biogenesis. By modulating the expression of these regulatory factors, miRNAs can indirectly control their own transcription and processing, establishing feedback loops that align their expression levels (Bell et al., 2010). In case of Arabidopsis, miR168 targets AGO1, a key component of the RNA-induced silencing complex. Interestingly, AGO1 also regulates miR168 biogenesis, establishing a feedback loop that tightly controls miRNA levels (Xie et al., 2003). Similarly, in rice, miR528 targets AGO18, while AGO18 reciprocally modulates miR528 abundance. These feedback mechanisms enable plants to regulate miRNA activities, allowing them to swiftly adapt to environmental cues and ensure precise gene regulation during growth and stress conditions.

4.5 Environmental regulation and MIR gene expression

Environmental regulation of microRNA (miRNA) gene expression is an important aspect of plant adaptation towards changing conditions and facilitating precise adjustments in gene expression for optimal survival. Drought stress triggers upregulation of miR169 in plants such as in Arabidopsis, which targets NF-YA transcription factors involved in drought responses. In response to water scarcity, plants activate signaling pathways that elevate miR169 expression. This microRNA plays a critical role by targeting and downregulating NF-YA transcription factors, important components in the drought stress response (Li et al., 2021). Regulatory levels of these transcription factors, miR169 helps plants optimize water usage, stomatal closure, and enhance overall drought tolerance. The upregulation of miR169 during drought stress exemplifies a sophisticated molecular adaptation mechanism in plants, enabling them to modulate gene expression for survival in challenging environmental conditions. Under the pathogen attack, plants organize a sophisticated defense strategy to regulate immune responses and elevate the expression of specific microRNAs (miRNAs) such as miR393 (Jiang et al., 2022). MiR393 targets genes encoding auxin receptors, disrupting auxin signaling pathways. This targeted suppression of auxin receptors serves a dual purpose: it not only helps in plant immune system by inhibiting auxin-mediated susceptibility but also enables a rapid reallocation of resources towards defense mechanisms. By modulating miR393 expression, plants effectively balance growth and defense priorities, enhancing their ability to fend off pathogens while minimizing the impact on overall developmental processes. This complex regulatory network enhance the adaptive capacity of plants and precise response to microbial threats.

Temperature fluctuations greatly impact microRNA (miRNA) expression, with notable implications for developmental processes in crops such as wheat. One illustrative case is the temperature sensitivity of miR156, a pivotal regulator of developmental transitions. In wheat, variations in temperature modulate miR156 levels, influencing critical phases such as flowering time. Elevated temperatures often result in reduced miR156 expression, triggering an acceleration of developmental transitions and, consequently, earlier flowering. Conversely, lower temperatures lead to increased miR156 abundance, delaying flowering. This temperature-dependent regulation of miR156 underscores its role as a molecular thermostat, fine-tuning the timing of developmental stages in response to environmental conditions.

5. Plant miRNA and gene knockdown, gene silencing and artificial microRNAs (amiRNAs)

Plant microRNAs (miRNAs) can be utilized for gene knockdown, a technique that reduces the expression of a specific gene of interest. Gene knockdown using miRNAs involves designing synthetic miRNA-like molecules, known as artificial miRNAs (amiRNAs), that target and degrade the mRNA of the desired gene. Gene silencing refers to the regulation of gene expression by suppressing the activity or expression of a specific gene. It can occur at various levels of gene expression including transcriptional post-transcriptional and epigenetic regulation.

Transcriptional gene silencing (TGS), involves the suppression of gene expression at the level of transcription (Betti et al., 2021). It typically occurs through the modification of chromatin structure, making the DNA less accessible to transcriptional machinery. TGS can be mediated by DNA methylation, histone modifications (such as histone methylation and acetylation), and the recruitment of chromatin remodeling complexes.

Post-transcriptional gene silencing (PTGS): Small RNA molecules, such as microRNAs (miRNAs) and small interfering RNAs (siRNAs), play crucial roles in PTGS. These small RNAs bind to complementary sequences on target mRNAs and guide the RNA-induced silencing complex (RISC) to induce mRNA degradation or translational repression. miRNAs contribute to gene silencing through post-transcriptional regulation, primarily by targeting messenger RNAs (mRNAs) for degradation or translational repression, developmental processes, stress responses, hormone signaling etc. Artificial microRNAs (amiRNAs) are synthetic molecules designed to mimic the function of endogenous microRNAs (miRNAs) in plants. They are engineered to regulate the expression of specific target genes, offering a powerful tool for functional genomics, crop improvement, and biotechnological applications. The designed amiRNA sequence is embedded within an artificial miRNA precursor structure, which mimics the structure and function of typical stem-loop structure found in endogenous miRNAs. The designed amiRNAs can be introduced into plant cells through genetic transformation methods, such as viral vectors (Agrobacterium-mediated transformation) or biolistic particle delivery or synthetic delivery systems. amiRNAs are valuable tools in functional genomics for specifically silencing genes of interest to study their functions (Tiwari et al., 2014). Expression of specific genes can be downregulated by using amiRNAs, allowing researchers to study the effects of gene silencing on plant phenotypes. Furthermore, amiRNAs can be designed to modify specific traits in crops, such as improving resistance to pests, diseases, or environmental stresses.

6. Biological function of miRNA in plants

MicroRNAs (miRNAs) are the key regulators of gene expression in plants, which also pays central role in biological functions. These small RNA molecules post-transcriptionally modulate gene activity, influencing developmental processes such as organ formation, leaf growth, root development and seedling formation. miRNAs also play crucial role in stress responses against various environmental challenges such as drought and pathogens. With their ability to balance gene expression, miRNAs stand as essential players in the dynamic and adaptive biology of plants, ensuring their resilience and survival in diverse ecological conditions (Chen, 2005; Chu et al., 2005) (Fig. 1.4).

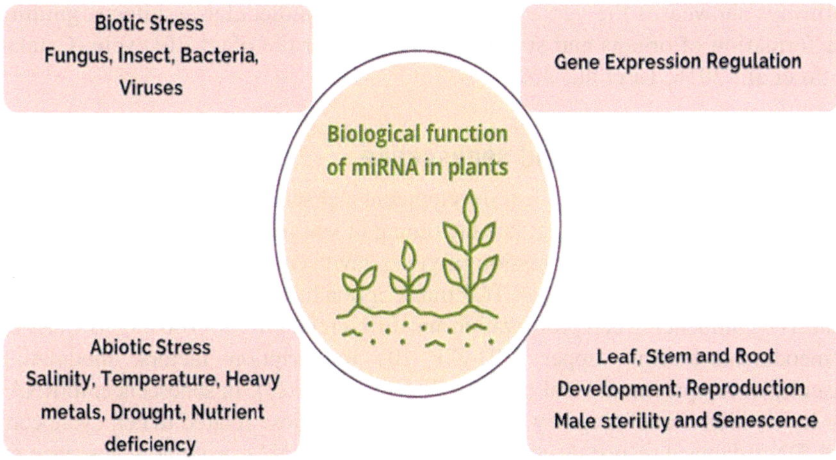

FIGURE 1.4

Roles of microRNAs in plant.

6.1 Gene expression regulation

MicroRNAs (miRNAs) play important role in post-transcriptional gene regulation in plants, exerting their influence by binding to the messenger RNAs (mRNAs) of target genes (Floyd and Bowman, 2004). This interaction leads to two major outcomes: inhibition of translation or promotion of mRNA degradation (Dong et al., 2022). During leaf development miR396 targets growth-regulating factor (GRF) genes, curbing their translation and influencing leaf size and shape. In the context of stress responses, miR398 downregulates copper-binding proteins, thereby reducing their translation and enhancing plant tolerance to oxidative stress (Li et al., 2022). Additionally, miR164 in root development targets NAC transcription factors, modulating lateral root development regulating gene expression.

6.2 Developmental processes

miRNAs are critical regulators of plant development. They control processes such as embryogenesis, organ development, and phase transitions (e.g., from vegetative to reproductive phases). Specific miRNAs are associated with the regulation of key developmental genes, ensuring proper growth and morphogenesis. One compelling example lies in leaf development, where miRNAs such as miR319 and miR166 target transcription factors, modulating leaf initiation, growth, and vascular patterning. Similarly, miR156 influences the transition from vegetative to reproductive phases by regulating SPL genes. In root development, miR160 and miR167 impact root architecture by regulating the expression of auxin response factors. The dynamic regulation of these developmental pathways showcases the versatility of miRNAs as molecular architects, guiding the formation of organs and structures essential for the plant life cycle (Catalanotto et al., 2016; Li et al., 2022).

6.3 Leaf development and senescence

Certain miRNAs play roles in leaf development and senescence (a state of irreversible cell cycle arrest). They regulate the timing of leaf initiation, growth, and senescence by modulating the expression of genes involved in these processes. During leaf development, miR319 targets TCP transcription factors, regulating leaf morphogenesis by influencing cell proliferation and differentiation. MiR166 targets Class III homeodomain-leucine zipper (HD-ZIP III) transcription factors, modulating vascular development and leaf polarity (Yang et al., 2023). In senescence, miRNAs are involved in modulating key pathways, including those related to cell cycle control, DNA damage response, and apoptosis. Some miRNAs promote senescence by targeting antisenescence factors, while others inhibit senescence by regulating prosenescence genes. As leaves undergo senescence, miRNAs such as miR164 target NAC (NAM, ATAF1/2, CUC2) transcription factors, contributing to the regulation of senescence-associated genes (Fang et al., 2014). Notably, miR396 influences senescence by targeting growth-regulating factor (GRF) genes, affecting cell cycle progression. The infraction between miRNAs and their target genes play crucial role in shaping plant architecture and optimizing resource allocation throughout the plant's life cycle. miR164 in root development targets NAC transcription factors, modulating lateral root development by centric regulation of gene expression. This dual functionality, inhibiting translation or promoting mRNA degradation, underscores the versatility of miRNAs in precisely sculpting the plant transcriptome.

6.4 Root development

miRNAs influence root architecture and development by regulating the expression of genes involved in root growth, lateral root formation, and response to environmental cues such as nutrient deficiency and pathogen infection (Xiao-xiao et al., 2022). MiR160 play key role in root growth, targeting Auxin Response Factors (ARFs) (ARF6, ARF8, ARF10, ARF16) and modulating the sensitivity of roots to auxin signals(Mallory et al., 2005). MiR164 regulates NAC (NAM, ATAF1/2, CUC2) transcription factors, impacting lateral root formation by modulating the expression of genes involved in this process. Additionally, miR167 targets ARF6 and ARF8, contributing to the regulation of lateral root initiation and development. Furthermore, miR393, targeting the Transport Inhibitor Response 1 (TIR1) and auxin receptor F-box (AFB) genes. miRNA regulates the auxin signaling pathway by targeting TIR1 and AFB genes. miR393 downregulate the TIR1 and AFB genes and modulates the sensitivity of plants to auxin signals and influences various aspects of plant development, including root architecture (Jiang et al., 2022). miR319 Targeting the TCP (*Teosinte Branched*1/Cycloidea/Proliferating Cell

Factor) transcription factors and influences cell proliferation and differentiation (Fang et al., 2021). Additionally, miR319's impact on root development further emphasizes its broad involvement in shaping plant architecture. The complex control exerted by miRNAs on these target genes ensures a balanced and responsive root system, crucial for nutrient uptake and improved resilience in diverse environmental conditions.

6.5 miRNA role in reproductive processes

miRNAs contribute to the control of reproductive processes, including pollen development, fertilization, and seed formation through regulating the expression of genes involved in these processes. Pollen Development: miR159 targets the MYB transcription factors involved in pollen development, pollen viability and maturation. Fertilization process is regulated by miR167, it targets the Auxin Response Factors (ARF6 and ARF8) and modulate the auxin signaling pathway, which directly influences the timing of fertilization and embryo development. During embryo development, miR396 regulates growth-regulating factor (GRF) genes and control the cell proliferation and differentiation during embryogenesis. The most crucial process of the plants is the floral and fruit development, which is regulated by miR172. This miRNA targets AP2-like transcription factors, influencing floral organ identity and proper formation of fruits. During seed development, miR156 regulates the expression of SPL (Squamosa Promoter Binding Protein-like) transcription factors, which impacts the timing of flowering and influences seed development and maturation. During seed germination, miR156 and miR159 collectively influence the timing of seed germination by regulating SPL and MYB genes, respectively.

miR164 participates in controlling flower development by targeting NAC domain transcription factors. miR166: Regulates the development of floral organs, including petals and stamens, by targeting class III homeodomain-leucine zipper (HD-Zip III) transcription factors. miR319: Involved in controlling leaf and flower development by targeting TCP (Teosinte Branched1/Cycloidea/Proliferating Cell Factor) transcription factors. Manipulating miRNAs offers potential applications in plant breeding for hybrid seed production (Dong et al., 2022).

7. miRNA and stress response

MiRNAs are centric regulators for gene expression in response to a spectrum of stresses both biotic and abiotic (Li et al., 2024). Here's an overview of the roles of some miRNAs in biotic and abiotic stress responses in plants.

7.1 Biotic stress response

Various miRNAs are involved in regulating the plant's defense response against pathogens, like bacteria, viruses, and fungi such as miR393 is known to target mRNAs encoding transport inhibitor response 1 (TIR1) and auxin signaling F-box (AFB) proteins. This regulation influences plant defense against pathogens. miR393, miR160, miR164 and miR167 react oppositely towards the pathogenic bacterium *Pseudomonas syringae* (Lee et al., 2017). miR159 and miR166 reported in cotton (*Gossypium hirsutum*), which show resistance to *Verticillium dahliae* (Zhang et al., 2016). Some of the miRNA such as miR482 and miR2118 are implicated in the regulation of nucleotide-binding site-leucine-rich repeat (NBS-LRR) and directly targeting resistance (Liao et al., 2022). In some of the host-pathogen relations Cross-Kingdom Communication also observed where, Plants can release miRNAs into the extracellular environment, and these miRNAs can be taken up by other organisms, potentially influencing interkingdom communication (Gualtieri et al., 2020).

7.2 Abiotic stress response

Plant Noncoding RNAs play crucial role in abiotic stresses such as such as drought, salinity, extreme temperatures, and nutrient deficiencies, can adversely affect plant growth and development. Two main classes of plant noncoding RNAs involved in abiotic stress responses are microRNAs (miRNAs) and long noncoding RNAs (lncRNAs) (Khraiwesh et al., 2012; Kar and Raichaudhuri, 2021).

microRNA 156: miR156 is known to target the SQUAMOSA PROMOTER BINDING PROTEIN-LIKE (SPL) transcription factors in plants such as Rice, Arabidopsis. These transcription factors contain a conserved SBP (SQUAMOSA Promoter-Binding Protein) domain and are involved in the control of various aspects of plant development (Preston and Hileman, 2013). The miR156 family is one of the well-studied microRNA families in plants, and it plays a crucial role in the regulation of various developmental processes, including the transition from the vegetative to the reproductive phase, flowering time and leaf development. Overexpression of miR156 enhances drought tolerance by downregulating SPL genes. Modulating SPL expression affects plant tolerance to drought stress and enhances drought tolerance. microRNA 159: miR159 reported in Cotton, Arabidopsis, it can modulate (downregulation) expression of MYB (myeloblastosis) transcription factors or HIC-15 and influences the plant's ability to cope with water deficit and salt stress. Some of the MYB proteins are known to interact with clock components. MYB transcription factors constitute a large family in plants, and they are involved in the regulation of various biological processes, including growth, development, and stress responses. Reduced levels of MYB transcription factors are associated with improved resilience to adverse environmental conditions such as osmotic regulation, ion homeostasis, antioxidant defense and also influence the circadian clock machinery. microRNA 160: miR160 regulates the stress responses by targeting ARF (Auxin Response Factor) genes involved in the auxin signaling pathway. This modulation contributes to the plant's ability to cope with drought stress and other environmental challenges and also essential for various processes including root development, seed germination.

microRNA 164: miR164 targeting NAC4/NAC60 (NAM, ATAF1/2, CUC2) transcription factors and leads to the degradation or translational inhibition of the target transcripts (Guo et al., 2005). The downregulation of NAC4 and NAC60 expression, facilitated by miR164, contributes to the plant's ability to cope with water stress.

microRNA167: miR167 plays a regulatory role in the plant's adaptive responses it is implicated in responses to adverse conditions such as water scarcity (drought) or high salt concentrations (salinity). The primary targets of miR167 are Auxin Response Factors (ARFs), specifically ARF6 and ARF8. During stress conditions, miR167 downregulates the expression of ARFs (ARF6 and ARF8) (Zhao et al., 2023). This downregulation influences the plant's response to stress by modulating the auxin signaling pathway, which is linked to the circadian clock. miR167 exhibit phase-specific expression patterns, indicating their involvement in specific phases of the circadian clock. This phase-specific regulation helps fine-tune the expression of target genes at different times of the day.

microRNA169: miR169 specifically targets NFYA transcription factors involved in nitrogen uptake and assimilation in plants. The regulation of NFYA by miR169 influences nitrogen use efficiency and maintains nitrogen homeostasis (Nguyen et al., 2015).

microRNA 319: miR319 is implicated in the plant's response to heat stress, a condition that poses challenges to various cellular processes and can adversely impact plant growth and development. The primary targets of miR319 are TCP (Teosinte Branched1/Cycloidea/Proliferating Cell Factor) transcription factors. The molecular mechanism involves the binding of miR319 to the mRNA of TCP transcription factors, resulting in either translational inhibition or mRNA degradation. TCP transcription factors are involved in the regulation of various developmental processes, including leaf development and growth. Impact: Under heat stress conditions, the expression of miR319 is often altered. The upregulation of miR319 leads to the downregulation of TCP transcription factors. Modulating TCP expression has cascading effects on downstream genes involved in growth and development. In the context of heat stress, this regulatory mechanism influenced by miR319 helps the plant redirect its resources from growth-related processes to stress response mechanisms (Zhao et al., 2016).

microRNA 395: miR395 is actively involved in Nutrient Stress such as sulfur deficiency. The primary target of miR395 is ATP sulfurylase, an enzyme involved in the initial step of sulfate assimilation. Sulfate assimilation is the process by which plants take up and incorporate sulfate into organic compounds, including amino acids and coenzymes. Under sulfur deficiency, miR395 expression is upregulated and lead to the downregulation of ATP sulfurylase, which is an essential enzyme in sulfur assimilation (Ai et al., 2016).

microRNA398: miR398 is known for its responsiveness to temperature changes and oxidative stress. Oxidative stress occurs when there is an imbalance between the production of reactive oxygen species (ROS) and the plant's ability to detoxify them. The main targets of miR398 is to downregulates the expression of copper-binding proteins can impact the redox state of the cell. This downregulation has a direct effect on copper homeostasis within the plant. Copper is an essential cofactor for copper superoxide dismutase (CSDs), enzymes involved in scavenging superoxide radicals (ROS) and by modulating the expression of genes involved in copper binding, miR398 influences the activity of these antioxidant enzymes. The fine-tuning of copper homeostasis, orchestrated by miR398, is crucial for the plant's ability to manage oxidative stress effectively. Further, Temperature compensation mechanisms are crucial for maintaining circadian rhythms, which optimize the plant adaptation to environmental changes and miR398 may play a role in this process. miR398 also targets the cytochrome *c* oxidase (CCO), a pivotal enzyme in the mitochondrial electron transport chain's final step transferring electrons to oxygen. Under stress, upregulated miR398 binds to CCO mRNA, potentially degrading it or inhibiting translation. Elevated miR398 levels lead to CCO downregulation, impacting mitochondrial function and electron transport. This influences the plant's energy generation through oxidative phosphorylation, subsequently affecting cellular energy status and stress response mechanisms (Li et al., 2022).

microRNA399: miR399 is specifically involved in the plant's response to phosphate deficiency, with its expression usually upregulated. The primary target of miR399 is the mRNA of PHO2 (Phosphate 2). PHO2 encodes an E2 ubiquitin-conjugating enzyme involved in ubiquitin-mediated degradation of phosphate transporters through ubiquitination, leading to subsequent proteasomal degradation (Hu et al., 2015).

microRNA408: miR408 is known to be responsive to various stresses, and one of its notable stress responses is to oxidative stress. Oxidative stress occurs when there is an imbalance between the production of reactive oxygen species (ROS) and the plant's antioxidant defense mechanisms. The primary targets of miR408 are plantacyanin genes and leading to the degradation of these mRNAs or translational inhibition. Plantacyanins are copper-containing proteins involved in various cellular processes, including electron transport, ROS scavenging and copper homeostasis. The reduced expression of plantacyanin genes affects the copper availability for these proteins, influencing their functionality. This modulation contributes to the plant's ability to manage oxidative stress by adjusting the levels of proteins involved in ROS scavenging. Increased miR408 expression under oxidative stress leads to downregulated plantacyanin genes, impacting the antioxidant system (Ma et al., 2015).

8. Future perspectives and conclusion

Research on plant miRNAs presents promising avenues for comprehending and manipulating plant development, stress responses, and crop enhancement. Identification of factors governing miRNA biogenesis, action, and degradation has progressed significantly. However, a comprehensive understanding of the mechanistic and cell biological details remains a current challenge. Unraveling the intricate regulation of plant development by miRNAs assumes paramount significance for progress in crop breeding. MiRNAs provide a nuanced and controlled approach to modifying gene expression, thereby enhancing plant development. Furthermore, exposure to biotic and abiotic stresses prompts plants to generate new small RNAs (sRNA), showcasing their adaptability in modulating genetic expression under environmental challenges. High-throughput sequencing technologies and bioinformatics tools have accelerated the identification and functional characterization of plant miRNAs. Utilizing this knowledge holds potential for developing miRNA-based biotechnological interventions, fostering crop improvement, pest resistance, and environmental adaptability. In conclusion, the burgeoning field of plant miRNA research not only enriches our understanding of plant biology but also unveils innovative strategies for sustainable agriculture, ensuring food security amid global challenges.

References

Ai, Q., et al., 2016. Control of sulfate concentration by miR395-targeted APS genes in *Arabidopsis thaliana*. Plant Divers. 38 (2), 92–100. Elsevier.

Ambros, V., et al., 2003. MicroRNAs and other tiny endogenous RNAs in *C. elegans*. Curr. Biol. 13 (10), 807–818. Elsevier.

Aravin, A.A., Hannon, G.J., Brennecke, J., 2007. The Piwi-piRNA pathway provides an adaptive defense in the transposon arms race. Science 318 (5851), 761–764. American Association for the Advancement of Science.

Bajczyk, M., et al., 2023. Recent insights into plant miRNA biogenesis: multiple layers of miRNA level regulation. Plants 12 (2), 342.

Baldrich, P., Hsing, Y.-I.C., San Segundo, B., 2016. Genome-wide analysis of polycistronic microRNAs in cultivated and wild rice. Genom. Biol. Evolut. 8 (4), 1104–1114. Oxford University Press.

Barik, S., et al., 2015. Coevolution pattern and functional conservation or divergence of miR167s and their targets across diverse plant species. Sci. Rep. 5 (1), 14611. Nature Publishing Group UK London.

Bartel, D.P., 2009. MicroRNAs: target recognition and regulatory functions. Cell 136 (2), 215–233. Elsevier.

Bell, M.L., Buvoli, M., Leinwand, L.A., 2010. Uncoupling of expression of an intronic microRNA and its myosin host gene by exon skipping. Mol. Cell Biol. 30 (8), 1937–1945. Taylor & Francis.

Betti, F., et al., 2021. Exogenous miRNAs induce post-transcriptional gene silencing in plants. Nat. Plant. 7 (10), 1379–1388. Nature Publishing Group UK London.

Borchert, G.M., Lanier, W., Davidson, B.L., 2006. RNA polymerase III transcribes human microRNAs. Nat. Struct. Mol. Biol. 13 (12), 1097–1101. Nature Publishing Group US New York.

Borsani, O., et al., 2005. Endogenous siRNAs derived from a pair of natural cis-antisense transcripts regulate salt tolerance in Arabidopsis. Cell 123 (7), 1279–1291.

Catalanotto, C., Cogoni, C., Zardo, G., 2016. MicroRNA in control of gene expression: an overview of nuclear functions. Int. J. Mole. Sci. 17 (10), 1712.

Chen, R., et al., 2014. Exploring microRNA-like small RNAs in the filamentous fungus *Fusarium oxysporum*. PLoS One 9 (8). Public Library of Science San Francisco, USA.

Chen, X., 2005. MicroRNA biogenesis and function in plants. FEBS Lett. 579 (26), 5923–5931. Elsevier.

Chen, Y., et al., 2021. Nuclear translocation of OsMFT1 that is impeded by OsFTIP1 promotes drought tolerance in rice. Mol. Plant 14 (8), 1297–1311. Elsevier.

Chithung, T.A., et al., 2023. Understanding the evolution of miRNA biogenesis machinery in plants with special focus on rice. Funct. Integr. Genom. 23 (1), 30. Springer.

Chiu, R.S., et al., 2012. The role of the Arabidopsis FUSCA3 transcription factor during inhibition of seed germination at high temperature. BMC Plant Biol. 12, 1–16. Springer.

Chu, C.-C., et al., 2005. A copper chaperone for superoxide dismutase that confers three types of copper/zinc superoxide dismutase activity in Arabidopsis. Plant Physiol. 139 (1), 425–436.

Connerty, P., Ahadi, A., Hutvagner, G., 2015. RNA binding proteins in the miRNA pathway. Int. J. Mole. Sci. 17 (1), 31.

Dang, Y., et al., 2011. RNA interference in fungi: pathways, functions, and applications', *Eukaryotic cell.* Am. Soc. Microbiol. 10 (9), 1148–1155.

Dong, Q., Hu, B., Zhang, C., 2022. microRNAs and their roles in plant development. Front. Plant Sci. 13, 824240.

Fang, Y., et al., 2021. Roles of miR319-regulated TCPs in plant development and response to abiotic stress. Crop J. 9 (1), 17–28. Elsevier.

Fang, Y., Xie, K., Xiong, L., 2014. Conserved miR164-targeted NAC genes negatively regulate drought resistance in rice. J. Exp. Bot. 65 (8), 2119–2135. Oxford University Press.

Floyd, S.K., Bowman, J.L., 2004. Ancient microRNA target sequences in plants. Nature 428 (6982), 485–486. Nature Publishing Group UK London.

Gao, F., et al., 2021. Characterization and comparative analysis of MicroRNAs in 3 representative red algae. Iran. J. Biotechnol. 19 (4), e2868.

Grigorova, B., et al., 2011. LEUNIG and SEUSS co-repressors regulate miR172 expression in Arabidopsis flowers. Development 138 (12), 2451–2456.

Gualtieri, C., Leonetti, P., Macovei, A., 2020. Plant miRNA cross-kingdom transfer targeting parasitic and mutualistic organisms as a tool to advance modern agriculture. Front. Plant Sci. 11, 531283.

Guo, H.-S., et al., 2005. MicroRNA directs mRNA cleavage of the transcription factor NAC1 to downregulate auxin signals for Arabidopsis lateral root development. Plant Cell 17 (5), 1376–1386.

Hu, B., et al., 2015. MicroRNA399 is involved in multiple nutrient starvation responses in rice. Front. Plant Sci. 6, 188. Frontiers Media SA.

Jia, T., et al., 2017. The Arabidopsis MOS4-associated complex promotes microRNA biogenesis and precursor messenger RNA splicing. Plant Cell 29 (10), 2626–2643.

Jiang, J., et al., 2022. The miR393-target module regulates plant development and responses to biotic and abiotic stresses. Int. J. Mol. Sci. 23 (16), 9477. MDPI.

Jones-Rhoades, M.W., Bartel, D.P., 2004. Computational identification of plant microRNAs and their targets, including a stress-induced miRNA. Mol. Cell 14 (6), 787–799. Elsevier.

Kar, M.M., Raichaudhuri, A., 2021. Role of microRNAs in mediating biotic and abiotic stress in plants. Plant Gene. 26, 100277. Elsevier.

Khraiwesh, B., Zhu, J.-K., Zhu, J., 2012. Role of miRNAs and siRNAs in biotic and abiotic stress responses of plants. Biochimica et Biophysica Acta (BBA)-Gene Regulat. Mechan. 1819 (2), 137–148. Elsevier.

Kozomara, A., Griffiths-Jones, S., 2014. miRBase: annotating high confidence microRNAs using deep sequencing data. Nucleic Acid. Res. 42 (D1), D68–D73. Oxford University Press.

Lagos-Quintana, M., et al., 2001. Identification of novel genes coding for small expressed RNAs. Science 294 (5543), 853–858. American Association for the Advancement of Science.

Lau, N.C., et al., 2001. An abundant class of tiny RNAs with probable regulatory roles in *Caenorhabditis elegans*. Science 294 (5543), 858–862.

Lee, M., et al., 2017. An Arabidopsis NAC transcription factor NAC4 promotes pathogen-induced cell death under negative regulation by microRNA164. New Phytol. 214 (1), 343–360. Wiley Online Library.

Lee, Y., et al., 2004. MicroRNA genes are transcribed by RNA polymerase II. EMBO J. 23 (20), 4051–4060. John Wiley & Sons, Ltd Chichester, UK.

Li, J., et al., 2021. The miR169n-NF-YA8 regulation module involved in drought resistance in *Brassica napus* L. Plant Sci. 313, 111062. Elsevier.

Li, J., et al., 2022. MicroRNA398: a master regulator of plant development and stress responses. Int. J. Mole. Sci. 23 (18), 10803.

Li, M., Yu, B., 2021. Recent advances in the regulation of plant miRNA biogenesis. RNA Biol. 18 (12), 2087–2096. Taylor & Francis.

Li, Z., et al., 2024. A review on the function of miRNAs in plant responses to stresses. Plant Stress 100347. Elsevier.

Liao, L., et al., 2022. New insight into the molecular mechanism of miR482/2118 during plant resistance to pathogens. Front. Plant Sci. 13. Frontiers.

Liu, J., Chang, C., 2021. Concerto on chromatin: interplays of different epigenetic mechanisms in plant development and environmental adaptation. Plants 10 (12), 2766.

Liu, W., et al., 2017. Characterization and function of microRNA* s in plants. Front. Plant Sci. 8, 319830.

Ma, C., Burd, S., Lers, A., 2015. Mi R 408 is involved in abiotic stress responses in Arabidopsis. Plant J. 84 (1), 169–187. Wiley Online Library.

Makhnevych, T., Houry, W.A., 2012. The role of Hsp90 in protein complex assembly. Biochim. Biophys. Acta, Mol. Cell Res. 1823 (3), 674–682. Elsevier.

Mallory, A.C., Bartel, D.P., Bartel, B., 2005. MicroRNA-directed regulation of Arabidopsis AUXIN RESPONSE FACTOR17 is essential for proper development and modulates expression of early auxin response genes. Plant Cell 17 (5), 1360–1375.

Marin, F.R., et al., 2022. RNA-seq, bioinformatic identification of potential MicroRNA-Like Small RNAs in the edible mushroom *Agaricus bisporus* and experimental approach for their validation. Int. J. Mol. Sci. 23 (9), 4923. MDPI.

Medley, J.C., Panzade, G., Zinovyeva, A.Y., 2021. microRNA strand selection: unwinding the rules. Wiley Interdiscipl. Rev.: RNA 12 (3), e1627. Wiley Online Library.

Mello, C.C., Conte Jr, D., 2004. Revealing the world of RNA interference. Nature 431 (7006).

Moazed, D., 2009. Small RNAs in transcriptional gene silencing and genome defence. Nature 457 (7228), 413–420. Nature Publishing Group UK London.

Narjala, A., et al., 2020. A conserved sequence signature is essential for robust plant miRNA biogenesis. Nucleic Acid. Res. 48 (6), 3103–3118. Oxford University Press.

Nguyen, G.N., et al., 2015. Role of microRNAs involved in plant response to nitrogen and phosphorous limiting conditions. Front. Plant Sci. 6, 629. Frontiers Media SA.

Okada, C., et al., 2009. A high-resolution structure of the pre-microRNA nuclear export machinery. Science 326 (5957), 1275–1279.

Patty, B.J., Hainer, S.J., 2020. Non-coding RNAs and nucleosome remodeling complexes: an intricate regulatory relationship. Biology 9 (8), 213.

Preston, J.C., Hileman, L.C., 2013. Functional evolution in the plant squamosa-promoter binding protein-like (SPL) gene family. Front. Plant Sci. 4, 80. Frontiers Media SA.

Qi, Y., et al., 2023. milR20 negatively regulates the development of fruit bodies in *Pleurotus cornucopiae*. Front. Microbiol. 14. Frontiers Media SA.

Raza, A., et al., 2023. miRNAs for crop improvement. Plant Physiol. Biochem. 107857. Elsevier.

Reinhart, B.J., Weinstein, E.G., Rhoades, M.W., Bartel, B., Bartel, D.P., 2002. MicroRNAs in plants. Genes Dev 16 (13), 1616–1626. https://doi.org/10.1101/gad.1004402. Erratum in: Genes Dev 2002 Sep 1;16(17):2313. PMID: 12101121; PMCID: PMC186362.

Samad, A.F.A., et al., 2017. MicroRNA and transcription factor: key players in plant regulatory network. Front. Plant Sci. 8, 565. Frontiers Media SA.

Shen, K., et al., 2019. The dual functional role of MicroRNA-18a (miR-18a) in cancer development. Clin. Transl. Med. 8 (1), 32. Springer.

Song, X., et al., 2019. MicroRNAs and their regulatory roles in plant–environment interactions. Annu. Rev. Plant Biol. 70, 489–525. Annual Reviews.

Taverna, S.D., Coyne, R.S., Allis, C.D., 2002. Methylation of histone h3 at lysine 9 targets programmed DNA elimination in tetrahymena. Cell 110 (6), 701–711.

Tiwari, M., Sharma, D., Trivedi, P.K., 2014. Artificial microRNA mediated gene silencing in plants: progress and perspectives. Plant Mol. Biol. 86, 1–18. Springer.

Vazquez, F., et al., 2004. Endogenous trans-acting siRNAs regulate the accumulation of Arabidopsis mRNAs. Mol. Cell 16 (1), 69–79. Elsevier.

Voshall, A., et al., 2017. miRNAs in the alga *Chlamydomonas reinhardtii* are not phylogenetically conserved and play a limited role in responses to nutrient deprivation. Sci. Rep. 7 (1), 5462. Nature Publishing Group UK London.

Wahba, L., Hansen, L., Fire, A.Z., 2021. An essential role for the piRNA pathway in regulating the ribosomal RNA pool in *C. elegans*. Develop. Cell 56 (16), 2295–2312.

Wang, H., et al., 2021. Genome-wide analysis and expression profiling of HD-ZIP III genes in three Brassica species. Diversity 13 (12), 684.

Wang, J., Mei, J., Ren, G., 2019. Plant microRNAs: biogenesis, homeostasis, and degradation. Front. Plant Sci. 10, 433008.

Wang, X., et al., 2018. Degradation of unmethylated miRNA/miRNA* s by a DEDDy-type 3′ to 5′ exoribonuclease Atrimmer 2 in Arabidopsis. Proc. Nat. Acad. Sci. 115 (28), E6659–E6667.

Xiao-xiao, Y., et al., 2022. The roles of microRNAs in regulating root formation and growth in plants. J. Integr. Agric. 21 (4), 901–916. ELSEVIER SCI LTD THE BOULEVARD, LANGFORD LANE, KIDLINGTON, OXFORD OX5 1GB.

Xie, Z., Kasschau, K.D., Carrington, J.C., 2003. Negative feedback regulation of Dicer-Like1 in Arabidopsis by microRNA-guided mRNA degradation. Curr. Biol. 13 (9), 784–789. Elsevier.

Xu, D., et al., 2021. Identification and physiological function of one microRNA (*Po-MilR-1*) in oyster mushroom *Pleurotus ostreatus*. Mycoscience 62 (3), 182–188. https://doi.org/10.47371/mycosci.2021.01.004.

Xu, Y., Chen, X., 2023. microRNA biogenesis and stabilization in plants. Fundam. Res. 3 (5), 707–717 (Elsevier).

Yang, M., et al., 2023. Fine mapping of a pleiotropic locus (BnUD1) responsible for the up-curling leaves and downward-pointing siliques in *Brassica napus*. Int. J. Mol. Sci. 24 (4), 3069. MDPI.

Yehu, M., et al., 2017. The evolutionary origin of plant and animal microRNAs. Nat. Ecol. Evolut. 1 (3). Nature Publishing Group.

Zhang, T., et al., 2016. Cotton plants export microRNAs to inhibit virulence gene expression in a fungal pathogen. Nat. Plant. 2 (10), 1–6. Nature Publishing Group.

Zhao, J., et al., 2016. Regulation of non-coding RNAs in heat stress responses of plants. Front. Plant Sci. 7, 1213.

Zhao, Y., et al., 2023. The miR167-OsARF12 module regulates rice grain filling and grain size downstream of miR159. Plant Communicat. 4 (5). Elsevier.

Zilberman, D., Cao, X., Jacobsen, S.E., 2003. ARGONAUTE4 control of locus-specific siRNA accumulation and DNA and histone methylation. Science 299 (5607), 716–719.

Zong, T., et al., 2021. tsRNAs: novel small molecules from cell function and regulatory mechanism to therapeutic targets. Cell Prolif. 54 (3), e12977. Wiley Online Library.

CHAPTER

What are microRNAs, their origin and functions in plants?

2

Nistala Shweta[1], Maravi Preeti[1], Singh Chauhan Nagendra[2], S. Keshavkant[3]

[1]*Department of Biotechnology, Bharti Vishwavidyalaya, Durg, Chhattisgarh, India;* [2]*Drug Testing Laboratory Avam Anusandhan Kendra, Raipur, Chhattisgarh, India;* [3]*School of Studies in Biotechnology, Pt. Ravishankar Shukla University, Raipur, Chhattisgarh, India*

1. Introduction

Gene regulation is an indispensable program observed in every organism. It is central to different developmental pathways and also responsive against environmental stimuli for providing better adaptability to organisms. Every organism is under strict control of multifactorial and multigenic regulatory systems during production of specific gene product. This regulatory control includes both modification and modulation in gene expression, which is majorly observed during transcription initiation, RNA processing, protein synthesis as well as at the time of post RNA and protein synthesis processes (Sunny et al., 2016). The control of gene expression at post transcriptional step by RNA silencing approach includes small, endogenous, noncoding miRNAs, also known as micro RNAs. It is found in most of the eukaryotes. In plants miRNAs are considered as one of the significant growth regulators involved in different growth processes as well as impose regulatory effects to combat against abiotic and biotic stresses. According to Griffiths-Jones et al. (2008) nearly 6000 premature miRNAs and 7000 fully developed miRNAs have been reported in a variety of plant species. These all are known to direct various physiological and developmental mechanisms in plants (Singh et al., 2016; Islam et al., 2018). The miRNAs usually show complementary base pairing with target messenger RNAs (mRNAs) and silence their expressions by one or more processes such as cleaving mRNA into pieces, shortening of the poly A tail to destabilize mRNA and repressing protein synthesis (Gielen et al., 2012). However, with the development of advance sequencing and genomics technologies many preserved and nonconserved miRNAs are quantified in different plants and are now used as a cost-effective approach to raise new varieties having better quality, enhanced production of biomass, and improved stress tolerance efficiency.

MicroRNA Advances and Application in Plant Biology. https://doi.org/10.1016/B978-0-443-21682-4.00016-6

2. What are microRNAs?

The RNA silencing pathways in most of the eukaryotes are mediated by various types of noncoding RNAs. The list includes small interfering RNAs (siRNAs), miRNAs, transfer RNA derived small RNAs (tsRNA), and PIWI-interacting RNA (piRNA): a small RNA reported in animals (Borges and Martienssen, 2015; Czech et al., 2018; Zhu et al., 2018; Treiber et al., 2019). The 24 nucleotides long siRNA, and 21 nucleotide miRNA shows their highest abundance while studying their expression among other similar type of RNAs in plants. Both these RNAs resembles with each other in structure except that the miRNAs are generated from the minute portions of single stranded RNA transcripts consisting of self-complementary hairpins precursors, whereas; siRNAs are produced from non-self-complementary regions of double stranded RNA (Borges and Martienssen, 2015). The very first miRNA recognized and discovered was lin-4 from *Caenorhabditis elegans,* a nematode (Lee et al., 1993; Wightman et al., 1993). This miRNA was found to control expression of those genes that are indulged in growth and development of nematode larva. This discovery changed the face of molecular biology, and after that many miRNAs were discovered from plants and other organisms, and then their function in controlling the pattern of gene expression has been scrutinized.

3. Biosynthesis of microRNAs in plants

Genome of plants consists of several hundreds of MIR genes encoding miRNAs, which controls expression of varied genes (Manavella et al., 2019). These genes are intergenic, present in middle of two protein coding genes or intronics, and present internal to introns (Srivastava and Suprasanna, 2021). The intergenic miRNAs are produced by transcription of MIR gene by the help of RNA Polymerase II, whereas; intronic are produced from introns during processing of RNA transcripts (Budak and Akpinar, 2015). Most of the miRNAs produced are intronic and processed from introns. Generation of miRNA through canonical pathway follows the sequential order as depicted in Fig. 2.1.

(1) Production of miRNA is primarily initiated by the RNA polymerase II mediated MIR genes transcription. Like all other genes, MIR genes also possess motifs for binding of transcription factor to a promoter region containing a TATA Box. Transcription of MIR gene is assisted by both general and specific transcription factors such as Mediator 17, Mediator 18, and Mediator 20A, which are a few specific proteins involved in recruiting RNA Polymerase II to their promoter site (Kim et al., 2011; Gao et al., 2020; Srivastava and Suprasanna, 2021). Other factors involved in various steps during transcription includes NOT2 {a subunit of carbon catabolite repression 4-negative on TATA-less (CCR4-NOT) complex} and CDC5 protein (Cell Division Cycle 5), which helps in transcriptional activity by interacting with polymerase II enzyme and initiating the

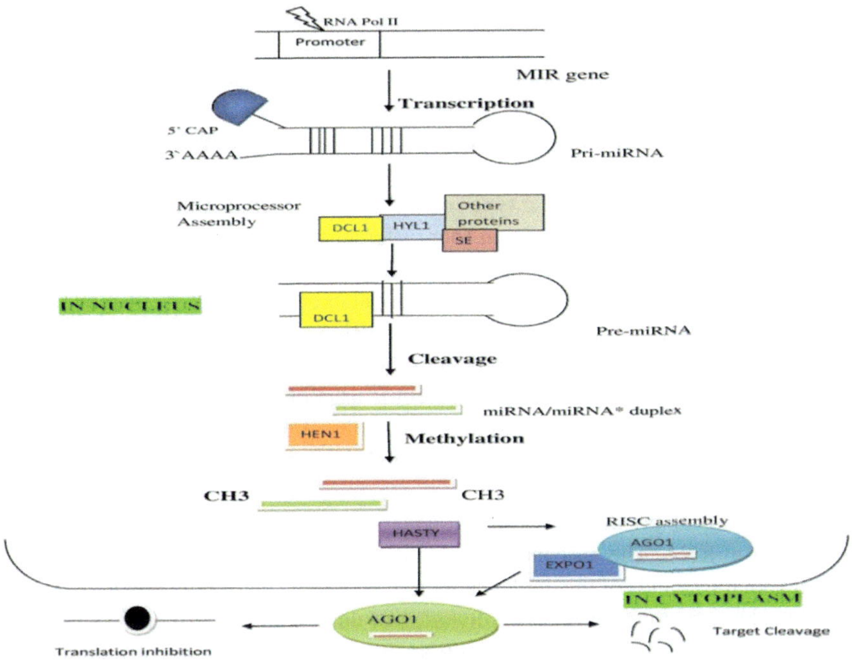

FIGURE 2.1

A diagrammatic demonstration of the important steps involved in the miRNA biosynthesis in plants. The sequence follows the MIR genes transcription by RNA polymerase II and production of pri-miRNA; conversion of pri-miRNA to pre-miRNA; and cleavage of pre-miRNA to produce duplex (miRNA/miRNA*) by dicing enzymes. Now, methylation of duplex by terminal methyltransferase HEN 1 takes place in the nucleus. Finally, the HASTY or EXPO 1 protein mediated translocation of methylated duplex to cytoplasm. The duplex is assembled with AGO1 protein to form RISC complex, which regulates splicing of target mRNA or translational restriction of mRNA.

process of transcription. Apart from initiation, elongation step of transcription is also crucial and requires involvement of conserved elongation factors such as ELP2 and ELP5 for transcription of miRNA in plants. Few other proteins such as the CDF2, known as cycling DOF (DNA binding with One Finger) Factor 2 further accelerates the transcription process (Sun et al., 2015). The Protein Phosphatase 4 (PP4), TPR1 and SNC1 (a disease resistant gene) are other such regulatory proteins required during transcription process (Cai et al., 2018; Wang et al., 2019).

(2) The very first product produced during transcription of MIR gene is known as primary miRNA abbreviated as pri-miRNA. A pri-miRNA is a spliced, 3′ polyadenylated and 5′ capped transcript (Rogers and Chen, 2013). It is usually a hairpin loop structure consisting of a lower stem, the miRNA/miRNA* region, upper stem, terminal loop, and two arms. The processing of this miRNA in plants requires a microprocessor assembly, which mainly consists of a Dicer-like RNase III endonuclease enzyme (DCLs), protein binding to double stranded RNA known as Hyponastic Leaves 1 (HYL1), and a Serrate (SE), which is a zinc-finger protein. The processing of pri- miRNAs takes place at subnuclear structures such as SmD3/SmB or dicing bodies (D-bodies) (Fang and Spector, 2007). Diverse types of DCLs are recognized in various plant species. In *Arabidopsis thaliana* four different types of DCL genes are found, out of which DCL1 is the first protein playing a major role in the pri-miRNA processing (Kurihara et al., 2004; Rajagopalan et al., 2006). Activation and deactivation of HYL1 necessary for pri-miRNA processing is controlled by light-dark and phosphorylation-dephosphorylation conditions. In dark and phosphorylated condition HYL1 protein is usually nonfunctional, whereas; light mediated dephosphorylation activates this protein and thus imposes regulatory control over processing of miRNA (Srivastava and Suprasanna, 2021).

(3) Processing of pri-miRNA leads to generation of pre-miRNA, known as pre-cursor miRNA. The nascent miRNA/miRNA* duplex consists of an overhang in both the strands of the duplex at their 3′ end. The length of overhang is about 2 nucleotides long. Further, methylation of duplex takes place at 2′ OH group located at 3′ end with the aid of an RNA methyltransferase enzyme commonly called as HUA enhancer 1 protein (HEN1) (Yu et al., 2005; Yang et al., 2006).

(4) Thereafter, transfer of methylated duplex (miRNA/miRNA*) to cytoplasm needs the help of a protein known as Hasty protein (HST), an analog of animal exportin 5 (EXPO5) protein (Park et al., 2005). Alternatively, transfer of methylated duplex also takes place with the help of exportin 1 protein (EXPO1) if already present as pre-assembled form in nucleus with a silencing complex commonly known as RISC (RNA-induced silencing complex) (Bologna et al., 2018).

(5) Now, the duplex transferred in cytoplasm binds with RISC for its final function. In the processed miRNA/miRNA* duplex form, one strand act as an active miRNA guide strand, while; another act as a passenger strand.

(6) Finally, the active miRNA guide strand selectively binds with the Argonaute (AGO) protein family associated with RISC. Different types of AGO proteins are available but the binding of majority of the miRNA takes place with AGO-1 associated protein (Mi et al., 2008). The AGO 1-RISC assembled miRNA then specifically binds to their target mRNAs for expression inhibition or cleavage, etc. On the other hand, the other strand in the double stranded structure is removed and degraded. The complete processing of miRNA and formation of functional RISC are under control of many positive and negative regulators. Proteins such as cyclophilin 40/Squint (CYP40/SQN), transportin 1 (TRN1) and a chaperone protein: heat shock protein 90 (HSP90) positively regulates the formation of functional RISC, while; its formation is down regulated by

Protein Phosphatase 5 (PP5) and Enhanced MiRNA Activity 1 (EMA1) proteins (Iki et al., 2012; Cui et al., 2016).

3.1 Approaches for identification of miRNA of plants

A diversity of miRNAs has now been identified from different plants, which has attracted immense interest due to their defensive role against biotic and abiotic stresses apart from their involvement in cellular growth and development. With the advent of experimental and computational approaches identification of miRNA has become quite easier these days (Verma et al., 2021). Their identification is also easier because of a few unique characteristics such as length of miRNA, which typically varies between 20 and 24 nucleotides in plants, and a well-defined stem loop hairpin structure (Devi et al., 2019). The identification of miRNA of plants can follow any of the four available approaches. They are mainly direct gene cloning, genetic screening, computational (sequence homology between known and unknown miRNAs), and expressed sequence tag analysis approaches (Reinhart et al., 2002; Devi et al., 2019).

The use of advance RNA sequencing method has also helped in identifying and functionally annotating various new miRNAs of different plant species (Zhang et al., 2022). According to the registered records, around 38,589 and 48,860 precursors and mature miRNAs belonging to more than 70 plant species including *Triticum aestivum, Oryza sativa, Arabidopsis thaliana, Zea mays, Glycine max, Solanum tuberosum, Nicotiana tabacum, Medicago truncatula, Sorghum bicolor, Brassica napus, etc.*, have been so far identified using experimental and computational approaches (http://www.mirbase.org, release 22). The methodological manner of identification of miRNA with any of the approaches includes selection of plant material, isolation of total RNA, enrichment of miRNA from total RNA, construction of cDNA library, sequencing and sequence processing of miRNA and their target prediction through bioinformatics tools such as psRNATarget, TargetMiner, RNAhybrid, miRanda, SVMicrO, and finally validation of miRNA and its target (Devi et al., 2019).

4. Functions of microRNAs in plants

4.1 Growth and development

The miRNA, a noncoding and endogenous RNA has been shown to regulate important functions such as normal developmental processes of plants (Singh et al., 2019). It induces expression of different regulatory genes. It has been determined to coordinate development of leaf, flower, seed, root, and stem. It maintains flowering time, assists in maintaining fruit quality and regulates plant fertility. It provides overall support to structure of plants. The miRNA has also been revealed to regulate phase changes, stimulates arbuscular mycorrhizal symbiosis, plant-microbe interaction, processes of metabolism and act as abiotic stress protector (Tungare et al., 2022).

In plants, the apical meristem is a collection of stem cells, which is important for self-replication, and differentiation of various organs (Maddala et al., 2023). The meristematic part of shoot known as shoot apical meristem (SAM) is important for development of organs exterior to ground. The shoot meristemless-Wuschel-Clavata (STM-WUS-CLV) pathway is determined to be involved in such meristematic activity (Dong et al., 2022). It was seen that the miRNAs control the expression of multiple genes involved in STM-WUS-CLV signaling pathway. The miR394 is reported as one of the key players regulating meristematic activity through STM-WUS-CLV pathway and maintain developmental processes of SAM. It has also been reported that during embryonic stage, production of miR394 takes place in L1 region of SAM, which further diffuses into organizing center (OC) through L3 layer. In L3, miR394 inhibits the expression of genes responsible for curling of leaf known as leaf curling responsiveness (LCR) gene and thereby negatively regulates the feedback-loop of CLV-WUS for development of SAM (Song et al., 2012; Knauer et al., 2013). On contrary, miR165/166 binds with a protein known as AGO10 to induce expression of transcription factors namely homeodomain leucine zipper belonging to Class III (HD-ZIP III TFs), which is important for development of SAM. It has been observed that the miR165/166 also helps in development of leaves from SAM. Similarly, miR390 and miR396 regulates expression of a number of other genes responsible for polarization of leaves. The miR396 imposes regulating effect on growth regulating factors (GRF) and thus affects dorsal-ventral polarity of leaves (Wang et al., 2011). The distribution and polarity of auxin is important for development of organs in plants. The miR160 has been revealed to target genes known as auxin response factor (ARF) such as ARF17, ARF16, and ARF10 thereby regulating auxin signaling pathways, which are found to be crucial in *Arabidopsis thaliana* somatic embryogenesis (Wojcik et al., 2017). One of the interesting miRNAs among others is miR156, which is a multifunctional miRNA. The protein namely Squamosa Promoter binding protein-Like (SPL protein) is the target of miR156 (Secic et al., 2021). It maintains overall architecture of plants by assisting in development of shoot and leaf, formation of vascular bundle, and flowering pattern (Zhang et al., 2022). The expression/formation of SPL transcription protein is down regulated by miR157 and miR156 at the early stages of development. Hence; decreased level of miR157 and miR156 increases the amounts of SBP/SPL protein, thus causing a change from growth to reproductive phase (He et al., 2018; Fouracre et al., 2021). It has been found that by regulating SPL protein, miR156 or miR157 changes morphology of leaves in *Arabidopsis thaliana*. The miR156-SPL also regulates shoot architecture and other agronomic traits. Flowering is an important phenomenon in plants. Development of flower mainly involves three different processes; flower induction, initiation of flower, and development of floral organs. These processes are also under the regulatory control of miRNAs. The over expression of miR156 inhibited the expression of SPL3/5, which later resulted in a delay in flowering time in *Arabidopsis thaliana* (Zhang et al., 2019). Similarly, the interactive roles of miR156 and SPL during development of flower in *Oryza sativa* have been reported (Xie et al., 2006). Another transcription factor AP2 is targeted by miR172 and thus regulates flowering time, flower morphology and development of floral organs (Aukerman and Sakai, 2003). The vascular bundle and secondary wall formation are regulated by miR156/miR157 by targeting HD-ZIP III transcription factor. The miRNA also plays a key role in controlling fruit and grain quality. The miR172 and its target transcription repressors namely AP2 and TOE3 are crucial for fruit growth (Patani et al., 2023). The ARF8 and ARF6 acts as activators of miR172. Thus, MADS/ARF-miR172-AP2 is considered as conserved regulatory pathway, which helps in morphogenesis of fruits. A study conducted with *Oryza sativa* showed that the grain quality of it is associated with regulation of osARF6-osAUX3. And, the miR167a maintains the grain weight and length by targeting osARF6 through promoter of osAUX3 (Qiao et al., 2021). Both regulation and maintenance of plant fertility are yet other role of miRNAs. Male sterility is important for production of hybrid seeds. A study done with anthers of *Arabidopsis thaliana* revealed that the transcription protein AtTTP helps in the maturation of miR160 (Shi et al., 2015). This miRNA controls the ARF17 expression, which is important for synthesis of callose and formation of pollen wall. Excessive expression of AtTTp leads to thin and reduced formation of callose wall, which is responsible for male sterility. Similarly, female plants are also sterilized with the help of miR167a (Liu et al., 2014). Hence, till now many different miRNAs are naturally expressed and identified, imparting their regulatory control over different target genes, which are directly or indirectly indulged in growth and developmental plant processes from juvenile to mature stage. The miRNA from different plant species, their target genes and responses are presented briefly in Table 2.1.

Table 2.1 A list of miRNAs from different species with their target genes and responses during growth and development of plants.

Plant species	Type of miRNA	Target genes	Responses	References
Growth and development				
Solanum lycopersicum	miR156	SPL family	Development of ovary and then fruit	Ferreira e Silva et al. (2014)
Arabidopsis thaliana	miR159 and miR319	MYB and TCP	Responsible for delayed flowering	Palatnik et al. (2007)
Dimocarpus longan	miR160	ARF17, ARF16, ARF10	Regulation of auxin responsible for leaf development; embryo development; signal transduction pathways	Lin et al. (2015), Mallory et al. (2005), Wang et al. (2005), Liu et al. (2007)

Continued

Table 2.1 A list of miRNAs from different species with their target genes and responses during growth and development of plants.—*cont'd*

Plant species	Type of miRNA	Target genes	Responses	References
Arabidopsis thaliana	miR164	NAC family	Over expression of it induces fusion of sepal and reduction in number of petal	Jung et al. (2009)
Cucumis sativus	miR164/ miR169	NAC family	Fruit spine development	Liu et al. (2018)
Arabidopsis thaliana	miR165 and miR166	HD-ZIPIII	Over expression it results in reduced development of floral organs	Zhong and Ye (2004), Jung and Park (2007)
Oryza sativa	miR167	ARFs	Development of male organ, root, stems, leaves flowers, seed, embryonic development; response towards stresses	Liu et al. (2012)
Solanum lycopersicum	miR167a	ARF6 and ARF8	Defects in floral developmental and female sterility	Liu et al. (2014)
Arabidopsis thaliana	miR160 and miR167	ARF6, ARF8, ARF17	Regulates initiation and development of adventitious roots through posttranscriptional control	Guttierrez et al. (2009)
Arabidopsis thaliana	miR172 and miR156	AP2-like genes; SM-LIKE 2, schnarchzapfen, target of early activation tagged 1/2/3	Regulates time of flowering and also floral organ formation by inhibiting particular target genes	Jung et al. (2007), Wollmann et al. (2010)
Solanum lycopersicum	miR319	TCP family	Induces a change in leaf size	Ori et al. (2007)
Arabidopsis thaliana	miR390	ARF4, ARF3, ARF2	Enhances lateral root growth	Marin et al. (2010)
Arabidopsis thaliana	miR394	Leaf curling responsiveness (LCRs)	Maintenance of leaf architecture; meristematic suppression	Knauer et al. (2013)
Oryza sativa	miR397	LAC	Controls size of grain and branching in panicle	Zhang et al. (2013)

Table 2.1 A list of miRNAs from different species with their target genes and responses during growth and development of plants.—*cont'd*

Plant species	Type of miRNA	Target genes	Responses	References
Oryza sativa	miR398	CSD1 and CSD2	Regulates seed development, panicle length, grain size and grain number	Zhang et al. (2017)

4.2 Abiotic stress tolerance

Growth and developmental processes of plants also depends on environmental factors including light, temperature, water content, mineral elements, etc., to a certain extent. However, extreme level of any of the environmental factors can result in the initiation of responses pertaining to abiotic stress (Zhang et al., 2022). The severe loss in the agricultural productivity is found to be accompanied by abiotic stresses. These abiotic stresses can direct the production of oxidative stress situation (Gangwar et al., 2023). The oxidative stress condition prevails due to the extreme temperature, high salinity, buildup of heavy metals and drought condition can influence the overall development of plant (Patani et al., 2024). It retards growth, affects seed germination, inhibits photosynthetic events, and minimizes crop yield (Kerchev et al., 2020). During such circumstances/in connection to abiotic pressures, plants do synthesize diverse forms of miRNAs, which play a chief role in controlling the expression of target genes and defend against the oxidative stress generated inside the plants (Pagano et al., 2021) (Table 2.2).

4.2.1 Cold and heat stress tolerance

Temperature is among the key physical factors and extreme cases of it impose severe threat to the plants. Both low and high temperature affects survival of plants, and leads to growth retardation, abnormal development, and loss of productivity (Parveen et al., 2023; Maddala et al., 2024). The adverse phenotypic symptoms include poor germination, yellowing of leaves, wilting, stunted growth, and even death of tissues. It also affects photosynthesis, nucleic acid and lipid metabolism, protein turnover, nitrogen metabolism, respiration, *etc.* (Zhang et al., 2022). The high and low temperature stress responsive miRNAs have been broadly studied to alleviate their adverse effects in plants (Zuo et al., 2021; Begum, 2022). The varied expression of miRNAs and their roles in providing defense against temperature stress have been considered in various species such as *Arabidopsis thaliana*, *Triticum aestivum*, *Zea mays*, *Oryza sativa*, *Gossypium herbaceum*, *Solanum tuberosum*, etc. (Zhang et al., 2022). A study conducted in a mutant namely auxin/indole acetic acid mutant (slr1) of *Arabidopsis thaliana* showed that one of the miRNAs viz.; miR169 and its forms miR169a/d/h were down regulated during continuous

Table 2.2 A lists of miRNAs from various plant species with their target genes and responses during abiotic stresses (heavy metal, drought, salt, cold, and heat).

Abiotic stresses	Plant species	miRNA	Target genes	Responses	References
Heavy metal	*Zea mays*	Aluminum: Zma-miR528; Cadmium: miR167, miR393	MATE Family GMZM2G148937	Up-regulates specific genes	Ding et al. (2013), Gao et al. (2019)
	Oryza sativa	Cadmium: miR268	NRAMP3	Over expression inhibits the growth of seedling	Ding et al. (2013, 2017)
	Oryza sativa	Cadmium: miR192	ATP- binding cassette (ABC) gene	Controls seed germination	Ding et al. (2013)
	Arabidopsis thaliana, Populus trichocarpa, and Vitis vinifera	Copper: miR397	–	Increases accumulation of copper; inhibit synthesis of plastocyanin or laccase	Wang et al. (2014)
	Arabidopsis thaliana	Copper: miR398	CSD1, CSD2, CCS1	Induces superoxide dismutase	Sunkar et al. (2006)
	Arabidopsis thaliana	Copper: miR857	SPL7	Copper concentration maintains activation and deactivation of SPL protein thus affecting the expression of Laccase enzyme and maintain Cu homeostasis	Yamasaki et al. (2009)
	Brassica juncea	Arsenic: miR838 and miR854	SULTR2; 1, SULTR2; 2, and serine acetyl transferase (SAT)	Controls sulfur metabolism	Srivastava et al. (2013)

Heat	*Solanum lycopersicum*	miR167, miR396, miR482, miR159, miR166, miR482	–	Up-regulation of miRNAs was not reported but long heat exposure could down-regulate the expression of these miRNAs.	Pan et al. (2017)
	Arabidopsis thaliana	miR160	–	The expression of heat shock proteins is under regulation	Lin et al. (2018)
	Zea mays	miR156, miR169, miR528, miR398 and Tae-miR818	–	Down regulates expression of genes	Raghupathy et al. (2016)
	Musa acuminata	miR156	SPL family	Correlation between miRNA156 and their response towards temperature stress	Zhu et al. (2019)
Cold stress	*Solanum tuberosum*	stu-novel-miR5125	StuABF8011 and StuAAO36826	Regulates abscisic acid metabolic pathways	Zhang et al. (2022)
	Arabidopsis thaliana	miR169a/miR169d/miR69h	NF-YA, YUC2	Regulates Auxin biosynthetic YUC2 gene and play a major role in providing cold tolerance	Aslam et el. (2020)
	Oryza sativa	miR319	OsTCP21 and OsPCF6	Enhances tolerance to cold	Wang et al. (2014)
	Oryza sativa	miR1320	AP2/ERF TF OsERF096	Repression of JA mediated signaling pathway regulates cold tolerance	Sun et al. (2022)
Drought stress	*Zea mays*	miR160	ARF10	Auxin production	Das and Mondal (2021)

Continued

Table 2.2 A lists of miRNAs from various plant species with their target genes and responses during abiotic stresses (heavy metal, drought, salt, cold, and heat).—*cont'd*

Abiotic stresses	Plant species	miRNA	Target genes	Responses	References
	Zea mays	miR528	POD	Reactive oxygen species detoxification	Das and Mondal (2021)
	Zea mays	miR167	PLD	Down regulated in *Zea mays and Oryza sativa* by ABA - induced signal pathway	Das and Mondal (2021)
	Zea mays	miR474	PDH	Regulates proline accumulation	Wei et al. (2009)
Salt stress	*Tamarix chinensis*	miR167	TcARF6	Down regulation of miRNA and specific gene in roots on exposure to 340 mM sodium chloride solution	Ye et al. (2020)
	Oryza sativa and Triticum aestivum	miR172	IDS1	It positively regulate salt tolerance; maintains ROS homeostasis	Cheng et al. (2021)
	Panicum virgatum	miR319	PvTCPs, PvPCF5	Promotes ethylene accumulation	Liu et al. (2019)

exposure to cold stress. The miR169 regulates biosynthesis of auxin by targeting YUC2 gene (flavin-binding monooxygenase family protein 2) through NF-YA (Nuclear Factor Y Subunit A) (Aslam et al., 2020). Hence, the miR169 has critical function in providing cold tolerance. Another example revealing importance of miRNA in cold stress resistance has been determined in the seedlings of *Solanum tuberosum*. Two miRNAs; stu-novel-miR5125 and stu-novel-miR10881 were identified to target expressions of Stu-ABF8011, Stu-AAO_36826 and GA_3ox_123158 genes respectively (Vora et al., 2023). The stu-novel-miR5125 down-regulates the expression of Stu-ABF8011 and Stu-AAO_36826, thus controlling hormonal signal transduction and carotenoid biosynthesis. This miRNA is crucial for the production of abscisic acid. While stu-novel-miR10881 negatively regulates GA_3ox_123158, which in turn regulates biosynthesis of gibberellic acid by imposing its control over gibberellin 3-beta-dioxygenase gene (GA3ox1) (Yan et al., 2021). Thus, cold resistance is provided by the regulation of gibberellic acid and abscisic acid signaling pathways. Similarly, increased expression of miR1320, Osa-miR528, Osa-miR319 and Osa-miR156 resulted in providing cold stress resistance in *Oryza sativa* plants by regulating hormonal signaling pathways with the help of transcription factors (Huo et al., 2022; Sun et al., 2022). Similarly, miRNA assisted regulation at post RNA synthesis level during increased temperature stress has been observed in *Zea mays*. The miR169 was the most differentially expressed miRNA followed by miR156, miR159, miR164, miR166, miR172, miR396, miR5381, etc. These miRNAs and their concerned targets such as miR169-SBP, miR159-MYB (transcription factor related to proto-oncogene), miR156-SBP/SPL, miR164-NAC (transcription factor), miR166- HD zip (homeodomain-zipper motif), miR172-AP2/ERF (ethylene response factor), miR396-GRF (growth-regulating factor1), and miR5381-SAC2 (adenylate cyclase) were negatively regulated under high temperature in *Zea mays* (Zhang et al., 2019). The miRNA assisted regulation under heat stress was seen to be developmental stage specific as was determined during development of anther in *Gossypium herbaceum* (Chen et al., 2020). The function of miRNAs such as miR9662, and miR528, which are heat responsive in *Triticum aestivum* was studied by Ravichandran et al. (2019), which revealed regulation of stress related proteins in mitochondria thereby antioxidant activity.

4.2.2 Drought stress tolerance

The condition of prolonged dry period due to the lack of precipitation and water shortage in the natural climate results in drought condition. It is a harsh abiotic stress factor, which influences the growth and developmental processes of plants, and reduces productivity drastically, hence; leading to a risk to global food protection (Fahlgren et al., 2007). The research on the drought responsive miRNA to understand its molecular mechanism in adjusting the plants with adverse condition and enhance their overall performance is on the way (Khare et al., 2017). Drought responsive different miRNAs produced in varied varieties of plants has been investigated by various research groups, which includes different monocotyledonous and dicotyledonous species such as *Arabidopsis thaliana* (Yang et al., 2019; Yu et al.,

2019), *Glycine* max (Zhou et al., 2020), *Zea mays* (Das and Mondal, 2021), and several other cereals (Geng et al., 2021; Singroha et al., 2021). The abscisic acid alone or in union with other hormones such as gibberellin, auxin, salicylic acid, jasmonic acid, ethylene, provides an adaptive role to plants under water depletion by regulating stomatal conductance. The abscisic acid signaling pathway is controlled by miRNAs. Different transcription factors namely NF−Y, MYB, HD-ZIP and NAC are the target genes controlled by miRNAs such as miR169, miR159, miR167, miR156, miR160, miR390 and miR393 respectively, either through abscisic acid dependent or independent signaling pathways (Singroha et al., 2021). It has been seen that the overexpression of gma-miR169c and gma-miR398c has resulted in lowered drought resistance in *Arabidopsis thaliana* by suppressing target genes (Yu et al., 2019; Zhou et al., 2020). However, heightened expression of gma-miR172c provided increased drought stress resistance (Li et al., 2016). Similarly, in *Zea mays*, miRNAs such as miR474, miR402, miR397b, miR393, miR389a, miR167, miR160, and miR159 were identified to regulate expressions of different target genes such as NAC, MYC/MYB, ZF-HD, PDH, DREB/CBF and ABRE involved in providing drought stress tolerance (Das and Mondal, 2021). In this series, few of the target genes such as miR402-MYB101/MYB33 and miR393-MYB101/MYB33 were seen to be up-regulated, whereas genes such as miR164-NAC and miR398a-MYB101/MYB33 were down-regulated. Similarly, the auxin response pathway is induced by up-regulation of miR160-ARF10, and ROS detoxification was controlled by down-regulation of miR528-POD (Das and Mondal, 2021).

4.2.3 Salt stress tolerance

Salinity is one more major stress as excess existence of salt in the soil depletes the water potential of it thereby reducing the intake of water by the roots. Salt stress imposes hyperosmoticity, ionic imbalance and toxicity in plants due to oxidative stress (Kumar et al., 2018). It also causes detrimental effects on developmental processes, and crop yield (Wani et al., 2020). Many evidences showed that the miRNAs act as posttranscriptional regulators and provides salt stress tolerance to plants. The miRNAs help to increase salt tolerance so as to plants cannot compromise in their performance and survival (Sun et al., 2021). Plants can resist the ill effects of salt stress by regulating the expressions of miRNAs and their associated target genes. The miR393 regulated salt stress tolerance in *Arabidopsis thaliana* by abscisic acid signaling pathway and through a scaffold protein (Denver and Ullah, 2019). The over appearance of miR156 suppresses the performance of its concerned gene SPL and thereby reduces uptake of sodium ions thus maintaining the homeostasis in high salinity stress (Arshad et al., 2017). Similarly, miR319 is another abiotic pressure regulator in C3 and C4 plants. The miR319 positively regulates its target genes and induces salt stress tolerance and promote ethylene production (Liu et al., 2019). The miR172 provided salt stress tolerance to *Oryza sativa* and *Triticum aestivum* (Cheng et al., 2021). It targets IDS1 (Indeterminate Spikelet 1) gene, which thereafter binds with promoter region of ROS scavenging gene for its repression.

Thus, maintaining a balance in ROS level under salt stress condition. The salt inducible miRNAs in different species such as *Arabidopsis thaliana*, *Medicago sativa*, *Gossypium herbaceum*, are the key regulating elements, which reduces building up of ROS by controlling expression of their target genes such as ARS, SPL, LAC (Laccase enzyme), AFB (Auxin F box protein), IDS1, and FSD1 (iron superoxide dismutase 1) (Arshad et al., 2017; Wang et al., 2019; Nguyen et al., 2020; Cheng et al., 2021), thereby providing salinity stress tolerance.

4.2.4 Heavy metal stress tolerance

One of the global serious and challenging problems is contamination of heavy metals. The accumulation and building up of heavy metals in various sites such as soil, sediment, and nearby water bodies results in contamination of surrounding environment and ecosystem. In fact, with the rapid progression in industrial establishments, different heavy metals are being released in industrial effluents, leading to pollution in different ecosystems. The excess application of heavy metals in terrestrial environment damages the structure of soil thus affecting growth, quality and yield of crops (Tiwari and Lata, 2018). A few of the toxic heavy metals such as cobalt, manganese, zinc nickel, and copper (Mn, Co, Ni, Zn, and Cu) are important and essentially required for proper maintenance of growth of plants but in minute concentration. However, higher concentrations of these lead to adverse impacts on health of plants (Noman et al., 2018). On the contrary, heavy metals such as mercury (Hg), lead (Pb), arsenic (As), cadmium (Cd), chromium (Cr), are severely toxic and lead to various symptoms such as plant growth reduction, damaging of metabolic pathways, morphological changes and also reduction in crop yield. Exposure to heavy metals also results in generation of ROS, such as superoxide anion (O^{2-}), hydrogen peroxide (H_2O_2), hydroxyl ion (OH-) (Sharma et al., 2012). Plants are well known to develop various defense mechanisms to withstand toxic impacts of heavy metals by reducing uptake of metal ions, sequestration of metals into the vacuoles, metal ions chelation, production of various antioxidant enzymes and proteins, activation of regulatory mechanisms including synthesis of miRNAs (Shahid et al., 2015). Heavy metals such as Cd, Pb, Cr, Hg, and As are considered as highly toxic metals/metalloids in the environmental context by “Agency for Toxic Substances and Disease Registry (ATSDR, 2003)”, and are also involved in causing human poisoning upon accumulation through food chain. Involvement of different miRNAs in providing tolerance and alleviating negative effects of various heavy metals are mentioned below in orderly manner.

4.2.4.1 miRNA and cadmium toxicity

Cadmium is a toxic heavy metal usually released out of different anthropogenic actions such as metallurgical work, use of synthetic phosphate-based fertilizers, mining, combustion of fossil fuel, sewage sludge, landfills, wastes from municipalities, use of Cd–Li batteries (Bigalke et al., 2017; Kubier et al., 2019). The fate of transportation of Cd in the environment can be local (industrial sites, mines) or diffused (atmospheric and agricultural activities). The higher concentration of Cd imposes severe indications in plants such as necrosis, chlorosis, reduced growth, stopping of carbon fixation, decreased photosynthetic activity, minimum rate of transpiration, impaired stomatal conductance, water shortage, excess production of ROS, reduced nutrient uptake (Haider et al., 2021). It also affects seed germination by hindering production of hormones and activities of enzymes such as amylases, proteases, and phosphatases, which are necessary for process of germination. Cadmium induces change in membrane lipid composition, inhibition in uptake of macro- and micro-nutrients, inhibition of electron transport chain, reduced activities of enzymes of the Calvin cycle, imbalance in cellular moisture content, etc. (Haider et al., 2021). Movement of Cd in the soil is relatively high and is readily absorbed by plant roots. It gets easily transported to edible parts of the plants and accumulates there thus contaminating the food chain and causing various health issues in human being (Yang et al., 2022). It also binds with different symplastic and apoplastic proteins such as ATPase (ATPase), Fe-superoxide dismutase (Fe-SOD), natural resistance-associated macrophage protein (NRAMP), and ATP binding cassette (ABC), and leads to oxidative stress condition by interfering with their activities (Ding and Zhu, 2009; Huang et al., 2009; Pegler et al., 2021).

The miRNAs have an important participation in providing defense against Cd toxicity. Different types of miRNAs are included in both up and down regulation of different target genes associated with Cd toxicity. It includes miR156, miR159, miR169, miR168, miR167, miR166, miR171, miR192, miR268, miR393, miR398, miR395, miR529 and miR528 (Yang et al., 2022). These miRNAs providing tolerance against Cd stress by regulating either transport or chelation of Cd or synthesis of hormones.

Presence of dedicated transporters in the cell membranes and organelles plays an important role in uptake, translocation and chelation of Cd thereby sustaining steady level of it inside the cell (Srivastava and Suprasanna, 2021). Among different transporters so far reported, NRAMP and ABC are most important ones. The NRAMP is an extremely conserved membranous protein found in most of the organisms. However, ABC transporters are present on the surface of the cell organelles and helps in chelation and absorption of metals in the vacuoles (Yang et al., 2022). It has been shown that different miRNAs are responsible for a variety of functions during Cd toxicity. Among them, the roles of miR167 and miR268 were found to be quite important. Both these miR167 and miR268 regulates expression of genes encoding NRAMP transporters, while miR159 and miR192 regulates ABC genes (Yang et al., 2022). While studying about control of these genes by miRNAs in *Oryza sativa* under Cd stress, it was observed that the miR167 and miR159 acts as positive regulators of NRAMP and ABC genes. But, over expression of miR268 and miR192 down regulates the expression of NRAMP and ABC genes. Apart, miR166 was also seen to regulate Cd translocation. The homeodomain containing protein 4 (OSHB4) is one of the targets of miR166. This gene encodes for HD-Zip protein, which responds against Cd toxicity. The level of expression of miR166 regulates performance of OSHB4 gene either positively or negatively. The over expression of miR166 reduces expression of OSHB4 gene thereby interferes with translocation of Cd from roots and diminishes plant's oxidative stress. Along with transport and translocation of Cd, miRNAs such as miR156 also functions as regulators of the synthetic pathways of antioxidants. It regulates expression of various enzyme based and non-enzyme-based antioxidant components, which includes catalase (CAT), SOD, ascorbate peroxidase (APX), as well as non-enzyme-based agents such as glutathione (GSH) and ascorbic acid (AsA). In addition, miR395 plays a key role in sulfur assimilation by regulating sulfur assimilating genes such as ATP sulfonylase 1, 3, 4 (APS 1, 3, 4) and sulfate transporter 2; 1 (SULTR2; 1) (Yang et al., 2022). Sulfur is an important macronutrient and provides defense against heavy metal stress. Expression of miR395 is under control of a transcription factor; sulfur-restricted 1 (SLIM1). The miR395 is present in between SLIM1 and SULTR2; 1 thus predicting their regulation as SLIM1-miR395-SULTR2; 1. It has been seen to control sulfate assimilation thereby alleviating Cd pressure (Zhang et al., 2013; Yang et al., 2022). Further, miR393 and miR168 targets transport inhibitor response 1 (TIR1) and lipoxygenase 1 (LOX1), which are involved in synthesis of hormones such as auxin and JA respectively. These hormones are also involved in improving tolerance against Cd toxicity (Takahashi, 2010; Ding et al., 2011; Noman and Aqeel, 2017).

4.2.4.2 miRNA and arsenic toxicity

Arsenic is widely distributed and highly toxic metalloid in nature. It is considered as an environmental pollutant of serious concern as it imposes severe effects on ecosystem and causes health hazards. Contamination of As is either natural; through volcanic eruption and erosion of rocks, or man-made such as smelting, mining, and frequent use of pesticides (Neumann et al., 2010; Tiwari and Lata, 2018). The accumulation of As in soil results in lesser productivity of crops and food chain contamination (Verma et al., 2016). In nature, As exists in two toxic forms namely; arsenate and arsenite [As (V) and As (III)]. Among these, As (III) is comparatively more lethal than the As (V). As (III) binds with proteins having sulfhydryl groups and hampers the activities of enzymes. On the contrary, As (V) inhibits phosphorylation reaction and synthesis of ATP (Khalid et al., 2017). The toxic impacts of As in plants includes inhibited seed germination, reduced rate of photosynthesis, transpiration, respiration, impaired carbon and nitrogen metabolism, etc. (Yang et al., 2022). Toxicity of As also results in generation of ROS, cell membrane damage and decreased actions of antioxidant enzymes. The disruption in cell membrane structure damages the transporters imbedded in it and involved in uptake and absorption of metal ions required by plants (Noman et al., 2019).

Exposure of As induces generation of miRNA, which later acts as regulator and helps in providing tolerance against stress and also helps in alleviation of harmful effects. Sharma et al. (2015) demonstrated differential expression of miRNAs in *Oryza sativa*. Expressions of miRNAs up-regulated in connection to As (V) and As (III) were miR408, miR399, miR396, miR1861, miR2907, miR2102, and miR528. While, miR529, miR395, miR820, miR164, miR171, miR1432, and miR1846 were seen to be down-regulated. As discussed earlier, one of the important functions of miRNA in answer to metal toxicity is to control expression of genes of various metal ion transporters, which are important for metal uptake and its translocation. The miR159 targets one of the transporters namely oligopeptide transporter 1 gene (OPT1) (Srivastava et al., 2013). In addition, miR838 targets a lipase encoding gene, which is involved in biosynthesis of oxylipin and JA. Similarly, miR528 is found to be closely involved in providing As stress tolerance (Liu et al., 2015). The miR528 targets few of the important genes viz.; gene encoding protein, which is like plastocyanin (LOC_Os07g38290.1), two genes encoding laccase (LOC_Os01g62600.1, LOC_Os01g44330.1), two genes encoding multicopper oxidase (LOC_Os01g03620.1, LOC_Os01g03640.1), and genes encoding enzyme of transferase family, galactosyltransferase (LOC_Os07g09690.1, LOC_Os12g41956.1) (Liu et al., 2015). Similarly, few other miRNAs such as miR408, miR159 and miR167 targets ascorbate oxidase, ATP-binding cassette, and NRAMP encoding genes respectively (Zhou et al., 2012a,b). The miR395 also regulates expressions of genes encoding ATP sulfurylases and sulfate transporters participating in sulfur uptake and its assimilation (Sharma et al., 2015).

4.2.4.3 miRNA and chromium toxicity

A number of anthropogenic activities serve as common sources of Cr contamination in the nature. Chromium occurs in two forms in nature; a stable Cr (III) form (trivalent form) and a Cr (VI) form (hexavalent form) (Cervantes et al., 2001). Later one is comparatively more toxic, mutagenic and carcinogenic. The solubility and movement of Cr hexavalent form in the soil are more as against to Cr (III) (Shahid et al., 2017). Chromium is released from various industries in water and leads to severe pollution. It is indeed needed in small amount for better growth, while higher amount of Cr causes severe ill effects in plants. It inhibits seed germination, disrupts nutrient balance, affects enzymes activities, reduces root growth, and induces oxidative stress thereby death of plants (Noman et al., 2019). Liu et al. (2015) determined differential expressions of 16 novel miRNAs and 54 known miRNAs in Cr stressed *Raphanus sativus*. Genes targeted by these miRNAs encodes different transcription factors (MYBs, SPLs, bZIPs and ERFs), which regulates induction of genes included in both uptake and homeostasis of Cr inside the plants. Similarly, in Cr treated *Oryza sativa*, 512 differentially expressed miRNAs have been determined. It included 13 different conserved miRNAs (miR159, miR156, miR166, miR160, miR171, miR444, miR397, miR408, miR169, miR396, miR5072, miR2877, and miR1883) (Dubey et al., 2020). Similarly, in *Nicotiana tabacum*, 53 identified miRNAs and 29 new miRNAs were reported. Out of which, miR156 and miR167 were seen to be the key players (Bukhari et al., 2015). Sequencing of miRNAs revealed presence of 104 preserved type miRNAs and 158 miRNAs of nonconserved type in *Miscanthus sinensis* (Nie et al., 2021). These miRNAs probably played important role in detoxification of Cr involving various transcription factors, ABC transporters, metal transporters, growth hormones and heat shock proteins (Nie et al., 2021).

4.2.4.4 miRNA and mercury toxicity

Mercury is one more toxic heavy metal, which under normal temperature and pressure exists in liquid form. In nature, it exists in either of three forms; as metallic, organic or inorganic (Guo et al., 2021). It exists in soil, water and atmosphere. The most accessible form of Hg in nature is Hg^{2+}, which is easily absorbed by plant roots and imposes toxicities such as generation of oxidative stress by ROS, deformity in cell shape by disrupting structure of cell membrane, inhibiting seed germination and affecting photosynthesis process (Wang and Greger, 2004; Azevedo and Rodriguez, 2012). It binds with sulfhydryl group of protein, one of the important biological macromolecules of cell membrane, and disrupts overall functioning of plasma membrane. Thus, to conquer toxic effects of Hg the miRNA plays a necessary role in post transcriptional gene regulation. In response to Hg stress, different miRNAs are expressed and involved in providing tolerance against it. A group of Hg stress responsive miRNAs were first reported in *Medicago truncatula* (Zhou et al., 2008). They recorded presence of about 210 known miRNAs and 54 potential miRNA candidates. Among which, expressions of miR319, miR393, and miR171 were up-regulated, while miR398 and miR166 were seen to be down regulated, however; expressions of miR395 and miR160 were not changed (Zhou et al., 2008). The miRNA usually targets the genes encoding enzymes or proteins, which are involved in giving resistance against stresses. The miR2681 targets a gene encoding a disease resistant protein (TIR-NBS-LRR); miR2708 targets the gene coding for salt-tolerant protein (TC114805), and miR2687 targets an enzyme involved in cell wall modification [hydrolase (XTH)/xyloglucan endotransglucosylase], which is involved in providing resistance to stresses by altering development of cell wall. An important protein; OsPDIL1 found in *Oryza sativa* is known as protein folding catalyst. The miR5144 (osa-miR5144-3p) targets OsPDIL1 protein and mediates formation of disulfide bonds in protein thereby improved resistance to Hg (Xia et al., 2018).

4.2.4.5 miRNA and lead toxicity

Lead is a well-known toxic and nonessential heavy metal. It is not important for any cellular and metabolic processes of plants, but is easily absorbed and accumulated in them. The higher amount of Pb in cells can produce lethal symptoms such as reduced plant growth and photosynthetic rate, impaired mineral uptake, blackening in roots (Hadi, 2015). Its exposure promotes up-regulation of the expressions of miR395, miR396, miR159, miR167, and miR162, while miR398, miR399, miR156, miR414, miR5658, and miR833 were seen to be down regulated in *Gossypium herbaceum* (Noman et al., 2019).

5. Applications of microRNAs in improvement of crops

Persistence of environmental problems for relatively longer duration can probably be tackled following various advance tools and technologies. One such tool is the use of miRNAs. However, the manipulation of naturally and endogenously expressed miRNA through advance technologies such as genetic engineering, can improve the quality, yield as well as nutritional values of crops, can induce higher production of biomass thereby biofuel, and can raise biotic and abiotic stress tolerance in plants (Basso et al., 2019; Begum, 2022; Raza et al., 2023). These are few of the supreme challenges for the world. Hence, the use of miRNAs is becoming one of the promising and novel tools for producing transgenic plants with better agronomic traits. Various strategies such as over expression and repression of miRNA, gene editing, RNA interference, and artificial miRNA may probably be used to produce miRNA based genetically engineered crops (Basso et al., 2019) (Fig. 2.2). The

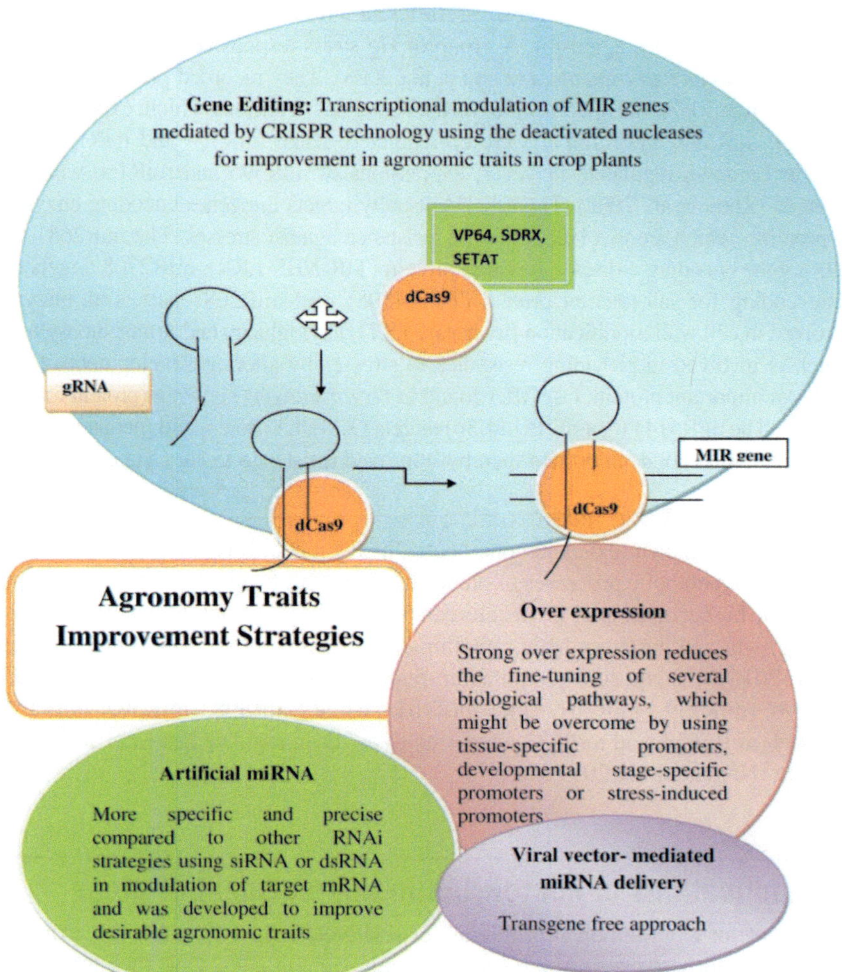

FIGURE 2.2

Different high throughput biotechnological approaches for manipulating miRNAs for improvising agronomic traits.

over expression of miRNAs negatively targets their associated genes and assists in producing stress resistant plants. The down regulation of miRNA is required if its target act as positive regulator. Alternatively, RNA interference technology can be used to target miRNAs to minimize pleiotropic effects produced due to their over expression. Even the artificially produced miRNAs can be utilized for crop improvement. The miRNA based genetic manipulation to induce improvements in crops can fall under any of the two categories; (a) improvements in architecture, growth of crops, and (b) improvements against abiotic as well as biotic stresses. Improvements in crop architecture include biomass production, grain yield, nutritional property, fruit and flower quality, plant fertility, etc. While, improvements in stress resistance includes biotic (viral, bacterial, fungal, insect) and abiotic (heat, cold, drought, salinity, heavy metal) stresses (Raza et al., 2023).

6. Conclusions and future prospects

The miRNAs are commonly distributed in most of the eukaryotes. It is transcribed from MIR genes, which are further modified to generate mature miRNA. The non-coding miRNAs are regarded as key controller of gene expression. It controls expression of specific mRNA at post-RNA synthesis level either by protein synthesis inhibition or cleavage. With the introduction of advance sequencing technologies and other advance biotechnological tools many miRNAs are being continuously reported from various plant species and their target genes are predicted. It is becoming the matter of great importance to identify different miRNAs from a variety of plant species and to study their roles in developmental processes. In this Chapter, miRNAs and their biosynthetic pathway has been discussed. It also included versatile role of miRNA in plant growth, and to provide resistance against physical stresses. The novel MIR gene-based tools, and miRNA mediated genetic manipulation have been evolved as effective and powerful approaches to improve several agronomic traits and develop new stress resistant varieties of plants. The over expression of specific miRNA gene or the use of artificially synthesized miRNAs to down regulate a particular gene are important strategies. Thus, exploitation of advance biotechnological tools and manipulating the expression of miRNA may possibly result in development of various stress resistant plants and improved products of agricultural importance.

Conflict of interest

The authors declare that they have no conflict of interest.

References

Arshad, M., Gruber, M.Y., Wall, K., Hannoufa, A., 2017. An insight into microRNA156 role in salinity stress responses of alfalfa. Front. Plant Sci. 8, 356. https://doi.org/10.3389/fpls.2017.00356.

Aslam, M., Sugita, K., Qin, Y., Rahman, A., 2020. Aux/IAA14 Regulates microRNA-mediated cold stress response in Arabidopsis roots. Int. J. Mol. Sci. 21, 8441. https://doi.org/10.3390/ijms21228441.

ATSDR, 2003. Agency for Toxic Substances and Disease Registry. Available from: http://www.atsdr.cdc.gov/.

Aukerman, M.J., Sakai, H., 2003. Regulation of flowering time and floral organ identity by a MicroRNA and its APETALA2-like target genes. Plant Cell 15, 2730–2741. https://doi.org/10.1105/tpc.016238.

Azevedo, R., Rodriguez, E., 2012. Phytotoxicity of mercury in plants: a review. J. Bot., Le 2012, 848614.

Basso, M.F., Ferreira, P.C.G., Kobayashi, A.K., Harmon, F.G., Nepomuceno, A.L., Molinari, H.B.C., Grossi-de-Sa, M.F., 2019. MicroRNAs and new biotechnological tools for its modulation and improving stress tolerance in plants. Plant Biotechnol. J. 17 (8), 1482–1500. https://doi.org/10.1111/pbi.13116.

Begum, Y., 2022. Regulatory role of microRNAs (miRNAs) in the recent development of abiotic stress tolerance of plants. Gene 821, 146283. https://doi.org/10.1016/j.gene.2022.146283.

Bigalke, M., Ulrich, A., Rehmus, A., Keller, A., 2017. Accumulation of cadmium and uranium in arable soils in Switzerland. Environ. Pollut. 221, 85–93.

Bologna, N.G., Iselin, R., Abriata, L.A., Sarazin, A., Pumplin, N., Jay, F., et al., 2018. Nucleo-cytosolic shuttling of ARGONAUTE1 prompts a revised model of the plant microRNA pathway. Mol. Cell. 69, 709–719.

Borges, F., Martienssen, R.A., 2015. The expanding world of small RNAs in plants. Nat. Rev. Mol. Cell Biol. 16, 727–741. https://doi.org/10.1038/nrm4085.

Budak, H., Akpinar, B.A., 2015. Plant miRNAs: biogenesis, organization and origins. Funct. Integr. Genom. 15, 523–531.

Bukhari, S.A., Shang, H., Zhang, S., Zheng, M., Zhang, W., Wang, G., Shamsi, T.Z., Wu, F., 2015. Genomewide identification of chromium stress-responsive micro RNAs and their target genes in tobacco (Nicotiana tabacum) roots. Environ. Toxicol. Chem. 34, 257–258. https://doi.org/10.1002/etc.3097.

Cai, Q., Liang, C., Wang, S.K., Hou, Y.N., Gao, L., Liu, L., et al., 2018. The disease resistance protein SNC1 represses the biogenesis of microRNAs and phased siRNAs. Nat. Commun. 9, 5080.

Cervantes, C., Campos-García, J., Devars, S., Gutiérrez-Corona, F., Loza-Tavera, H., Torres-Guzmán, J.C., Moreno-Sánchez, R., 2001. Interactions of chromium with microorganisms and plants. FEMS Microbiol. Rev. 25, 335–347.

Chen, Q.J., Deng, B.H., Gao, J., Zhao, Z.Y., Chen, Z.L., Song, S.R., et al., 2020. A miRNA-encoded small peptide, vvi-miPEP171d1, regulates adventitious root formation. Plant Physiol. 183, 656–670. https://doi.org/10.1104/pp.20.00197.

Cheng, X., He, Q., Tang, S., Wang, H., Zhang, X., Lv, M., et al., 2021. The miR172/IDS1 signaling module confers salt tolerance through maintaining ROS homeostasis in cereal crops. New Phytol. 230, 1017–1033. https://doi.org/10.1111/nph.1721.

Cui, Y.W., Fang, X.F., Qi, Y.J., 2016. TRANSPORTIN1 promotes the association of microRNA with ARGONAUTE1 in Arabidopsis. Plant Cell 28, 2576–2585.

Czech, B., Munafo, M., Ciabrelli, F., Eastwood, E.L., Fabry, M.H., Kneuss, E., et al., 2018. piRNA-guided genome defense: from biogenesis to silencing. Annu. Rev. Genet. 52, 131–157. https://doi.org/10.1146/annurev-genet-120417-031441.

Das, R., Mondal, S.K., 2021. Plant miRNAs: biogenesis and its functional validation to combat drought stress with special focus on maize. Plant Gene 27, 100294. https://doi.org/10.1016/j.plgene.2021.100294.

Denver, J.B., Ullah, H., 2019. miR393s regulate salt stress response pathway in *Arabidopsis thaliana* through scaffold protein RACK1A mediated ABA signaling pathways. Plant Signal. Behav. 14, 1600394. https://doi.org/10.1080/15592324.2019.1600394.

Devi, K., Dey, K.K., Singh, S., Mishra, S.K., Modi, M.K., Sen, P., 2019. Identification and validation of plant miRNA from NGS data-an experimental approach. Brief Funct. Genomics 18 (1), 13–22. https://doi.org/10.1093/bfgp/ely034.

Ding, H., Gao, J., Luo, M., Peng, H., Lin, H., Yuan, G., 2013. ScienceDirect identification and functional analysis of MiRNAs in developing kernels of a viviparous mutant in maize. Crop J. 1, 115–126.

Ding, Y., Chen, Z., Zhu, C., 2011. Microarray-based analysis of cadmium-responsive MicroRNAs in rice (*Oryza sativa*). J. Exp. Bot. 62, 3563–3573.

Ding, Y., Wang, Y., Jiang, Z., Wang, F., Jiang, Q., Sun, J., Chen, Z., Zhu, C., 2017. MicroRNA268 overexpression affects rice seedling growth under cadmium stress. J. Agric. Food Chem. 65, 5860–5867.

Ding, Y.F., Zhu, C., 2009. The role of MicroRNAs in copper and cadmium homeostasis. Biochem. Biophys. Res. Commun. 386, 6–10.

Dong, Q., Hu, B., Zhang, C., 2022. microRNAs and their roles in plant development. Front. Plant Sci. 13, 824240. https://doi.org/10.3389/fpls.2022.824240.

Dubey, S., Saxena, S., Chauhan, A.S., Mathur, P., Rani, V., Chakrabaroty, D., 2020. Identification and expression analysis of conserved microRNAs during short and prolonged chromium stress in rice (*Oryza sativa*). Environ. Sci. Pollut. Res. Int. 27, 380–390. https://doi.org/10.1007/s11356-019-06760-0.

Fahlgren, N., Howell, M.D., Kasschau, K.D., Chapman, E.J., Sullivan, C.M., Cumbie, J.S., et al., 2007. High-throughput sequencing of Arabidopsis microRNAs: evidence for frequent birth and death of MIRNA genes. PLoS One 2, e219. https://doi.org/10.1371/journal.pone.0000219.

Fang, Y.D., Spector, D.L., 2007. Identification of nuclear dicing bodies containing proteins for microRNA biogenesis in living Arabidopsis plants. Curr. Biol. 17, 818–823.

Ferreira e Silva, G.F., Silva, E.M., Azevedo Mda, S., Guivin, M.A., Ramiro, D.A., Figueiredo, C.R., et al., 2014. microRNA156-targeted SPL/SBP box transcription factors regulate tomato ovary and fruit development. Plant J. 78, 604–618. https://doi.org/10.1111/tpj.12493.

Fouracre, J.P., He, J., Chen, V.J., Sidoli, S., Poethig, R.S., 2021. VAL genes regulate vegetative phase change via miR156-dependent and independent mechanisms. PLoS Genet. 17, e1009626. https://doi.org/10.1371/journal.pgen.1009626.

Gangwar, P., Johri, P., Datta, R., Singh, S., Trivedi, M., 2023. Overview on 1 phytopharmaceuticals and biotechnology of herbal plants. In: Phytopharmaceuticals and Biotechnology of Herbal Plants. CRC Press, pp. 1–17.

Gao, J., Luo, M., Peng, H., Chen, F., Li, W., 2019. Characterization of cadmium-responsive MicroRNAs and their target genes in maize (*Zea mays*) roots. BMC Mol. Biol. 20, 14.

Gao, Z., Nie, J., Wang, H., 2020. MicroRNA biogenesis in plant. Plant Growth Regul. 2020, 1–12. https://doi.org/10.1007/s10725-020-00654-9.

Geng, Z., Liu, J., Li, D., Zhao, G., Liu, X., Dou, H., et al., 2021. A conserved miR394-targeted F-Box gene positively regulates drought resistance in foxtail millet. J. Plant Biol. 64, 243–252. https://doi.org/10.1007/s12374-021-09303-8.

Gielen, H., Remans, T., Vangronsveld, J., Cuypers, A., 2012. MicroRNAs in metal stress: specific roles or secondary responses? Int. J. Mol. Sci. 13, 15826–15847.

Griffiths-Jones, S., Saini, H.K., Van Dongen, S., Enright, A.J., 2008. miRBase: tools for microRNA genomics. Nucleic Acids Res. 36, D154–D158. https://doi.org/10.1093/nar/gkm952.

Guo, P., Du, H., Wang, D., Ma, M., 2021. Effects of mercury stress on methylmercury production in rice rhizosphere, methylmercury uptake in rice and physiological changes of leaves. Sci. Total Environ. 765, 142682.

Gutierrez, L., Bussell, J.D., Pacurar, D.I., Schwambach, J., Pacurar, M., Bellini, C., 2009. Phenotypic plasticity of adventitious rooting in Arabidopsis is controlled by complex regulation of AUXIN RESPONSE FACTOR transcripts and MicroRNA abundance. Plant Cell 21, 3119–3132.

Hadi, F., 2015. A mini review on lead (Pb) toxicity in plants. J. Biol. Life Sci. 6, 91–101.

Haider, F.U., Liqun, C., Coulter, J.A., Cheema, S.A., Wu, J., Zhang, R., Wenjun, M., Farooq, M., 2021. Cadmium toxicity in plants: impacts and remediation strategies. Ecotoxicol. Environ. Saf. 211, 111887. https://doi.org/10.1016/j.ecoenv.2020.111887.

He, J., Xu, M., Willmann, M.R., McCormick, K., Hu, T., 2018. Thresholddependent repression of SPL gene expression by miR156/miR157 controls vegetative phase change in *Arabidopsis thaliana*. PLoS Genet. 14, e1007337. https://doi.org/10.1371/journal.pgen.1007337.

Huang, S.Q., Peng, J., Qiu, C.X., Yang, Z.M., 2009. Heavy metal-regulated new MicroRNAs from rice. J. Inorg. Biochem. 103, 282–287.

Huo, C., Zhang, B., Wang, R., 2022. Research progress on plant noncoding RNAs in response to low-temperature stress. Plant Signal. Behav. 17, 2004035. https://doi.org/10.1080/15592324.2021.2004035.

Iki, T., Yoshikawa, M., Meshi, T., Ishikawa, M., 2012. Cyclophilin 40 facilitatesHSP90- mediated RISC assembly in plants. EMBO J. 31, 267–278.

Islam, W., Noman, A., Qasim, M., Wang, L., 2018. Plant responses to pathogen attack: small RNAs in focus. Int. J. Mol. Sci. 19 (2). https://doi.org/10.3390/ijms19020515.

Jung, J.H., Seo, P.J., Park, C.M., 2009. MicroRNA biogenesis and function in higher plants. Plant Biotechnol. Rep. 3, 111–126. https://doi.org/10.1007/s11816-009-0085-8.

Jung, J.H., Seo, Y.H., Seo, P.J., Reyes, J.L., Yun, J., 2007. The GIGANTEA-regulated microRNA172 mediates photoperiodic flowering independent of CONSTANS in Arabidopsis. Plant Cell 19, 2736–2748. https://doi.org/10.1105/tpc.107.054528.

Jung, J.H., Park, C.M., 2007. MIR166/165 genes exhibit dynamic expression patterns in regulating shoot apical meristem and floral development in Arabidopsis. Planta 225, 1327–1338.

Kerchev, P., Van Der Meer, T., Sujeeth, N., Verlee, A., Stevens, C.V., Van Breusegem, F., et al., 2020. Molecular priming as an approach to induce tolerance against abiotic and oxidative stresses in crop plants. Biotechnol. Adv. 40, 107503. https://doi.org/10.1016/j.biotechadv.2019.107503.

Khalid, S., Shahid, M., Niazi, N.K., Rafiq, M., Bakhat, H.F., Imran, M., Abbas, T., Bibi, I., Dumat, C., 2017. Arsenic behaviour in soil-plant system: biogeochemical reactions and chemical speciation influence. In: Enhancing Cleanup of Environmental Pollutants, vol. 2. Springer, Cham, Switzerland, pp. 97–139.

Khare, P., Kishore, K., Sharma, D.K., 2017. A study on the standardization parameters of Cassia angustifolia. Asian J. Pharmaceut. Clin. Res. 10 (7), 329–332.

Kim, Y.J., Zheng, B.L., Yu, Y., Won, S.Y., Mo, B.X., Chen, X.M., 2011. The role of Mediator in small and long noncoding RNA production in *Arabidopsis thaliana*. EMBO J. 30, 814–822.

Knauer, S., Holt, A.L., Rubio-Somoza, I., Tucker, E.J., Hinze, A., 2013. A protodermal miR394 signal defines a region of stem cell competence in the Arabidopsis shoot meristem. Dev. Cell 24, 125–132. https://doi.org/10.1016/j.devcel.2012.12.009.

Kubier, A., Wilkin, R.T., Pichler, T., 2019. Cadmium in soils and groundwater: a review. Appl. Geochem. 108, 1–16. https://doi.org/10.1016/j.apgeochem.2019.104388.

Kumar, V., Khare, T., Shriram, V., Wani, S.H., 2018. Plant small RNAs: the essential epigenetic regulators of gene expression for salt-stress responses and tolerance. Plant Cell Rep. 37, 61–75. https://doi.org/10.1007/s00299-017-2210-4.

Kurihara, Y., Watanabe, Y., 2004. Arabidopsis micro-RNA biogenesis through Dicer-like 1 protein functions. P. Natl. Acad. Sci. USA. 101, 12753–12758.

Lee, R.C., Feinbaum, R.L., Ambros, V., 1993. The *C. elegans* heterochronic gene lin-4 encodes small RNAs with antisense complementarity to lin-14. Cell 75, 843–854. https://doi.org/10.1016/0092-8674(93)90529-Y.

Li, W., Wang, T., Zhang, Y., Li, Y., 2016. Overexpression of soybean miR172c confers tolerance to water deficit and salt stress, but increases ABA sensitivity in transgenic *Arabidopsis thaliana*. J. Exp. Bot. 67, 175–194. https://doi.org/10.1093/jxb/erv450.

Lin, J.S., Kuo, C.C., Yang, I.C., Tsai, W.A., Shen, Y.H., Lin, C.C., et al., 2018. MicroRNA160 modulates plant development and heat shock protein gene expression to mediate heat tolerance in Arabidopsis. Front. Plant Sci. 9, 68. https://doi.org/10.3389/fpls.2018.00068.

Lin, Y., Lai, Z., Tian, Q., Lin, L., Lai, R., Yang, M., et al., 2015. Endogenous target mimics down-regulate miR160 mediation of ARF10,−16, and−17 cleavage during somatic embryogenesis in Dimocarpus longan Lour. Front. Plant Sci. 6, 956. https://doi.org/10.3389/fpls.2015.00956.

Liu, H., Jia, S., Shen, D., Liu, J., Li, J., 2012. Four AUXIN RESPONSE FACTOR genes downregulated by microRNA167 are associated with growth and development in *Oryza sativa*. Funct. Plant Biol. 39, 736–744. https://doi.org/10.1071/FP12106.

Liu, N., Wu, S., Van Houten, J., Wang, Y., Ding, B., 2014. Downregulation of AUXIN RESPONSE FACTORS 6 and 8 by microRNA 167 leads to floral development defects and female sterility in tomato. J. Exp. Bot. 65, 2507–2520. https://doi.org/10.1093/jxb/eru141.

Liu, P.P., Montgomery, T.A., Fahlgren, N., Kasschau, K.D., Nonogaki, H., 2007. Repression of AUXIN RESPONSE FACTOR10 by microRNA160 is critical for seed germination and post-germination stages. Plant J. 52, 133–146. https://doi.org/10.1111/j.1365-313X.2007.03218.x.

Liu, Q., Hu, H., Zhu, L., Li, R., Feng, Y., 2015. Involvement of MiR528 in the regulation of arsenite tolerance in rice (*Oryza sativa* L.). Agric. Food Chem. 63, 8849–8861.

Liu, W., Xu, L., Wang, Y., Shen, H., Zhu, X., Zhang, K., Chen, Y., Yu, R., Limera, C., Liu, L., 2015. Transcriptome-wide analysis of chromium-stress responsive microRNAs to explore miRNAmediated regulatory networks in radish (*Raphanus sativus* L.). Sci. Rep. 5, 1–7. https://doi.org/10.1038/srep14024.

Liu, X., Wang, T., Bartholomew, E., Black, K., Dong, M., Zhang, Y., et al., 2018. Comprehensive analysis of NAC transcription factors and their expression during fruit spine development in cucumber (*Cucumis sativus* L.). Hortic. Res. 5, 31. https://doi.org/10.1038/s41438-018-0036-z.

Liu, Y., Li, D., Yan, J., Wang, K., Luo, H., Zhang, W., 2019. MiR319 mediated salt tolerance by ethylene. Plant Biotechnol. J. 17, 2370–2383. https://doi.org/10.1111/pbi.13154.

Maddala, V.K.S., Singh, S., 2023. Essential oils from plants and their role in nanomaterial synthesis characterization and applications. In: Secondary Metabolites Based Green Synthesis of Nanomaterials and Their Applications. Springer Nature Singapore, Singapore, pp. 191–207.

Maddala, V.K.S., Singh, S., Bandi, J., 2024. Soil properties influencing nutrient signaling in plant soil systems. In: Essential Minerals in Plant-Soil Systems. Elsevier, pp. 373–393.

Mallory, A.C., Bartel, D.P., Bartel, B., 2005. MicroRNA-directed regulation of Arabidopsis AUXIN RESPONSE FACTOR17 is essential for proper development and modulates expression of early auxin response genes. Plant Cell 17, 1360–1375. https://doi.org/10.1105/tpc.105.031716.

Manavella, P.A., Yang, S.W., Palatnik, J., 2019. Keep calm and carry on: miRNA biogenesis under stress. Plant J. 99, 832–843.

Marin, E., Jouannet, V., Herz, A., Lokerse, A.S., Weijers, D., Vaucheret, H., et al., 2010. miR390. ArabidopsisTAS3 tasiRNAs their auxin response factor targets define an autoregulatory network quantitatively regulating lateral root growth. Plant Cell 22, 1104–1117.

Mi, S.J., Cai, T., Hu, Y.G., Chen, Y., Hodges, E., Ni, F.R., et al., 2008. Sorting of small RNAs into Arabidopsis argonaute complexes is directed by the 5'terminal nucleotide. Cell 133, 116–127.

Neumann, R.B., Ashfaque, K.N., Badruzzaman, A.B.M., Ali, M.A., Shoemaker, J.K., Harvey, C.F., 2010. Anthropogenic influences on groundwater arsenic concentrations in Bangladesh. Nat. Geosci. 3, 46–52. https://doi.org/10.1038/ngeo685.

Nguyen, D.Q., Brown, C.W., Pegler, J.L., Eamens, A.L., Grof, C.P.L., 2020. Molecular manipulation of microRNA397 abundance influences the development and salt stress response of *Arabidopsis thaliana*. Int. J. Mol. Sci. 21, 7879. https://doi.org/10.3390/ijms21217879.

Nie, G., Liao, Z., Zhong, M., Zhou, J., Cai, J., Liu, A., Wang, X., Zhang, X., 2021. MicroRNA-mediated responses to chromium stress provide insight into tolerance characteristics of *Miscanthus sinensis*. Front. Plant Sci. 2021, 12.

Noman, A., Ali, Q., Maqsood, J., Iqbal, N., Javed, M.T., Rasool, N., Naseem, J., 2018. Deciphering physio-biochemical, yield, and nutritional quality attributes of water-stressed radish (*Raphanus sativus* L.) plants grown from Zn-Lys primed seeds. Chemosphere 195, 175–189.

Noman, A., Aqeel, M., 2017. miRNA-based heavy metal homeostasis and plant growth. Environ. Sci. Pollut. Res. Int. 24, 10068–10082. https://doi.org/10.1007/s11356-017-8593-5.

Noman, A., Sanaullah, T., Khalid, N., Islam, W., Khan, S., Irshad, M.K., Aqeel, M., 2019. Crosstalk between plant miRNA and heavy metal toxicity. In: Sablok, G. (Ed.), Plant Metallomics and Functional Omics. Springer, Cham, pp. 145–168. https://doi.org/10.1007/978-3-030-19103-0_7.

Ori, N., Cohen, A.R., Etzioni, A., Brand, A., Yanai, O., Shleizer, S., Menda, N., Amsellem, Z., Efroni, I., Pekker, I., Alvarez, J.P., Blum, E., Zamir, D., Eshed, Y., 2007. Regulation of LANCEOLATE by miR319 is required for compound-leaf development in tomato. Nat. Genet. 39, 787–791.

Pagano, L., Rossi, R., Paesano, L., Marmiroli, N., Marmiroli, M., 2021. miRNA regulation and stress adaptation in plants. Environ. Exp. Bot. 184, 104369. https://doi.org/10.1016/j.envexpbot.2020.104369.

Palatnik, J.F., Wollmann, H., Schommer, C., Schwab, R., Boisbouvier, J., 2007. Sequence and expression differences underlie functional specialization of Arabidopsis microRNAs miR159 and miR319. Dev. Cell 13, 115–125. https://doi.org/10.1016/j.devcel.2007.04.012.

Pan, C., Ye, L., Zheng, Y., Wang, Y., Yang, D., Liu, X., et al., 2017. Identification and expression profiling of microRNAs involved in the stigma exsertion under high-temperature stress in tomato. BMC Genom. 18, 843. https://doi.org/10.1186/s12864-017-4238-9.

Park, M.Y., Wu, G., Gonzalez-Sulser, A., Vaucheret, H., Poethig, R.S., 2005. Nuclear processing and export of microRNAs in Arabidopsis. Proc. Natl. Acad. Sci. U.S.A. 102, 3691–3696.

Parveen, R., Khare, N., Singh, S., Johri, P., 2023. An in silico outlook for the detection and surveillance of evolving and persistent plant pathogens. In: Detection, Diagnosis and Management of Soil-Borne Phytopathogens. Springer Nature Singapore, Singapore, pp. 17–40.

Patani, A., Patel, A., Prajapati, D., Khare, N., Singh, S., 2023. Various agriculture crop plant-based bioactive compounds and their use in nanomaterial synthesis and applications. In: Nanomaterials from Agricultural and Horticultural Products. Springer Nature Singapore, Singapore, pp. 223–241.

Patani, A., Prajapati, D., Shukla, K., Patel, M., Patani, P., Patel, A., Singh, S., 2024. Environmental stress–induced alterations in the micro-and macronutrients status of plant. In: Essential Minerals in Plant-Soil Systems. Elsevier, pp. 181–195.

Pegler, J.L., Oultram, J.M.J., Nguyen, D.Q., Grof, C.P.L., Eamens, A.L., 2021. Microrna-mediated responses to cadmium stress in *Arabidopsis thaliana*. Plants 10, 130.

Qiao, J., Jiang, H., Lin, Y., Shang, L., Wang, M., Li, D., et al., 2021. A novel miR167a-OsARF6-OsAUX3 module regulates grain length and weight in rice. Mol. Plant 14, 1683–1698. https://doi.org/10.1016/j.molp.2021.06.023.

Ragupathy, R., Ravichandran, S., Mahdi, M.S., Huang, D., Reimer, E., Domaratzki, M., et al., 2016. Deep sequencing of wheat sRNA transcriptome reveals distinct temporal expression pattern of miRNAs in response to heat, light and UV. Sci. Rep. 6, 39373. https://doi.org/10.1038/srep39373.

Rajagopalan, R., Vaucheret, H., Trejo, J., Bartel, D.P., 2006. A diverse and evolutionarily fluid set of microRNAs in *Arabidopsis thaliana*. Gene Dev. 20, 3407–3425.

Ravichandran, S., Ragupathy, R., Edwards, T., Domaratzki, M., Cloutier, S., 2019. MicroRNA-guided regulation of heat stress response in wheat. BMC Genom. 20, 488. https://doi.org/10.1186/s12864-019-5799-6.

Raza, A., Charagh, S., Karikari, B., Sharif, R., Yadav, V., Mubarik, M.S., Habib, M., Zhuang, Y., Zhang, C., Chen, H., Varshney, R.K., Zhuang, W., 2023. miRNAs for crop improvement. Plant Physiol. Biochem. 201, 107857. https://doi.org/10.1016/j.plaphy.2023.107857.

Reinhart, B.J., Weinstein, E.G., Rhoades, M.W., Bartel, B., Bartel, D.P., 2002. MicroRNAs in plants. Genes Dev. 16, 1616–1626. https://doi.org/10.1101/gad.1004402.

Rogers, K., Chen, X.M., 2013. Biogenesis, turnover, and mode of action of plant microRNAs. Plant Cell 25, 2383–2399.

Secic, E., Kogel, K.H., Ladera-Carmona, M.J., 2021. Biotic stressassociated microRNA families in plants. J. Plant Physiol. 263, 153451. https://doi.org/10.1016/j.jplph.2021.153451.

Shahid, M., Khalid, S., Abbas, G., Shahid, N., Nadeem, M., Sabir, M., et al., 2015. Heavy metal stress and crop productivity. In: Hakeem, K.R. (Ed.), Crop Production and Global Environmental Issues. Springer International Publishing, Cham, pp. 1–25.

Shahid, M., Shamshad, S., Rafiq, M., Khalid, S., Bibi, I., Niazi, N.K., Dumat, C., Rashid, M.I., 2017. Chromium speciation, bioavailability, uptake, toxicity and detoxification in soil-plant system: a review. Chemosphere 178, 513–533.

Sharma, D., Tiwari, M., Lakhwani, D., Tripathi, R.D., Trivedi, P.K., 2015. Differential expression of MicroRNAs by arsenate and arsenite stress in natural accessions of rice. Metallomics 7, 174–187.

Sharma, P., Jha, A.B., Dubey, R.S., Pessarakali, M., 2012. Reactive oxygen species, oxidative damage and antioxidant defense mechanism in plants under stressful conditions. J. Bot., Le 2012, 217037.

Shi, Z.-H., Zhang, C., Xu, X.-F., Zhu, J., Zhou, Q., Ma, L.-J., et al., 2015. Overexpression of AtTTP affects ARF17 expression and leads to male sterility in Arabidopsis. PLoS One 10 (3), e0117317. https://doi.org/10.1371/journal.pone.0117317.

Singh, A., Sengar, R.S., Singh, A.K., Singh, S., Kumar, A., Ahmad, K., 2019. Impact of genetic engineering on agriculture food crops. Biotech Today: Int. J. Biol. Sci. 9 (1), 17–24.

Singh, S., Parihar, P., Singh, R., Singh, V.P., Prasad, S.M., 2016. Heavy metal tolerance in plants: role of transcriptomics, proteomics, metabolomics and ionomics. Front. Plant Sci. 6, 1143. https://doi.org/10.3389/fpls.2015.01143.

Singroha, G., Sharma, P., Sunkur, R., 2021. Current status of microRNAmediated regulation of drought stress responses in cereals. Physiol. Plantarum 172, 1808–1821. https://doi.org/10.1111/ppl.13451.

Song, J.B., Huang, S.Q., Dalmay, T., Yang, Z.M., 2012. Regulation of leaf morphology by microRNA394 and its target LEAF CURLING RESPONSIVENESS. Plant Cell Physiol. 53, 1283–1294. https://doi.org/10.1093/pcp/pcs080.

Srivastava, S., Srivastava, A.K., Suprasanna, P., D'Souza, S.F., 2013. Identification and profiling of arsenic stress-induced microRNAs in Brassica juncea. J. Exp. Bot. 64 (1), 303–315. https://doi.org/10.1093/jxb/ers333.

Srivastava, S., Suprasanna, P., 2021. MicroRNAs: tiny, powerful players of metal stress responses in plants. Plant Physiol. Biochem. 166, 928–938.

Sun, M., Shen, Y., Chen, Y., Wang, Y., Cai, X., Yang, J., et al., 2022. Osa-miR1320 targets the ERF transcription factor OsERF096 to regulate cold tolerance via JA-mediated signaling. Plant Physiol. 189, 2500–2516. https://doi.org/10.1093/plphys/kiac208.

Sun, Y., Xiong, X., Wang, Q., Zhu, L., Wang, L., He, Y., et al., 2021. Integrated analysis of small RNA, transcriptome, and degradome sequencing reveals the MiR156, MiR5488 and MiR399 are involved in the regulation of male sterility in PTGMS rice. Int. J. Mol. Sci. 22, 2260. https://doi.org/10.3390/ijms22052260.

Sun, Z., Guo, T., Liu, Y., Liu, Q., Fang, Y., 2015. The roles of Arabidopsis CDF2 in transcriptional and post-transcriptional regulation of primary microRNAs. PLoS Genet. 11, e1005598.

Sunkar, R., Kapoor, A., Zhu, J.K., 2006. Posttranscriptional induction of two Cu/Zn superoxide dismutase genes in Arabidopsis is mediated by down regulation of miR398 and important for oxidative stress tolerance. Plant Cell 18, 2051–2065.

Sunny, J.S., Snijesh, V.P., Singh, S., 2016. Potential role of mirnas in rheumatoid arthritis. Austin Arthritis 1 (2), 1009.

Takahashi, H., 2010. Regulation of sulfate transport and assimilation in plants. Int. Rev. Cell Mol. Biol. 281, 129–159.

Tiwari, S., Lata, C., 2018. Heavy metal stress, signaling, and tolerance due to plant-associated microbes: an overview. Front. Plant Sci. 9, 452. https://doi.org/10.3389/fpls.2018.00452.

Treiber, T., Treiber, N., Meister, G., 2019. Regulation of microRNA biogenesis and its crosstalk with other cellular pathways. Nat. Rev. Mol. Cell Biol. 20, 5–20. https://doi.org/10.1038/s41580-018-0059-1.

Tungare, K., Shahu, R., Zambare, V., Agarwal, P., Jobby, R., Nisar, N., Jha, P., 2022. Toxicity mitigation of textile dye reactive blue 4 by hairy roots of Helianthus annuus and testing its effect in in vivo model systems. BioMed Res. Int. 2022.

Verma, N., Sao, P., Srivastava, A., Singh, S., 2021. Physiological and molecular responses to drought, submergence and excessive watering in plants. In: Harsh Environment and Plant Resilience: Molecular and Functional Aspects, pp. 305–321.

Verma, S., Verma, P.K., Meher, A.K., Dwivedi, S., Bansiwal, A.K., Pande, V., et al., 2016. A novel arsenic methyltransferase gene of Westerdykella aurantiaca isolated from arsenic contaminated soil: phylogenetic, physiological, and biochemical studies and its role in arsenic bioremediation. Metallomics 8, 344–353. https://doi.org/10.1039/c5mt00277j.

Vora, Z., Pandya, J., Vaikuntapu, P.R., Singh, S., Asha, S., Purama, R.K., 2023. Genome-wide association of defense hormone crosstalk in plants. In: Hormonal Cross-Talk, Plant Defense and Development. Academic Press, pp. 353–371.

Wang, B., Cheng, D., Chen, Z., Zhang, M., Zhang, G., Jiang, M., et al., 2019. Bioinformatic exploration of the targets of xylem sap miRNAs in maize under cadmium stress. Int. J. Mol. Sci. 20, 1474.

Wang, J.W., Wang, L.J., Mao, Y.B., Cai, W.J., Xue, H.W., 2005. Control of root cap formation by MicroRNA-targeted auxin response factors in Arabidopsis. Plant Cell 17, 2204–2216. https://doi.org/10.1105/tpc.105.033076.

Wang, L., Gu, X., Xu, D., Wang, W., Wang, H., 2011. MiR396-targeted AtGRF transcription factors are required for coordination of cell division and differentiation during leaf development in Arabidopsis. J. Exp. Bot. 62, 761–773. https://doi.org/10.1093/jxb/erq307.

Wang, S.T., Sun, X.L., Hoshino, Y., Yu, Y., Jia, B., Sun, Z.W., et al., 2014. MicroRNA319 positively regulates cold tolerance by targeting OsPCF6 and OsTCP21 in rice (*Oryza sativa* L.). PLoS One 9, e91357. https://doi.org/10.1371/journal.pone.0091357.

Wang, W., Liu, D., Chen, D., Cheng, Y., Zhang, X., Song, L., et al., 2019. MicroRNA414c affects salt tolerance of cotton by regulating reactive oxygen species metabolism under salinity stress. RNA Biol. 16, 362–375. https://doi.org/10.1080/15476286.2019.1574163.

Wang, Y., Greger, M., 2004. Plant and environment interactions clonal differences in mercury tolerance, accumulation, and distribution in willow. J. Environ. Qual. 33, 1779–1785.

Wani, S.H., Kumar, V., Khare, T., Guddimalli, R., Parveda, M., Solymosi, K., et al., 2020. Engineering salinity tolerance in plants: progress and prospects. Planta 251, 76. https://doi.org/10.1007/s00425-020-03366-6.

Wei, L., Zhang, D., Xiang, F., Zhang, Z., 2009. Differentially expressed miRNAs potentially involved in the regulation of defense mechanism to drought stress in maize seedlings. Int. J. Plant Sci. 170, 979–989. https://doi.org/10.1086/605122.

Wightman, B., Ha, I., Ruvkun, G., 1993. Posttranscriptional regulation of the heterochronic gene lin-14 by lin-4 mediates temporal pattern formation in *C. elegans*. Cell 75, 855–862. https://doi.org/10.1016/0092-8674(93)90530-4.

Wojcik, A.M., Nodine, M.D., Gaj, M.D., 2017. miR160 and miR166/165 contribute to the LEC2-mediated auxin response involved in the somatic embryogenesis induction in Arabidopsis. Front. Plant Sci. 8, 2024. https://doi.org/10.3389/fpls.2017.02024.

Wollmann, H., Mica, E., Todesco, M., Long, J.A., Weigel, D., 2010. On reconciling the interactions between APETALA2, miR172 and AGAMOUS with the ABC model of flower development. Development 137, 3633–3642.

Xia, K., Zeng, X., Jiao, Z., Li, M., Xu, W., Nong, Q., Mo, H., Cheng, T., Zhang, M., 2018. Formation of protein disulfide bonds catalyzed by OsPDIL1;1 is mediated by MicroRNA5144-3p in rice. Plant Cell Physiol. 59, 331–342.

Xie, K., Wu, C., Xiong, L., 2006. Genomic organization, differential expression, and interaction of SQUAMOSA promoter-binding-like transcription factors and microRNA156 in rice. Plant Physiol. 142, 280–293. https://doi.org/10.1104/pp.106.084475.

Yamasaki, H., Hayashi, M., Fukazawa, M., Kobayashi, Y., Shikanai, T., 2009. SQUAMOSA promoter binding protein-like7 is a central regulator for copper homeostasis in Arabidopsis. Plant Cell 21, 347–361.

Yan, C., Zhang, N., Wang, Q., Fu, Y., Wang, F., Su, Y., et al., 2021. The effect of low temperature stress on the leaves and microRNA expression of potato seedlings. Front. Ecol. Evol. 9, 727081. https://doi.org/10.3389/fevo.2021.727081.

Yang, L., Liu, Z.Q., Lu, F., Dong, A.W., Huang, H., 2006. SERRATE is a novel nuclear regulator in primary microRNA processing in Arabidopsis. Plant J. 47, 841–850.

Yang, T., Wang, Y., Teotia, S., Wang, Z., Shi, C., Sun, H., et al., 2019. The interaction between miR160 and miR165/166 in the control of leaf development and drought tolerance in Arabidopsis. Sci. Rep. 9, 2832. https://doi.org/10.1038/s41598-019-39397-7.

Yang, Y., Huang, J., Sun, Q., Wang, J., Huang, L., Fu, S., Qin, S., Xie, X., Ge, S., Li, X., et al., 2022. microRNAs: key players in plant response to metal toxicity. Int. J. Mol. Sci. 23, 8642. https://doi.org/10.3390/ijms23158642.

Yang, Z., Yang, F., Liu, J.L., Wu, H.T., Yang, H., Shi, Y., Liu, J., Zhang, Y.F., Luo, Y.R., Chen, K.M., 2022. Heavy metal transporters: functional mechanisms, regulation, and application in phytoremediation. Sci. Total Environ. 809, 151099.

Yang, Z., Hui, S., Lv, Y., Zhang, M., Chen, D., Tian, J., Zhang, H., Liu, H., Cao, J., Xie, W., et al., 2022. MiR395-Regulated sulfate metabolism exploits pathogen sensitivity to sulfate to boost immunity in rice. Mol. Plant 15, 671–688.

Ye, Y., Wang, J., Wang, W., Xu, L.A., 2020. ARF family identification in Tamarix chinensis reveals the salt responsive expression of TcARF6 targeted by miR167. PeerJ 8, e8829. https://doi.org/10.7717/peerj.8829.

Yu, B., Yang, Z.Y., Li, J.J., Minakhina, S., Yang, M.C., Padgett, R.W., et al., 2005. Methylation as a crucial step in plant microRNA biogenesis. Science 307, 932–935. https://doi.org/10.1126/science.1107130.

Yu, Y., Ni, Z., Wang, Y., Wan, H., Hu, Z., Jiang, Q., et al., 2019. Overexpression of soybean miR169c confers increased drought stress sensitivity in transgenic *Arabidopsis thaliana*. Plant Sci. 285, 68–78. https://doi.org/10.1016/j.plantsci.2019.05.003.

Zhang, F., Yang, J., Zhang, N., Wu, J., Si, H., 2022. Roles of microRNAs in abiotic stress response and characteristics regulation of plant. Front. Plant Sci. 13, 919243. https://doi.org/10.3389/fpls.2022.919243.

Zhang, H., Zhang, J., Yan, J., Gou, F., Mao, Y., Tang, G., et al., 2017. Short tandem target mimic rice lines uncover functions of miRNAs in regulating important agronomic traits. Proc. Natl. Acad. Sci. U.S.A. 114, 5277–5282. https://doi.org/10.1073/pnas.1703752114.

Zhang, H., Zhang, L., Han, J., Qian, Z., Zhou, B., 2019. The nuclear localization signal is required for the function of squamosa promoter binding protein-like gene 9 to promote vegetative phase change in Arabidopsis. Plant Mol. Biol. 100, 571–578. https://doi.org/10.1007/s11103-019-00863-5.

Zhang, L.W., Song, J.B., Shu, X.X., Zhang, Y., Yang, Z.M., 2013. MiR395 is involved in detoxification of cadmium in Brassica napus. J. Hazard Mater. 250–251, 204–211.

Zhang, M., An, P., Li, H., Wang, X., Zhou, J., Dong, P., et al., 2019. The miRNAmediated post-transcriptional regulation of maize in response to high temperature. Int. J. Mol. Sci. 20, 1754. https://doi.org/10.3390/ijms20071754.

Zhong, R., Ye, Z.H., 2004. Amphivasal vascular bundle 1, a gain-of-function mutation of the IFL1/REV gene, is associated with alterations in the polarity of leaves, stems and carpels. Plant Cell Physiol. 45, 369–385. https://doi.org/10.1093/pcp/pch051.

Zhou, Y., Liu, W., Li, X., Sun, D., Xu, K., Feng, C., et al., 2020. Integration of sRNA, degradome, transcriptome analysis and functional investigation reveals gma-miR398c negatively regulates drought tolerance via GmCSDs and GmCCS in transgenic Arabidopsis and soybean. BMC Plant Biol. 20, 190. https://doi.org/10.1186/s12870-020-02370-y.

Zhou, Z.S., Huang, S.Q., Yang, Z.M., 2008. Bioinformatic identification and expression analysis of new MicroRNAs from Medicago truncatula. Biochem. Biophys. Res. Commun. 374, 538–542.

Zhou, Z.S., Song, J.B., Yang, Z.M., 2012a. Genome wide identification of Brassica napus microRNAs and their targets in response to cadmium. J. Exp. Bot. 63, 4597–4613. https://doi.org/10.1093/jxb/ers136.

Zhou, Z.S., Zeng, H.Q., Liu, Z.P., Yang, Z.M., 2012b. Genome-wide identification of Medicago truncatula microRNAs and their targets reveals their differential regulation by heavy metal. Plant Cell Environ. 35, 86–99.

Zhu, H., Zhang, Y., Tang, R., Qu, H., Duan, X., Jiang, Y., 2019. Banana sRNAome and degradome identify microRNAs functioning in differential responses to temperature stress. BMC Genom. 20, 33. https://doi.org/10.1186/s12864-018-5395-1.

Zhu, L., Ow, D.W., Dong, Z.C., 2018. Transfer RNA-derived small RNAs in plants. Sci. China Life Sci. 61, 155–161. https://doi.org/10.1007/s11427-017-9167-5.

Zuo, Z.F., He, W., Li, J., Mo, B., Liu, L., 2021. Small RNAs: the essential regulators in plant thermotolerance. Front. Plant Sci. 12, 726762. https://doi.org/10.3389/fpls.2021.726762.

CHAPTER

Genetic modification of plants using microRNAs

3

Soumya Singh[1], Arpan Modi[2], Megha Barot[3,4], Sachidanand Singh[5], Tanvi Jain[1]

[1]*Faculty of Biotechnology, Institute of Biosciences and Technology, Shri Ramswaroop Memorial University, Barabanki, Uttar Pradesh, India;* [2]*Axenic Agritech Pvt Ltd, Ahmedabad, Gujarat, India;* [3]*Research and Development Cell, Parul University, Vadodara, Gujarat, India;* [4]*Department of Environmental Science, Parul Institute of Applied Sciences, Parul University, Vadodara, Gujarat, India;* [5]*Department of Biotechnology, School of Energy Technology, Pandit Deendayal Energy University, Gandhinagar, Gujarat, India*

1. Introduction

1.1 miRNA

A noncoding RNA molecule is known as microRNA, or miRNA. It is a tiny, single-stranded RNA molecule with a length of roughly 22—24 nucleotides. It contributes to the control of both gene silencing and gene expression. It is found naturally in certain viruses, plants, and mammals.

The discovery of miRNA was made first by Ambros, Lee, Feinbaum, and the team. Many kinds of mammalian cells include miRNAs. Additionally, extracellular fluids including blood and cerebrospinal fluid are shown to contain them in circulation. By base matching with the complementary mRNA molecule, miRNA, which is complementary to mRNA, silences genes. By cleaving or destabilizing mRNA or by decreasing translation efficiency, they prevent the expression of genes.

1.2 Biosynthesis of micro RNA and their mode of action

RNA polymerase II typically transcribes miRNA genes that have been extracted from different genomic loci into lengthy primary transcripts, or pri-miRNAs. Following this, these primary transcripts are cleaved onto the precursor mRNA and various proteins such as serrate and hyponastic leaves one by RNase III-like enzyme which is known as DICER-LIKE (DCLI).

The DCLI enzymes further cleave the defective stem-loop composition of the pre-miRNA hairpin-like structure to produce the miRNA duplexes. HUA enhancer I (HEN I), an S-adenosyl-L-methionine-dependent RNA methyltransferase, methylates these duplexes. Enzyme HASTY then exports the methylated duplexes to the cytoplasm, where they are loaded into an RNA-induced silencing complexes which contains ARGONAUTE protein. After the miRNA is fully matured, it is placed into the RISC along with the ARGONAUTE proteins. The AGO protein unwinds these

MicroRNA Advances and Application in Plant Biology. https://doi.org/10.1016/B978-0-443-21682-4.00010-5

complexes, leaving one strand of miRNA that is pointed toward exosome for the degradation by small-RNA degrading nucleases. The mature miRNA then joins RISC and together they move toward mRNAs that are complementary to the miRNA. If there is an abnormal base pairing among the miRNA and the targeted mRNA, it affects the translational repression or mRNA. However, if there is broad homology among the miRNA and the target mRNA, it results in the site-specific slicing of mRNA.

Many researches have shown the evolutionary protection of multiple miRNAs in each major plant lineages, together with the mosses, gymnosperms, eudicots, and monocots (Cuperus et al., 2011; Shen et al., 2012). Two groups of mi RNA units can be distinguished based upon the conservation and diversity of miRNA over the course of the plant kingdom's evolution.

While little miRNAs are produced at relatively small levels are simply generated in the specific conditions, which are typically created in a midget number of the species, they are evolutionarily nonconserved. In contrast, ancient miRNAs are ofttimes substantially expressed and have shown to be conserved across time.

1.3 Regulatory function of miRNAs in plants

These short RNAs are diverse enough to precisely match almost any RNA encoded in a genome, controlling nearly each facet of plant biotechnology and frequently impacting transcription factor (TF) control. A budding body of research has examined the mechanisms via which miRNAs work as significant controller of the gene expression (Sanan-Mishra and Mukherjee, 2007; Lelandais-Brière et al., 2010). Plant miRNAs are small RNAs that are typically 20 to 24 nt long. They function by binding to specific mRNAs, resulting in the repression of the gene expression at the posttranscriptional stage. It has been found that miRNAs pair with their target mRNAs with high complementarity, and imperfect pairing results in translational repression. On the other hand, when a miRNA binds optimally with its corresponding target mRNA, it leads in the destruction of the targeted mRNA (Sun, 2012). The loss of poly(A) tail caused by cleavage causes the target mRNAs to become unstable and to decay.

Cereal Crops:

1. Most of the world's community depends on the rice as the main food. Similar to another crops, rice faces a number of environmental stressors that might hinder its growth and productivity, like salt, cold, drought, heat, and other nutrient inadequacies. In 2004, first 20 miRNAs were identified in the rice crop with the assistance of a cDNA library. These miRNAs were expected to target the genes that are engaged in several processes such as transcription (miR-4), disease resistance (miR-10), transport (miR-4, miR-15), and metabolism (miR-5), among others (Wang et al., 2004). In a study by Sunkar et al. (2005), 35 rice miRNAs were discovered, out of which 14 were resulted to target the genes which were involved in different physiological processes of

rice (Sunkar et al., 2005). As of right now, 592 rice miRNA sequences have been acknowledged; these can be found in miRbase.

2. The second significant food crop worldwide is maize which is utilized for fuel generation, as well as for food, feed, and pasture. Apart from its significance in agriculture and the economy, maize has been utilized as an exemplary plant used for scientific studies (Bennetzen and Hake, 2009). The identified and characterized five miRNAs, namely the miR156, miR160, miR166, miR167, and miR169 groups were involved in the development of maize, its growth, and their responses against biotic and abiotic stresses, have their putative target genes—auxin response factor, squamosa promoter-binding protein, HAP-2-like proteins, HD-ZIP TF, and CCAAT-binding factor, respectively—identified and characterized (Mica et al., 2006).
3. Since it has high nutritional value and significant impact on global food security, wheat is considered as one of the most significant cereal crops. After considerable work, 58 wheat miRNAs were identified by sequencing-based subgenomic miRNA research in wheat. Twenty of these families were new, and 20 were conserved (Yao et al., 2007). They should target genes participated in the wheat growth and several physiological processes, according to Yao et al. (2007).
4. In terms of productivity, barley is another significant crop that is produced across the world. Barley grain is fed to cattle and utilized for human consumption (Curaba et al., 2012). Apart from its significance in agriculture, it serves as a model plant which has been well explored in terms of genetics, breeding, and genomics (Sreenivasulu et al., 2008). Approximately 100 miRNAs were found in barley for the first time after utilizing deep sequencing. Here, up to 44 appeared to be expressed exclusively in barley, whereas 56 had orthologs that were known to be expressed in wheat or rice. Similar to several major crops, barley had generated curiosity in miRNA research since, in 2012, 84 preserved miRNAs and seven novel miRNAs were identified and then characterized, along with 96 acknowledged miRNA target genes expected to be involved in various purposes such as cell differentiation, defense response, photosynthesis, carbohydrate translocation, and phytohormone signaling pathways.

Legume Crops:

5. Soybean is the most widely used oil crop globally. It is significant for both humans and animals and has a high oil (20%) and seed protein (40%) content. The identification of 55 families of miRNAs in 2008 marked the beginning of the study of miRNAs in soybeans. Of these, 20 were and 35 were novel (Subramanian et al., 2008).
6. Grown in tropical areas of Africa, Asia, Central America, and Southern America that be semiarid or desert, cowpeas are valuable legume crops that provide both nutrient-rich feed for cattle and sustenance for people. Cowpeas' drought tolerance makes them a fantabulous model to examine the genetic basis of

abiotic stress like drought response. Understanding genetic components, drought forbearance mechanisms, drought-associated miRNAs have been studied (Barrera-Figueroa et al., 2011). In the beginning, 47 miRNAs total from 13 miRNA families were found using computational techniques (Lu and Yang, 2010).

7. Growing widely, groundnuts, also known as peanuts, are among the most vital oil crops in the world.

2010 saw the use of high-efficiency Solexa sequencing on peanuts for first miRNA-related study. There are now 75 conserved miRNAs that go to 22 families and 14 novel miRNAs families, 13 of which are peanut precise (Zhao et al., 2010). Targets of 14 preserved miRNAs and 7 newly identified peanut miRNAs were predicted by Zhao et al. (2010). These included mRNAs for protein kinase, auxin signaling protein, growth regulator factor, resveratrol synthase, endonuclease, protein synthesis proteins, transport inhibitor response, and the GRAS family of TFs and protein synthesis proteins. It was proposed that the discovered miRNAs were crucial for development, growth, and stress response in the environment.

Tuber Crops:

8. After rice, wheat, and maize, potatoes rank like the fourth most significant food crop in the world. According to Zhang et al. (2009), 48 putative potato miRNAs be initially discovered and anticipate to target 186 genes related to signal transduction, metabolism pathways, flora, leaves, roots, and stems development, as well as stress responses.
9. A common perennial shrub, cassava is farmed for both animal feed and human consumption. In some parts of Latin America, Africa, and Asia, this is a key supply of energy for about semibillion people (El-Sharkawy, 2004). Apart from its conventional significance, cassava starch is becoming more and more in demand as a raw ingredient in a variety of industries, including textile, paper, glue, and biofuels (Patanun et al., 2013). Various computational techniques were used to identify 159 putative cassava miRNAs, which are members of 34 miRNA families (Patanun et al., 2013). The recently discovered cassava miRNAs aim a variety of enzymes, such as acyl-protein thioesterase, 5-methyl-tetrahydrofolate homocysteine methyltransferase, laccase, multicopper oxidases, sulfate adenylyltransferase, and short chain alcohol dehydrogenase. These enzymes are known to be essential for a number of metabolic processes (Patanun et al., 2013).
10. The perennial dicotyledonous sweet potato is cultured in temperate, subtropical, and tropical regions worldwide and is utilized as a food source due to its high content of proteins, carbs, carotenoids, fiber, vitamins A, B, and C, and minerals (calcium, potassium, and iron). Fibrous roots and storage roots are the two types of roots that sweet potatoes develop. The part of the plant that humans eat is the store roots, which have high starch content and sweet flavor. Sweet potatoes are used as staple meal, source of starch, an ingredient in animal feed and raw material in industrial processes such as alcohol

manufacture. Like other crops, sweet potatoes were found to contain a huge number of miRNAs (Lin et al., 2012).

Fruit Crops:

11. The most commercially significant and extensively grown fruit crop in the world, grapevine has excellent processing and nutritional qualities. Since it is the first fruit crop with a fully sequenced genome, scientists are very interested in learning more about miRNAs (Jaillon et al., 2007). A prior study found that because miRNAs regulate fruit development, there are totally 24 preserved miRNA families, 26 nonconserved miRNAs, and 21 new main miRNAs.
12. Apples are grown all throughout the world as economically significant fruit trees, and people eat a lot of their fruit. It is abundant in phenolic chemicals, such as flavonoids, which may be important in lowering the danger of chronic illness in people (Boyer and Liu, 2004). The recognition and classification of 413 miRNAs (miRBase 21) from apples has been completed, which contributes to our understanding of their crucial regulatory mechanisms.
13. Oranges can be eaten raw, their juice, which has an elevated vitamin C concentration, is frequently drunk as beverage, and their skin can be used for skin care or added to home cleaning solutions. Citrus is a valuable fruit economically, but it's too a great source of investigational matter to understand the roles and directive pathways of miRNAs because of its sequenced genome, abundance of ESTs, and other factors (Xu et al., 2013). In miRbase 21, roughly 124 orange miRNAs were found.
14. One of the most beloved and extensively consumed vegetable crops is the tomato. Its molecular tractability and hereditary nature attracted researchers to study the role of miRNA during fruit growth by focusing on genes related to the ethylene pathway and fruit ripening. According to Giovannoni (2007), they are also a perfect plant for researching senescence and the ripening of plump fruit. More than 350–700 short RNAs were found in tomato fruit and leaf, respectively, and most of them are said to be crucial to fruit growth (Itaya et al., 2008).

Biofuel Crops:

15. About 75% of the sugar produced worldwide comes from sugarcane, a crop of significant economic importance. According to Azevedo et al. (2011), it is becoming more and more important into the development of replenishable energy sources like ethanol. Due to its high biomass yield and inherent confinement, it has gained attention as a target for improving the production of sustainable biomaterials due to the relevance of biofuel (Birch, 2007). As per genomic survey sequence database, sugarcane possesses the most complex polyploid genomes that hasn't been sequenced yet (Jannoo et al., 2007). Beyond miRNA mature sequence, many sugarcane miRNA originators exhibit strong similarity with their putative sorghum orthologs (Zanca et al., 2010).

16. A temperate-seasonal perennial grass which is from North America, a switchgrass is a gifted crop for herbaceous lignocellulosic biofuels. It is not a food crop. Switchgrass's wide range of adaptation and quick growth rate offer a reliable and abundant contributor of biomass used in the manufacturing of the biofuels (McLaughlin et al., 2006). As a result, research on switchgrass has increased, focusing on miRNAs in an effort to increase biomass yield. Switchgrass has 154 miRNAs that have been found and their targets described through the use of bioinformatics techniques (Matts et al., 2010). According to a study, one effective way to alter the synthesis of the lignin and cell wall phenolic esters group in switchgrass and increase the efficiency of bioenergy production is to manipulate the regulatory roles of miRNAs (Shen et al., 2012).
17. A valuable and nutrient-dense cereal crop, sorghum is extremely resilient to external stressors like flooding and drought. Fodder, food, and the source of natural material used in the synthesis of ethanol, biofuels, and starch are among its applications (Prasad et al., 2008, Mutegi et al., 2010). Being a C4 plant species, it can flourish on poor and marginal lands, in locations prone to high temperatures and drought, and thanks to its high photosynthetic competence, which converts solar energy into biomass. There are now 446 sorghum miRNAs listed in miRbase 21.

Beverage Crops:

18. Coffee ranks second on worldwide trade exchanges and is one of the most significant beverage crops as well as a significant and profitable agricultural commodity. Only two species of coffee plants, *Coffea arabica* and *Coffea canephora*, are significant since an economic standpoint; they account for 64% and 36% of global production, respectively. The examination of miRNA in *C. canephora* and *C. arabica* was made possible by the sequencing, characterization, and analysis of their transcripts (Mondego et al., 2011). According to Akter et al. (2014), the first publication on the detection of miRNAs in coffee (*C. arabica*) was just released. For the first time, expressed sequence tags (EST)-based homology search was utilized in this study along with a sophisticated, potent, and comparative computational strategy to identify the big miRNA family miR393 with the proper fold-back structures.
19. Grown in various agroclimatic zones, the tea plant is an eternally green woody perennial which is a major global source of commercial beverage crops. Due to the presence of polyphenols, catechins, caffeine, theanine, saponin, and volatile oils in its leaves, tea has been shown to provide numerous health benefits to people (Prabu and Mandal, 2010; Shi et al., 2011). The tea leaf or plant is very interesting for miRNA study because of its economic and health benefits as well as the advancements in functional genomics studies, gene cloning, and analysis. The initial study on the tea miRNAs found four miRNAs that belong to four miRNA families and had a full of 30 probable target genes using computational techniques (Prabu and Mandal, 2010). These four miRNAs that have been discovered belong to two specialized miRNA families,

miR1863 and miR 1846, and two extremely conserved miRNA families, miR169 and miR164.

Thirty expected target genes, comprising six TF-encoding genes, six cell development genes, seven metabolic pathway genes (three for protein metabolism, four for carbohydrate metabolism), three stress response genes, and eight hypothetical proteins, are expected to be controlled by the newly discovered miRNAs (Prabu and Mandal, 2010).

Nonfood Crops:

20. Cotton is a highly significant crop for the economy since it can be used to make edible oil, natural textiles, and fiber. Apart from being the most significant crop for fiber, cotton is the second most significant source of this plant protein, fifth most significant crop for oil, also an ideal species for research on plant polyploidization and the production of cell walls and cellulose (Wang et al., 2012). Many miRNAs (158 in miRbase 21) have been found and described since the first publications on the finding of miRNAs in cotton (Zhang et al., 2007; Barozai et al., 2008).
21. As per Andreinov et al. (2010), tobacco is a significant nonfood crop used for economic purposes worldwide. It has also been looked into as a possible crop for biofuel. It is a model plant used to research secondary metabolites of alkaloids and basic biological processes. Additionally, the origin of the cell line of the said plant that was essential to the study of plant molecular biology. The tobacco draft genome was just made public (Sierro et al., 2014), and the public database now contains the sequences from the genome survey sequences (GSSs), which allowed for the completion of several transcriptome, genomic, and miRNA studies. Initially, Frazier et al. (2010) utilized the computational techniques used for the prediction of 259 likely miRNAs of tobacco that fall into 65 groups and 1225 potential miRNA target gene of tobacco, which include TFs, DNA replication protein, and metabolic enzyme essential to tobacco growth and development.

2. Techniques for miRNA gene manipulation

Extensive survey has reported that miRNAs include serious role in growth and development of the plants, stress response, and could render worth genetic sources for various plants investigations. These are broadly categorized into two types.

2.1 Gain-of-function analysis techniques

Based on gain-of-function systems, there are various methods for purposefully characterizing miRNAs. For miRNA overexpression, one can employ the whole cDNA of MIR gene (Yang et al., 2013) and precursor miRNAs (pre-miRNAs,). Furthermore, endogenous miRNAs can be successfully overexpressed using a vector system that incorporates a two-hit artificial miRNA in the *Arabidopsis* miR168a backbones. According to Ji et al. (2011), this method introduces endogenous miRNAs by inserting complementary endogenous miRNA* that has mismatches on sites analogous to site I and site XII of the strand of miRNA. Gain-of-function effects are commonly achieved by utilizing a strong constitutive 35S promoter; however, since the expression level and pattern of miRNAs in vivo are misrepresentative, data drawn from this approach should be interpreted with caution. Furthermore, considering the role that miRNAs play in a variety of intricate regulatory networks, distinct transgenesis approaches must be used by utilizing particular promoters, like those that are tissue-specific, developmental stage-specific, stress-induced, which can prevent pleiotropic effects and correspond to particular research objectives.

2.2 Loss-of-function analysis technique

Methods aiming at genome sequence of miRNAs: In the past years, many gene-editing techniques were developed based on various sequence-specific technically made-up endonucleases (SSNs), few of which were used for the functional analysis of miRNAs in plants. Zinc Finger Nucleases (ZFNs) and Transcription Activator-like Effector Nucleases (TALENs) identify the sequences of the target using protein motifs that limit their widespread adoption by requiring researchers to assemble specific proteins for every target (Gaj et al., 2013). Apart from that, the CRISPR/Cas method identifies sites of the genomic target via base complementary pairing among the single-guide RNA (SG RNA) and the target DNA, significantly simplifying its usage (Xu et al., 2013). A single study created TALEN constructs for miRNA gene editing which led to transmissible alterations in plants, despite the fact that many miRNAs have been described by TALENs in mammals (Bi et al., 2020). Soon after its creation, the CRISPR/Cas9 system was effectively employed to do protein-coding gene editing in rice (Shan et al., 2013), *Nicotiana benthamiana* (Nekrasov et al., 2013), and *Arabidopsis* (Li et al., 2013). Many miRNAs have been fruitfully knocked out in a variety of plant species recently through the use of CRISPR/Cas9 nonhomologous end joining (NHEJ) to initiate indels at MIR genes (Damodharan et al., 2018; Hou et al., 2019; Miao et al., 2019). Also, the full-length deletion and knock-in of MIR genes may be accomplished by CRISPR/Cas9 editing via homology-directed repair (HDR; 20). Furthermore, by combining a disabled Cas9 nuclease (dCas9) with the transcriptional regulators like various activators (Li et al., 2017) and repressors (Tang et al., 2017), the CRISPR/Cas9 system can be used to alter the gene expression through the transcriptional repression or activation of the target genes. The small length of these MIR genes restricts its ability to build gRNA target which target helps in targeting MIR genes, despite the fact that the CRISPR/Cas9 system can be an effective tool for changing the sequence of miRNA genes and creating miRNA null mutant plants. Furthermore, the ability of CRISPR/Cas9 to concurrently beat the full MIR gene family may be restricted by the presence of several miRNA family members of the genome.

a) Methods aiming pri-/pre-miRNAs: Functional investigations of miRNAs have embraced a number of methods that are frequently used to target transcripts and mute protein-coding genes. RNA interference (RNAi), which is frequently employed to knock down coding genes, can also prevent miRNAs from building up. RNAi constructs that target both the primary miRNA transcripts factors and their promoters were successful in blocking miR163 and miR171a (Vaistij et al., 2010). The variety of siRNAs were generated via RNAi that may have off-target effects. An amiRNA method was devised to purposely mute targets by expressing the amiRNA with miRNA precursors as back, hence avoiding off-target effects (Carbonell et al., 2016). Genes of interest could be effectively silenced by substituting amiRNA/amiRNA* intended to target a particular mRNA for the original miRNA/miRNA* (Warthmann et al., 2008). The meaningful validation of the individual MIR loci in the genome is aided by the fact that it has been reported with the aim of an amiRNA designed to target the mature sequence of a miRNA can silence all members, whereas amiRNA is created to target the nonconserved stem-loop regions of the precursor transcripts which can just silence individual member (Eamens et al., 2011). The primary difficult task involved in using amiRNA is selecting precise and adequate pre- and amiRNA sequences to efficiently and specifically mute targets of interest (Carbonell et al., 2016; Peng et al., 2018).

Moreover, CRISPR/Cas13A may target and cleave single-stranded RNA directed by gRNA using class II type VI-A endoribonuclease. This mechanism could also be utilized to target pri-/pre-miRNA transcripts of MIR genes (Aman et al., 2018). Plant miRNA silencing using the CRISPR/Cas13a system which has not been tested yet.

b) Methods aiming at mature miRNAs: Both endogenous target mimics and exogenous synthetic AMOs have the ability to deceive mature miRNAs. Originally employed to reduce miRNA activity in animals, AMOs are chemically customized antisense oligonucleotides that are engineered to join up with and block mature microRNAs by sequence complementarity. According to current reports, AMO can effectively suppress miRNAs in rice protoplasts and intact leaves through sucrose-mediated delivery. AMOs can quickly examine the characterization of miRNAs in plants by acting in a sequence-specific method against selected miRNAs and transiently inducing miRNA blockages. According to a report by Franco-Zorrilla et al., endogenous target imitates have the ability to suppress miRNA function by obstructing the interaction between miRNAs and their targets. It was discovered that the endogenous long noncoding *RNA INDUCED BY PHOSPHATE STARVATION* 1 (IPS1) in Arabidopsis possessed a miR399-binding site with a middle "bulge" composed of three nucleotides (CUA) which could efficiently trap miR399 and prevent miR399-guided cleavage (Frazier et al., 2010). Artificial TMs of interest were used to decoy miRNAs in plants by substituting the endogenous miR399 target site of IPS1. To prevent miRNA activity in SP systems, four copies of the miRNA-binding sites were employed in animal cells, with two central mismatches on the cleavage site connected by 4-nt spacers. Using SPs to block miRNAs in plants has proven successful in recent years (Tang et al., 2017). A sponge's ability to obstruct the function of miRNA in plants is enhanced by the utilization of more miRNA-binding sites—up to 15—in the sponge. However, creating a lengthy SP with numerous tandem repeats is challenging, which might restrict the use of SPs. TMs and SPs served as the foundation for the original development of STTM. A 48–88 nt RNA spacer connects two miRNA-binding sites in the modified artificial RNA structure known as STTM. Similar to TMs, STTMs' miRNA-binding sites feature mismatches at the miRNA cleavage sites, which aid in the sequestration of miRNAs by STTMs without allowing them to be cleaved. According to Yang et al. (2013), the spacer that separates the two miRNA-binding sites generates a little "stem" that is essential for resisting Dicer attack and maintaining the stability of the expressed STTM in cells. STTMs have the ability to repress the expression of a whole family of miRNAs and block or eliminate particular endogenous short RNA functions in plants (Yang et al., 2013). Even though the exact mechanism behind STTM-induced miRNA degradation is not known, it is partially reliant on SDN-mediated miRNA degradation (Yang et al., 2013). Furthermore, the F-box protein HAWAIIAN SKIRT (HWS) is involved in the crucial process of miRNA degradation that is initiated by the TM/STTM system. This could be attributed to its role in eliminating nonoptimal RISC that is produced by mimicking targets. It was observed that the silencing efficiency of TMs, SPs, and STTMs changes among distinct miRNA families, disrupting the role of miRNAs as miRNA decoys. While TM more successfully targets low-abundance miRNAs (e.g., miR159), STTMs were shown to be effective in disrupting the activities of greatly abundant miRNAs and SPs serve as a backup for miRNAs that neither TM nor STTM can effectively block. To achieve the intended result, it is therefore advised to use a variety of decoy techniques.

c) Methods for simulating the loss of function of miRNAs: Upregulating the targets of miRNAs by overexpressing the cleavage-resistant target genes is another method for simulating the loss-of-function status of a miRNA (Li et al., 2013). Cleavage-resistant targets were produced by altering the miRNA cleavage site through the introduction of synonymous mutations, as overexpressed original targets are still susceptible to miRNA cleavage. Numerous transgenic plants that overexpress cleavage-resistant targets have been developed.

3. Conclusion

As demonstrated by the case studies that were covered, miRNAs are now thought to be some of the most significant gene regulators. Plant miRNA analysis and characterization have advanced significantly in the last several years, and there are an increasing number of studies stressing the critical role miRNAs take part in agricultural plants. According to Sun (2012), these tiny nucleic acids showed a vital role in many processes which includes hormone signaling, plant development and growth, transitions from the vegetative to the reproductive phases, homeostasis reactions and signal transduction pathways, to various biotic and abiotic environmental stressors. Additionally, a number of latest researches had demonstrated the part of the plant miRNAs as molecular regulators of defense and immunological responses in plants. Growing evidence of the importance of miRNAs in crop plants means that these molecules emerged as the future generation of targets for recombinant and genetic engineering intended to improve crop agronomic traits.

4. Future scope

Plant miRNA approaches for agriculture have been found to be generally successful; however, the production of complex multigenic traits such as yield may require varying the expression of multiple genes at various stages of plant growth.

The concept and strategy for enhancing the aforementioned crop improvement approaches would be further enhanced by a thorough understanding of miRNA-mediated target control in crops and model plants.

Another challenge, as previously noted, is that large target gene knockouts and miRNA overexpression typically result in phenotypes that are extremely similar, which is typically opposite to what is perceived in plants with decreased miRNA activity. Another strategy for reaching a desired result is the capacity of designed decoys to modify miRNA activity and hence regulate miRNA regulatory networks. Through the development of a desired set of plant phenotypes and precise control over local miRNA targets, this technique offers the chance to alter the sequence of the miRNA trap sites, thereby attenuating the effects of miRNA inactivation.

The promise of miRNA alteration as a viable method for improving plant crops has been recognized, but it's crucial to remember that these genetic modification techniques may have unforeseen negative effects.

Changes in the appearance of the target gene or a miRNA may cause unintentional pleiotropic modification in the development of the plants and its morphology.

Thus, it is essential to comprehend how miRNA regulates plant development and growth, as well as how plants react to different biotic and abiotic stimuli. This will make it easier to create methods that will produce the desired features in the modified crops with the least amount of trade-off.

References

Akter, A., Islam, M.M., Mondal, S.I., Mahmud, Z., Jewel, N.A., Ferdous, S., et al., 2014. Computational identification of miRNA and targets from expressed sequence tags of coffee (Coffea arabica). Saudi J. Biol. Sci. 21, 3–12. https://doi.org/10.1016/j.sjbs.2013.04.007.

Aman, R., Ali, Z., Butt, H., Mahas, A., Aljedaani, F., et al., 2018. RNA virus interference via CRISPR/Cas13a system in plants. Genome Biol. 19, 1−9. https://doi.org/10.1186/s13059-017-1381-1.

Andrianov, V., Borisjuk, N., Pogrebnyak, N., Brinker, A., Dixon, J., Spitsin, S., et al., 2010. Tobacco as a production platform for biofuel: overexpression of Arabidopsis DGAT and LEC2 genes increases accumulation and shifts the composition of lipids in green biomass. Plant Biotechnol. J. 8, 277−287. https://doi.org/10.1111/j.1467-7652.2009.00458.x.

Azevedo, R.A., Carvalho, R.F., Cia, M.C., Gratão, P.L., 2011. Sugarcane under pressure: an overview of biochemical and physiological studies of abiotic stress. Trop. Plant Biol. 4, 42−51. https://doi.org/10.1007/s12042-011-9067-4.

Barozai, M.Y.K., Irfan, M., Yousaf, R., Ali, I., Qaisar, U., Maqbool, A., et al., 2008. Identification of micro-RNAs in cotton. Plant Physiol. Biochem. 46, 739−751. https://doi.org/10.1016/j.plaphy.2008.05.009.

Barrera-Figueroa, B.E., Gao, L., Diop, N.N., Wu, Z., Ehlers, J.D., Roberts, P.A., et al., 2011. Identification and comparative analysis of droughtassociated microRNAs in two cowpea genotypes. BMC Plant Biol. 11−127. https://doi.org/10.1186/1471-2229-11-127.

Bennetzen, J.L., Hake, S.C., 2009. Handbook of Maize, its Biology. Springer, New York, NY.

Bi, H., Fei, Q., Li, R., Liu, B., Xia, R., Char, S.N., et al., 2020. Disruption of miRNA sequences by TALENs and CRISPR/Cas9 induces varied lengths of miRNA production. Plant Biotechnol. J. 18, 1526−1536. https://doi.org/10.1111/pbi.13315.

Birch, R.G., 2007. Metabolic engineering in sugarcane: assisting the transition to a bio-based economy. In: Verpoorte, R., Alfermann, A.W., Johnson, T.S. (Eds.), Applications of Plant Metabolic Engineering. Springer, pp. 249−281.

Boyer, J., Liu, R., 2004. Apple phytochemicals and their health benefits. Nutr. J. 3, 5. https://doi.org/10.1186/1475-2891-3-5.

Carbonell, A., Carrington, J.C., Daros, J.-A., 2016. Fast-forward generation of effective artificial small RNAs for enhanced antiviral defense in plants. RNA Dis. 3 (1), e1130. https://pubmed.ncbi.nlm.nih.gov/26925463/.

Cuperus, J.T., Fahlgren, N., Carrington, J.C., 2011. Evolution and functional diversification of miRNA genes. Plant Cell 23, 431−442. https://doi.org/10.1105/tpc.110.082784.

Curaba, J., Spriggs, A., Taylor, J., Li, Z., Helliwell, C., 2012. miRNA regulation in the early development of barley seed. BMC Plant Biol. 12−120. https://doi.org/10.1186/1471-2229-12-120.

Damodharan, S., Corem, S., Gupta, S.K., Arazi, T., 2018. Tuning of SlARF10A dosage by sly-miR160a is critical for auxin-mediated compound leaf and flower development. Plant J. 96, 855−868. https://doi.org/10.1111/tpj.14073.

Eamens, A.L., Agius, C., Smith, N.A., Waterhouse, P.M., Wang, M.-B., 2011. Efficient silencing of endogenous microRNAs using artificial microRNAs in *Arabidopsis thaliana*. Mol. Plant 4 (1), 157−170. https://doi.org/10.1093/mp/ssq061.

El-Sharkawy, M.A., 2004. Cassava biology and physiology. Plant Mol. Biol. 56, 481−501. https://doi.org/10.1007/s11103-005-2270-7.

Frazier, T.P., Xie, F., Freistaedter, A., Burklew, C.E., Zhang, B., 2010. Identification and characterization of microRNAs and their target genes in tobacco (Nicotiana tabacum). Planta 232, 1289−1308. https://doi.org/10.1007/s00425-010-1255-1.

Gaj, T., Gersbach, C.A., Fau-Barbas, C.F., Barbas, C.F., 2013. ZFN, TALEN, and CRISPR/Cas-based methods for genome engineering. Trends Biotechnol. 31, 397−405. https://doi.org/10.1016/j.tibtech.2013.04.004.

Giovannoni, J.J., 2007. Fruit ripening mutants yield insights into ripening control. Curr. Opin. Plant Biol. 10, 283–289. https://doi.org/10.1016/j.pbi.2007.04.008.

Hou, N., Cao, Y., Li, F., Yuan, W., Bian, H., Wang, J., et al., 2019. Epigenetic regulation of miR396 expression by SWR1-C and the effect of miR396 on leaf growth and developmental phase transition in Arabidopsis. J. Exp. Bot. 70, 5217–5229. https://doi.org/10.1093/jxb/erz285.

Itaya, A., Bundschuh, R., Archual, A.J., Joung, J.G., Fei, Z.J., Dai, X.B., et al., 2008. Small RNAs in tomato fruit and leaf development. Biochim. Biophys. Acta. 1779, 99–107. https://doi.org/10.1016/j.bbagrm.2007.09.003.

Jaillon, O., Aury, J.M., Noel, B., Policriti, A., Clepet, C., Casagrande, A., et al., 2007. The grapevine genome sequence suggests ancestral hexaploidization in major angiosperm phyla. Nature 449, 463–465. https://doi.org/10.1038/nature06148.

Jannoo, N., Grivet, L., Chantret, N., Garsmeur, O., Glaszmann, J.C., Arruda, P., et al., 2007. Orthologous comparison in a gene-rich region among grasses reveals stability in the sugarcane polyploidy genome. Plant J. 50, 574–585. https://doi.org/10.1111/j.1365-313X.2007.03082.x.

Ji, L., Liu, X., Yan, J., Wang, W., Yumul, R., Kim, Y., et al., 2011. ARGONAUTE10 and ARGONAUTE1 regulate the termination of floral stem cells through two microRNAs in Arabidopsis. PLoS Genet. 7. https://doi.org/10.1371/journal.pgen.1001358.

Lelandais-Brière, C., Sorin, C., Declerck, M., Benslimane, A., Crespi, M., Hartmann, C., 2010. Small RNA diversity in plants and its impact in development. Curr. Genom. 11, 14–23. https://doi.org/10.2174/138920210790217918.

Li, J.-F., Norville, J.E., Aach, J., McCormack, M., Zhang, D., Bush, J., et al., 2013. Multiplex and homologous recombination-mediated genome editing in Arabidopsis and Nicotiana benthamiana using guide RNA and Cas9. Nat. Biotechnol. 31, 688–691. https://doi.org/10.1038/nbt.2654.

Li, Z., Zhang, D., Xiong, X., Yan, B., Xie, W., Sheen, J., et al., 2017. A potent Cas9-derived gene activator for plant and mammalian cells. Nat. Plants 3, 930–936. https://doi.org/10.1038/s41477-017-0046-0.

Lin, J.S., Lin, C.C., Lin, H.H., Chen, Y.C., Jeng, S., 2012. MicroR828 regulates lignin and H2O2 accumulation in sweet potato on wounding. New Phytol. 196, 427–440. https://doi.org/10.1111/j.1469-8137.2012.04277.x.

Lu, Y., Yang, X., 2010. Computational identification of novel microRNAs and their targets in Vigna unguiculata. Comp. Funct. Genom. 2010, 128297. https://doi.org/10.1155/2010/128297.

Matts, J., Jagadeeswaran, G., Roe, B.A., Sunkar, R., 2010. Identification of microRNAs and their targets in switchgrass, a model biofuel plant species. J. Plant Physiol. 167, 896–904. https://doi.org/10.1016/j.jplph.2010.02.001.

McLaughlin, S.B., Kiniry, J.R., Taliaferro, C.M., Ugarte, D.D., 2006. Projecting yield and utilization potential of switchgrass as an energy crop. Adv. Agron. 90, 267–297. https://doi.org/10.1016/S0065-2113(06)90007-8.

Miao, C., Wang, Z., Zhang, L., Yao, J., Hua, K., Liu, X., et al., 2019. The grain yield modulator miR156 regulates seed dormancy through the gibberellin pathway in rice. Nat. Commun. 10, 3822. https://doi.org/10.1038/s41467-019-11830-5.

Mica, E., Gianfranceschi, L., Pe, M.E., 2006. Characterization of five microRNA families in maize. J. Exp. Bot. 57, 2601–2612. https://doi.org/10.1093/jxb/erl013.

Mondego, J.M., Vidal, R.O., Carazzolle, M.F., Tokuda, E.K., Parizzi, L.P., Costa, G.G., et al., 2011. An EST-based analysis identifies new genes and reveals distinctive gene expression features of Coffea arabica and Coffea canephora. BMC Plant Biol. 11, e30. https://doi.org/10.1186/1471-2229-11-30.

Mutegi, E., Fabrice, S., Moses, M., Ben, K., Bernard, R., Caroline, M., et al., 2010. Ecogeographical distribution of wild, weedy and cultivated Sorghum bicolor (L.) Moench in Kenya: implications for conservation and crop-to-wild gene flow. Genet. Resour. Crop Evol. 57, 243–253. https://doi.org/10.1007/s10722-009-9466-7.

Nekrasov, V., Staskawicz, B., Weigel, D., Jones, J.D.G., Kamoun, S., 2013. Targeted mutagenesis in the model plant Nicotiana benthamiana using Cas9 RNA-guided endonuclease. Nat. Biotechnol. 31, 691–693. https://doi.org/10.1038/nbt.2655.

Patanun, O., Lertpanyasampatha, M., Sojikul, P., Viboonjun, U., Narangajavana, J., 2013. Computational identification of microRNAs and their targets in Cassava (Manihot esculenta Crantz.). Mol. Biotechnol. 53, 257–269. https://doi.org/10.1007/s12033-012-9521-z.

Peng, T., Qiao, M., Liu, H., Teotia, S., Zhang, Z., et al., 2018. A resource for inactivation of micrornas using short tandem target mimic technology in model and crop plants. Mol. Plant 11 (11), 1400–1417. https://doi.org/10.1016/j.molp.2018.09.003.

Prabu, G.R., Mandal, A.K., 2010. Computational identification of miRNAs and their target genes from expressed sequence tags of tea (Camellia sinensis). Dev. Reprod. Biol. 8, 113–121. https://doi.org/10.1016/S1672-0229(10)60012-5.

Prasad, P.V.V., Pisipati, S.R., Mutava, R.N., Tuinstra, M.R., 2008. Sensitivity of grain sorghum to high temperature stress during reproductive development. Crop Sci. 48, 1911. https://doi.org/10.2135/cropsci2008.01.0036.

Sanan-Mishra, N., Mukherjee, S.K., 2007. A peep into the plant miRNA world. Open Plant Sci. J. 1, 1–9. https://doi.org/10.2174/1874294700701010001.

Shan, Q., Wang, Y., Li, J., Zhang, Y., Chen, K., Liang, Z., et al., 2013. Targeted genome modification of crop plants using a CRISPR-Cas system. Nat. Biotechnol. 31, 686–688. https://doi.org/10.1038/nbt.2650.

Shen, H., He, X., Poovaiah, C.R., Wuddineh, W.A., Ma, J., Mann, D.G., et al., 2012. Functional characterization of the switchgrass (Panicum virgatum) R2R3-MYB transcription factor PvMYB4 for improvement of lignocellulosic feedstocks. New Phytol. 193, 121–136. https://doi.org/10.1111/j.1469-8137.2011.03922.x.

Shi, C.Y., Yang, H., Wei, C.L., Yu, O., Zhang, Z.Z., Jiang, C.J., et al., 2011. Deep sequencing of the Camellia sinensis transcriptome revealed candidate genes for major metabolic pathways of tea specific compounds. BMC Genom. 12, 131. https://doi.org/10.1186/1471-2164-12-131.

Sierro, N., Battey, J.N.D., Ouadi, S., Bakaher, N., Bovet, L., Willig, A., et al., 2014. The tobacco genome sequence and its comparison with those of tomato and potato. Nat. Commun. 5–3833. https://doi.org/10.1038/ncomms4833.

Sreenivasulu, N., Graner, A., andWobus, U., 2008. Barley genomics: an overview. Int. J. Plant Genom. 2008, 486258. https://doi.org/10.1155/2008/486258.

Subramanian, S., Fu, Y., Sunkar, R., Barbazuk, W.B., Zhu, J.K., Yu, O., 2008. Novel and nodulation-regulated microRNAs in soybean roots. BMC Genom. 9, 160. https://doi.org/10.1186/1471-2164-9-160.

Sun, G., 2012. MicroRNAs and their diverse functions in plants. Plant Mol. Biol. 80, 17–36. https://doi.org/10.1007/s11103-011-9817-6.

Sunkar, R., Girke, T., Jain, P.K., Zhu, J.K., 2005. Cloning and characterization of microRNA from rice. Plant Cell 17, 1397–1411. https://doi.org/10.1105/tpc.105.031682.

Tang, X., Lowder, L.G., Zhang, T., Malzahn, A.A., Zheng, X., Voytas, D.F., et al., 2017. A CRISPR-Cpf1 system for efficient genome editing and transcriptional repression in plants. Nat. Plants 3, 17103. https://doi.org/10.1038/s41477-017-0001-0.

Vaistij, F.E., Elias, L., George, G.L., Jones, L., 2010. Suppression of microRNA accumulation via RNA interference in *Arabidopsis thaliana*. Plant Mol. Biol. 73, 391–397. https://doi.org/10.1007/s11103-010-9625-4.

Wang, J.F., Zhou, H., Chen, Y.Q., Luo, Q.J., Qu, L.H., 2004. Identification of 20 microRNAs from Oryza sativa. Nucleic Acids Res. 32, 1688–1695. https://doi.org/10.1093/nar/gkh332.

Wang, M., Wang, Q., Wang, B., 2012. Identification and characterization of microRNAs in Asiatic cotton (Gossypium arboretum L.). PLoS One 7, e33696. https://doi.org/10.1371/journal.pone.0033696.

Warthmann, N., Chen, H., Ossoswki, S., Weigel, D., Herve, P., 2008. Highly specific gene silencing by artificial miRNAs in rice. PLos One 3 (3), e1829. https://doi.org/10.1371/journal.pone.0001829.

Xu, Q., Chen, L.L., Ruan, X., Chen, D., Zhu, A., Chen, C., et al., 2013. The draft genome of sweet orange (Citrus sinensis). Nat. Genet. 45, 59–66. https://doi.org/10.1038/ng.2472.

Yang, C., Li, D., Mao, D., Liu, X., Ji, C., Li, X., et al., 2013. Overexpression of microRNA319 impacts leaf morphogenesis and leads to enhanced cold tolerance in rice (Oryza sativa L.). Plant Cell Environ. 36, 2207–2218. https://doi.org/10.1111/pce.12130.

Yao, Y., Guo, G., Ni, Z., Sunkar, R., Du, J., Zhu, J.K., et al., 2007. Cloning and characterization of microRNAs from wheat (*Triticum aestivum* L.). Genome Biol. 8, R96. https://doi.org/10.1186/gb-2007-8-6-r96.

Zanca, A.S., Vicentini, R., Ortiz-Morea, F.A., Bem, L.E.D., da Silva, M.J., Vincentz, M., et al., 2010. Identification and expression analysis of microRNAs and targets in the biofuel crop sugarcane. BMC Plant Biol. 10–260. https://doi.org/10.1186/1471-2229-10-260.

Zhang, B.H., Wang, Q.L., Wang, K.B., Pan, X.P., Liu, F., Guo, T.L., et al., 2007. Identification of cotton microRNAs and their targets. Gene 397, 26–37. https://doi.org/10.1016/j.gene.2007.03.020.

Zhang, W., Luo, Y., Gong, X., Zeng, W., Li, S., 2009. Computational identification of 48 potato microRNAs and their targets. Comput. Biol. Chem. 33, 84–93. https://doi.org/10.1016/j.compbiolchem.2008.07.006.

Zhao, C.Z., Xia, H., Frazier, T.P., Yao, Y.Y., Bi, Y.P., Li, A.Q., et al., 2010. Deep sequencing identifies novel and conserved microRNAs in peanuts (*Arachis hypogaea* L.). BMC Plant Biol. 10–13. https://doi.org/10.1186/1471-2229-10-3.

CHAPTER

Plant development: What role do microRNAs play?

4

Kotte Bhargava[1], E. Abhishek[1], P. Bhavana[2]

[1]*ICAR-Indian Agricultural Research Institute, New Delhi, India;* [2]*ICAR-Research Complex for Eastern Region, Farming Systems Research Centre Hill and Plateau Region, Ranchi, Jharkhand, India*

1. Introduction

MicroRNA (miRNA) is a noncoding form of RNA. It is a small, single-stranded RNA (22 nucleotides in length). It aids in the silencing of RNA and gene expression regulation. By creating stable base pairs with its complementary messenger RNA (mRNA), miRNA suppresses gene expression. They inhibit gene expression by inhibiting translation, cleaving, or destabilizing mRNA. The term "microRNA" was used in 2001; however, they were initially reported in 1993 by Lee and colleagues (Lee et al., 1993). The role of miRNAs in plant biology is well established at this point. They were first found in plants during the last 20 years. In 1993, two independent research groups led by Ambros and Ruvkun found that a little 22 nt RNA made from a gene called lin-4 suppressed the expression of lin-14, therefore regulating the time of *Caenorhabditis elegans* larval development (Lee et al., 1993; Wightman et al., 1993). When discussing the microRNA family tree, lin-4 is often cited as the progenitor (Lee and Ambros, 2001). Hundreds of additional miRNAs from a wide variety of organisms, including *C. elegans*, *Drosophila melanogaster*, fish, mouse, rat, and human, have been found since lin-4 was initially recognized as a miRNA in 2001 (Lai, 2003). These 872 miRNAs have been discovered so far, and they have been located in 71 different plant species and 42 different miRNA families. Unfortunately, only 391 of them have been reported to the miRNA Registry Database (Rfam 6.0, published in April 2005). This view is shared by others (Griffiths-Jones, 2004). Over the last 3 years, there has been a dramatic increase in the total number of miRNAs due to the fast development of technologies for discovering miRNAs unique to each species. MiRNAs have a crucial role in the development of the leaf, flower, plant embryo, and meristem, as well as in the plant's reaction to biotic and abiotic stress (D'Ario et al., 2017).

MicroRNA Advances and Application in Plant Biology. https://doi.org/10.1016/B978-0-443-21682-4.00001-4

2. Identification of plant microRNAs

Methods for identifying miRNAs are based on their major characteristics:

Approach	Procedure
Genetic screening	The procedure was very identical to the one that is used in the identification of conventional genes. This method did not lead to the discovery of all miRNAs; however, it did lead to the discovery of several, such as lin-4 and lin-7. But the method is a significant amount of effort and financial investment.
Direct cloning after isolation of small RNAs	The very first stage of this technique involves the isolation of minute RNA molecules depending on their size. Following this step, the 5 and 3V ends of the short RNAs are linked to RNA adapters. After that, their cDNA goes through amplification and sequencing, after which it is reverse-transcribed (Lu et al., 2005a). A separation of extremely tiny RNA molecules based on their size is the first step in this methodology. The 3 and 5V ends of short RNAs are connected to RNA adapters, respectively. Their cDNA undergoes reverse transcription, which is followed by amplification and sequencing (Lu et al., 2005a). When it comes to isolating miRNAs, this method is better to conventional genetic screening since it just involves the extraction and analysis of short RNAs. This strategy has been used to the investigation of new miRNAs in a wide variety of organisms, including mammals (Fu et al., 2005) and plants (Sunkar et al., 2005). Lu et al. (2005a) have recently improved upon this technique for the research of Arabidopsis miRNAs by using massively parallel signature sequencing (MPSS). This method may be used for the detection of plant miRNAs, and it also has the capacity of quantifying the amount of miRNA in a sample.
Computational strategy	This strategy is based on the sequencing of a genome (Wang et al., 2005). Several research institutions have developed computational programmes, such as MIRscan (http://genes.mit.edu/mirscan/) (Lim et al., 2003a, b) and MiRAlign (http://bioinfo.au.tsinghua.edu.cn/miralign/) (Wang et al., 2005), and have utilized these programs to successfully predict miRNA genes in Arabidopsis (Lim et al., 2003a). The bulk of the microRNAs that are now stored in miRNA databases, such as "The miRNA Registry" (http://www.sanger.ac.uk/Software/Rfam/mirna/index.shtml) and "The Arabidopsis Small RNA Project" (http://asrp.cgrb.oregonstate.edu/), were discovered via the use of computational methods. Nevertheless, the actual computational technique that was applied was only modestly efficient and was by no means thorough. Experiments, such as cloning and Northern blotting, need to be performed in order to validate the anticipated miRNAs.

Approach	Procedure
Expressed sequence tags (ESTs) analysis	It is known that some miRNAs have been maintained over the course of evolutionary history and species (Llave et al., 2002). This demonstrates very strongly that there is a way to anticipate homologs or orthologs of known miRNAs. As an example, Weber (2005) used this methodology to find 45 unique mouse putative miRNA genes in addition to 35 novel human miRNA genes. In addition, this technique is very useful for predicting miRNA across species, particularly in species for which the genome is not known. The development of an EST analysis approach for determining miRNA homologies has been the primary focus of recent work in our laboratory. We searched through an EST collection that was open to the public and discovered 481 miRNA homologs that were only previously thought to be present in a single plant species (Arabidopsis). This suggests that EST analysis is a workable approach for locating miRNAs, particularly for species whose genomes are not well characterized. This is especially true for species like these. On the other hand, this method can only find miRNAs that have significant levels of conservation in their sequences. It is challenging to find these genes by using the EST method due to the fact that the overwhelming majority of miRNAs are not likely to have been saved.

3. Origin and diversity of miRNAs in plants

A great number of miRNAs, such as miR7 and miR18, have been passed down from the nematode worm to the human species (Lagos-Quintana et al., 2001). According to Pasquinelli et al. (2000), the first appearance of miRNAs in animals likely occurred at least 420 million years ago, during the time that the last common ancestor of metazoans was alive. On the other hand, mature miRNAs are the only ones that are preserved in plants, but in mammals, both mature miRNAs and their precursors are retained (Llave et al., 2002). Experimental and computational work reveals that many plant miRNAs and their targets are conserved between monocots and dicots, based on sparse data for Arabidopsis and rice (Bonnet et al., 2004). Given that monocots and dicots diverged from the same parent around 125 million years ago, it is reasonable to assume that miRNAs were also present at that time. Axtell and Bartel (2005) examined the expression of a variety of miRNAs across a wide variety of plant species by making use of the technology known as microarrays. They found that numerous miRNAs are present not just in dicots and monocots but also in ferns, lycopods, and mosses in addition to those two plant types. Our EST data also shown that microRNAs may be found in over 71 different plant taxonomic groups, ranging from mosses to higher flowering plants (Zhang et al., 2005). Floyd and Bowman (2004) demonstrated that miR166 has conserved target sites in the class-III homeodomain-leucine zipper (HD-Zip) genes. Despite the fact that other regions have far lower nucleotide conservation, they were able to demonstrate that miR166 possesses these target areas. There is evidence that plant microRNAs have a deep evolutionary origin, extending back to the period of the last common ancestor of bryophytes and seed plants, which is provided by the conservation of both the microRNAs and the targets of those microRNAs (Floyd and Bowman, 2004). The regulation of gene expression by microRNAs is a very ancient evolutionary technique that has been used for a very long time. It is an essential component of the global systems that regulate gene expression in all organisms. This demonstrates that microRNAs were already actively controlling gene activity in plants more than 425 million years ago. When compared to the regulation of miRNA in metazoans, the process of miRNA regulation in plants is almost the same age. However, as of right now, there is no evidence to imply that miRNAs found in plants and miRNAs found in animals have a similar ancestry. There are 117 distinct miRNAs that have been identified in Arabidopsis, and the miRNA database has information on all of them (Griffiths-Jones, 2004). These microRNAs may be placed into one of 42 families by analyzing the nucleotide sequences they have in common with those families. The vast majority of miRNA families have been found in other species, such as rice, maize, and sorghum (Griffiths-Jones, 2004). The genomes of many distinct plant species are only differentiated by a few number of nucleotide differences; nonetheless, some of the 42 families of plant miRNAs have remained very conserved over the course of evolution (Zhang et al., 2005). Axtell and Bartel found evidence of members of the miR159/319 family in 10 distinct plant species via their research (2005). There are a possible nine distinct plant species that include members of the miRNA families 156/157 and 165/166. These miRNA families can be found in all plant phyla. According to the findings of Zhang et al., almost one third of miRNA families were discovered to be present in at least 10 different plant species (2005). A member of one of these miRNA families called MiR156 was found in 31 different plant species (Zhang et al., 2005). According to their degree of genetic similarity to other miRNAs, microRNAs have been divided into the following four categories: highly conserved, moderately conserved, poorly conserved, and nonconserved miRNAs (Zhang et al., 2005). Jones-Rhoades and Bartel (2004) found that some miRNAs, such as miR163 and miR158, did not exhibit conservation. Despite this, the dbEST investigation found that there were 5 and 4, respectively, different plant species that contained homologs of these two miRNAs. As a result of this, we need to certainly redefine them as families of miRNAs that have a low level of conservation (Zhang et al., 2005). It is likely that the categorization of those miRNAs that are now labeled as nonconserved or poorly conserved miRNAs may alter as a result of the discovery of more miRNAs in the future. It may be deduced what a miRNA does by looking at how well its sequence has been preserved throughout the course of history. The regulation of genes that are engaged in critical activities, including the formation of leaves and flowers and the transmission of signals, is an important function performed by conserved microRNAs. It is probable that nonconserved miRNAs have more specialized activities in certain plant species, such as in the differentiation and elongation of cotton fibers. This notion is supported by the observation that nonconserved miRNAs are present in all plant species. The duplication of protein-coding sequences is one of the aspects that contributes to the complex interaction of factors that leads to the formation of new gene regulatory sources and their networks (Levine and Tjian, 2003). After researching a large number of sequences in Arabidopsis that were connected to microRNAs, Allen et al. (2004) developed a theory to explain their origin. They hypothesize that inverted duplications producing plant pre-miRNAs are transcribed directly from their target genes without undergoing any further modification (Allen et al., 2004). A large number of genomic regions that were previously believed to be devoid of features were shown to be loci of microRNAs, demonstrating that microRNAs may originate anywhere in the plant genome (Lu et al., 2005a, b). Based on these findings, it seems that the mechanisms of miRNA origin, such as inversion and duplication, are more complicated than was previously thought.

4. Biogenesis of microRNAs

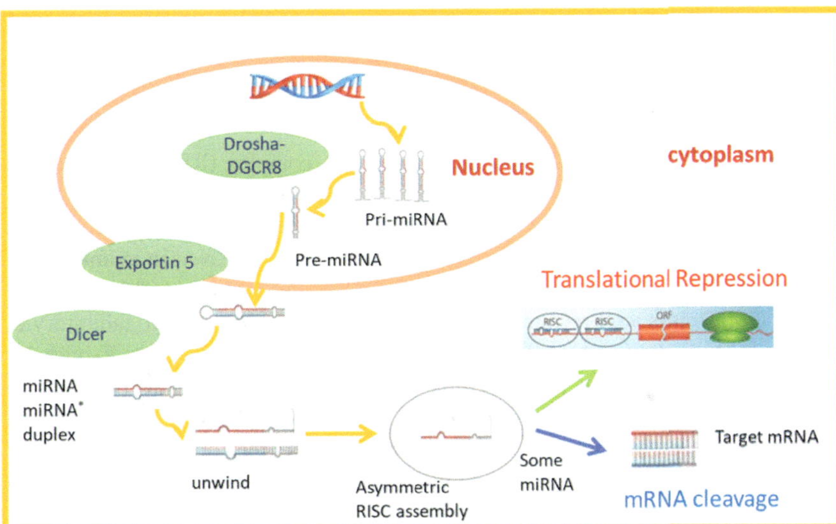

The genes responsible for encoding miRNAs are much longer than the processed miRNA molecule due to the fact that miRNAs are encoded by bigger genes. miRNAs that are located in the introns of their pre-mRNA host genes have many features, such as expression patterns, regulatory elements, and main transcripts. A limited quantity of fully characterized primary transcripts has been identified for the other miRNA genes that are synthesized from their respective promoters. To synthesize a microRNA, the first step involves the transcription of a long RNA precursor called a pri-miRNA by RNA polymerase II. According to Lee et al. (2004), the pri-miRNA has a 5′ cap and a poly-A tail. The nucleus contains a microprocessor complex that consists of the RNase III enzyme Drosha (Han, 2004) and the double-stranded RNA-binding protein Pasha/DGCR8 (Lund, 2004) responsible for processing the pri-miRNAs. The pre-miRNAs, consisting of around 70 nucleotides, eventually undergo folding into suboptimal stem-loop structures. Subsequently, the transportation of pre-miRNAs into the cytoplasm is facilitated by the karyopherin exportin 5 (Exp5) and Ran-GTP complex (Yi, 2003). Ran, or Ras-related nuclear protein, is a small protein that binds to GTP and belongs to the RAS superfamily. The transportation of RNA and proteins via the nuclear pore complex is of utmost importance. According to Yi (2003), the Ran GTPase engages with the Exp5 protein to produce a nuclear heterotrimer that encompasses pre-miRNAs. Dicer, an RNAse III enzyme situated in the cytoplasm, performs further processing on the pre-miRNAs. This results in the synthesis of miRNA, a double-stranded RNA with a length of around 22 nucleotides (Bernstein et al., 2001). The dicer plays a crucial role in the first phases of the RISC assembly process. Hammond's study indicates that the RISC is responsible for the suppression of genes resulting from the generation of miRNA and RNA interference (RNAi) (Hammond et al., 2000).

5. Mechanism of mi RNA gene regulation in plants

The role of miRNAs in the RNA-induced silencing complex (RISC) involves their ability to attach to mRNA in a manner that is nearly perfect complementarity. This binding process leads to the suppression of gene expression (Bartel, 2004). According to Almeida and Allshire (2005), this phenomenon results in RNAi in mammals and posttranscriptional gene silencing (PTGS) in plants. According to Carrington and Ambros (2003), it is well recognized that microRNAs in animals have inhibitory effects on gene expression by selectively binding to mRNA targets that exhibit low complementarity at various places within the 3V untranslated regions (UTRs). The regulation of gene expression in plants is governed by processes that have resemblance to those seen in mammals. On the other hand, it is noteworthy that a significant proportion of target mRNAs present in plants include a solitary miRNA complementary site. Furthermore, the majority of miRNAs that align with these sites often exhibit perfect complementarity, hence facilitating the cleavage of the target mRNAs (Bartel, 2004). The presence of complementary sites for miRNAs in plants is not restricted only to the 3-UTR region of the target mRNA. Instead, these sites may be located at various locations along the mRNA. In contrast to the scenario seen in mammals, it is essential that the complimentary sites be situated inside the 3-UTR. A recent study has shown a new method for controlling miRNA in plants. Although some microRNAs, such as miR172, may exhibit a harmonious complementarity with the mRNAs they target, they nonetheless possess the capacity to modulate gene expression by suppressing gene translation, perhaps by slowing ribosome mobility (Aukerman and Sakai, 2003). In contrast, recent research has shown that miR172 exhibits effective regulation of the cleavage process of its target RNAs (Schwab et al., 2005). Based on the aforementioned data, it can be inferred that the role of miRNAs in plants exhibits a higher level of complexity compared to miRNAs in mammals in the regulation of gene expression. Plant miRNAs have the ability to not only restrict the translation of mRNA but also have posttranscriptional effect on gene expression by direct mRNA cleavage. This is an additional mechanism via which plant microRNAs regulate gene expression. In their study, Schwab et al. (2005) used genome-wide expression profiling to examine the characteristics associated with miRNA-guided mRNA cleavage. The researchers' results yielded distinctive perspectives on the mechanism behind plant miRNA activity. Transgenic technology has facilitated the upregulation of certain plant miRNAs, such as miR156, miR159, miR164, and miR319/JAW. The purpose of this research was to investigate the impact of various circumstances on the process of mRNA cleavage. Furthermore, consideration was given to the quantity of mismatches, the extent to which the sequences of the miRNA and its target mRNA exhibit complementarity, and the free energy associated with miRNA pairings with potential target sites. Schwab et al. (2005) used these data to develop a collection of empirical variables for identifying miRNA targets. Furthermore, they used these data to deduce that plant miRNAs exhibit greater specificity compared to siRNAs. Additionally, the research findings indicate that the impact of miRNAs on the stable levels of target mRNAs is intricately influenced by feedback regulation. This discovery was identified by the researchers. Schwab et al. (2005) propose the use of feedback regulation of miRNA targets as a viable approach for fine-tuning the expression of target genes.

6. miRNA role in plant development

MicroRNAs play a vital role in the development of plants. Plant stem cells, similar to animal stem cells, are a type of cell population that can undergo various types of differentiation. This is because they have two key characteristics: the ability to renew themselves in a pluripotent state and the ability to differentiate into specialized cell types (Weigel and Jurgens, 2002). Plant stem cells are found in the confined habitats of shoot apical meristems (SAMs) and root apical meristems (RAMs). According to Weigel and Jurgens (2002) and Aida and Tasaka (2006), the formation of SAM takes place in higher plants during embryogenesis from the axil of cotyledon(s), and again during postembryogenesis from the axil of leaves. The study of meristems is consistently popular due to the pivotal function that stem cells play in the process of plant growth. Recent studies have contributed to a more comprehensive understanding of the mechanisms by which miRNAs regulate gene expression and facilitate intercellular communication in the context of meristem differentiation.

6.1 miRNA-regulated SAM development

Singh and Bhalla (2006) assert that the precise spatiotemporal regulation of SAM stem cells, which act as progenitors for various aerial functional cells, is pivotal for determining cell fate and organ development in higher plants. Evidence suggests that specific miRNAs operate as mobile signaling molecules, crucial for maintaining stem cell adhesion mechanisms within the SAM by directly regulating key SAM-related genes at the posttranscriptional level (Zhang and Zhang, 2012). The SAM comprises three distinct cell layers: the outer cell layer L1 (protoderm), the middle layer L2, and the inner layer L3 (organizing center, OC) (Satina et al., 1940). The pluripotent nature of ovarian cancer (OC) cells is dependent on the expression of the homeodomain transcription factor WUS (WUSCHEL) protein (Knauer et al., 2013). The development of rice tillers, or shoot branches, is a critical determinant of rice grain yield, with WUS proteins playing a significant role in regulating this process due to their central involvement in SAM formation (Wang et al., 2014).

AGO10, also known as PINHEAD and ZWILLE, has been identified as a key component in maintaining SAM in Arabidopsis, as revealed by genetic screenings (Zhang and Zhang, 2012; Moussian et al., 1998). Mutation in the AGO10 gene results in the absence of embryonic meristem, producing a pinhead phenotype characterized by an empty apex (McConnell and Barton, 1995). Further research indicates that AGO10 regulates WUS protein activity independently, sustaining SAM via AGO10-mediated mechanisms (Tucker et al., 2008). The translocation of AGO10 across cells appears to be facilitated by certain signal molecules, as the protein itself cannot move (Tucker et al., 2008). Genetic analysis by Zhu et al. (2011) reveals that the ago10 mutant shows elevated levels of miR165 and miRNA166 synthesis, miRNAs that differ by a single nucleotide in their mature sequences (Liu et al., 2009). Reducing miR165/166 expression in ago10 mutant plants alleviates some abnormal shoot apex characteristics (Liu et al., 2009). Both miR165 and miR166 target the class III homeodomain-leucine zipper (HD-ZIP III) family of transcription factors, essential for SAM development, including apical embryo patterning, SAM establishment and maintenance, and lateral organ polarity. Molecular analysis by Zhu et al. (2011) shows that AGO10 has a specific affinity for miR165/166, positioning them as prime mobile signal molecules for AGO10-mediated noncell-autonomous regulation. AGO10 acts as a decoy for miR165/166, preventing their repressive effect on HD-ZIP III genes and thus maintaining SAM synthesis (Zhu et al., 2011).

In maize, the rolled leaf1 (Rld1) mutation alters the adaxial/abaxial (dorsoventral) polarity of the leaf (Nelson et al., 2002). Genetic research shows that Rld1 encodes a gene in the maize HD-ZIP III family (rev1), with 70% similarity to one of the five Arabidopsis HD-ZIP III members (REV) (Juarez et al., 2004; Zhang and Zhang, 2012). In situ hybridization studies by Juarez et al. (2004) demonstrate that, like the Arabidopsis HD-ZIP III genes, the maize rev1 gene is most highly expressed near the terminal region of SAM and SAM-related areas. Additionally, the conserved miRNA165/166 complementary site in maize rev1 mirrors that in Arabidopsis, suggesting a common regulatory mechanism between the two species.

Trans-acting siRNAs (ta-siRNAs), 21-nucleotide short RNAs with posttranscriptional regulatory roles similar to miRNAs, are implicated in SAM development in cereal plants like maize and rice (Mallory and Bouche, 2008). The maize mutant leafbladeless1 (lbl1) exhibits leaf polarity defects due to faulty primordia development (Timmermans et al., 1998). Sequence analysis by Nogueira et al. (2007) reveals that the protein suppressor of gene silencing3 (SGS3), encoded by LBL1, is involved in ta-siRNA production. Loss of SGS3 induces miR166c, miR166i, and miR166h expression at the SAM base, reducing maize HD-ZIP III gene rolled leaf2 (rld2) expression (Nogueira et al., 2007). Satoh et al. (1999) identify the rice shoot organization (SHO) and SHOOTLESS (SHL) gene families as critical for SAM initiation and maintenance. Mutations in SHO1, SHO2, and SHL2 result in loss or impairment of rice SAM function (Liu et al., 2007). SHO1 encodes DICER-like 4, SHO2 encodes ARGONAUTE 7, and SHL2 encodes RNA-dependent RNA polymerase 6 (Liu et al., 2007). Mallory and Bouche (2008) identify the full gene repertoire for ta-siRNA biosynthesis, highlighting ta-siRNAs' significant role in SAM development. The miR165-HD-ZIP III regulatory mechanism in rice underpins abnormal SAM production in shl mutants (Nagasaki et al., 2007). Increased miR166 expression in sho mutant tissues and suppressed rice HD-ZIP III genes OSHB1 and OSHB2 further indicate miRNA and ta-siRNAs' collaborative role in SAM maintenance (Nagasaki et al., 2007).

Researchers employed an ethyl methanesulfonate (EMS) mutant screen to identify enhancer loci of the weak ago10-1 stem cell mutant, uncovering that the enh146 mutant enhances meristem termination in the ago10-1 background (Knauer et al., 2013). Molecular evidence suggests that the ENH146 locus encodes miR394, with suppression impacting stem cell preservation (Knauer et al., 2013). Baumann (2013) provides evidence that miR394 functions as a mobile signal from the outer cell layer L1 (protoderm) to maintain SAM region stem cell competence.

6.2 miRNA-mediated postembryonic shoot meristem regulation

Higher plants derive their principal axis and fundamental structure from the SAM, which predominantly forms during embryogenesis (McSteen and Leyser, 2005; Hibara et al., 2006). Postembryonic shoot meristems, also known as axillary shoot meristems, are pivotal in initiating shoot branching and inflorescence structure formation, both of which are critical for crop yield enhancement. These meristems can develop multiple times throughout a plant's lifecycle (Komatsu et al., 2003).

Notably, scientific investigations have pinpointed the lateral suppressor (Ls or Las) gene as the key regulator of axillary meristem growth in the tomato (*Solanum lycopersicum*). Mutant plants deficient in this gene exhibit pronounced defects in axillary bud initiation and inflorescence formation (Schumacher et al., 1999). Analogous phenotypes are observed in rice, where mutations in the MONOCULM 1 (MOC1) gene lead to a spectrum of lateral organ developmental issues, including a reduction in the number of shoot laterals (tillers), rachis laterals, and spikelet laterals (Li et al., 2003). The MOC1 gene in rice shares a 44% sequence similarity with the tomato Las gene, indicating their orthologous nature (Li et al., 2003).

Research by Greb et al. (2003) has elucidated that the Arabidopsis Las gene is integral to axillary shoot production, a process regulated in a conserved manner. Genetic evidence from Kasschau et al. (2003) suggests that miR164 modulates Las expression in Arabidopsis through posttranscriptional regulation of the CUP-SHAPED COTYLEDON genes CUC1 and CUC2. These genes are implicated in diverse developmental roles, including the initiation of embryonic meristems, border size control, and cotyledon establishment. This is underscored by the array of developmental anomalies observed in single or double mutants of cuc1 and cuc2 (Long et al., 1996).

Moreover, the miR164-CUC1/CUC2 regulatory mechanism is potentially involved in axillary shoot meristem initiation via LAS. This hypothesis is supported by the accumulation of CUC1 and CUC2 mRNA in the axils of leaf primordia, underscoring their significant role in establishing axillary meristems through an LAS-mediated pathway (Hibara et al., 2006).

6.3 miRNAs and leaf development

The leaf, serving as the primary locus for photosynthetic processes, is indispensable in the production of biomass and agricultural yields. Leaf formation is a complex orchestration of developmental stages culminating in a fully mature leaf. Leaf primordia production initiates from undifferentiated cells located at the SAM periphery (Byrne, 2005). These border cells are pivotal in establishing leaf polarity and demarcating leaf primordia from the SAM. Rubio-Somoza and Weigel (2011) have demonstrated that the genes CUC1, CUC2, and their master regulator miR164 play a crucial role in leaf development by delineating organ boundaries.

In agricultural contexts, productivity is intrinsically linked to photosynthetic efficiency, which is, in turn, influenced by the dimensions and morphology of the foliage. The Arabidopsis mutant jaw-D, characterized by serrated leaves, was identified using activation-tagging techniques, revealing that the JAW locus harbors the miR319, a well-known noncoding RNA (Weigel et al., 2000). MiR319 is a critical regulator of leaf development, a role conserved across species. Ectopic overexpression of miR319 in tomatoes results in significant alterations in leaf size and shape (Ori et al., 2007). Similarly, overexpression of miR319 in rice and creeping bentgrass markedly increases leaf blade size (Yang et al., 2013). MiR319 modulates leaf shape by targeting TCP transcription factors, as evidenced by Palatnik et al. (2003). TCP4, identified as a target of miR319, plays a significant role in leaf shape development. EMS mutagenesis screening has shown that mutations impairing the binding of miR319 to TCP4 can partially restore normal leaf phenotypes in jaw-D plants (Palatnik et al., 2007). The tomato mutant lanceolate (La) exhibits reduced leaf complexity due to a mutation in the La gene, which disrupts miR319-mediated inhibition, further substantiating miR319's role in leaf development.

Genetic mapping has confirmed that La encodes a TCP transcription factor with a miR319-binding site (Ori et al., 2007).

Growth-regulating factors (GRFs) are another group of transcription factors linked to leaf growth regulation. Overexpression of AtGRF1 and AtGRF2 in Arabidopsis results in larger leaves and cotyledons, whereas the AtGRF1-AtGRF3 triple mutant exhibits significantly smaller leaves and cotyledons (Kim et al., 2003). MiR396 selectively targets GRF genes posttranscriptionally, demonstrating evolutionary conservation across species. Upregulation of miR396 in Arabidopsis leads to narrow-leaf phenotypes and suppression of six GRF targets, likely by reducing cellular division and proliferation during leaf development. Overexpression of miR396 in rice also results in narrow-leaf features. Beyond GRF target genes, miR396 interacts with several nonconserved bHLH transcription factors. Debernardi et al. (2012) identified bHLH74 as a novel miR396 target involved in margin and vein pattern formation in Arabidopsis. In *Medicago truncatula*, miR396 negatively regulates the expression of six MtGRF genes and two bHLH79-like targets, as demonstrated by Bazin et al. (2013). This multifaceted regulatory network underscores the sophisticated interplay of miR396 in leaf development across diverse plant species.

6.4 miRNAs and root system development

The root system is esteemed as the second most critical component of a plant, fulfilling essential roles in nutrient and water absorption, providing structural support, synthesizing hormones and secondary metabolites, and contributing to the overall vitality of the plant (Meng et al., 2010). Water and mineral nutrients are indispensable for plant survival and development, as well as for the adaptive capacity and biomass production of agricultural plants in their natural habitats. Consequently, substantial research over the past several decades has focused on the genetic mechanisms underpinning the construction of plant root systems. The genesis of a plant's root system is primarily influenced by both embryonic and postembryonic phases of development, paralleling the origin of the SAM (Rogers and Benfey, 2015). Root structures such as primary and seminal roots develop during embryogenesis, whereas lateral, crown, and brace roots typically emerge postembryonically (Rogers and Benfey, 2015). Additionally, the development of a plant's root system is profoundly affected by the interactions between its roots and the surrounding environment, both above and below the soil surface. Plants possess the remarkable ability to alter their root architecture in response to various environmental stressors, such as drought, phosphorus (P) deficiency, and nitrogen (N) scarcity, to enhance their nutrient uptake capabilities (Malamy and Ryan, 2001).

Accordingly, the configuration of a plant's root system is modulated by the interplay between internal signaling pathways and external conditions. Auxin, a pivotal phytohormone, plays a substantial role in regulating root growth (Gutierrez et al., 2012). Recent studies have elucidated that certain miRNAs are instrumental in modulating root architecture via posttranscriptional modifications of genes involved in the auxin signaling pathway (Wang et al., 2005). The TIR1 protein in Arabidopsis is directly involved in the degradation of AUXIN/INDOLE ACETIC ACID (Aux/IAA) transcriptional repressors in response to auxin signals (Kepinski and Leyser, 2005). TIR1, a member of the AUXIN SIGNALING F-BOX gene family, is complemented by five additional members (AFB1-5), each possessing distinct biochemical and biological functions. miR393 acts as a repressor of TIR1 and AFB2, genes integral to seedling root development. In rice, two miRNAs, OsmiR393a and OsmiR393b, significantly influence root growth, particularly main root elongation and the number of adventitious roots, in overexpression lines (Bian et al., 2012). Further, OsmiR393 targets two rice homologs of TIR1 and TIR2, namely OsTIR1 and OsAFB2.

Recent research substantiates the role of the IAA/ARF transcriptional repressor IAA28 in lateral root development in Arabidopsis. The iaa28-1 mutant displays a marked deficiency in initiating lateral root growth. Studies have shown a link between the crucial transcription factor GATA23, which regulates lateral root founder cell selection, and lateral root development controlled by IAA28. De Rybel et al. (2010) provide evidence from yeast two-hybrid analyses supporting IAA28's involvement in auxin-driven lateral root growth, mediated by interactions with five ARF proteins: ARF5, ARF6, ARF7, ARF8, and ARF19. Wang and Guo (2015) have presented molecular data indicating that IAA28 has cleavage sites partially complementary to miR847. The overexpression of miR847 mimics the iaa28-1 mutant's developmental abnormalities in lateral root formation, bolstering this claim (Wang and Guo, 2015).

The primary responsibility for auxin signal transduction and subsequent growth and development lies with the auxin response factor (ARF) gene family. The *Arabidopsis thaliana* ARF family comprises 23 members (Remington et al., 2004). miR160 selectively targets ARF10, ARF16, and ARF17. Overexpression of miR160c results in alterations in root production, including a reduction in primary roots and an increase in lateral roots. Pro35S:MIR160c seed-derived roots exhibit reduced gravitropism and characteristic curled main roots associated with root cap abnormalities (Wang et al., 2005). Further studies indicate that the Pro35S:MIR160c plant's agravitropic root phenotype is observed in arf10-2 and arf16-2 double mutant plants, suggesting miR160s role in root cap development regulation by cleaving ARF10 and ARF16 transcripts (Wang et al., 2005). Meng et al. (2010) observed that miR160 overexpression in *M. truncatula* and rice might result in root phenotype mutations. However, upregulating miR160 expression in soybean (*Glycine max*) did not significantly alter root development rate or lateral root density despite affecting auxin sensitivity. miR167 has also been linked to root development regulation via its interaction with ARF8 and ARF6. Unlike miR160, miR167 positively influences adventitious root formation (Gutierrez et al., 2009, 2012). Gutierrez et al. (2009) suggest that adventitious root development regulation may involve a complex feedback loop comprising miR160, miR167, and their specific targets ARF17, ARF6, and ARF8. Additionally, miRNAs may collaborate with other short RNA molecules, such as ta-siRNAs, to regulate certain biological processes.

miR173 and miR390 facilitate ta-siRNA synthesis by cleaving transcripts from the TAS1 and TAS2 loci and the TAS3 locus by cleaving transcripts from miR390, while miR828 targets the TAS4 locus (Axtell et al., 2006).

For instance, TAS3 ta-siRNA targets ARF2, ARF3, and ARF4, transcription factors involved in developmental timing and leaf development. Expression analysis data demonstrating overlapping expression regions for miR390 and TAS3, which govern lateral root initiation, suggest the miR390-TAS3 pathway's involvement in lateral root formation. Recent research underscores the critical role of miRNAs in regulating plant responses to nutrient scarcity and drought (Ferdous et al., 2015). These findings indicate that miRNAs and their targets establish an intricate molecular network to integrate environmental and cellular signals.

6.5 Plant floral transitions and their miRNAs

Following the development of a plant's mature structural components, such as leaves and roots, during the vegetative stage, the plant transitions into the reproductive stage, often termed the floral transition. This shift from flowering to fruiting plants holds significance beyond the immediate lifecycle of the plant. Through the examination of the annual model plant Arabidopsis, researchers have successfully delineated five distinct processes that regulate the timing of the plant's blooming phase. These mechanisms encompass the gibberellic acid (GA) pathway, the autonomous pathway, the age pathway, the photoperiod pathway, and the vernalization pathway (Wang, 2014). A plethora of floral integrative regulators, including SUPPRESSOR OF OVEREXPRESSION OF CO1 (SOC1), FLOWERING LOCUS T (FT), LEAFY (LFY), and PETALA1 (AP1), are activated by these pathways, forming an intricate molecular network that mediates the transmission of both endogenous and environmental cues related to flowering time (Komeda, 2004).

miR156 and miR172 are two extensively studied miRNAs involved in floral regulation. The identification of conserved activities across monocotyledons and dicotyledons has elevated miRNAs to a promising breeding tool for agricultural plants. miR156 and miR172, while independent entities, exhibit an interconnected relationship. An inverse correlation is often observed between miR172 and miR156 concerning their expression patterns and regulatory functions. miR172 expression starts at a low level in juvenile plants, progressively increasing as they age. Additionally, the overexpression of miR172 results in an earlier onset of flowering in both monocotyledons and dicotyledons (Zhu and Helliwell, 2011). Conversely, miR156 expression levels decline from the seedling stage to maturity, with excessive miR156 production impeding the blooming rate.

According to Park et al. and Chen, the mature 21-nucleotide miR172 exhibits binding affinity toward the 3′ end of AP2 domain transcription factors within the coding region of their respective target genes. Numerous plant-specific transcription factors, including SCHLAFMUTZE (SMZ), SCHNARCHZAPFEN (SNZ), TARGET OF EAT1 (TOE1), and TOE2, have inhibitory roles in regulating flowering time. Previous studies have shown that miR156 targets genes belonging to the

SQUAMOSA PROMOTER BINDING LIKES (SPLs) family, such as SPL3 and SPL9. Intriguingly, some targets of miR156, like SPL9 and SPL10, redundantly regulate miR172's transcription by binding to its promoter region. The interplay of miR156 and miR172 forms a molecular network that governs several developmental phases (Wu et al., 2009). miR156 and miR172 are significant contributors to the plant age-dependent floral pathway due to their time-varying expression patterns and crucial roles in transitioning from the juvenile to adult phases of plants (Bergonzi and Albani, 2011). Furthermore, the modulation of interactions between the age pathway and other floral pathways appears to be influenced by miR156 and miR172. The concept of "vernalization" refers to the requirement of certain winter plant ecotypes for a specific period of cold treatment to ensure optimal flowering timing. However, Arabidopsis' perennial relatives, *Arabis alpina* and *Cardamine flexuosa*, respond to this induction only after a minimum of 5 weeks of cultivation (Wang et al., 2011). These perennial plants necessitate a period of vegetative growth to achieve reproductive maturity. miR156 and miR172, which exhibit distinct molecular regulation, play a critical role in the age-dependent vernalization process.

GA functions as a floral activator in Arabidopsis by destabilizing a cluster of GA repressors known as DELLA proteins. According to a study by Yu et al. (2012), molecular evidence suggests that the DELLA protein RGA can physically interact with SPL9, thereby impeding its transcriptional activities on miR172, SOC1, and FRUITFULL (FUL). GIGANTEA (GI) plays a vital role in regulating photoperiodic blooming within the photoperiod pathway by controlling the transcription of CONSTANS (CO), the primary component responsible for differentiating between day and night. The work by Jung et al. (2007) indicates that GI also employs a miR172-dependent mechanism for photoperiodic induction, independent of CO but requiring a functional FT, the target of CO. Several SPL genes targeted by miR156 can also respond to changes in photoperiod. The shared molecular function of miR156 and miR172 in monocotyledons and dicotyledons suggests that these miRNAs may have a more significant impact on enhancing agricultural productivity than previously believed.

6.6 miRNA and vascular development

HD-ZIP proteins control not only meristem and lateral organ polarity but also vascular development. Members of the HD-ZIP family, such as ATHB15, are highly expressed in vascular tissues, which raises the possibility that it contributes to the formation of plant blood vessels. As one of its targets, ATHB15 may be manipulated by miR166 (Rhoades et al., 2002). Reduced ATHB15 mRNA levels and rapid differentiation of cambial/procambial cells into vascular cells due to miR166a overexpression led to a remodeled vascular system characterized by increased xylem tissue and an interfascicular area. All vascular plant species may make use of this regulatory system (Floyd and Bowman, 2004; Kim et al., 2005). Cotton fiber formation and the manufacture of cell wall metabolites (Lu et al., 2005b) are two other potential roles for miRNAs (Zhang et al., 2005) (Fig. 4.1, Table 4.1).

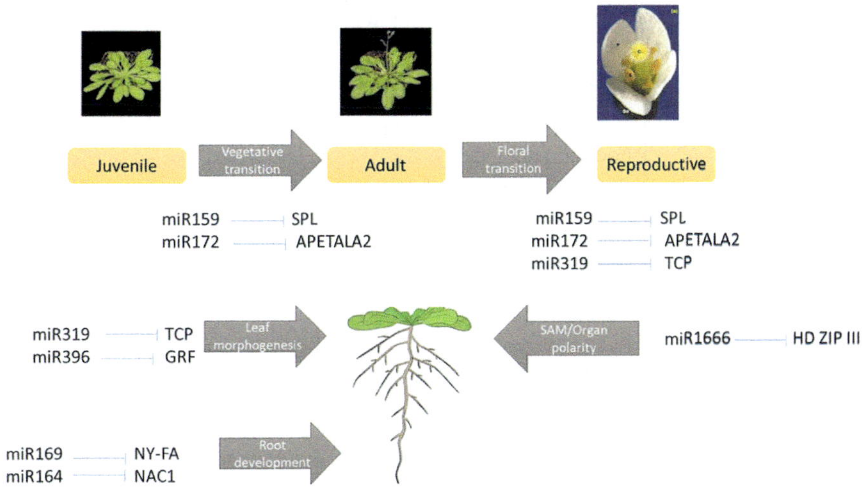

FIGURE 4.1

Schematic representation of different miRNAs mechanism in plant development.

Table 4.1 Different miRNAs and their roles in plant development (Dong et al.,.2022).

mi RNA	Target	Target function	Crop	References
miR156	SPL	The factors that influence blooming in *Zea mays* include plastochron duration, leaf development, root development, secondary metabolism, abiotic stress, tillering, and corn growth.	Arabidopsis and maize	Aukerman and Sakai (2003)
miR159	GAMYB or GAMYB-like gene	This study focuses on the development of reproductive systems, vegetative tissues, and seed development.	Arabidopsis	Allen et al. (2007)
miR160	ARFs	Development of the embryo, leaf, and root, as well as elongation of the hypocotyl.	Arabidopsis, maize, and Medicago	Bustos-Sanmamed et al. (2013)
miR164	NAC family	Identification of meristem boundaries, establishment of auxiliary meristems, growth of leaves and flowers, beginning of lateral roots.	Arabidopsis, maize, and rice	Li et al. (2003)
miR165/166	HD-ZIP III	The preservation of meristematic cells, the adaxial identity of leaves, the formation of lateral roots, and the identity of the procambium.	Arabidopsis	Williams et al. (2005)

Continued

Table 4.1 Different miRNAs and their roles in plant development (Dong et al.,.2022).—*cont'd*

mi RNA	Target	Target function	Crop	References
miR167	ARFs	The process of male organ development includes the growth of roots, stems, leaves, and flowers, as well as the timing of blooming, embryonic development, seed production, stress response, and defensive mechanisms against infections.	Arabidopsis and Oryza	
miR169	CBF and NY FI family	Facilitates the transcription of C homeotic genes and improves root architecture.	Arabidopsis, and maize	Cartolano et al. (2007)
miR171	SCL	The process of chlorophyll production, phase transitions, and the determinants of floral meristems.	Arabidopsis, and barley	Curaba et al. (2013)
miR172	AP2 family	Inhibits the process of blooming, alteration of floral meristem identity and patterning; alteration of vegetative phase, growth of carpel and stamen; opening of flowers, tuberization, and tolerance to salt.	Arabidopsis, maize, rice, *H. vulgare*, and *S. tuberosum*	
miR319	TCP family	Development and aging of leaves, curvature of organs, and the production and communication of hormones.	Arabidopsis and tomato	Ori et al. (2007)
miR390	TAS3	The synthesis of ta-siRNA for the suppression of ARF and the indirect control of miR165/166 in relation to lateral root development and leaf patterning.	Arabidopsis	Fahlgren et al. (2006)
miR393	TIR1 and AFB	Homeostasis of auxin, development of lateral roots, and shape/number of leaves.	Arabidopsis and rice	Parry et al. (2009)
miR394	LCR	The downregulation of WUS, leaf inclination, and architecture leads to the suppression of meristematic identity.	Arabidopsis	Baumann (2013)
miR396	GRF	Leaf cell proliferation, disease resistance, somatic embryogenesis, grain size, and panicle branching are all important factors to consider.	Arabidopsis, Medicago, and rice	Debernardi et al. (2012)
mir397	OsLAC	Grain yield and panicle branches.		Zhang et al. (2013)
TAS3	ARF3/4	Development of blood vessels, polarity of leaves, and phase transition.		Fahlgren et al. (2006)

7. Conclusion and future prospects

Thus far, by using genetic screenings, high-throughput screening, and bioinformatics, a considerable number of miRNAs have been detected in plants. Significant advancements have also been achieved in the characterization of the biological functions of pivotal miRNAs in diverse plant species. Nevertheless, the comprehensive analysis of the regulatory network of miRNAs in plant development remains incomplete. The precise roles and functions of many miRNAs remain ambiguous, and the intricate mechanisms by which these miRNAs collaborate to orchestrate organ growth are inadequately comprehended.

The investigation of the developmental processes of specific miRNAs, as well as the examination of the interactions between various miRNA-related pathways, is a significant area of inquiry within the field. Prior research has indicated various avenues that warrant further exploration. These include the potential for miRNA targets, typically encoding transcription factors, to directly modulate the transcription of another miRNA, thereby initiating a signaling cascade. Additionally, it is worth considering whether miRNA targets can impact histone modifications, which frequently influence the expression of downstream miRNA genes. Furthermore, it is worth exploring the possibility that the target gene products of multiple miRNAs may interact with one another and operate in a cooperative manner. Moreover, the outcomes of these inquiries may also prompt us toward a more intricate objective of comprehending the manner in which these interactions amalgamate inside a network to govern a certain developmental process, and ultimately how these distinct processes interconnect to shape the entirety of the plant.

References

Aida, M., Tasaka, M., 2006. Genetic control of shoot organ boundaries. Curr. Opin. Plant Biol. 9 (1), 72–77. https://doi.org/10.1016/j.pbi.2005.11.011.

Allen, R.S., Li, J., Stahle, M.I., Dubroue, A., Gubler, F., 2007. Genetic analysis reveals functional redundancy and the major target genes of the Arabidopsis miR159 family. Proc. Natl. Acad. Sci. U.S.A. 104, 16371–16376. https://doi.org/10.1073/pnas.0707653104.

Allen, E., Xie, Z., Gustafson, A.M., Sung, G.-H., Spatafora, J.W., Carrington, J.C., 2004. Evolution of microRNA genes by inverted duplication of target gene sequences in *Arabidopsis thaliana*. Nat. Genet. 36 (12), 1282–1290. https://doi.org/10.1038/ng1478.

Almeida, R., Allshire, R.C., 2005. RNA silencing and genome regulation. Trends Cell Biol. 15 (5), 251–258. https://doi.org/10.1016/j.tcb.2005.03.006.

Aukerman, M.J., Sakai, H., 2003. Regulation of flowering time and floral organ identity by a microRNA and its APETALA2-like target genes. Plant Cell 15 (11), 2730–2741. https://doi.org/10.1105/tpc.016238.

Axtell, M.J., Bartel, D.P., 2005. Antiquity of microRNAs and their targets in land plants. Plant Cell 17 (6), 1658–1673. https://doi.org/10.1105/tpc.105.032185.

Axtell, M.J., Jan, C., Rajagopalan, R., Bartel, D.P., 2006. A two-hit trigger for siRNA biogenesis in plants. Cell 127, 565–577.

Bartel, D.P., 2004. MicroRNAs: genomics, biogenesis, mechanism, and function. Cell 116 (2), 281–297. https://doi.org/10.1016/s0092-8674(04)00045-5.

Baumann, K., 2013. Plant cell biology: mobile miRNAs for stem cell maintenance. Nat. Rev. Mol. Cell Biol. 14 (3), 128. https://doi.org/10.1038/nrm3529.

Bazin, J., Khan, G.A., Combier, J.-P., Bustos-Sanmamed, P., Debernardi, J.M., Rodriguez, R., Sorin, C., Palatnik, J., Hartmann, C., Crespi, M., Lelandais-Brière, C., 2013. MiR396 affects mycorrhization and root meristem activity in the Legume Medicago truncatula. Plant J. 74 (6), 920–934. https://doi.org/10.1111/tpj.12178.

Bergonzi, S., Albani, M.C., 2011. Reproductive competence from an annual and a perennial perspective. J. Exp. Bot. 62 (13), 4415–4422. https://doi.org/10.1093/jxb/err192.

Bernstein, E., Caudy, A.A., Hammond, S.M., Hannon, G.J., 2001. Role for a bidentate ribonuclease in the initiation step of RNA interference. Nature 409 (6818), 363–366. https://doi.org/10.1038/35053110.

Bian, H., Xie, Y., Guo, F., Han, N., Ma, S., Zeng, Z., Wang, J., Yang, Y., Zhu, M., 2012. Distinctive expression patterns and roles of the miRNA393/TIR1 homolog module in regulating flag leaf inclination and primary and crown root growth in rice (*Oryza sativa*). New Phytol. 196 (1), 149–161. https://doi.org/10.1111/j.1469-8137.2012.04248.x.

Bonnet, E., Wuyts, J., Rouzé, P., Van de Peer, Y., 2004. Detection of 91 potential conserved plant microRNAs in *Arabidopsis thaliana* and *Oryza sativa* identifies important target genes. Proc. Natl. Acad. Sci. U.S.A. 101 (31), 11511–11516. https://doi.org/10.1073/pnas.0404025101.

Bustos-Sanmamed, P., Mao, G., Deng, Y., Elouet, M., Khan, G.A., Mie Bazin, J.R., Turner, M., Subramanian, S., Oliver, Y., Crespi, M., Re, C.L.-B., 2013. Overexpression of miR160 affects root growth and nitrogen-fixing nodule number in *Medicago truncatula*. Funct. Plant Biol. 40 (12), 1208–1220. https://doi.org/10.1071/FP13123.

Byrne, M.E., 2005. Networks in leaf development. Curr. Opin. Plant Biol. 8 (1), 59–66. https://doi.org/10.1016/j.pbi.2004.11.009.

Carrington, J.C., Victor, A., 2003. Role of microRNAs in plant and animal development. Science 301 (5631), 336–338. https://doi.org/10.1126/science.1085242.

Cartolano, M., Castillo, R., Efremova, N., Kuckenberg, M., Zethof, J., 2007. A conserved microRNA module exerts homeotic control over *Petunia hybrida* and *Antirrhinum majus* floral organ identity. Nat. Genet. 39, 901–905. https://doi.org/10.1038/ng2056.

Curaba, J., Talbot, M., Li, Z., Helliwell, C., 2013. Over-expression of microRNA171 affects phase transitions and floral meristem determinancy in barley. BMC Plant Biol. 13, 6. https://doi.org/10.1186/1471-2229-13-6.

D'Ario, M., Griffiths-Jones, S., Kim, M., 2017. Small RNAs: big impact on plant development. Trends Plant Sci. 22 (12), 1056–1068. https://doi.org/10.1016/j.tplants.2017.09.009.

De Rybel, B., Vassileva, V., Parizot, B., Demeulenaere, M., Grunewald, W., Audenaert, D., Van Campenhout, J., Overvoorde, P., Jansen, L., Vanneste, S., et al., 2010. A novel aux/IAA28 signaling cascade activates GATA23-dependent specification of lateral root founder cell identity. Curr. Biol. 20, 1697–1706.

Debernardi, J.M., Rodriguez, R.E., Mecchia, M.A., Palatnik, J.F., 2012. Functional specialization of the plant miR396 regulatory network through distinct microRNA-target interactions. PLoS Genet. 8 (1). https://doi.org/10.1371/journal.pgen.1002419 e1002419.

Dong, Q., Hu, B., Zhang, C., 2022. microRNAs and their roles in plant development. Front. Plant Sci. 13, 824240. https://doi.org/10.3389/fpls.2022.824240.

Fahlgren, N., Montgomery, T.A., Howell, M.D., Allen, E., Dvorak, S.K., Alexander, A.L., Carrington, J.C., 2006. Regulation of auxin response factor3 by TAS3 Ta-siRNA affects developmental timing and patterning in Arabidopsis. Curr. Biol. 16 (9), 939–944. https://doi.org/10.1016/j.cub.2006.03.065.

Ferdous, J., Hussain, S.S., Shi, B.-J., 2015. Role of microRNAs in plant drought tolerance. Plant Biotechnol. J. 13 (3), 293–305. https://doi.org/10.1111/pbi.12318.

Floyd, S.K., Bowman, J.L., 2004. Ancient microRNA target sequences in plants. Nature 428, 485–486.

Fu, H., Tie, Y., Xu, C., Zhang, Z., Zhu, J., Shi, Y., Jiang, H., Sun, Z., Zheng, X., 2005. Identification of human fetal liver miRNAs by a novel method. FEBS (Fed. Eur. Biochem. Soc.) Lett. 579 (17), 3849–3854. https://doi.org/10.1016/j.febslet.2005.05.064.

Greb, T., Clarenz, O., Schafer, E., Muller, D., Herrero, R., Schmitz, G., Theres, K., 2003. Molecular analysis of the lateral suppressor gene in Arabidopsis reveals a conserved control mechanism for axillary meristem formation. Gene Dev. 17 (9), 1175–1187. https://doi.org/10.1101/gad.260703.

Griffiths-Jones, S., 2004. The microRNA Registry. Nucleic Acids Res. 32, D109–D111. https://doi.org/10.1093/nar/gkh023. Database issue.

Gutierrez, L., Bussell, J.D., Pacurar, D.I., Schwambach, J., Pacurar, M., Bellini, C., 2009. Phenotypic plasticity of adventitious rooting in Arabidopsis is controlled by complex regulation of auxin response factor transcripts and microRNA abundance. Plant Cell 21, 3119–3132.

Gutierrez, L., Mongelard, G., Floková, K., Pacurar, D.I., Novák, O., Staswick, P., Kowalczyk, M., Pacurar, M., Demailly, H., Geiss, G., Bellini, C., 2012. Auxin controls Arabidopsis adventitious root initiation by regulating Jasmonic acid homeostasis. Plant Cell 24 (6), 2515–2527. https://doi.org/10.1105/tpc.112.099119.

Hammond, S.M., Bernstein, E., Beach, D., Hannon, G.J., 2000. An RNA-directed nuclease mediates post-transcriptional gene silencing in Drosophila cells. Nature 404 (6775), 293–296. https://doi.org/10.1038/35005107.

Hibara, K., Karim, Md R., Takada, S., Taoka, K., Furutani, M., Aida, M., Tasaka, M., 2006. Arabidopsis cupshaped cotyledon3 regulates postembryonic shoot meristem and organ boundary formation. Plant Cell 18 (11), 2946–2957. https://doi.org/10.1105/tpc.106.045716.

Jones-Rhoades, M.W., Bartel, D.P., 2004. Computational identification of plant microRNAs and their targets, including a stress-induced miRNA. Mol. Cell 14 (6), 787–799. https://doi.org/10.1016/j.molcel.2004.05.027.

Juarez, M.T., Kui, J.S., Thomas, J., Heller, B.A., Timmermans, M.C.P., 2004. MicroRNA-mediated repression of rolled leaf1 specifies maize leaf polarity. Nature 428 (6978), 84–88. https://doi.org/10.1038/nature02363.

Jung, J.-H., Seo, Y.-H., Seo, P.J., Reyes, J.L., Yun, J., Chua, N.-H., Park, C.-M., 2007. The GIGANTEA-regulated microRNA172 mediates photoperiodic flowering independent of CONSTANS in Arabidopsis. Plant Cell 19 (9), 2736–2748. https://doi.org/10.1105/tpc.107.054528.

Kasschau, K.D., Xie, Z., Allen, E., Llave, C., Chapman, E.J., Krizan, K.A., Carrington, J.C., 2003. P1/HC-Pro, a viral suppressor of RNA silencing, interferes with Arabidopsis development and miRNA unction. Dev. Cell 4 (2), 205–217. https://doi.org/10.1016/s1534-5807(03)00025-x.

Kepinski, S., Ottoline, L., 2005. The Arabidopsis F-box protein TIR1 is an auxin receptor. Nature 435 (7041). https://doi.org/10.1038/nature03542, 446–217.

Kim, J.H., Choi, D., Hans, K., 2003. The AtGRF family of putative transcription factors is involved in leaf and cotyledon growth in Arabidopsis. Plant J. 36 (1), 94–104. https://doi.org/10.1046/j.1365-313x.2003.01862.x.

Kim, J., Jung, J.H., Reyes, J.L., Kim, Y.S., Kim, S.Y., Chung, K.S., Kim, J.A., Lee, M., Lee, Y., Kim, V.N., Chua, N.H., Park, C.M., 2005. microRNA directed cleavage of ATHB15 mRNA regulates vascular development in Arabidopsis inflorescence stems. Plant J. 42, 84–94.

Knauer, S., Holt, A.L., Rubio-Somoza, I., Tucker, E.J., Hinze, A., Pisch, M., Javelle, M., Timmermans, M.C., Tucker, M.R., Laux, T., 2013. A protodermal miR394 signal defines a region of stem cell competence in the Arabidopsis shoot meristem. Dev. Cell 24 (2), 125–132. https://doi.org/10.1016/j.devcel.2012.12.009.

Komatsu, K., Maekawa, M., Ujiie, S., Satake, Y., Furutani, I., Okamoto, H., Shimamoto, K., Kyozuka, J., 2003. LAX and SPA: major regulators of shoot branching in rice. Proc. Natl. Acad. Sci. U.S.A. 100 (20), 11765–11770. https://doi.org/10.1073/pnas.1932414100.

Komeda, Y., 2004. Geneticregulation of time to flowerin *Arabidopsis thaliana*. Annu. Rev. Plant Biol. 55, 521–535.

Lagos-Quintana, M., Rauhut, R., Lendeckel, W., Tuschl, T., 2001. Identification of novel genes coding for small expressed RNAs. Science 294 (5543), 853–858. https://doi.org/10.1126/science.1064921.

Lai, E.C., 2003. microRNAs: runts of the genome assert themselves. Curr. Biol. 13 (23), R925–R936. https://doi.org/10.1016/j.cub.2003.11.017.

Lee, R.C., Ambros, V., 2001. An extensive class of small RNAs in *Caenorhabditis elegans*. Science 294 (5543), 862–864. https://doi.org/10.1126/science.1065329.

Lee, R.C., Feinbaum, R.L., Ambros, V., 1993. The *C. elegans* heterochronic gene lin-4 encodes small RNAs with antisense complementarity to lin-14. Cell 75 (5), 843–854. https://doi.org/10.1016/0092-8674(93)90529-y.

Lee, Y., Kim, M., Han, J., Yeom, K.-H., Lee, S., Baek, S.H., Kim, V.N., 2004. MicroRNA genes are transcribed by RNA polymerase II. EMBO J. 23 (20), 4051–4060. https://doi.org/10.1038/sj.emboj.7600385.

Levine, M., Tjian, R., 2003. Transcription regulation and animal diversity. Nature 424 (6945), 147–151. https://doi.org/10.1038/nature01763.

Li, X., Qian, Q., Fu, Z., Wang, Y., Xiong, G., Zeng, D., Wang, X., Liu, X., Teng, S., Hiroshi, F., Yuan, M., Luo, D., Han, B., Li, J., 2003. Control of tillering in rice. Nature 422 (6932), 618–621. https://doi.org/10.1038/nature01518.

Lim, L.P., Glasner, M.E., Yekta, S., Burge, C.B., Bartel, D.P., 2003a. Vertebrate microRNA genes. Science 299 (5612), 1540. https://doi.org/10.1126/science.1080372.

Lim, L.P., Lau, N.C., Weinstein, E.G., Abdelhakim, A., Yekta, S., Rhoades, M.W., Burge, C.B., Bartel, D.P., 2003b. The microRNAs of *Caenorhabditis elegans*. Gene Dev. 17 (8), 991–1008. https://doi.org/10.1101/gad.1074403.

Liu, B., Chen, Z., Song, X., Liu, C., Cui, X., Zhao, X., Fang, J., Xu, W., Zhang, H., Wang, X., Chu, C., Deng, X., Xue, Y., Cao, X., 2007. *Oryza sativa* dicer-like4 reveals a key role for small interfering RNA silencing in plant development. Plant Cell 19, 2705–2718.

Liu, Q., Yao, X., Pi, L., Wang, H., Cui, X., Huang, H., 2009. The ARGONAUTE10 gene modulates shoot apical meristem maintenance and establishment of leaf polarity by repressing miR165/166 in Arabidopsis. Plant J. 58 (1), 27–40. https://doi.org/10.1111/j.1365-313X.2008.03757.x.

Llave, C., Xie, Z., Kasschau, K.D., Carrington, J.C., 2002. Cleavage of scarecrow-like mRNA targets directed by a class of Arabidopsis miRNA. Science 297 (5589), 2053–2056. https://doi.org/10.1126/science.1076311.

Long, J.A., Moan, E.I., Medford, J.I., Barton, M.K., 1996. A member of the KNOTTED class of homeodomain proteins encoded by the STM gene of Arabidopsis. Nature 379 (6560), 66–69. https://doi.org/10.1038/379066a0.

Lu, C., Singh Tej, S., Luo, S., Haudenschild, C.D., Meyers, B.C., Green, P.J., 2005a. Elucidation of the small RNA component of the transcriptome. Science 309 (5740), 1567–1569. https://doi.org/10.1126/science.1114112.

Lu, S., Sun, Y.-H., Shi, R., Clark, C., Li, L., Chiang, V.L., 2005b. Novel and mechanical stress-responsive MicroRNAs in *Populus trichocarpa* that are absent from Arabidopsis. Plant Cell 17 (8), 2186–2203. https://doi.org/10.1105/tpc.105.033456.

Malamy, J.E., Ryan, K.S., 2001. Environmental regulation of lateral root initiation in Arabidopsis. Plant Physiol. 127 (3), 899–909. https://doi.org/10.1104/pp.010406.

Mallory, A.C., Nicolas, B., 2008. MicroRNA-directed regulation: to cleave or not to cleave. Trends Plant Sci. 13 (7), 359–367. https://doi.org/10.1016/j.tplants.2008.03.007.

McConnell, J.R., Barton, M.K., 1995. Effect of mutations in the PINHEAD gene of Arabidopsis on the formation of shoot apical meristems. Dev. Genet. 16, 358–366.

McSteen, P., Leyser, O., 2005. Shoot branching. Annu. Rev. Plant Biol. 56, 353–374. https://doi.org/10.1146/annurev.arplant.56.032604.144122.

Meng, Y., Ma, X., Chen, D., Wu, P., Chen, M., 2010. MicroRNA-mediated signaling involved in plant root development. Biochem. Biophys. Res. Commun. 393 (3), 345–349. https://doi.org/10.1016/j.bbrc.2010.01.129.

Moussian, B., Schoof, H., Haecker, A., Jürgens, G., Laux, T., 1998. Role of the ZWILLE gene in the regulation of central shoot meristem cell fate during Arabidopsis embryogenesis. EMBO J. 17 (6), 1799–1809. https://doi.org/10.1093/emboj/17.6.1799.

Nagasaki, H., Itoh, J., Hayashi, K., Hibara, K., Satoh-Nagasawa, N., Nosaka, M., Mukouhata, M., Ashikari, M., Kitano, H., Matsuoka, M., Nagato, Y., Sato, Y., 2007. The small interfering RNA production pathway is required for shoot meristem initiation in rice. Proc. Natl. Acad. Sci. U.S.A. 104 (37), 14867–14871. https://doi.org/10.1073/pnas.0704339104.

Nelson, J.M., Lane, B., Freeling, M., 2002. Expression of a mutant maize gene in the ventral leaf epidermis is sufficient to signal a switch of the leaf's dorsoventral axis. Development 129 (19), 4581–4589. https://doi.org/10.1242/dev.129.19.4581.

Nogueira, F.T.S., Madi, S., Chitwood, D.H., Juarez, M.T., Timmermans, M.C.P., 2007. Two small regulatory RNAs establish opposing fates of a developmental Axis. Gene Dev. 21 (7), 750–755. https://doi.org/10.1101/gad.1528607.

Ori, N., Cohen, A.R., Etzioni, A., Brand, A., Yanai, O., Shleizer, S., Menda, N., Amsellem, Z., Efroni, I., Pekker, I., Alvarez, J.P., Blum, E., Zamir, D., Eshed, Y., 2007. Regulation of LANCEOLATE by miR319 is required for compound-leaf development in tomato. Nat. Genet. 39 (6), 787–791. https://doi.org/10.1038/ng2036.

Palatnik, J.F., Allen, E., Wu, X., Schommer, C., Schwab, R., Carrington, J.C., Weigel, D., 2003. Control of leaf morphogenesis by microRNAs. Nature 425 (6955), 257–263. https://doi.org/10.1038/nature01958.

Palatnik, J.F., Wollmann, H., Schommer, C., Schwab, R., Boisbouvier, J., Rodriguez, R., Warthmann, N., Allen, E., Dezulian, T., Huson, D., Carrington, J.C., Weigel, D., 2007. Sequence and expression differences underlie functional specialization of Arabidopsis microRNAs miR159 and miR319. Dev. Cell 13, 115–125.

Parry, G., Calderon-Villalobos, L.I., Prigge, M., Peret, B., Dharmasiri, S., Itoh, H., Lechner, E., Gray, W.M., Bennett, M., Estelle, M., 2009. Complex regulation of the TIR1/AFB family of auxin receptors. Proc. Natl. Acad. Sci. U.S.A. 106 (52), 22540–22545. https://doi.org/10.1073/pnas.0911967106.

Pasquinelli, A.,E., Reinhart, B.J., Slack, F., Martindale, M.Q., Kuroda, M.I., Maller, B., Hayward, D.C., Ball, E.E., Degnan, B., Müller, P., Spring, J., Srinivasan, A., Fishman, M., Finnerty, J., Corbo, J., Levine, M., Leahy, P., Davidson, E., Ruvkun, G., 2000. Conservation of the sequence and temporal expression of Let-7 heterochronic regulatory RNA. Nature 408 (6808), 86–89. https://doi.org/10.1038/35040556.

Remington, D.L., Vision, T.J., Guilfoyle, T.J., Reed, J.W., 2004. Contrasting modes of diversification in the Aux/IAA and ARF gene families. Plant Physiol. 135 (3), 1738–1752. https://doi.org/10.1104/pp.104.039669.

Rhoades, M.W., Reinhart, B.J., Lim, L.P., Burge, C.B., Bartel, B., Bartel, D.P., 2002. Prediction of plant microRNA targets. Cell 110 (4), 513–520. https://doi.org/10.1016/s0092-8674(02)00863-2.

Rogers, E.D., Benfey, P.N., 2015. Regulation of plant root system architecture: implications for crop advancement. Curr. Opin. Biotechnol. 32, 93–98. https://doi.org/10.1016/j.copbio.2014.11.015.

Rubio-Somoza, I., Weigel, D., 2011. MicroRNA networks and developmental plasticity in plants. Trends Plant Sci. 16 (5), 258–264. https://doi.org/10.1016/j.tplants.2011.03.001.

Satina, S., Blakeslee, A.F., Avery, A.G., 1940. Demonstration of the three germ layers in the shoot apex of datura by means of induced polyploidy in periclinal chimeras. Am. J. Bot. 27 (10), 895–905. https://doi.org/10.1002/j.1537-2197.1940.tb13952.x.

Satoh, N., Hong, S.K., Nishimura, A., Matsuoka, M., Kitano, H., Nagato, Y., 1999. Initiation of shoot apical meristem in rice: characterization of four SHOOTLESS genes. Development 126 (16), 3629–3636. https://doi.org/10.1242/dev.126.16.3629.

Schumacher, K., Schmitt, T., Rossberg, M., Schmitz, G., Theres, K., 1999. The lateral suppressor (Ls) gene of tomato encodes a new member of the VHIID protein family. Proc. Natl. Acad. Sci. U.S.A. 96 (1), 290–295. https://doi.org/10.1073/pnas.96.1.290.

Schwab, R., Palatnik, J.F., Riester, M., Schommer, C., Schmid, M., Weigel, D., 2005. Specific effects of microRNAs on the plant transcriptome. Dev. Cell 8 (4), 517–527. https://doi.org/10.1016/j.devcel.2005.01.018.

Singh, M.B., Bhalla, P.L., 2006. Plant stem cells carve their own Niche. Trends Plant Sci. 11 (5), 241–246. https://doi.org/10.1016/j.tplants.2006.03.004.

Sunkar, R., Girke, T., Zhu, J.-K., 2005. Identification and characterization of endogenous small interfering RNAs from rice. Nucleic Acids Res. 33 (14), 4443–4454. https://doi.org/10.1093/nar/gki758.

Timmermans, M., Schultes, N.P., Jankovsky, J.P., Nelson, T., 1998. Leafbladeless1 is required for dorsoventrality of lateral organs in maize. Development 125 (15), 2813–2823. https://doi.org/10.1242/dev.125.15.2813.

Tucker, M.R., Hinze, A., Tucker, E.J., Takada, S., Jürgens, G., Laux, T., 2008. Vascular signalling mediated by ZWILLE potentiates WUSCHEL function during shoot meristem stem cell development in the Arabidopsis embryo. Development 135 (17), 2839–2843. https://doi.org/10.1242/dev.023648.

Wang, L., Gu, X., Xu, D., Wang, W., Wang, H., Zeng, M., Chang, Z., Huang, H., Cui, X., 2011. MiR396 targeted AtGRF transcription factors are required for coordination of cell division and differentiation during leaf development in Arabidopsis. J. Exp. Bot. 62, 761–773.

Wang, J.-J., Guo, H.-S., 2015. Cleavage of indole-3-acetic ACID INDUCIBLE28 mRNA by microRNA847 upregulates auxin signaling to modulate cell proliferation and lateral organ growth in Arabidopsis. Plant Cell 27 (3), 574–590. https://doi.org/10.1105/tpc.15.00101.

Wang, W., Li, G., Zhao, J., Chu, H., Lin, W., Zhang, D., Wang, Z., Liang, W., 2014. DWARF TILLER1, a WUSCHEL-related homeobox transcription factor, is required for tiller growth in rice. PLoS Genet. 10 (3). https://doi.org/10.1371/journal.pgen.1004154 e1004154.

Wang, X.W., Zhang, J., Li, F., Gu, J., He, T., Zhang, X.G., Li, Y.D., 2005. MicroRNA identification based on sequence and structure alignment. Bioinformatics 21 (18), 3610–3614. https://doi.org/10.1093/bioinformatics/bti562.

Weber, M.J., 2005. New human and mouse microRNA genes found by homology search. FEBS J. 272 (1), 59–73. https://doi.org/10.1111/j.1432-1033.2004.04389.x.

Weigel, D., Ahn, J.H., Blázquez, M.A., Borevitz, J.O., Christensen, S.K., Fankhauser, C., Ferrándiz, C., Kardailsky, I., Malancharuvil, E.J., Neff, M.M., Nguyen, J.T., Sato, S., Wang, Z.Y., Xia, Y., Dixon, R.A., Harrison, M.J., Lamb, C.J., Yanofsky, M.F., Chory, J., 2000. Activation tagging in Arabidopsis. Plant Physiol. 122 (4), 1003–1013. https://doi.org/10.1104/pp.122.4.1003.

Weigel, D., Jürgens, G., 2002. Stem cells that make stems. Nature 415 (6873), 751–754. https://doi.org/10.1038/415751a.

Wightman, B., Ha, I., Ruvkun, G., 1993. Posttranscriptional regulation of the heterochronic gene lin-14 by lin-4 mediates temporal pattern formation in *C. elegans*. Cell 75, 855–862.

Williams, L., Grigg, S.P., Xie, M., Christensen, S., Fletcher, J.C., 2005. Regulation of Arabidopsis shoot apical meristem and lateral organ formation by microRNA miR166g and its AtHD-ZIP target genes. Development 132 (16), 3657–3668. https://doi.org/10.1242/dev.01942.

Wu, G., Park, M.Y., Conway, S.R., Wang, J.-W., Weige, l.D., Poethig, R.S., 2009. The sequential action of miR156 and miR172 regulates developmental timing in *Arabidopsis*. Cell 138, 750–759.

Yang, C., Li, D., Mao, D., Liu, X., Ji, C., Li, X., Zhao, X., Cheng, Z., Chen, C., Zhu, L., 2013. Overexpression of microRNA319 impacts leaf morphogenesis and leads to enhanced cold tolerance in rice (*Oryza sativa* L.). Plant Cell Environ. 36 (12), 2207–2218. https://doi.org/10.1111/pce.12130.

Yi, R., in, Y., Macara, I.G., Cullen, B.R., 2003. Exportin-5 mediates the nuclear export of pre-microRNAs and short hairpin RNAs. Gene Dev. 17 (24), 3011–3016. https://doi.org/10.1101/gad.1158803.

Yu, S., Galvão, V.C., Zhang, Y.C., Horrer, D., Zhang, T.Q., Hao, Y.H., Feng, Y.Q., Wang, S., Schmid, M., Wang, J.W., 2012. Gibberellin regulates the Arabidopsis floral transition through miR156 targeted SQUAMOSA PROMOTER BINDING-LIKE transcription factors. Plant Cell 24, 3320–3332.

Zhang, B.H., Pan, X.P., Wang, Q.L., Cobb, G.P., Anderson, T.A., 2005. Identification and characterization of new plant microRNAs using EST analysis. Cell Res. 15 (5), 336–360. https://doi.org/10.1038/sj.cr.7290302.

Zhang, Y.C., Yu, Y., Wang, C.Y., Li, Z.Y., Liu, Q., Xu, J., Liao, J.Y., Wang, X.J., Qu, L.H., Chen, F., Xin, P., Yan, C., Chu, J., Li, H.Q., Chen, Y.Q., 2013. Over expression of microRNA OsmiR397 improves rice yield by increasing grain size and promoting panicle branching. Nat. Biotechnol. 31, 848–852.

Zhang, Z., Zhang, X., 2012. Argonautes compete for miR165/166 to regulate shoot apical meristem development. Curr. Opin. Plant Biol. 15 (6), 652–658. https://doi.org/10.1016/j.pbi.2012.05.007.

Zhu, H., Hu, F., Wang, R., Zhou, X., Sze, S.-H., Wen Liou, L., Barefoot, A., Dickman, M., Zhang, X., 2011. Arabidopsis Argonaute10 specifically sequesters miR166/165 to regulate shoot apical meristem development. Cell 145 (2), 242–256. https://doi.org/10.1016/j.cell.2011.03.024.

Zhu, Q.-H., Helliwell, C.A., 2011. Regulation of flowering time and floral patterning by mi R172. J. Exp. Bot. 62 (2), 487–495. https://doi.org/10.1093/jxb/erq295.

CHAPTER

Root architecture and microRNAs: A key role for microRNAs

5

Noopur Khare[1,2], Pragati Khare[3], Sachidanand Singh[4], Megha Barot[5,6], Tanvi Jain[1]

[1]*Faculty of Biotechnology, Institute of Biosciences and Technology, Shri Ramswaroop Memorial University, Barabanki, Uttar Pradesh, India;* [2]*Bhai Gurdas Institute of Engineering and Technology, Sangrur, Punjab, India;* [3]*Department of Pharmacy, Bhagwant University, Ajmer, Rajasthan, India;* [4]*Department of Biotechnology, School of Energy and Technology, Pandit Deendayal Energy University, Gandhinagar, Gujarat, India;* [5]*Research and Development Cell, Parul University, Vadodara, Gujarat, India;* [6]*Department of Environmental Science, Parul Institute of Applied Sciences, Parul University, Vadodara, Gujarat, India*

1. Introduction

A key factor in determining the general health and production of a plant is its root system, which anchors nutrient intake and offers stability. Root architecture is dynamically regulated to adjust to different environmental conditions. It is characterized by its spatial organization and distribution (Downie et al., 2015). The complex regulatory networks controlling root architecture have been clarified by recent study, and microRNAs (miRNAs) have emerged as important participants in this process. Small noncoding RNA molecules target messenger RNAs for translational suppression or destruction in order to control gene expression posttranscriptionally. Their importance in plant development is further shown by their role in modifying root architecture. This introduction lays the groundwork for investigating the interactions between miRNAs and root architecture, clarifying their functions in forming root systems and emphasizing the possible agricultural applications (Santosh et al., 2015). Gaining insight into the mechanisms behind the regulation of root architecture by miRNAs has the potential to increase crop resilience and production in a changing environment. The spatial arrangement and distribution of roots within the soil, which forms the basis for plant anchoring, intake of water and nutrients, and interactions with the surrounding environment, is referred to as root architecture. Together with several specialized components such as root hairs and lateral roots, the root system is made up of primary, secondary, and tertiary roots. Because of their extremely flexible and dynamic root design, plants can adapt to a wide range of environmental pressures and soil conditions. Genetic predisposition, environmental cues including soil moisture and nutrient availability, and interactions with symbiotic bacteria are some of the factors that affect root design (Kou et al., 2022). Plants can effectively explore the soil volume thanks to their complex

MicroRNA Advances and Application in Plant Biology. https://doi.org/10.1016/B978-0-443-21682-4.00008-7

network of root systems, which maximizes resource acquisition and promotes above-ground growth and development. Gaining an understanding of root architecture is crucial for increasing agricultural productivity since it has a direct impact on resistance to biotic and abiotic stressors, water use efficiency, and nutrient uptake efficiency (Homulle et al., 2021).

Small, endogenous RNA molecules known as microRNAs (miRNAs) are essential for the posttranscriptional control of genes in a variety of species, including plants. These noncoding RNAs, which are usually between 21 and 24 nucleotides long, work by attaching themselves to messenger RNAs (mRNAs) and causing translational suppression or mRNA destruction. The complex biogenesis route that produces miRNAs involves transcription, processing, and maturation. The genome's endogenous loci or intergenic regions are frequently the source of miRNAs (Betti et al., 2021). When miRNAs reach maturity, they direct the RNA-induced silencing complex (RISC) to target mRNAs, which modifies gene expression and affects many biological processes, such as stress responses, differentiation, and development. Numerous facets of plant growth and development, such as root architecture, leaf shape, and flowering timing, have been linked to miRNAs. Deciphering the roles and regulatory processes of miRNAs is crucial for resolving the intricacies of gene regulatory networks and has potential uses in biotechnology and agriculture (Tiwari and Rajam, 2022). Root architecture is largely regulated by microRNAs (miRNAs), which have fine control over many aspects of root growth and development. Through their targeting of particular genes involved in root morphogenesis, elongation, and patterning, miRNAs regulate the distribution and spatial arrangement of roots in the soil (Parveen et al., 2023). miRNAs shape root architectures and maximize resource uptake efficiency by influencing critical processes such cell division, differentiation, and elongation through their capacity to modify gene expression posttranscriptionally (Barrera-Rojas et al., 2021).

Furthermore, miRNAs help plants adapt to shifting growth conditions and preserve the ideal root system architecture by coordinating the root responses to environmental stimuli and stress conditions. The complex regulatory network that miRNAs orchestrate guarantees the appropriate growth and functionality of roots, which in turn affects the general growth, uptake of nutrients, and resilience to stress in plants (Vives-Peris et al., 2020). Comprehending the role that miRNAs play in controlling root architecture offers important insights into the biology of plants and possible ways to improve crop resilience and production in farming systems. This chapter aims to investigate the critical function of microRNAs (miRNAs) in controlling root architecture. It will explore the processes behind the regulation of root development by miRNA, the methods used in experiments to research the function of miRNA, and the consequences for crop improvement and agriculture as seen in Fig. 5.1 (Joshi et al., 2022).

2. Root architecture

A crucial component of plant biology is root architecture, which includes the morphology, distribution, and spatial organization of roots in the soil. It provides

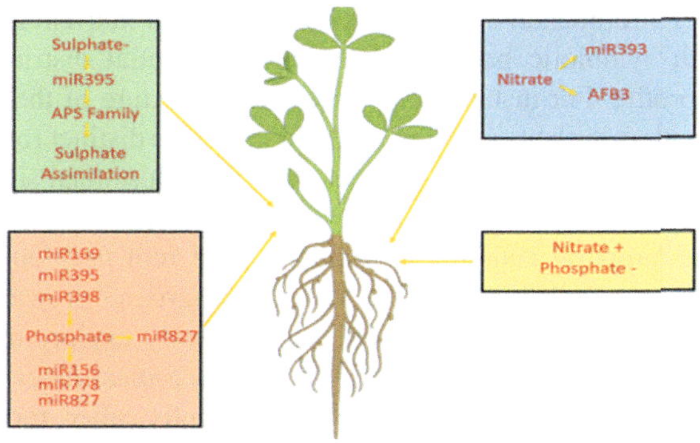

FIGURE 5.1

Root architecture and microRNAs.

the framework for interactions with the surrounding environment, water and nutrient intake, and plant anchoring as seen in Fig. 5.2. The primary, secondary, and tertiary roots that make up the root system each have different roles and traits (Hodge et al., 2009). The embryonic radicle gives rise to primary roots, which are the principal axis of root growth and give the plant structural support and anchoring. The lateral roots, or secondary roots, grow from the primary root and are essential for the root system's expansion, enabling it to explore a greater area of the soil (Vishwakarma et al., 2024). The intricacy and effectiveness of the root system are increased by tertiary roots, which split off from secondary roots even more. Because of their extremely dynamic and malleable root architecture, plants can adapt to a wide range of environmental pressures and soil conditions. Numerous factors, including genetic

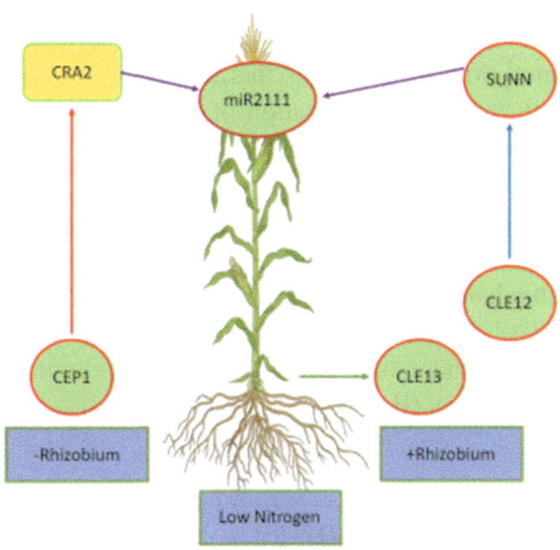

FIGURE 5.2

Root architecture.

predisposition, environmental cues such soil moisture and nutrient availability, and interactions with symbiotic bacteria, determine the spatial distribution of roots within the soil profile. For instance, the structure and texture of the soil can affect root branching patterns; plants growing in sandy soils have distinct root architectures than those growing in clayey soils (Viana et al., 2022). Hormonal signals including strigolactones, cytokinins, and auxins also control root growth and branching by coordinating the temporal and spatial components of root development. In order to maximize resource collection and facilitate above-ground growth and development, root architecture is essential. Plants can efficiently explore the soil volume thanks to their complex network of root systems, which maximizes their access to nutrients and water. Thin protrusions of epidermal cells called root hairs greatly enhance the surface area available for absorbing water and nutrients, which increases plant output (Jung and McCouch, 2013). Additionally, root architecture affects erosion management and soil stabilization, especially in habitats that are vulnerable to environmental degradation (Löbmann et al., 2020).

Enhancing crop production and agricultural sustainability requires an understanding of root architecture. Key factors that determine nutrient uptake efficiency, water use efficiency, and resilience to biotic and abiotic stressors are root characteristics such root length, branching density, and root depth (Koevoets et al., 2016). When crops are bred for desired root features, they can perform better even in harsh environments such as drought, salt, and nutrient shortage. For instance, breeding initiatives to increase root biomass and depth have been successful in producing crop types that are more stable in terms of output and are resistant to drought (Wasaya et al., 2018). The capacity to examine root architecture in detail has been transformed by developments in imaging methods and root phenotyping technology. Researchers can precisely assess root features by using noninvasive imaging techniques such as magnetic resonance imaging (MRI) and X-ray computed tomography (CT), which provide high-resolution visualization of the root structures in soil. Furthermore, new root phenotyping technologies that allow for controlled environmental conditions and high-throughput investigation of root properties include rhizotrons and root growth chambers (Hou et al., 2022).

A complicated characteristic, root architecture is essential to a plant's ability to grow, develop, and adapt to its surroundings. Untangling the complexity of plant biology and enhancing crop yield and sustainability in agriculture require an understanding of the mechanisms controlling root development and the variables influencing root design. The delicate relationship between root architecture and plant performance can be better understood with the help of developments in root imaging and phenotyping technologies (Bodner et al., 2021).

3. Characteristics of microRNAs

A class of short, noncoding RNA molecules known as microRNAs (miRNAs) is essential to the control of posttranscriptional genes in a wide range of organisms, including fungi, mammals, and plants as seen in Fig. 5.3 miRNAs are produced from endogenous transcripts by a multistep biogenesis process, and they are usually 21−24 nucleotides long (Sharma et al., 2022). The capacity of miRNAs to control gene expression by attaching to complementary sequences in messenger RNAs (mRNAs) and causing translational repression or mRNA destruction is one of their distinguishing features. Through this mode of action, miRNAs can regulate gene expression and a range of biological processes, such as stress responses, differentiation, and development. Sequence specificity is another feature of miRNAs that allows them to target numerous mRNAs that have partial complementarity to their seed region. Because of this property, a single miRNA can control the expression of several different genes, resulting in the formation of intricate regulatory networks that manage various biological processes. Furthermore, miRNAs display expression patterns that are particular to tissues and developmental stages, indicating their dynamic functions in determining the identity and function of cells (Pasquinelli, 2012).

miRNAs play a crucial role in the regulatory circuits that control basic biological functions such as metabolism, apoptosis, and cell division. miRNAs have a critical role in human health and disease, as seen by the numerous diseases, including cancer, neurological disorders, and cardiovascular diseases, that have been linked to dysregulation of miRNA expression (Zhang et al., 2015). miRNAs have a variety

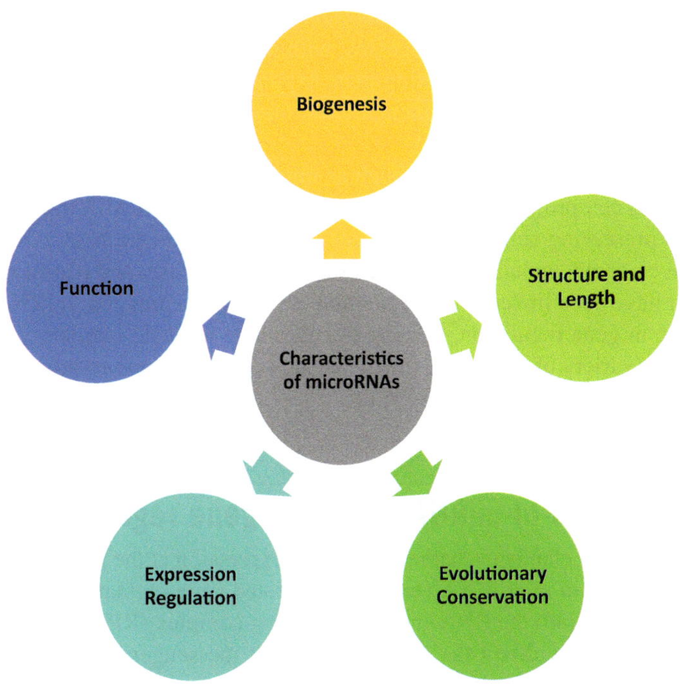

FIGURE 5.3

Characteristics of microRNAs.

of roles in gene regulation and cellular physiology in various organisms, which are supported by their properties, which include their tiny size, sequence specificity, and regulatory versatility (Gebert and MacRae, 2019).

4. Role of miRNAs in gene regulation

MicroRNAs (miRNAs) play a crucial role in gene regulation by modulating the expression of target genes at the posttranscriptional level. Messenger RNAs (mRNAs) have a 3′ untranslated region (UTR) where miRNAs bind to complementary sequences to either degrade or suppress translation of the target mRNA. This is the main mechanism by which miRNAs regulate expression of mRNAs. The RNA-induced silencing complex (RISC) is a ribonucleoprotein complex that is the site of this interaction between the miRNA and its target mRNA (Pu et al., 2019). The mechanism of regulation is determined by the degree of complementarity between the miRNA and its target mRNA. Generally, complete or nearly perfect complementarity leads to mRNA cleavage and degradation, whereas imperfect complementarity results in translational repression without mRNA degradation (Denzler et al., 2016). The sequence complementarity between the miRNA and its target mRNA dictates the specificity of miRNA-mediated gene regulation many mRNAs can be targeted by a single miRNA, and vice versa. This can result in complex regulatory networks that govern different biological processes, with many miRNAs controlling a single mRNA. For the purpose of regulating gene expression during development, differentiation, and reaction to environmental cues, these regulatory networks are essential (Correia de Sousa et al., 2019).

Numerous biological processes, such as immunological responses, metabolism, apoptosis, and cell proliferation, have been linked to miRNAs. MiRNAs play a critical role in preserving cellular homeostasis and organismal health, as evidenced by the several diseases, including cancer, cardiovascular disease, and neurological disorders that have been linked to dysregulation of miRNA expression. miRNAs play a critical role in gene regulation, helping to coordinate intricate regulatory networks that control a variety of biological functions and support the preservation of cellular identity and function (Gupta et al., 2022).

5. Mechanisms of miRNA-mediated gene regulation

Multiple critical mechanisms that regulate gene expression at the posttranscriptional stage are involved in microRNA (miRNA)-mediated gene regulation. Mature miRNAs are integrated into the RNA-induced silencing complex (RISC) during biogenesis. From there, they direct the complex to target messenger RNAs (mRNAs) by initiating base pairing interactions between the complementary sequences in the 3′ untranslated region (UTR) of the mRNA and the miRNA seed region (Michlewski and Cáceres, 2019). The mode of regulation depends on the degree of complementarity between the miRNA and its target mRNA. Perfect or nearly perfect complementarity causes mRNA degradation through RISC-enabled cleavage by the Argonaute (AGO) proteins, whereas imperfect complementarity leads to translational repression without mRNA degradation. Furthermore, miRNAs can affect mRNA stability by decapping and deadenylating mRNA, which can cause mRNA degradation and instability as seen in Fig. 5.4. Due to the precise control over gene expression made possible by these modes of miRNA-mediated gene regulation, miRNAs are essential for a variety of biological processes, such as development, differentiation, and response to environmental cues (Afonso-Grunz and Müller, 2015).

6. Importance of miRNAs in plant development

Plant development is significantly impacted by microRNAs (miRNAs), which have an impact on several aspects of growth, differentiation, and morphogenesis. miRNAs fine-tune the expression of target genes involved in important developmental processes by regulating gene expression at the posttranscriptional level throughout the plant life cycle (Singh et al., 2019). For example, during embryogenesis and

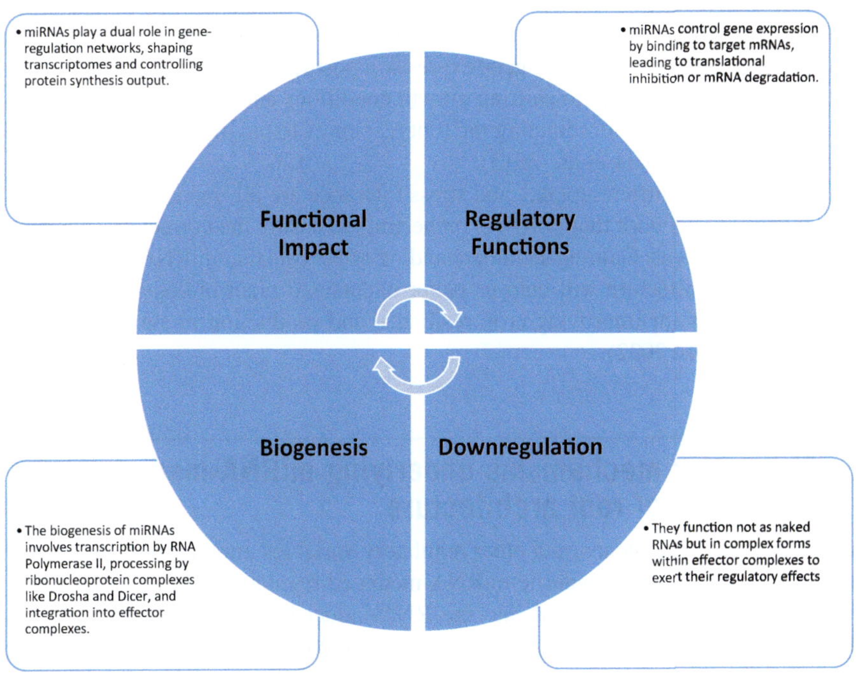

FIGURE 5.4

Mechanisms of miRNA-mediated gene regulation.

organogenesis, miRNAs control the expression of transcription factors, components of hormone signaling, and regulatory proteins that coordinate cell division, differentiation, and patterning (Liu et al., 2018).

Furthermore, miRNAs control root architecture, floral patterning, and leaf development, all of which affect the plant's overall morphology. Furthermore, miRNAs are crucial for coordinating how an organism responds to environmental stimuli and challenges in order to ensure healthy growth and adaptation (Singh and Storey, 2021). The significance of miRNAs in plant development is highlighted by the fact that dysregulation of miRNA expression can result in growth problems and aberrant developmental patterns. In general, miRNAs play a crucial role in controlling plant development by helping to create intricate developmental programs that support a plant's ability to grow and adapt to its surroundings (Song et al., 2019).

7. The role of microRNAs in regulating root architecture

Root architecture is largely regulated by microRNAs (miRNAs), which have fine control over many aspects of root growth and development. Through their targeting of particular genes involved in root morphogenesis, elongation, and patterning, miRNAs regulate the distribution and spatial arrangement of roots in the soil (Yan et al., 2022). miRNAs shape root architectures and maximize resource uptake efficiency by influencing critical processes such cell division, differentiation, and elongation through their capacity to modify gene expression posttranscriptionally. Furthermore, miRNAs help plants adapt to shifting growth conditions and preserve the ideal root system architecture by coordinating the root responses to environmental stimuli and stress conditions (Bazin et al., 2013).

Plant growth, nutrient uptake, and stress tolerance are all impacted by the complex regulatory network that miRNAs orchestrate to ensure the correct development and function of roots. Gaining an understanding of the role that miRNAs play in controlling root architecture will help us better understand plant biology and open up new possibilities for improving crop resilience and production in agricultural systems (Islam et al., 2022).

8. Molecular mechanisms underlying miRNA-mediated regulation of root architecture

miRNAs, their target genes, and other regulatory variables interact intricately in the molecular pathways that underlie miRNA-mediated regulation of root architecture. Mature miRNAs are integrated into the RNA-induced silencing complex (RISC) during biogenesis, and through base pairing interactions, they direct the complex to target messenger RNAs (mRNAs) (Gautam et al., 2021). miRNAs target genes encoding transcription factors, signaling proteins, and regulatory factors involved in root morphogenesis, elongation, and patterning in the context of root development. miRNAs control the expression of genes by binding to complementary sequences in the 3′ untranslated region (UTR) of target mRNAs and causing translational repression or mRNA destruction. Because of their exact control over gene expression, miRNAs can influence important biological processes such as differentiation, elongation, and cell division, which helps to shape root structures and maximize the efficiency of resource uptake (Ma et al., 2020).

Additionally, in order to coordinate root development in response to shifting growth conditions and challenges, miRNAs may interact with other regulatory mechanisms, including as hormone signaling and environmental reactions. Comprehending these molecular pathways offers valuable understanding of the intricate regulatory networks that dictate root design. It also presents possible avenues for modifying root features to enhance crop resilience and productivity (Singh et al., 2020a,b).

9. Interactions between miRNAs and other regulatory factors in shaping root architecture

Root architecture is shaped by interactions between microRNAs (miRNAs) and other regulatory factors, which integrate several signaling pathways and coordinate developmental processes (Patani et al., 2023). miRNAs frequently work in tandem with hormones, transcription factors, and environmental cues to control root development and regulate gene expression. For instance, miRNAs may specifically target genes that encode transcription factors important for root growth and patterning, adjusting their activity and affecting root architecture (Liebsch and Palatnik, 2020).

Furthermore, miRNAs can control how roots grow and develop in response to shifting environmental conditions by interacting with hormone signaling pathways such as auxin, cytokinin, and gibberellin. Moreover, miRNAs may be involved in posttranslational modifications, epigenetic modifiers, feedback loops, and regulatory networks including other small RNAs, which would support the resilience and adaptability of root development. Insights into the mechanisms controlling root design can be gained by comprehending the intricate interactions between miRNAs and other regulatory elements. This also presents prospects for the targeted modification of root features to increase crop productivity and resilience in agricultural systems (Ahmad et al., 2022).

10. Experimental approaches to studying MicroRNA regulation of root architecture

Utilizing a combination of molecular, genetic, and imaging techniques, experimental approaches to researching microRNA (miRNA) regulation of root architecture allow researchers to clarify the roles of miRNAs in influencing root development. A frequently employed method is short RNA sequencing, which enables the detection and measurement of miRNAs expressed in particular root tissues or developmental phases (Khan et al., 2011).

Furthermore, transcriptome studies such as microarray profiling and mRNA sequencing can shed light on the downstream targets of miRNAs and the regulatory networks that support them. Through the use of genetic modification techniques, such as transgenic plants and gene editing tools such as CRISPR/Cas9, researchers can target specific genes or modify the amounts of miRNA expression to examine their effects on root architecture. The phenotypic effects of miRNA-mediated gene regulation in roots can be studied with the help of loss-of-function mutants and overexpression lines (Tseng et al., 2019).

Furthermore, real-time visualization of root structures and dynamic analysis of changes in root shape and growth patterns are made possible by sophisticated imaging techniques including confocal microscopy and live-cell imaging. The expression patterns of miRNAs and their target genes can be tracked in vivo using fluorescent protein markers and reporter constructs, which offer temporal and spatial details regarding the regulatory actions of these molecules throughout root development. Through the utilization of these experimental methodologies, scholars can acquire a thorough comprehension of the molecular mechanisms that underpin miRNA regulation of root architecture. This can clarify the complex networks that oversee root development and offer valuable perspectives on plausible tactics for modifying root characteristics to enhance crop productivity and sustainability (Clark et al., 2020).

11. Techniques for miRNA detection and quantification

The challenges of investigating these small RNA molecules have led to an evolution in techniques for microRNA (miRNA) detection and quantification as seen in Fig. 5.5. Quantitative real-time PCR (qRT-PCR) is a popular technique that allows precise determination of miRNA expression levels by utilizing reverse transcription and PCR amplification using primers unique to miRNAs (Jet et al., 2021). Northern blotting, which involves isolating miRNAs based on size by gel electrophoresis and identifying them using complementary probes, is another often used approach. The ability to thoroughly profile miRNA expression patterns throughout the transcriptome has enabled high-throughput sequencing technologies, including small RNA sequencing, to completely transform miRNA detection in more recent times. Furthermore, many miRNAs can be analyzed in parallel in a single experiment thanks to microarray-based methods. The detection of miRNAs within cells or tissues can be done spatially resolvedly using methods such as fluorescence in situ hybridization (FISH) and in situ PCR. Each of these methods has pros and cons, and in order to obtain reliable and thorough miRNA detection and quantification for

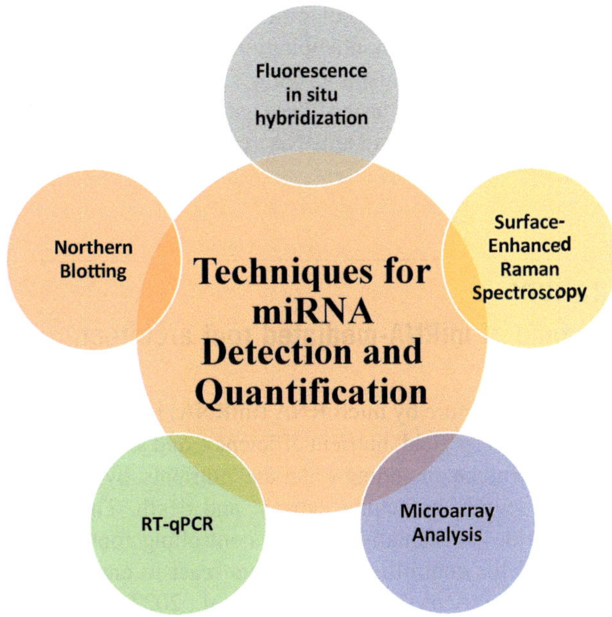

FIGURE 5.5

Techniques for miRNA detection and quantification.

investigating their functions in controlling diverse biological processes, such as root architecture, researchers frequently combine methodologies (Reon and Dutta, 2016).

12. Genetic and molecular tools for studying miRNA function in root development

Studying the role of microRNAs (miRNAs) in root formation has grown increasingly dependent on genetic and molecular methods. It is possible to investigate the involvement of certain miRNAs in root architecture by using transgenic plants that have been modified to either overexpress or inhibit those miRNAs. In a similar vein, loss-of-function mutants produced by gene editing techniques such as CRISPR/Cas9 allow scientists to analyze the phenotypic effects of miRNA depletion (Lima et al., 2012).

Researchers can observe the spatial and temporal expression patterns of miRNAs and their target genes during root development by using reporter constructs, which are comprised of reporter genes fused with miRNA target sequences. This makes it easier to visualize miRNA activity in vivo. Furthermore, protein complexes and RNA-binding proteins implicated in miRNA-mediated gene regulation can be identified thanks to sophisticated molecular methods such as chromatin immunoprecipitation (ChIP) and RNA immunoprecipitation (RIP). Researchers can gain insights into the molecular mechanisms driving root design and identify possible targets for crop improvement by using these genetic and molecular methods to disentangle the complex networks of miRNA regulation underlying root development (Wójcik, 2020).

13. Applications

13.1 Significance of miRNA-mediated root architecture regulation in agriculture

The control of root architecture by microRNA (miRNA) is important for agriculture because it can improve crop yield, nutrient efficiency, and stress tolerance. miRNAs can optimize resource intake, including water and nutrients, by modifying root characteristics including length, branching density, and depth. This enhances overall plant growth and yield. Additionally, through controlling root shape and growth, miRNAs are essential for controlling how plants react to environmental stressors such salt, drought, and nutrient shortage (Islam et al., 2022).

Comprehending the molecular processes that underlie the regulation of root architecture by miRNA presents prospects for focused agricultural enhancement tactics. Through the manipulation of miRNA expression or targeting of particular miRNA-regulated genes, scientists may be able to develop crops with improved root systems that are more resilient to difficult growing environments. In the end, generating robust and high-yielding crop varieties that can maintain agricultural production in the face of shifting environmental stressors and resource restrictions may be possible by using the regulatory potential of miRNAs in root architecture (Singh et al., 2020a,b).

14. Potential applications in crop improvement and stress tolerance

MicroRNA (miRNA)-mediated regulation has a wide range of potential uses in stress tolerance and crop development. Through the targeting of particular genes involved in root architecture, miRNAs present a potent tool for augmenting resistance and yield in crops. For instance, modifying the amounts of miRNA expression or focusing on the genes that regulate root characteristics might result in the development of crops that are more efficient at absorbing nutrients, using water, and withstanding environmental challenges including salt, drought, and nutrient shortage (Begum et al., 2022).

Additionally, by focusing on genes involved in defense systems, miRNAs can be used to control how plants respond to biotic challenges such as pathogen infections.

Moreover, miRNA-based approaches have the potential to improve agricultural output stability and nutritional quality, thereby addressing global issues with sustainable agriculture and food security. In summary, the utilization of miRNA-mediated regulation has promising prospects for the development of crops possessing improved agronomic characteristics and resistance to various environmental obstacles, ultimately enhancing the sustainability and adaptability of agricultural systems (Waheed et al., 2021).

15. Future directions and challenges in research on miRNA-Mediated regulation of root architecture

Future studies on the regulation of root architecture by microRNA (miRNA) have enormous potential to further our knowledge of plant development and increase agricultural yield. To understand the intricacies of root development, one approach is to clarify the regulatory networks including miRNAs and their target genes, as well as their interactions with other regulatory variables. Furthermore, investigating the activities of recently identified miRNAs and how they contribute to particular root features may reveal new regulatory pathways (Mandal et al., 2021).

The development of effective methods for regulating miRNA expression in crop plants, as well as the requirement for more accurate functional characterization of miRNAs and their target genes, present challenges. Moreover, a thorough understanding of miRNA-mediated regulatory networks would need the integration of omics techniques, including transcriptomics, proteomics, and genomes. It will take interdisciplinary teams and creative approaches to overcome these obstacles, but in the end, doing so will open the door for miRNA research to be applied practically in crop enhancement and sustainable agriculture (Sun, 2012).

16. Conclusion

In summary, the complex interactions between root architecture and microRNAs (miRNAs) highlight the critical role miRNAs play in influencing plant growth and environmental adaptation. MiRNAs have a significant impact on root elongation, branching, and patterning, among other features of root architecture, by precisely controlling gene expression at the posttranscriptional level. miRNAs enhance resource uptake efficiency, water and nutrient intake, and stress tolerance by focusing on particular genes involved in root development and growth. This eventually affects total plant resilience and productivity. Beyond basic plant biology, miRNA-mediated modulation of root architecture has practical uses in agriculture. Utilizing miRNA-based techniques has promising prospects for agricultural advancement, such as the creation of crops with improved agronomic characteristics and resistance to environmental perturbations. There are still a number of obstacles to overcome, though, such as the requirement for a thorough comprehension of miRNA regulatory networks, accurate functional characterization of miRNAs and their target genes, and the creation of effective instruments for modifying miRNA expression in agricultural plants. Our understanding of miRNA-mediated regulation of root architecture will grow as a result of addressing these issues through interdisciplinary cooperation and creative approaches. Additionally, there is tremendous potential to increase food security, sustainability, and the ability of agricultural systems to withstand environmental challenges worldwide by putting miRNA research into useful applications for crop improvement. The significance of miRNAs in plant biology and agriculture is underscored by their critical role in controlling root architecture. This will influence future research paths and applications that aim to increase agricultural yield and sustainability.

References

Afonso-Grunz, F., Müller, S., 2015. Principles of miRNA−mRNA interactions: beyond sequence complementarity. Cell. Mol. Life Sci. 72, 3127−3141.

Ahmad, H.M., Wang, X., Ijaz, M., Oranab, S., Ali, M.A., Fiaz, S., 2022. Molecular aspects of microRNAs and phytohormonal signaling in response to drought stress: a review. Curr. Issues Mol. Biol. 44 (8), 3695−3710.

Barrera-Rojas, C.H., Otoni, W.C., Nogueira, F.T.S., 2021. Shaping the root system: the interplay between miRNA regulatory hubs and phytohormones. J. Exp. Bot. 72 (20), 6822−6835.

Bazin, J., Khan, G.A., Combier, J.P., Bustos-Sanmamed, P., Debernardi, J.M., Rodriguez, R., Lelandais-Brière, C., 2013. miR396 affects mycorrhization and root meristem activity in the legume M edicago truncatula. Plant J. 74 (6), 920−934.

Begum, Y., 2022. Regulatory role of microRNAs (miRNAs) in the recent development of abiotic stress tolerance of plants. Gene 821, 146283.

Betti, F., Ladera-Carmona, M.J., Weits, D.A., Ferri, G., Iacopino, S., Novi, G., Perata, P., 2021. Exogenous miRNAs induce post-transcriptional gene silencing in plants. Nat. Plants 7 (10), 1379−1388.

Bodner, G., Mentler, A., Keiblinger, K., 2021. Plant Roots for Sustainable Soil Structure Management in Cropping Systems. The Root Systems in Sustainable Agricultural Intensification, pp. 45−90.

Clark, N.M., Van den Broeck, L., Guichard, M., Stager, A., Tanner, H.G., Blilou, I., Sozzani, R., 2020. Novel imaging modalities shedding light on plant biology: start small and grow big. Annu. Rev. Plant Biol. 71, 789−816.

Correia de Sousa, M., Gjorgjieva, M., Dolicka, D., Sobolewski, C., Foti, M., 2019. Deciphering miRNAs' action through miRNA editing. Int. J. Mol. Sci. 20 (24), 6249.

Denzler, R., McGeary, S.E., Agarwal, V., Bartel, D.P., Stoffel, M., 2016. Impact of microRNA levels, target-site complementarity, and cooperativity on competing endogenous RNA-regulated gene expression. Mol. Cell 64 (3), 565−579.

Downie, H.F., Adu, M.O., Schmidt, S., Otten, W., Dupuy, L.X., White, P.J., Valentine, T.A., 2015. Challenges and opportunities for quantifying roots and rhizosphere interactions through imaging and image analysis. Plant Cell Environ. 38 (7), 1213−1232.

Gautam, V., Gupta, P., Salvi, P., Sharma, A., Kumar, D., Dwivedi, A., 2021. miRNA mediated signaling involved in *Arabidopsis thaliana* root development. In: Rhizobiology: Molecular Physiology of Plant Roots, pp. 93–113.

Gebert, L.F., MacRae, I.J., 2019. Regulation of microRNA function in animals. Nat. Rev. Mol. Cell Biol. 20 (1), 21–37.

Gupta, M., Akhtar, J., Sarwat, M., 2022. MicroRNAs: regulators of immunological reactions in hepatocellular carcinoma. In: Seminars in Cell and Developmental Biology, vol. 124. Academic Press, pp. 127–133.

Hodge, A., Berta, G., Doussan, C., Merchan, F., Crespi, M., 2009. Plant Root Growth, Architecture and Function.

Homulle, Z., George, T.S., Karley, A.J., 2021. Root traits with team benefits: understanding belowground interactions in intercropping systems. Plant Soil 1–26.

Hou, L.H., Gao, W., Weng, Z.H., Doolette, C.L., Maksimenko, A., Hausermann, D., Kopittke, P.M., 2022. Use of X-ray tomography for examining root architecture in soils. Geoderma 405, 115405.

Islam, W., Tauqeer, A., Waheed, A., Zeng, F., 2022. MicroRNA mediated plant responses to nutrient stress. Int. J. Mol. Sci. 23 (5), 2562.

Jet, T., Gines, G., Rondelez, Y., Taly, V., 2021. Advances in multiplexed techniques for the detection and quantification of microRNAs. Chem. Soc. Rev. 50 (6), 4141–4161.

Joshi, S., Chinnusamy, V., Joshi, R., 2022. Root system architecture and omics approaches for belowground abiotic stress tolerance in plants. Agriculture 12 (10), 1677.

Jung, J.K., McCouch, S.R., 2013. Getting to the roots of it: genetic and hormonal control of root architecture. Front. Plant Sci. 4, 42810.

Khan, G.A., Declerck, M., Sorin, C., Hartmann, C., Crespi, M., Lelandais-Brière, C., 2011. MicroRNAs as regulators of root development and architecture. Plant Mol. Biol. 77, 47–58.

Koevoets, I.T., Venema, J.H., Elzenga, J.T., Testerink, C., 2016. Roots withstanding their environment: exploiting root system architecture responses to abiotic stress to improve crop tolerance. Front. Plant Sci. 7, 215317.

Kou, X., Han, W., Kang, J., 2022. Responses of root system architecture to water stress at multiple levels: a meta-analysis of trials under controlled conditions. Front. Plant Sci. 13, 1085409.

Liebsch, D., Palatnik, J.F., 2020. MicroRNA miR396, GRF transcription factors and GIF co-regulators: a conserved plant growth regulatory module with potential for breeding and biotechnology. Curr. Opin. Plant Biol. 53, 31–42.

Lima, J.C.D., Loss-Morais, G., Margis, R., 2012. MicroRNAs play critical roles during plant development and in response to abiotic stresses. Genet. Mol. Biol. 35, 1069–1077.

Liu, H., Yu, H., Tang, G., Huang, T., 2018. Small but powerful: function of microRNAs in plant development. Plant Cell Rep. 37, 515–528.

Löbmann, M.T., Geitner, C., Wellstein, C., Zerbe, S., 2020. The influence of herbaceous vegetation on slope stability–a review. Earth Sci. Rev. 209, 103328.

Ma, J., Zhao, P., Liu, S., Yang, Q., Guo, H., 2020. The control of developmental phase transitions by microRNAs and their targets in seed plants. Int. J. Mol. Sci. 21 (6), 1971.

Mandal, K., Boro, P., Chattopadhyay, S., 2021. Micro-RNA based gene regulation: a potential way for crop improvements. Plant Gene 27, 100312.

Michlewski, G., Cáceres, J.F., 2019. Post-transcriptional control of miRNA biogenesis. Rna 25 (1), 1–16.

Parveen, R., Khare, N., Singh, S., Johri, P., 2023. An in Silico Outlook for the detection and surveillance of evolving and persistent plant pathogens. In: Detection, Diagnosis and

Management of Soil-Borne Phytopathogens. Springer Nature Singapore, Singapore, pp. 17–40.

Pasquinelli, A.E., 2012. MicroRNAs and their targets: recognition, regulation and an emerging reciprocal relationship. Nat. Rev. Genet. 13 (4), 271–282.

Patani, A., Patel, A., Prajapati, D., Khare, N., Singh, S., 2023. Various agriculture crop plant-based bioactive compounds and their use in nanomaterial synthesis and applications. In: Nanomaterials from Agricultural and Horticultural Products. Springer Nature Singapore, Singapore, pp. 223–241.

Pu, M., Chen, J., Tao, Z., Miao, L., Qi, X., Wang, Y., Ren, J., 2019. Regulatory network of miRNA on its target: coordination between transcriptional and post-transcriptional regulation of gene expression. Cell. Mol. Life Sci. 76, 441–451.

Reon, B.J., Dutta, A., 2016. Biological processes discovered by high-throughput sequencing. Am. J. Pathol. 186 (4), 722–732.

Santosh, B., Varshney, A., Yadava, P.K., 2015. Non-coding RNAs: biological functions and applications. Cell Biochem. Funct. 33 (1), 14–22.

Sharma, Y., Sharma, A., Madhu, S., Singh, K., Upadhyay, S.K., 2022. Long non-coding RNAs as emerging regulators of pathogen response in plants. Non-coding RNA 8 (1), 4.

Singh, G., Storey, K.B., 2021. MicroRNA cues from nature: a roadmap to decipher and combat challenges in human health and disease? Cells 10 (12), 3374.

Singh, A., Sengar, R.S., Singh, A.K., Singh, S., Kumar, A., Ahmad, K., 2019. Impact of genetic engineering on agriculture food crops. Biotech Today Int. J. Biol. Sci. 9 (1), 17–24.

Singh, A., Gandhi, N., Mishra, V., Yadav, S., Rai, V., Sarkar, A.K., 2020a. Role of abiotic stress responsive miRNAs in Arabidopsis root development. J. Plant Biochem. Biotechnol. 29, 733–742.

Singh, R.K., Prasad, A., Muthamilarasan, M., Parida, S.K., Prasad, M., 2020b. Breeding and biotechnological interventions for trait improvement: status and prospects. Planta 252, 1–18.

Song, X., Li, Y., Cao, X., Qi, Y., 2019. MicroRNAs and their regulatory roles in plant–environment interactions. Annu. Rev. Plant Biol. 70, 489–525.

Sun, G., 2012. MicroRNAs and their diverse functions in plants. Plant Mol. Biol. 80, 17–36.

Tiwari, R., Rajam, M.V., 2022. RNA-and miRNA-interference to enhance abiotic stress tolerance in plants. J. Plant Biochem. Biotechnol. 31 (4), 689–704.

Tseng, C.C., Wu, L.Y., Tsai, W.C., Ou, T.T., Wu, C.C., Sung, W.Y., Yen, J.H., 2019. Differential expression profiles of the transcriptome and miRNA interactome in synovial fibroblasts of rheumatoid arthritis revealed by next generation sequencing. Diagnostics 9 (3), 98.

Viana, W.G., Scharwies, J.D., Dinneny, J.R., 2022. Deconstructing the root system of grasses through an exploration of development, anatomy and function. Plant Cell Environ. 45 (3), 602–619.

Vishwakarma, S.K., Ilyas, T., Shahid, M., Malviya, D., Kumar, S., Singh, S., Singh, H.V., 2024. Bacillus spp.: nature's Gift to agriculture and humankind. In: Applications of Bacillus and Bacillus Derived Genera in Agriculture, Biotechnology and Beyond. Springer Nature Singapore, Singapore, pp. 1–36.

Vives-Peris, V., Lopez-Climent, M.F., Perez-Clemente, R.M., Gomez-Cadenas, A., 2020. Root involvement in plant responses to adverse environmental conditions. Agronomy 10 (7), 942.

Waheed, S., Anwar, M., Saleem, M.A., Wu, J., Tayyab, M., Hu, Z., 2021. The critical role of small RNAs in regulating plant innate immunity. Biomolecules 11 (2), 184.

Wasaya, A., Zhang, X., Fang, Q., Yan, Z., 2018. Root phenotyping for drought tolerance: a review. Agronomy 8 (11), 241.

Wójcik, A.M., 2020. Research tools for the functional genomics of plant miRNAs during zygotic and somatic embryogenesis. Int. J. Mol. Sci. 21 (14), 4969.

Yan, X.X., Liu, X.Y., Hong, C.U.I., Zhao, M.Q., 2022. The roles of microRNAs in regulating root formation and growth in plants. J. Integr. Agric. 21 (4), 901–916.

Zhang, H.M., Kuang, S., Xiong, X., Gao, T., Liu, C., Guo, A.Y., 2015. Transcription factor and microRNA co-regulatory loops: important regulatory motifs in biological processes and diseases. Briefings Bioinf. 16 (1), 45–58.

CHAPTER

Tolerance-enhancing microRNA: A new therapeutic target for plants

6

Sagnik Nag[1], Garima Tripathi[2], Anamika Mishra[2], Kasturi Banerjee[2], Saloni Koli[2], Sai Anisha Bhupathiraju[2], Rohan Dutta[2], Shubhrima Ghosh[3,4]

[1]*Jeffrey Cheah School of Medicine and Health Sciences, Monash University Malaysia, Selangor, Malaysia;* [2]*Department of Bio-Sciences, School of Bio-Sciences & Technology (SBST), Vellore Institute of Technology (VIT), Vellore, Tamil Nadu, India;* [3]*Trinity Translational Medicine Institute, School of Medicine, Trinity College Dublin, Dublin, Ireland;* [4]*School of Biological, Health and Sports Sciences, Technological University Dublin, Dublin, Ireland*

1. Introduction

Plants, as stationary organisms are continuously subjected to adverse climatic conditions. Extreme climatic conditions cause different types of stress in the plants, leading to economic losses (Malhi et al., 2021). With the increasing population worldwide, the demand for agriculture is rising. In addition to this, agriculture contributes significantly to the economy of a nation (Calicioglu et al., 2019). Due to the advent of subsequent recent viral outbreaks of COVID-19 (Chatterjee et al., 2023) and the monkeypox virus (Bhattacharya et al., 2022), individuals have shifted their focus on sustainable green produce supplemented with conventional diets to yield better immunity and to give protection against fatal morbidities such as cancer, diabetes, cardiovascular ailments and others (Ghosh et al., 2023; Hoque et al., 2023). Therefore, much of the focus is diverted to enhancing the quality of the plants cultivated and providing the ever-growing population with high-quality food (Sunny et al., 2016).

Majorly, stress in plants is classified into two distinct categories namely abiotic stress and biotic stress (Gull et al., 2019). Abiotic stress comprises of nonliving components such as temperature, radiation, thermal energy, salinity, and drought (Motos et al., 2022). Evidence exists that depicts the negative effect of abiotic stress in the form of seed germination inhibition and reduction in chlorophyll synthesis (Sharma et al., 2021). Additionally, abiotic stress is demonstrated to cause developmental alterations due to changes at a genetic level (Miryeganeh, 2021). Abiotic stress due to extreme environmental changes is one of the major causes of yield loss, which leads to immense economic loss for a nation. In contrast, biotic stress incorporates pathogenic attacks on plants (Tarfeen et al., 2022). These pathogens include bacteria,

MicroRNA Advances and Application in Plant Biology. https://doi.org/10.1016/B978-0-443-21682-4.00005-1

fungi, viruses, nematodes as well as insects. Biotic stress is responsible for the lack of growth sustenance of plants due to pathogenic attacks. Both biotic and abiotic stress are known to cause several negative impacts at physiological, cellular and even molecular levels (Meddya et al., 2023).

In the case of biotic stress, fungicide and pesticide have been utilized to repulse the fungus and pests, respectively (Abubakar et al., 2019). However, such chemical treatment not only harms the fungi and pests but also became incorporated into the plants (Ons et al., 2020). Consumption of such plants by humans has led to immediate as well as long-term negative health impacts (Alengebawy et al., 2021). In addition, chemical treatment with pesticides, herbicides, fungicides and insecticides led to bioaccumulation in plants, which altered their yield and numerous metabolic pathways (Wijntjes et al., 2022). Moreover, environmental pollution became a significant issue with the application of chemical sprays. Therefore, further research focussed towards alternative methods to strengthen the tolerance of plants towards various stresses.

Since the impact of the stresses is at the genetic level, several significant genes are observed to respond to the stresses. Overexpression of these genes in the plant species has been shown to enhance the stress tolerance capacity of the plants (Hang et al., 2021). However, many agricultural practices remain adhered to traditional methods of breeding, which have been applied to obtain high-yielding crop varieties (Luo et al., 2022). However, it is often labor-intensive and time-consuming to breed crops through traditional methods over a long period of time. Therefore, research is directed to new technology that can facilitate the stress tolerance ability of plants.

One of such techniques is genetic engineering, which has been playing a significant role in introducing desirable traits in plants including enhancing their tolerance to different kinds of stress (Zafar et al., 2020; Verma et al., 2021). Novel technologies, for instance, microarrays and next-generation sequencing revealed several genes that are associated with environmental stresses (Saidi et al., 2020). However, plenty of areas remain unexplored such as the mechanism of these gene regulations and the genetic networks responsible for stress tolerance. Research has highlighted the ability of a small regulatory RNA molecule named microRNA (miRNA), which plays a significant part in enhancing the stress tolerance ability of plants (Wani et al., 2020).

In the field of genetic engineering, the concept of miRNAs emerged as an effective solution to yield crops with high-stress tolerance. miRNAs are long noncoding regulatory RNAs, which contain 20–24 nucleotides (Wu et al., 2020). Essentially, the functional activity of miRNAs is to regulate various biological processes through the silencing of genes at a transcriptional level (Mewalal et al., 2019). In the case of plants, miRNAs are believed to be highly conserved and distributed widely (Singh et al., 2019). However, studies over the years have elucidated that plants contain species-specific miRNAs as well (Liu et al., 2021). This is indicative of the fact that the miRNA holds the capability to influence common plant traits such as growth, morphology and development. miRNAs, both conserved and species-specific, facilitate abiotic and biotic stress tolerance (Patra et al., 2023). miRNA aids in the understanding of plants' ability to sense and respond to stresses such as salinity, temperature fluctuation and heat. This poses a deep impact on plant breeding and plant development (Tungare et al., 2022).

Decades of studies have depicted the various role of miRNA in the metabolic as well as regulatory processes in plants (Adjei et al., 2021). This is detailed in Fig. 6.1, which shows the functions of miRNA in plants for enhanced stress tolerance. Based on the figure, primary miRNA (pri-miRNA) conforms to a loop-like structure, which inhibits certain protein translations by pairing with target mRNA transcripts (Mencia et al., 2023). As a result, the resistance to pathogenic attacks and tolerance to stress enhances. The response of miRNA to stresses is highly dependent upon miRNA type, stress type, plant organ/tissues and genetic composition (Xing et al., 2021; Patani et al., 2023). Stress-driven response of miRNA is observed either as a down-regulation mechanism of negative regulators or an upregulation of positive regulators that confer high-stress tolerance (Kumar et al., 2021). Regulation also correlates with the interaction between plants and beneficial microorganisms

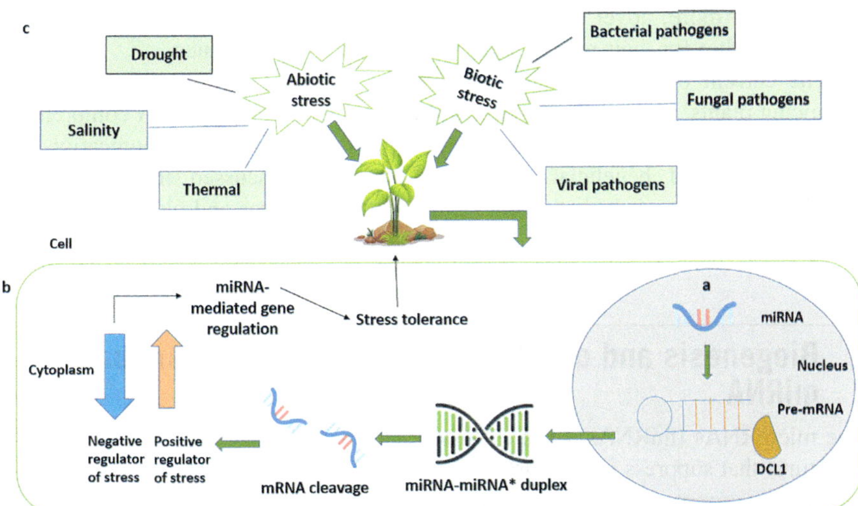

FIGURE 6.1 Multimodal function of plant-based miRNA (A). Biosynthesis of miRNA and miRNA-miRNA* duplex in the nucleus: In the nucleus, the miRNA gene is transcribed into the preliminary miRNA, which is single-stranded. This preliminary miRNA is converted to precursor-miRNA (pre-miRNA), characterized by the loop structure. DCL1 or dicer-like 1 acts on the pre-miRNA to produce miRNA-miRNA* duplex, which then travels to the cytoplasm. (B) miRNA-mediated gene regulation in cytoplasm: In the cytoplasm, in association with AGO1, the duplex is broken down, producing single strand of miRNA and miRNA*. AGO1 processes the mature miRNA by loading it onto RISC, forming a complex. The complex either cleaves the target strand or stops the translation, thereby regulating gene expression. (C) Enhanced stress tolerance: The miRNA-mediated gene expression enables the plant, exposed to abiotic and biotic stress, to depict stress tolerance and resistance).

(Bukhat et al., 2020). For instance, it was observed that mycorrhizal plants depict different miRNA expression when compared with plants without mycorrhiza (Mewalal et al., 2019; Gangwar et al., 2023). Furthermore, detailed scrutiny of miRNA-mRNA pairs is highly important in regulating or silencing gene expression, metabolic activity, plant-microbe interaction and signaling (Qiu et al., 2020).

Recent research is directed towards understanding the target site and validation of the stress-driven response of miRNA and their functional activity under different stresses. This is made possible due to technology involving high throughput sequencing and other advanced online tools. However, an increasing understanding of miRNA activity has highlighted the contributory role of miRNA networks in biotic and abiotic stresses. The regulatory network associated with miRNA in plants depicts several mechanisms to mediate regulation during transcription and posttranscription to form a synergistic miRNA regulatory network (Pu et al., 2019). In addition, miRNA functional studies and epigenetic studies are conducted to elucidate the regulatory networks and pathways that trigger stress tolerance in plants (Zhang et al., 2022). Additionally, the technology associated with genome editing is being exploited to utilize miRNA to its maximum capacity and provision of desired plant traits. Furthermore, novel pathways and loci origins of miRNA remain yet to be explored further for their role in biotic and abiotic stress tolerance.

The current book chapter is conducted to highlight the role of miRNA in enhancing plants' ability to withstand both abiotic and biotic stress. A comprehensive summary of existing work and insights into the symbiotic communication between plants and beneficial microbes have been discussed in this review. Furthermore, this review sheds light on the specific stresses and plants' capability to tolerate those stresses through different miRNA networks.

2. Biogenesis and operational mechanism of plant based miRNA

The microRNAs (miRNAs) form a cluster of nucleotide endogenous minute RNA structures that suppress gene expression. They drive the gene expression by encoding specific transcription factors, responsive stress-mediated proteins, and protein domains that regulate plant physiology and development. Plant miRNA synthesis is similar to that of miRNA synthesis in animals, with respect to the two primary steps of miRNA synthesis, in which the action of RNAase III like enzymes incorporates the synthesized miRNA in a RISC protein complex (Zhang et al., 2022). However, the plant miRNAs are further processed through methylation on their last nucleotide with the help of miRNA methyltransferase HEN1 (Chen et al., 2019). The maturation of the miRNAs plays a preparative part in the biogenesis of the same. TOUGH (TGH) G patch domain protein and SERRATE (SE) help in the pri-miRNA metabolism and synthesis, along with the further accumulation of the processed miRNA (Kansal et al., 2021). The pri-miRNA is later processed with the help of enzymes - Drosha and Dicer. They help in the cleavage of the hairpin loop of the pre-miRNA at the site of the3′ overhang and further process the formation of the duplex structure of small RNAs. The 5′ end of the RNA being less stable is then incorporated into RISC. This is the place of gene regulation in plants, which helps further metabolism and regulation in the plant body (Rani et al., 2022). The DCL genes play a central role in this kind of regulation. Thus, DCL 1 gene is the most essential factor that helps in the synthesis of dicer and drosha. The miRNA duplex formation is proven by the viral RNA silencing suppressors that stabilized the transient duplex form during the miRNA biogenesis. DCL 1 protein is seen to be localized at the nuclear center of epidermal cells (Tyagi et al., 2019). Protein-protein interactivity alongwith structural conformations of the pri-miRNA stimulate the formation of DCL1 complex. Some of the proteins have a dual influence in deploying the DCL1 complex or spliceosomes to the pri-miRNA. HYPONASTIC LEAVES1 (HYL1) plays a pivotal role in the duplex binding and RISC loading on the Dicer. The upregulation or downregulation of the phosphorylation event impacts the interactivity of DCL1 with DDL and favors the processing of DCL1 complex by the action of HYL1. Consequent phosphorylation of the protein structures can impact the succeeding stages in pri-miRNA processing. The miRNA biogenesis thus likely occurs in the nucleus since p19 leads to the reduced production of miRNAs in the cell. HASTY (HST) facilitates the migration of miRNA duplex outside the nucleus to the cytoplasm. miRNA are found in single-stranded form in both nuclear and cytoplasmic compartments. However, there is ambiguity related to the occurrence of RISC assembly occurrence in the cytoplasm or nucleus (Cambiagno et al., 2021). HEN1 is involved in the methylation of the miRNA duplex strands involving the C terminal methyltransferase domain. The methylation enables the protection of the hydroxyl group in the nucleotide end of the miRNA (Yang et al., 2021). The aggregated protein complex comprising DCL1, TGH, SE, HYL1 and CPL1 is recognized in the subnuclear foci, which additionally involves prorich protein SIC to help in mature miRNA accumulation. The involvement of DCL1 complex and pri-miRNAs in the pri-miRNA processing led to these getting ascertained as dicing bodies (Zhan et al., 2012). Furthermore, the overlapping between miRNA and mRNA biogenesis results in pri-miRNA specific splicing. This involves the process of mi-RNA mediated gene splicing, wherein the pri-miRNA are produced inside the nucleus and then the mature miRNAs are moved to cytoplasm. Consequently, the insertion of the miRNAs occurs inside the AGO protein complexes and regulates the posttranscriptional gene-splicing mechanism through the action of splicing and translational surrendering. Despite the little knowledge that is available concerning the cytoplasmic locations in which RISC loading and miRNA targeted slicing takes place, research has showcased that translational inhibition arises in the Endoplasmic reticulum. P-body constituents play a key factor in translational repression, importing the AGO complexes inside the nucleus and then starting off the cascade producing translational silencing of targeted gene expression (Chen et al., 2013).

There are various transcription factors that enable the biogenesis of miRNA family groups in plant systems (Bajczyk et al., 2023). POWDERESS regulates the miR-172 family members and the FUSCA3 transcription factor regulates the accumulation of miR156 family members. Other few of metabolic factors that are involved in the transcriptional pathways of different miRNAs are - SHORT VEGETATIVE PHASE, which acts like a transcriptional repressor of miR-172, MYB2 helps in the transcription of miR-399f, during phosphate starvation and SQUAMOUS PROMOTER BINDING PROTEIN LIKE7 acts as a transcriptional factor of miR-398b during copper deficient conditions. Cup-shaped cotyledon 1 (CUC1) and CUC2 act as the transcription factors of the embryonic and floral development of the plant.

The miRNAs work in the regulatory pathways of the plant systems and consequently help in the upregulation and downregulation of the required factors for development. They help in the release of phytohormones and their regulation. The release of these miRNAs is also dependent on environmental factors (Dong et al., 2022). Further knowledge and understanding of the miRNA function can help us to explore how they can be beneficial in the plant growth system by increasing resistance, crop growth and quality to various biotic and abiotic stresses.

3. miRNA network associated symbiotically with plant growth promoting microbes for enhanced stress response

MicroRNAs are a class of endogenous RNA that plays a vital role in governing and regulation of gene expressions in various developmental and signaling metabolic pathways. From the studies conducted it is concluded that abiotic stress avail in the induction of expression of various miRNAs, implicating that miRNAs can be explored as a new way to genetically modify plant tolerance to atmospheric stress. The research also adds to our understanding of how miRNAs respond to environmental stress in a manner that depends on miRNA type, genotype, tissue, and the specific stress involved. The control of the signaling pathways and regulation of the root development target genes that are present in the miRNA during abiotic stress (Lü et al., 2022). Negative regulation of stress-induced miRNAs leads to the inhibition of genes that play a role as inhibitors of stress tolerance, while stress-suppressed miRNAs facilitate the activation of genes that function as promoters of stress resilience. At present, the primary emphasis in miRNA investigation centers on identifying miRNAs that react to various stress conditions and analyzing changes in their expression patterns in such situations. These studies have largely been facilitated by advanced sequencing technologies and other expression analysis techniques such as quantitative real-time PCR. Further, it will be imperative to delve deeper into the functional aspects and expression patterns of miRNAs to unravel the shared regulatory mechanisms orchestrated by miRNAs that enhance resilience against various environmental stressors. To understand the specific functions of individual miRNAs reacting to environmental pressures, techniques such as altering artificial miRNAs, increasing their expression, and conducting gene knockout or downregulation experiments targeting both miRNAs and their target genes will be extremely valuable (Ho et al., 2017).

There is a variety of bidirectional communication as well as cross-communication between the microbes and host plants. These interactions are beneficial for the host plants and also for the microbes. The miRNA present in plants shows a quick response to both pathogenic and beneficial microorganisms when they interact and enter the plant tissue, the interaction occurs between the miRNA of plants and miRNA-like RNA (milRNA) and effects the health of the plants. The response refers to a rapid and crucial molecular reaction that serves as a mediator in the intricate interplay between plants and microbes. This response encompasses several aspects, including the monitoring role of microRNAs (miRNAs) in bolstering the resistance of host plant against diseases in the context of plant-microbe interactions. Additionally, miRNAs play a pivotal regulatory function in enhancing host development and proliferation when plants engage in beneficial interactions with probiotic microorganisms. Furthermore, miRNAs exhibit a fascinating cross-kingdom regulatory profile, exerting their influence on host plants while interacting with microbes from different kingdoms. These multifaceted roles of miRNAs underscore their significance in orchestrating the delicate balance between plants and microbes, shaping the outcomes of these interactions in diverse ways (De Vries et al., 2019).

When infective or beneficial microbes enter plant tissues, plant miRNA reacts quickly, and microbes also generate miRNA-like RNA (milRNA) that has an impact on plant health. These internal mechanisms have improved in recent years, which indicates that miRNA or milRNA plays a significant role as a quick-acting molecular intermediary in plant-microbe interactions (Harris et al., 2022). The regulatory functions of miRNA in the interactions between plants, pathogens, and probiotics aree summarized in this chapter. In-depth discussions would focus on the interspecies regulatory functions of milRNA in the host systems, the regulatory functions of milRNA in facilitating host growth and advancement in plant-probiotic interactions, and the regulatory functions of milRNA in conferring resistance to disease of plant species during plant-pathogen interactions (Sharma et al., 2021).

For instance, research has revealed that miRNAs are involved in facilitating communication between rhizobia, which are nitrogen-fixing bacteria residing in the root nodules of leguminous plants, and these plants. The gene expression involved in the symbiotic relationship, including those governing the development of root nodules, is believed to be partially regulated by miRNAs (Lü et al., 2022).

The communication among plants and mycorrhizal fungi, that also aid in the absorption of nutrients by plants-like phosphorus from the soil, has also been found to involve miRNAs. The expression of gene that is involved in the growth of fungal hyphae, the constructions that enhance exchange of nutrients between both the plant and the fungus, is thought to be regulated by the miRNAs. Overall, miRNAs are essential for controlling gene expression and enhancing stress response in the mutually beneficial relationship among plants and advantageous microbes (Parveen et al., 2023). Additional study in this field may result in the creation of fresh ways to improve productivity and development of plants while also enhancing interactions between plants and microbe (De Vries et al., 2019).

The RNA silencing machinery are employed in plant-microbe symbiotic regulatory mechanism. Targets of pathogens include the transcription factor TCP14 and the RNA silencing machinery, which are conserved parts of the immune-like system residing in plants (Vora et al., 2023). The convergent evolution of these effector molecules in various pathogens suggests their significance in undermining the plant's defenses. TCP14 is a transcription factor that is present in the genomes of both land plants and streptophyte algae. It controls immune signaling in plants. TCP14 is a target that pathogens have developed to attack so as to thwart the immune response shown by the plants (Rajwanshi et al., 2019).

Another target of pathogens is the RNA silencing apparatus. Plants use RNA silencing as a natural defense against viral infections. Some pathogens have developed suppressors of RNA silencing, which obstruct this defense mechanism (Song et al., 2019). The suppressors target multiple steps in the RNA silencing process, rather than a single component, making them effective at subverting the plant's defenses. Overall, this passage highlights the importance of understanding the mechanisms by which pathogens evade the plant immune system in order to develop effective strategies for crop protection. Plant pathogens such as *Phytophthora sojae* and *Pseudomonas syringae* control the RNA silencing system of plants to inhibit their immune responses. The RNA silencing apparatus is targeted by these pathogens' effector molecules, which have a wide range of effects on plant stress responses. Nevertheless, some RNA silencing suppressors have the potential to unintentionally activate NBS-LRRs, which are involved in plant pathogen defense. The passage also notes that some RNA silencing suppressors, such as PSR2 in the *Phytophthora* genus, Remain conserved among a broader spectrum of similar pathogens and that adaptive mutations in particular amino acid motifs may permit functional conservation while enabling variation. Further to the function in refining immune response, the passage also suggests that miR482/2118-mediated RNA silencing can act as a sensor for pathogen (De Vries et al., 2019). Fig. 6.2 suggest the mechanism in an illustrative manner.

Table 6.1 provides information about the miRNA that play significant role in association with growth-promoting bacteria that work by different mechanisms in various plants. This helps to know about the miRNA application in various plants and with growth-promoting bacteria. The interactions between plants and the microbial communities that live there, particularly the phyllosphere, rhizosphere, and endophytic microbiomes. The passage also discusses how these microbial communities are shaped by plant genotype and root exudates. In numerous plant species, including *Arabidopsis thaliana* and maize, QTL analysis, GWAS, and candidate gene approaches have been used to show the effects of genetic variation in plants on the composition and abundance of microbial communities. Additionally, it was discovered that the constitution of root exudates differs between different plant

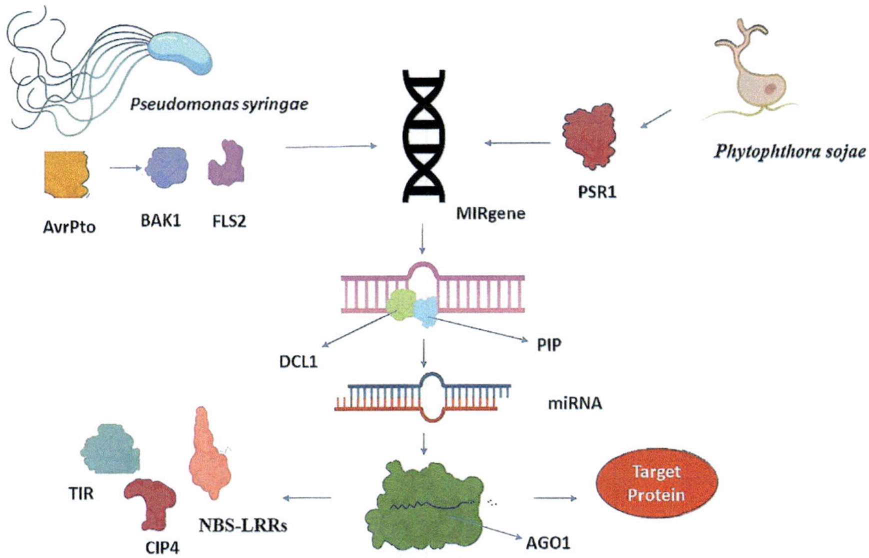

FIGURE 6.2 Pathogens evading the plant immune system, through targeting the RNA silencing apparatus.

Plants employ RNA silencing as an innate safeguard mechanism to microbial infections. These suppressors disrupt several steps in the RNA silencing process, making them highly effective at subverting the plant's immune defenses. *Phytophthora sojae* and *Pseudomonas syringae* are specifically mentioned for their ability to manipulate the RNA silencing system in plants, thereby inhibiting the plant's immune responses. These pathogens utilize effector molecules to target the RNA silencing apparatus, causing various effects on plant stress responses. Interestingly, some of these RNA silencing suppressors have the unintended potential to activate NBS-LRRs, which are components engaged in plant defense against pathogens.

species, with each having a unique microbial community that it attracts and shapes in the rhizosphere. Examples of how certain root exudates, such as organic acids and amino acids, affect the development of soil-borne pathogens as well as the colonization and biofilm formation of rhizobacteria that promote plant growth is shown here .

4. miRNA networks and underlying mechanisms for management of abiotic stress

4.1 Drought stress and salinity stress

Drought stress stands as a prominent abiotic challenge impacting plant growth and agricultural productivity significantly. In response to drought-induced challenges, plants have evolved a range of adaptive strategies, including the crucial role of

Table 6.1 miRNA associated with the plants and growth-promoting bacteria along with the mechanism of their action.

miRNA	Plant involved	Growth-promoting bacteria	Mechanism	References
miR156	Arabidopsis	*Rhizobium* sp.	Enhances root elongation and lateral branching	Wani et al. (2020)
miR167	Tomato	*Bacillus subtilis*	Improves nutrient uptake and root development	Zhang et al. (2022)
miR390	Rice	*Azospirillum brasilense*	Stimulates lateral root growth	Dong et al. (2022)
miR399	Arabidopsis	*Pseudomonas putida*	Enhances phosphate uptake and transport	Yang et al. (2021)
miR164	Maize	*Azotobacter vinelandii*	Promotes root hair elongation	Dexheimer and Cochella (2020)
miR172	Soybean	*Bradyrhizobium japonicum*	Regulates nodulation and symbiosis	Šečić et al. (2021)
miR160	Arabidopsis	*Agrobacterium tumefaciens*	Enhances auxin signaling and root formation	Meng et al. (2021)
miR396	Wheat	*Pseudomonas fluorescens*	Stimulates cell proliferation and growth	Zhang et al. (2022)
miR395	Brassica	*Sinorhizobium meliloti*	Improves sulfur assimilation and plant growth	Dexheimer and Cochella (2020)
miR319	Cotton	*Azospirillum lipoferum*	Enhances lateral shoot development	Deng et al. (2022)

regulating gene expression through small RNA molecules, notably microRNAs (miRNAs). As previously mentioned, miRNAs, which are short, noncoding RNA molecules, govern gene expression posttranscriptionally by either prompting the degradation of targeted messenger RNAs (mRNAs) or impeding their translation (Khare et al., 2017). Under the influence of drought stress, plants exhibit the remarkable ability to adjust the expression of specific miRNAs, thereby fine-tuning target gene expression that participate in stress response pathways. For instance, miR156, miR159, miR164, miR167, and miR169 have been identified as miRNAs exhibiting varying levels of expression in reaction to drought stress across a range of plant species, encompassing extensively researched models such as *Arabidopsis*, along with significant crops such as rice and maize. This dynamic regulatory network mediated by miRNAs allows plants to modulate their genetic responses to drought stress, ultimately enhancing their chances of survival and adaptation in water-limited environments. Understanding the complex interaction between miRNAs and their target genes in the context of drought stress response is crucial for enhancing our comprehension of plant resilience. Moreover, it holds significance for enhancing drought tolerance in crops through genetic engineering and breeding strategies (Singh et al., 2021).

Indeed, miRNAs play a vital role in regulating a plant's reaction to drought stress by precisely managing the expression of key genes involved in stress-responsive pathways. miR156 has been elucidated to play a crucial role in orchestrating the plant's reaction to drought stress. It achieves this by targeting a repertoire of transcription factors that wield authority over gene expression pivotal in responding to stress-induced conditions. Through this mechanism, miR156 finely tunes the plant's ability to cope with drought-related challenges. In a similar vein, miR159 takes on a central role by regulating MYB transcription factors, which are integral components of drought stress response pathways. Through regulating the expression of MYB genes, miR159 directly impacts the plant's ability to respond and adjust to drought stress conditions. miR164, on the other hand, emerges as a key player in the regulation of NAC transcription factors, pivotal actors in the plant's response to drought stress. Through its control over NAC expression, miR164 significantly Contributes to refining the plant's reaction to challenges induced by drought. Furthermore, miR167 and miR169 each have a unique role to play in this complex regulatory network. miR167 targets auxin response factors (ARFs), critical for modulating various elements of plant growth and development, encompassing reactions to drought stress. Simultaneously, miR169 steps in to regulate nuclear factor Y (NF−Y) transcription factors, which wield substantial influence over the plant's responses to drought stress (Banerjee, 2020).

Two of the most important and harmful abiotic stressors that impair plant growth and development are salinity and drought stress. To tackle these stressors, plants develop numerous coping strategies, among which is the regulation of gene expression through pathways mediated by miRNA. MiRNAs are essential regulators of plant stress responses because they suppress gene expression at the posttranscriptional level. In response to drought stress, plants adjust miRNA activity to regulate the expression of target genes associated with signaling pathways and mechanisms for adaptation to stress (Gull et al., 2019). For instance, a study in rice found that overexpressing the miR397a gene reduced water loss during drought stress and increased root biomass by controlling the laccase genes responsible for lignin biosynthesis. Similar to how miR396 affects target genes involved in cell division and growth, miR396 is reported to contribute to drought tolerance. The transcript of the F-box protein family gene, AUXIN SIGNALING F-BOX, is the target of miR393, one of the most thoroughly researched miRNAs involved in the response to salinity stress (AFB). The plant can detect and react to changes in its environment thanks to the role played by AFB as an auxin receptor. MiR393 is upregulated under salt stress conditions, which results in decreased expression of AFB and modifications to auxin signaling that support tolerance to salinity stress. MiR399 is a significant miRNA that plays a role in the reaction to salinity stress. The PHOSPHATE2 (PHO2) gene, which is involved in the control of phosphate uptake and homeostasis in plants, is expressed under the control of this miRNA. Under conditions of salt stress, miR399 is downregulated, which causes an increase in PHO2 expression and a decrease in phosphate uptake, enabling the plant to conserve resources and energy. Other miRNAs involved in the response to salinity stress include, among others, miR156, miR164, miR167, miR169, miR172, miR390, and miR393. These miRNAs control how plants respond to abiotic stress, such as salinity, as well as various features of their growth and development (Gahlaut et al., 2018; Yadav et al., 2020).

4.2 Thermal stress

Heat stress is recognized to be more fatal as it causes extensive damage to most of the organs in plants. High temperatures disturb cells' regular functioning and cease growth and development. The immediate result of stress-induction is the accumulation of harmful compounds such as protease inhibitors, phytohormones, chaperones, reactive oxygen species and late embryogenesis related proteins. Season-wide long thermal stress decreases seed production and weight as a result of reduced accumulation of proteins, carbohydrates and gluten. Plants during different phenophases possess different genetic mechanisms to tolerate the effects caused by heat based on its duration as well as intensity (Prasad et al., 2017).

Current studies have identified the importance of miRNA in thermal stress management by regulating the function of target genes through cleaving transcripts of targeted genes or inhibiting translation and controlling signaling pathways (Singh et al., 2021). When faced with thermal stress, various signaling pathways are triggered, leading to the varying production of multiple phytohormones. These hormones also regulate developmental processes, a distinctive trait of phytohormones. Although the transduction and metabolic pathways of phytohormones and miRNAs are different, miRNAs interplay with the transduction of phytohormones under thermal stress conditions. During thermal stress, miRNAs will be either up- or down-regulated; this differential expression with varied outputs was seen in different plant species. Generally, heat stress upregulated miRNAs downregulates the target gene; similarly, suppressed miRNAs function as positive regulators and result in upregulation of target genes. Different miRNAs are studied to understand their role in overcoming heat stress (Wani et al., 2020).

In plants, certain genes causes modulations in metabolic and physiological adaptations during heat stress, such as nuclear transcription factor Y subunit alpha (NFY A5), laccase, sulfate transporter and superoxide dismutase are targeted by certain classes of miRNAs (miR169, miR397, miR395 and miR398 respectively). Another group of miRNAs are involved in the morphological adaptations (leaf shape, size changes and root, shoot development) such as miR156, miR160/167, miR399, miR164 and miR319, which target Squamosa promoter binding-like (SPL), auxin response factor (ARF) TF, PHO2, NAC TF genes and teosinte branched/cycloidea protein (TCP), respectively (Gahlaut et al., 2018).

miRNA169 is evolutionarily conserved and widespread in plants. Scientifically proven target genes for miRNA169 are nuclear factors (Gahlaut et al., 2018). A nuclear transcription factor having NF-YA, NF-YB and NF-YC subunits functions in controlling the expression of 25% of genes. MiRNA169 shows its significant function in physiological adaptations of plants by maintaining transcription rate and controlling stomatal opening and closing (Ni et al., 2013). The sequence analysis of miRNA169 elucidated the presence of two proximate dehydration responsive elements. Upregulation of miRNA169 under heat stress is observed in wheat crops; besides, downregulation of miRNA169 is observed in *P. tomentose*. This alternative up-and-down regulation of miRNA 169 allows the differential expression of mRNAs of nuclear transcription factor YA5 (Zhou et al., 2016). miRNA397 negatively regulates its target gene, which is laccase (LAC). LAC gene codes for multicopper containing glycoprotein, which has a role in the maintenance of cell wall structure and improving cell-cell adhesion. During heat stress, downregulation of miRNA 397 will help increase the levels of glycoprotein (Roy et al., 2017). miRNA156 regulates the juvenile to young stage transition by modulating the time of flowering. It down regulates the expression of SPL, which results in the induction of genes such as FRUITFUL (FUL), LOCUS T (FT), and LEAFY (LFY) (Stief et al., 2014). miRNA156 exhibits contrasting expression patterns, where upregulation is seen in wheat, which down regulates its target SPL gene. In rice, a decrease in activity or expression is noted in response to heat stress (Sailaja et al., 2014). This class of miRNAs also possesses memory of heat stress response in Arabidopsis plants by ensuring sustainable expression of the target genes (Stief et al., 2014). miRNA164 plays a role in managing heat stress by controlling the expression of NAC TF genes, which are responsible for regulating various growth processes in plants such as root and shoot development, as well as leaf senescence (Jia et al., 2021). During heat stress, miRNA164 was reported to be upregulated and its target NAC family TFs were found to be downregulated in Arabidopsis (Li et al., 2014).

4.3 Oxidative stress

In case of plants, reactive oxygen species such as hydroxyl and superoxide radicals, as well as hydrogen peroxide are frequently generated as by-products during aerobic metabolism in cellular organelles such as mitochondria and chloroplasts; those are further neutralized by plant-specific defense systems, i.e., enzymatic and nonenzymatic antioxidant systems. Different external conditions will change the equilibrium between antioxidant enzymes and reactive oxygen species. This in turn results in the damage of macromolecules such as proteins, nucleic acids and lipids, which ultimately causes oxidative stress (Chaki et al., 2020). Other responses to this stress include redox imbalances. The repair for this condition is carried out by redox signaling networks, which adjust the redox homeostasis or reduce oxidative damage by modulating the target gene expression (Panda et al., 2015).

Recent studies have discovered that genome-encoded miRNAs are emerging as prominent regulators of target gene expression that are important in helping plants to acclimatize with stress. The first line of defense against ROS produced in response to oxidative stress is provided by superoxide dismutase enzymes. These enzymes work in the conversion of the most dangerous superoxide radicals into relatively stable and less toxic hydrogen peroxide. Cytosolic Cu/Zn superoxide dismutase (CSD1) and chloroplastic Cu/Zn superoxide dismutase (CSD2) enzymes are induced in response to oxidative stress (Yadav et al., 2020). miRNA398 plays a vital role in the regulatory pathway of these enzymes. This miRNA398 gets down regulated transcriptionally as a result of the oxidative burst, which leads to the accumulation of CSD 1 and 2 mRNAs, which leads to improved tolerance of oxidative stress (Iyer et al., 2012).

In a study conducted on rice seedlings that are subjected to hydrogen peroxide induced oxidative stress, five miRNAs were reported to be upregulated and two were downregulated. Upregulated miRNA subgroups include miRNA397, miRNA1425, miRNA169, miRNA827 and miRNA408-5p. miRNA397, which are highly expressed in the undifferentiated embryonic callus rather than in mature organs such as stems and leaves, are found to be upregulated in response to many environmental changes such as high temperatures, cold, and NaCl. miRNA397 downregulates its target gene corresponding to laccase, involved in copper homeostasis and lignin biosynthesis. These functions were comparatively nonessential and the energy will be used for defense purposes. miRNA1425 targets the PPR gene, which is seen in the higher plants and is involved in the processes of posttranscriptional modifications and organellar gene expression. While miRNA1425 is known to target only a small part of the rice PPR gene, it is efficient in controlling the biogenesis of organelles. Induction of this miRNA suppresses the expression of PPR genes, leading to reduced organellar biogenesis. miRNA827 and miRNA408-5P target the genes that encode for major facilitator superfamily proteins that allow different substrates across cytoplasmic membranes and the SPX domain protein that plays a role in nutrition homeostasis, respectively. Upregulation of miRNA827 and miRNA408-5P suppresses the levels of the following proteins, thereby reducing nutrient transport and consumption. Induction of miRNA169 downregulates the expression of HAP2-type transcription factor subunits. HAP2 shows its role in cell differentiation and respiration. Upregulation of miRNA169 suppresses or slows down the metabolism and developmental processes such as respiration and cell differentiation (Li et al., 2011).

In other instance, two miRNAs were downregulated in rice crops in action to hydrogen peroxide, which are miRNA319a.2 and miRNA528. MiRNA319a.2 targets the gene that codes for putative meta caspase. Putative meta caspase participate in the activation of PCD in plants. Reduction in the expression of miRNA319a.2 upregulates meta caspases, thereby activating PCD. miRNA528 regulates the gene encoding for IAR1 proteins. These proteins show their role in maintaining cellular free auxin, mainly indole-3 acetic acid. Since IAA is sensitive and readily prone to oxidative damage, suppression of miRNA528 induces IAR1 proteins. Further they cause the hydrolysis of IAA amino acid conjugates and maintain the levels of free IAA in the cells (Li et al., 2011).

4.4 Metal ion stress

The exceeding concentrations have led to heavy metal pollution in various areas including cultivation sites. Certain heavy metals are essential for the growth and development of plants but beyond a certain tolerable amount, they prove to be a major abiotic stress factor for plants. Heavy metal toxicity has been known to cause several hazardous effects in plants majorly by altering their metabolic mechanisms (Izzo et al., 2011). Some research has indicated that exposure to heavy metals can hinder or reduce the expression of a significant number of proteins, potentially resulting in complications. such as leaf senescence, weakened photosynthesis, suppressed root development, and necrosis (Singh et al., 2016). Heavy metals such as copper (Cu), Manganese (Mn), and zinc (Zn) in lower concentrations play essential roles in the biochemical reactions of plants. However, when these essential metals are present at higher concentrations than required, they are toxic for the plant's survival. Whereas, the nonessential metal ions such as aluminum (Al), mercury (Hg) and cadmium (Cd) are toxic for the plants even in lower concentrations (Gielen et al., 2012). To avoid and overcome the destructive complications caused by metal ions stress, plants have developed their own coping mechanisms (Singh et al., 2016). One of such mechanism is the role of miRNAs as major regulators of the posttranscriptional expression of genes. Recent findings have found that miRNAs alter the numerous gene expressions during the growth cycle as a response to the abiotic stress caused by metal ion toxicity (Shriram et al., 2016).

To ascertain the functions and interactions of their targets, it is imperative to examine specific miRNAs' roles. Metal transporters are important complexes in plants, which are responsible for the uptake and translocation of metals (Clemens, 2001). Studies have revealed that the absorption and movement of cadmium (Cd) in plants are facilitated by particular sets of metal transporters that are controlled by miRNAs. Research indicates that a total of 84 miRNAs, both conserved and nonconserved, participate in this regulatory process. The miRNAs reportedly target specific genes of the transporter leading to their cleavage and thus suppressing the uptake of Cadmium (Zhou et al., 2012). Zhou et al. discovered that miR159 and miR167 targeted the genes encoding an ABC transporter involved in Cd uptake in *B. napus*. Another mode of action of miRNAs involves metal chelation. The chelation of heavy metals by ligands is a key mechanism for the tolerance and further detoxification of heavy metals in plants. Sequencing performed in plants such as *B. napus* has identified certain genes, which are a part of the metal chelation pathway and were targeted by various miRNAs (Min Yang and Chen, 2013). An example of such miRNAs is miR156 whose expression is upregulated in response to Cd toxicity. It targeted the gene coding for glutathione-γ-glutamyl cysteinyl transferase (GGT) that forms metal complexes, which are transported and detoxified in the vacuole for detoxification (Yang et al., 2013). It has also been known that heavy metal stress is linked with the accumulation of reactive oxygens species (ROS) causing destructive effects on proteins, nucleic acids and lipids (Foyer et al., 1994). There is emerging evidence that shows that miRNAs play a vital role in regulating posttranscriptional genes associated with antioxidant response pathways (Sunkar et al., 2006). miR398 is an extensively studied miRNA, which is known to target copper (Cu) and zinc (Zn) superoxide dismutase that function in the conversion of ROS to H_2O_2 (Sunkar et al., 2006). The miR398 family in Arabidopsis is known to have crucial roles in copper (Cu) ion toxicity. It inhibits Cu uptake by using mechanisms such as transcript cleavage and translational inhibition (Sunkar et al., 2006). With the use of technologies such as RT PCR and biological information prediction, it was discovered that a particular set of miRNAs, namely miR171, miR319 and miR393 participated in mercury (Hg) stress response (Zhou et al., 2008).

4.5 Radiation stress

Sunlight is a necessity for the growth and survival of plants. But some intense radiations such as ultraviolet and gamma (γ) rays are harmful to plants. Radiation stress includes the abiotic stress caused mainly due to ultraviolet (UV) radiations. These radiations induce various plant reactions, including alterations in yield and growth, disruptions in cellular metabolism, and modifications in gene expression (Singh et al., 2020). The major defect caused by UV radiations is damage to nucleic acids such as DNA or RNA and buildup of ROS. Ultraviolet-B (UV−B) radiation, a type of ultraviolet radiation is most harmful to plants. Being a component of solar radiations, higher exposure levels of UV-B causes defects in the growth and development of plants (Gill et al., 2015); while a low dose exposure has been shown to enhance the production of secondary metabolites in plants (Gu et al., 2010). Recently, miRNAs responsive to UV-B exposures have been discovered in wheat. One novel miR6000 and others known as miR156, miR159, miR164, miR167, miR171, and miR395 were identified to be responsive to treatments of UV-B radiations. The expression of some miRNAs was upregulated while some were downregulated in response to UV-B radiations. Their expression levels can then be correlated with the cellular mechanisms they play a role in (Wang et al., 2013). In another study performed on Chrysanthemum leaves revealed a total of 137 miRNAs, which showed differential expression when exposed to UV-B radiation (Yang et al., 2020). Out of the 137 miRNAs identified, 67 were novel. By predicting the biological processes, the miRNAs were involved in, it was found that their major target was three main glycolysis genes, pyruvate kinase, glucan endo-1,3-beta-D-glucosidase and glyceraldehyde-3-phosphate dehydrogenase. The major mechanism by which the miRNAs suppressed the stress caused by radiation was related to the synthesis of bioactive ingredients of pathways. Other ways by which miRNAs functioned was related to biosynthesis of flavonoids, which suggests that low dose of UV-B radiations are required for accumulation of flavonoids (Yang et al., 2020). Gamma (γ) rays are potentially more harmful when compared with UV radiations as they can easily penetrate through various living cells including plant cells. Thus, they have a more dangerous or deleterious effect on organisms. γ- IR have been previously used on plants to create mutation, which enhance traits such as rate of seed germination, yield percentage, and biochemical content. Current investigations suggest that a low dose irradiation has energy enhancing effects on plants when compared with high-dose irradiations (Zaichkina et al., 2004). During this exposure of harmful irradiation, the known mechanisms opted by plants is altering the biochemical or physiological processes in order to alter the expression of certain genes (Seki et al., 2001). In a study performed by Subburaj (2017), they found 37 miRNAs, which showed altered expression under γ- IR exposure in *Tradescantia.* mR164, miR166k, miR169b, and miR172d showed upregulation in their concentrations when exposed to γ- IR rays. The miRNAs found in this study were also found to be highly conserved across several species of plants with 1–3 nucleotide variations suggesting common ancestry.

5. miRNA networks in biotic stress management

Biotic stress is a significant concern, which hinders plant growth due to interaction with plant pathogens. Studies have shown that during a pathogen attack, differential expression of several miRNA networks is triggered (Kulshrestha, 2020). Plant evolution shows that they have developed several mechanisms for defense against pathogens. One such mechanism is through miRNA networking that enhances the combatting power of plants against pathogenic attacks (Yang et al., 2021). As shown in Table 6.2, a variety of miRNA networks are involved in the management of several pathogenic attacks. Pathogenic attacks on plants are generally bacterial, fungal and viral. Bacterial plant pathogens are observed to cause immense economic losses. The pathogenic manifestation corresponds to spots, pustules, hormonal imbalance, the proliferation of plant cells or mosaic patterns (Muñoz et al., 2021). Evidence suggests that a decrease in miRNA population is a natural depiction during a bacterial pathogenic attack (Wang et al., 2021). However, the miRNA network of miR397, miR398 and miR408 families increased during exposure to certain bacterial pathogens such as *Agrobacterium* (Carbonell et al., 2019). miR408 is a conserved miRNA in plants that have existed for a long duration and holds a key role in biotic stress response (Gao et al., 2022). The tolerance to stress including aversion to biotic stress is regulated by miRNA through its ability to downregulate the target genes and increase the antioxidant formation in plants (Banerjee, 2020). The most common plant pathogens are fungi that cause significant damage to plants. However, miRNA networks play a key role in managing biotic stress. One of the largest miRNA regulatory networks is the *NBS-LRR*-miR482/2118 network, which is involved in enhancing plant immunity by strengthening the plant's ability to resist the disease (Pastor et al., 2019; Jiang et al., 2019). Therefore, the extent of the miRNA network expression impacts the degree of resistance depicted by different plant species at different development stages. One of the most damaging fungal diseases observed in plants is the stripe rust disease whose causative agent is *Puccinia striiformis.* The network of the miR408 family, being upregulated during the fungal attack, is associated with the initiation of the lignin biosynthetic pathway, which strengthens the epidermal layer of the plant, preventing its rupture (Feng et al.,

Table 6.2 Different miRNA signatures associated as a defense mechanism against different pathogens involving multistep molecular mechanisms.

miRNA	Pathogen	Mechanism	References
miR398	*Pseudomonas syringae*	Activation of defense genes	Campo et al. (2021)
miR398	*Beet necrotic yellow vein virus*	Gene silencing and regulating reactive oxygen intermediate production	Liu et al. (2020)
miR408	*Puccinia striiformis*	Activates lignin biosynthetic pathway	Feng et al. (2022)
miR164	*Verticillium dahliae*	Breaks down mRNA NAC 100 to gain disease resistance	Hu (2020)
miR159	Sugarcane mosaic virus	Strengthens immunity	Ling et al. (2019)
miR159, miR838, miR854, miR5658 and csa-miRn6-3p	Cucumber green mottle mosaic virus	Target gene silencing	Liang et al. (2019)
miR159	Sugarcane yellow leaf virus	RNAi-based gene silencing	Ashraf et al. (2021)
ath-miR164c	*Pseudomonas syringae*	Regulation of proline metabolism	Gupta et al. (2020)
miR171b	*L. asiaticus*	Downregulation of target genes to promote resistance	
miR164 and miR397	*Verticillium dahliae*	Expression suppression of *OsAPS1*, leading to broad-spectrum resistance	Mei et al. (2022)

2022). The defense against fungal pathogens is not dependent upon strengthening the epidermal layer but also enhancing immunity through regulatory miRNA networks (Rose et al., 2019). The plant hormone involved in enhancing immunity corresponds to auxin signaling pathways, which are regulated by the miRNA network of miR160, miR167 and miR393 (Luo et al., 2022). With regards to vascular plants, the miRNA network of miR159 and miR166 depicts immunity against the fungal plant pathogen *Verticillium dahliae* by modulating the plant's immune response (Gabriel et al., 2019).

Plants respond to biotic stress through auxin homeostasis (Devireddy et al., 2021). One of the major miRNAs involved in auxin homeostasis is miR164, which interacts with five different members of the NAC domain associated with transcription factors of various plants (Iqbal et al., 2022). This is further reflected in the study by Hu et al., which depicted upregulation by miR164 in cotton plant when infected with *Verticillium dahliae* (Hu, 2020). The mechanism corresponded to miR164 breaking the mRNA NAC100 as a response to the infection, which conferred the plant with resistance to *Verticillium dahliae*, thus strengthening the defense of the plant. In an investigation by Sanz-Carbonell et al., the heat map of a melon plant infected by HSVd revealed that out of the 20 miRNAs, 15 were down-regulated (Carbonell et al., 2020). Therefore, overexpression of the 15 down-regulated miRNA will lead to susceptibility to viral infection. The study by Liu et al., deals with species-specific miRNA network that involves strengthening the immunity of the plant against viral attack (Liu et al., 2020). The study discovered that the upregulation of miR398 acts as the first line of defense against viral infection. The target gene of miR398 corresponds to *umecyanin*, which is knocked down with the activation of miR398. This leads to the plant acquiring systematic resistance against the pathogen and defend itself against the virus attack. In addition to conferring disease resistance, the study showed that the regulatory network of miR398 regulates the production of reactive oxygen intermediates such as O_2^-. The production of reactive intermediates in high concentrations damages the viral nucleic acids, thus preventing viral replication. miR398 regulates production by lowering the plant's ability to inhibit the reactive oxygen intermediates. Therefore, miRNA networks are essential in facilitating the plant to fight against biotic stress caused by viral attacks. In contrary to usual beliefs, ROS is formed in the host due to a bacterial or fungal infection (Silwal et al., 2020). Plants, being susceptible to pathogenic attacks are prone to ROS formation that damages the cells and leads to plant death. Studies have depicted that the miRNA network involving miR398 is down-regulated as a response to *Pseudomonas syringae* (Li et al., 2022). In other words, overexpression of this miRNA renders plants such as Arabidopsis exposed to *Pseudomonas syringae* infections. However, when a plant is infected by *M. oryzae,* miR398 responds by being upregulated and thus, overexpression of the miRNA provided plants such as *Oryzae sativa* with resistance against the fungal pathogen as well as activated defense genes (Campo et al., 2021).

6. Future prospectives and conclusion

Certainly, the study of microRNAs (miRNAs) in plants as integral players in stress management has seen remarkable progress in recent years. This advancement has been made possible through the utilization of cutting-edge technologies, along with significant developments in bioinformatics and genome editing. These breakthroughs have deepened our comprehension of how miRNAs regulate stress responses in plants and hold great promise for agricultural enhancement. Amidst evolving environmental circumstances. One of the key insights gained is that miRNAs are central to plants' ability to cope with various stressors, acting as crucial regulators of stress-responsive genes and signaling pathways. To counter environmental stressors, plants exhibit specific miRNAs that are either induced or suppressed. These stress-responsive miRNAs function as master regulators, precisely managing the expression of target genes pivotal for stress tolerance and defense mechanisms. Through binding to complementary sequences in messenger RNAs (mRNAs), miRNAs can either inhibit translation or promote mRNA degradation. This fine-tuned regulatory mechanism enables miRNAs to change the expression of genes involved in stress-response, thus facilitating plants in effectively dealing with challenging environmental conditions. These regulatory functions span diverse aspects of plant responses to stress, including ion transport, osmotic adjustment for salinity stress, activation of heat shock and other stress-related proteins under heat stress, modulation of cold-responsive genes to enhance freezing tolerance, and control of genes associated with defense mechanisms to repel pathogen invasions(Carbonell et al., 2019).

The discovery and characterization of stress-responsive miRNAs and their target genes have yielded invaluable insights into the intricate networks governing plant stress responses. This knowledge has the potential to be harnessed for the development of stress-tolerant crops through genetic engineering or selective breeding programs. By manipulating miRNA expression or activity, scientists aim to bolster plant resilience and productivity in challenging environmental conditions. Recent strides in miRNA research in plants have been catalyzed by advances in high-throughput sequencing technologies, such as next-generation sequencing. These technologies empower researchers to comprehensively profile miRNAs on a genome-wide scale, offering a holistic view of miRNA expression patterns and their regulatory networks under stress conditions. Moreover, bioinformatics tools and algorithms have greatly simplified the characterization and analysis of miRNA sequencing data. These tools facilitate the prediction of miRNA target genes and the identification of enriched functional pathways associated with stress responses. Integration of miRNA expression data with omics datasets, related to transcriptomics and proteomics, has further enriched the comprehension of the intricate regulatory networks underpinning plant stress responses. Recent studies have delved deeper into elucidating the molecular mechanisms of miRNA-mediated stress responses. Researchers have pinpointed specific transcription factors and other regulatory proteins that govern the expression of stress-responsive miRNAs. This deeper understanding sheds light on the intricate gene regulatory networks involved in stress tolerance and opens up new avenues for precise manipulation of miRNA expression to improve stress tolerance in plants (Luo et al., 2022).

Advancements in genome editing technologies, particularly CRISPR-Cas9, have enabled the precise manipulation of miRNA genes in plants. This revolutionary technology allows scientists to investigate the functional roles of specific miRNAs by creating knockouts or overexpression lines. These experiments offer concrete proof of how miRNAs influence stress responses, affirming their potential as targets for enhancing crop quality. Furthermore, the application of miRNAs as biotechnological tools for enhancing tolerance to stress in crops represents a burgeoning area of research. Scientists are exploring the use of synthetic miRNAs or miRNA mimics to modulate the expression of stress-responsive genes and improve plant performance under adverse conditions. This innovative approach holds great promise for developing stress-tolerant crop varieties with increased yields and resilience to environmental challenges (Jiang et al., 2019).

In conclusion, the study of miRNAs in plants for stress management has witnessed extraordinary advancements in recent years. High-throughput sequencing technologies, bioinformatics tools, molecular mechanisms, genome editing, and biotechnological applications have collectively propelled our understanding of miRNAs' roles in regulating stress responses. These innovations have deepened our insights into the complex regulatory networks governing plant stress tolerance, paving the way for targeted miRNA pathway manipulation to improve stress tolerance in crops (Ghosh et al., 2022). Furthermore, the application of synthetic miRNAs or miRNA mimics holds the potential to revolutionize stress management in agriculture. By fine-tuning the expression of stress-responsive genes, it becomes possible to augment plant performance and yield under challenging environmental conditions. Overall, the recent strides in miRNA research in plants for stress management have brought us closer to realizing the potential of these small RNA molecules in improving crop quality and promoting sustainability in agriculture. Continued research in this field, alongside the integration of emerging technologies, will undoubtedly contribute to the creation of crops resilient to stress, capable of flourishing in a shifting climate, and securing food provisions for forthcoming generations.

Acknowledgments

Sagnik Nag, as the leading author, wants to acknowledge the Graduate Research Excellence Scholarship (GRES) conferred by the management of Monash University (Malaysia Campus) and Jeffrey Cheah School of Medicine and Health Sciences (JCSMHS), supporting his overall academic and research pursuits. The authors also wish to express gratitude to their respective departments and parent institutions for their overall support.

References

Abubakar, Y., Tijjani, H., Egbuna, C., Adetunji, C.O., Kala, S., Kryeziu, T.L., Patrick-Iwuanyanwu, K.C., 2019. Pesticides, history, and classification. In: Natural Remedies for Pest, Disease and Weed Control, pp. 29−42. https://doi.org/10.1016/B978-0-12-819304-4.00003-8.

Alengebawy, A., Abdelkhalek, S.T., Qureshi, S.R., Wang, M.Q., 2021. Heavy metals and pesticides toxicity in agricultural soil and plants: ecological risks and human health implications. Toxics 9 (3), 1−34. https://doi.org/10.3390/TOXICS9030042.

Álvarez, S., Acosta-Motos, J.R., 2022. Miscellaneous sets of abiotic stresses and plant strategies to cope with them. Agronomy 12 (11), 2727. https://doi.org/10.3390/AGRONOMY12112727.

Ashraf, M.A., Ashraf, F., Feng, X., Hu, X., Shen, L., Khan, J., Zhang, S., 2021. Potential targets for evaluation of sugarcane yellow leaf virus resistance in sugarcane cultivars: in silico sugarcane miRNA and target network prediction. Biotechnol. Biotechnol. Equip. 35 (1), 1990−2001. https://doi.org/10.1080/13102818.2022.2041483/SUPPL_FILE/TBEQ_A_2041483_SM6157.ZIP.

Bajczyk, M., Jarmolowski, A., Jozwiak, M., Pacak, A., Pietrykowska, H., Sierocka, I., Swida-Barteczka, A., Szewc, L., Szweykowska-Kulinska, Z., 2023. Recent insights into plant miRNA biogenesis: multiple layers of miRNA level regulation. Plants 12 (2). https://doi.org/10.3390/PLANTS12020342.

Banerjee, P., 2020. Plant abiotic stress responses and MicroRNAs. Adv. Agric. https://doi.org/10.30954/NDP-ADVAGR.2020.6.

Bhattacharya, M., Chatterjee, S., Nag, S., Dhama, K., Chakraborty, C., 2022. Designing, characterization, and immune stimulation of a novel multi-epitopic peptide-based potential vaccine candidate against monkeypox virus through screening its whole genome encoded proteins: an immunoinformatics approach. Trav. Med. Infect. Dis. 50. https://doi.org/10.1016/J.TMAID.2022.102481.

Bukhat, S., Imran, A., Javaid, S., Shahid, M., Majeed, A., Naqqash, T., 2020. Communication of plants with microbial world: exploring the regulatory networks for PGPR mediated defense signaling. Microbiol. Res. 238. https://doi.org/10.1016/J.MICRES.2020.126486.

Calicioglu, O., Flammini, A., Bracco, S., Bellù, L., Sims, R., 2019. The future challenges of food and agriculture: an integrated analysis of trends and solutions. Sustainability 11 (1), 222. https://doi.org/10.3390/SU11010222.

Cambiagno, D.A., Giudicatti, A.J., Arce, A.L., Gagliardi, D., Li, L., Yuan, W., Lundberg, D.S., Weigel, D., Manavella, P.A., 2021. HASTY modulates miRNA biogenesis by linking pri-miRNA transcription and processing. Mol. Plant 14 (3), 426–439. https://doi.org/10.1016/J.MOLP.2020.12.019.

Campo, S., Sánchez-Sanuy, F., Camargo-Ramírez, R., Gómez-Ariza, J., Baldrich, P., Campos-Soriano, L., Soto-Suárez, M., San Segundo, B., 2021. A novel Transposable element-derived microRNA participates in plant immunity to rice blast disease. Plant Biotechnol. J. 19 (9), 1798–1811. https://doi.org/10.1111/PBI.13592.

Canto-Pastor, A., Santos, B.A.M.C., Valli, A.A., Summers, W., Schornack, S., Baulcombe, D.C., 2019. Enhanced resistance to bacterial and oomycete pathogens by short tandem target mimic RNAs in tomato. Proc. Natl. Acad. Sci. U.S.A. 116 (7), 2755–2760. https://doi.org/10.1073/PNAS.1814380116.

Chaki, M., Begara-Morales, J.C., Barroso, J.B., 2020. Oxidative stress in plants. Antioxidants 9 (6). https://doi.org/10.3390/ANTIOX9060481.

Chatterjee, S., Bhattacharya, M., Nag, S., Dhama, K., Chakraborty, C., 2023. A detailed overview of SARS-CoV-2 omicron: its sub-variants, mutations and pathophysiology, clinical characteristics, immunological landscape, immune escape, and therapies. Viruses 15 (1). https://doi.org/10.3390/V15010167.

Chen, S., Ren, G., 2019. Analysis of methylation status of plant MicroRNAs. Methods Mol. Biol. (Clifton, N.J.) 1932, 285–293. https://doi.org/10.1007/978-1-4939-9042-9_21.

Clemens, S., 2001. Molecular mechanisms of plant metal tolerance and homeostasis. Planta 212 (4), 475–486. https://doi.org/10.1007/S004250000458.

De Vries, S., De Vries, J., Rose, L.E., 2019. The elaboration of miRNA regulation and gene regulatory networks in Plant–Microbe interactions. Genes 10 (4). https://doi.org/10.3390/GENES10040310.

Deng, C., Liang, X., Zhang, N., Li, B., Wang, X., Zeng, N., 2022. Molecular mechanisms of plant growth promotion for methylotrophic Bacillus aryabhattai LAD. Front. Microbiol. 13. https://doi.org/10.3389/FMICB.2022.917382.

Devireddy, A.R., Zandalinas, S.I., Fichman, Y., Mittler, R., 2021. Integration of reactive oxygen species and hormone signaling during abiotic stress. Plant J. : Cell. Mol. Biol. (Sarreguemines, Fr., Online) 105 (2), 459–476. https://doi.org/10.1111/TPJ.15010.

Dexheimer, P.J., Cochella, L., 2020. MicroRNAs: from mechanism to organism. Front. Cell Dev. Biol. 8. https://doi.org/10.3389/FCELL.2020.00409.

Dong, Q., Hu, B., Zhang, C., 2022. microRNAs and their roles in plant development. Front. Plant Sci. 13. https://doi.org/10.3389/FPLS.2022.824240.

Feng, Y., Qi, N., Lei, P., Wang, Y., Xuan, Y., Liu, X., Fan, H., Chen, L., Duan, Y., Zhu, X., 2022. Gma-miR408 enhances soybean cyst nematode susceptibility by suppressing reactive oxygen species accumulation. Int. J. Mol. Sci. 23 (22). https://doi.org/10.3390/IJMS232214022.

Foyer, C.H., Descourvières, P., Kunert, K.J., 1994. Protection against oxygen radicals - an important defense-mechanism studied in transgenic plants. Plant Cell Environ. 17 (5), 507−523. https://doi.org/10.1111/J.1365-3040.1994.TB00146.X.

Gabriel, A.F., Costa, M.C., Enguita, F.J., Leitão, A.L., 2019. Si vis pacem para bellum: A prospective in silico analysis of miRNA-based plant defenses against fungal infections. Plant Sci. 288. https://doi.org/10.1016/J.PLANTSCI.2019.110241.

Gahlaut, V., Baranwal, V.K., Khurana, P., 2018. miRNomes involved in imparting thermotolerance to crop plants. 3 Biotech 8 (12). https://doi.org/10.1007/S13205-018-1521-7.

Gangwar, P., Johri, P., Datta, R., Singh, S., Trivedi, M., 2023. Overview on 1 phytopharmaceuticals and biotechnology of herbal plants. In: Phytopharmaceuticals and Biotechnology of Herbal Plants. CRC Press, pp. 1−17.

Gao, Y., Feng, B., Gao, C., Zhang, H., Wen, F., Tao, L., Fu, G., Xiong, J., 2022. The evolution and functional roles of miR408 and its targets in plants. Int. J. Mol. Sci. 23 (1). https://doi.org/10.3390/IJMS23010530.

Ghosh, A., Nag, S., Gomes, A., Gosavi, A., Ghule, G., Kundu, A., Purohit, B., Srivastava, R., 2022. Applications of smart material sensors and soft electronics in healthcare wearables for better user compliance. Micromachines 14 (1). https://doi.org/10.3390/MI14010121.

Ghosh, N., Chacko, L., Bhattacharya, H., Vallamkondu, J., Nag, S., Dey, A., Karmakar, T., Reddy, P.H., Kandimalla, R., Dewanjee, S., 2023. Exploring the complex relationship between diabetes and cardiovascular complications: understanding diabetic cardiomyopathy and promising therapies. Biomedicines 11 (4). https://doi.org/10.3390/BIOMEDICINES11041126.

Gielen, H., Remans, T., Vangronsveld, J., Cuypers, A., 2012. MicroRNAs in metal stress: specific roles or secondary responses? Int. J. Mol. Sci. 13 (12), 15826−15847. https://doi.org/10.3390/IJMS131215826.

Gill, S.S., Anjum, N.A., Gill, R., Jha, M., Tuteja, N., 2015. DNA damage and repair in plants under ultraviolet and ionizing radiations. Sci. World J. 2015. https://doi.org/10.1155/2015/250158.

Gu, X.D., Sun, M.Y., Zhang, L., Fu, H.W., Cui, L., Chen, R.Z., Zhang, D.W., Tian, J.K., 2010. UV-B induced changes in the secondary metabolites of Morus alba L. leaves. Molecules (Basel, Switzerland) 15 (5), 2980−2993. https://doi.org/10.3390/MOLECULES15052980.

Gull, A., Lone, A.A., Wani, N.U.I., Gull, A., Lone, A.A., Wani, N.U.I., 2019. Biotic and abiotic stresses in plants. In: Abiotic and Biotic Stress in Plants. https://doi.org/10.5772/INTECHOPEN.85832.

Gupta, A., Patil, M., Qamar, A., Senthil-Kumar, M., 2020. ath-miR164c influences plant responses to the combined stress of drought and bacterial infection by regulating proline metabolism. Environ. Exp. Bot. 172, 103998. https://doi.org/10.1016/J.ENVEXPBOT.2020.103998.

Hang, N., Shi, T., Liu, Y., Ye, W., Taier, G., Sun, Y., Wang, K., Zhang, W., 2021. Overexpression of Os-microRNA408 enhances drought tolerance in perennial ryegrass. Physiol. Plantarum 172 (2), 733−747. https://doi.org/10.1111/PPL.13276.

Harris, J.M., Bede, J., Tsuda, K., 2022. Focus on the role of the abiotic environment on interactions between plants and microbes. Mol. Plant Microbe Interact. : MPMI (Mol. Plant-Microbe Interact.) 35 (7), 510. https://doi.org/10.1094/MPMI-04-22-0099-FI.

Ho, Y.-N., Mathew, D.C., Huang, C.-C., 2017. Plant-microbe ecology: interactions of plants and symbiotic microbial communities. In: Plant Ecology - Traditional Approaches to Recent Trends. https://doi.org/10.5772/INTECHOPEN.69088.

Hoque, S., Dhar, R., Kar, R., Mukherjee, S., Mukherjee, D., Mukerjee, N., Nag, S., Tomar, N., Mallik, S., 2023. Cancer stem cells (CSCs): key player of radiotherapy resistance and its clinical significance. Biomarkers: Biochemical Indicators of Exposure, Response, and Susceptibility to Chemicals 28 (2), 139–151. https://doi.org/10.1080/1354750X.2022.2157875.

Hu, 2020. The ghr-miR164 and GhNAC100 modulate cotton plant resistance against *Verticillium dahlia*. Plant Sci. 293, 110438.

Iqbal, A., Bocian, J., Hameed, A., Orczyk, W., Nadolska-Orczyk, A., 2022. Cis-regulation by NACs: a promising frontier in wheat crop improvement. Int. J. Mol. Sci. 23 (23). https://doi.org/10.3390/IJMS232315431.

Iyer, N.J., Jia, X., Sunkar, R., Tang, G., Mahalingam, R., 2012. microRNAs responsive to ozone-induced oxidative stress in *Arabidopsis thaliana*. Plant Signal. Behav. 7 (4), 484–491. https://doi.org/10.4161/PSB.19337.

Jia, D., Jiang, Z., Fu, H., Chen, L., Liao, G., He, Y., Huang, C., Xu, X., 2021. Genome-wide identification and comprehensive analysis of NAC family genes involved in fruit development in kiwifruit (Actinidia). BMC Plant Biol. 21 (1). https://doi.org/10.1186/S12870-020-02798-2.

Jiang, N., Cui, J., Yang, G., He, X., Meng, J., Luan, Y., 2019. Comparative transcriptome analysis shows the defense response networks regulated by miR482b. Plant Cell Rep. 38 (1), 1–13. https://doi.org/10.1007/S00299-018-2344-Z.

Kansal, S., Panwar, V., Mutum, R.D., Raghuvanshi, S., 2021. Investigations on regulation of MicroRNAs in rice reveal [Ca2+]cyt signal transduction regulated MicroRNAs. Front. Plant Sci. 12, 2239. https://doi.org/10.3389/FPLS.2021.720009/BIBTEX.

Khare, P., Kishore, K., Sharma, D.K., 2017. A study on the standardization parameters of Cassia angustifolia. Asian J. Pharmaceut. Clin. Res. 10 (7), 329–332.

Kulshrestha, 2020. Elucidating micro RNAs role in different plant–pathogen interactions. Mol. Biol. Rep. 47 (10), 8219–8227.

Kumar, A., Kaur, R., Rajam, M.V., 2021. Regulatory role of micro-RNAs in plants under challenging environmental conditions with special focus on drought and salinity. In: Harsh Environment and Plant Resilience, pp. 121–140. https://doi.org/10.1007/978-3-030-65912-7_6.

Li, J., Song, Q., Zuo, Z.F., Liu, L., 2022. MicroRNA398: a master regulator of plant development and stress responses. Int. J. Mol. Sci. 23 (18). https://doi.org/10.3390/IJMS231810803.

Li, S., Liu, J., Liu, Z., Li, X., Wu, F., He, Y., 2014. HEAT-INDUCED TAS1 TARGET1 mediates thermotolerance via HEAT STRESS TRANSCRIPTION FACTOR A1a-directed pathways in Arabidopsis. Plant Cell 26 (4), 1764–1780. https://doi.org/10.1105/TPC.114.124883.

Li, T., Li, H., Zhang, Y.X., Liu, J.Y., 2011. Identification and analysis of seven H_2O_2-responsive miRNAs and 32 new miRNAs in the seedlings of rice (*Oryza sativa* L. ssp. indica). Nucleic Acids Res. 39 (7), 2821–2833. https://doi.org/10.1093/NAR/GKQ1047.

Liang, C., Liu, H., Hao, J., Li, J., Luo, L., 2019. Expression profiling and regulatory network of cucumber microRNAs and their putative target genes in response to cucumber green mottle mosaic virus infection. Arch. Virol. 164 (4), 1121–1134. https://doi.org/10.1007/S00705-019-04152-W.

Ling, H., Huang, N., Xu, L., Peng, Q., Liu, F., Yang, Y., Que, Y., 2019. Suitable reference genes/miRNAs for qRT-PCR normalization of expression analysis in sugarcane under Sorghum mosaic virus infection. Sugar Tech. 21 (5), 780–793. https://doi.org/10.1007/S12355-019-00712-1.

Liu, J., Fan, H., Wang, Y., Han, C., Wang, X., Yu, J., Li, D., Zhang, Y., 2020. Genome-wide microRNA profiling using oligonucleotide microarray reveals regulatory networks of microRNAs in Nicotiana benthamiana during beet necrotic yellow vein virus infection. Viruses 12 (3). https://doi.org/10.3390/V12030310.

Liu, J., Liu, X., Zhang, S., Liang, S., Luan, W., Ma, X., 2021. TarDB: an online database for plant miRNA targets and miRNA-triggered phased siRNAs. BMC Genom. 22 (1). https://doi.org/10.1186/S12864-021-07680-5.

Lü, X., Zhang, W., Zhang, H., Liang, Z., Chen, H., 2022. Advances of miRNA-mediated regulatory roles in plant-microbe interaction. Sheng Wu Gong Cheng Xue Bao/Chin. J. Biotechnol. 38 (5), 1695–1705. https://doi.org/10.13345/J.CJB.210580.

Luo, G., Najafi, J., Correia, P.M.P., Trinh, M.D.L., Chapman, E.A., Østerberg, J.T., Thomsen, H.C., Pedas, P.R., Larson, S., Gao, C., Poland, J., Knudsen, S., DeHaan, L., Palmgren, M., 2022. Accelerated domestication of new crops: yield is key. Plant Cell Physiol. 63 (11), 1624–1640. https://doi.org/10.1093/PCP/PCAC065.

Malhi, G.S., Kaur, M., Kaushik, P., 2021. Impact of climate change on agriculture and its mitigation strategies: a review. Sustainability 13 (3), 1318. https://doi.org/10.3390/SU13031318.

Meddya, S., Meshram, S., Sarkar, D., Datta, R., Singh, S., Avinash, G., Thulasinathan, T., 2023. Plant stomata: an unrealized possibility in plant defense against invading pathogens and stress tolerance. Plants 12 (19), 3380.

Mei, J., Wu, Y., Niu, Q., Miao, M., Zhang, D., Zhao, Y., Cai, F., Yu, D., Ke, L., Feng, H., Sun, Y., 2022. Integrative analysis of expression profiles of mRNA and MicroRNA provides insights of cotton response to Verticillium dahliae. Int. J. Mol. Sci. 23 (9). https://doi.org/10.3390/IJMS23094702.

Mencia, R., Gonzalo, L., Tossolini, I., Manavella, P.A., 2023. Keeping up with the miRNAs: current paradigms of the biogenesis pathway. J. Exp. Bot. 74 (7). https://doi.org/10.1093/JXB/ERAC322.

Meng, X., Li, A., Yu, B., Li, S., 2021. Interplay between miRNAs and lncRNAs: mode of action and biological roles in plant development and stress adaptation. Comput. Struct. Biotechnol. J. 19, 2567–2574. https://doi.org/10.1016/J.CSBJ.2021.04.062.

Mewalal, R., Yin, H., Hu, R., Jawdy, S., Vion, P., Tuskan, G.A., Tacon, F. Le, Labbé, J.L., Yang, X., 2019. Identification of populus small RNAs responsive to mutualistic interactions with mycorrhizal fungi, laccaria bicolor and rhizophagus irregularis. Front. Microbiol. 10 (MAR). https://doi.org/10.3389/FMICB.2019.00515.

Min Yang, Z., Chen, J., 2013. A potential role of microRNAs in plant response to metal toxicity. Metallomics: Integrated Biometal Science 5 (9), 1184–1190. https://doi.org/10.1039/C3MT00022B.

Miryeganeh, M., 2021. Plants' epigenetic mechanisms and abiotic stress. Genes 12 (8). https://doi.org/10.3390/GENES12081106.

Muñoz, M., Cieniewicz, E., Faust, J.E., 2021. Diseases and disease management. In: Cut Flowers and Foliages, pp. 258–315. https://doi.org/10.1079/9781789247602.0006.

Ni, Z., Hu, Z., Jiang, Q., Zhang, H., 2013. GmNFYA3, a target gene of miR169, is a positive regulator of plant tolerance to drought stress. Plant Mol. Biol. 82 (1–2), 113–129. https://doi.org/10.1007/S11103-013-0040-5.

Ons, L., Bylemans, D., Thevissen, K., Cammue, B.P.A., 2020. Combining biocontrol agents with chemical fungicides for integrated plant fungal disease control. Microorganisms 8 (12), 1–19. https://doi.org/10.3390/MICROORGANISMS8121930.

Owusu Adjei, M., Zhou, X., Mao, M., Rafique, F., Ma, J., 2021. MicroRNAs roles in plants secondary metabolism. Plant Signal. Behav. 16 (7). https://doi.org/10.1080/15592324.2021.1915590.

Panda, S.K., Sunkar, R., 2015. Nutrient- and other stress-responsive microRNAs in plants: role for thiol-based redox signaling. Plant Signal. Behav. 10 (4), 1–3. https://doi.org/10.1080/15592324.2015.1010916.

Parveen, R., Khare, N., Singh, S., Johri, P., 2023. An in silico outlook for the detection and surveillance of evolving and persistent plant pathogens. In: Detection, Diagnosis and Management of Soil-Borne Phytopathogens. Springer Nature Singapore, Singapore, pp. 17–40.

Patani, A., Patel, A., Prajapati, D., Khare, N., Singh, S., 2023. Various agriculture crop plant-based bioactive compounds and their use in nanomaterial synthesis and applications. In: Nanomaterials from Agricultural and Horticultural Products. Springer Nature Singapore, Singapore, pp. 223–241.

Patra, G.K., Gupta, D., Rout, G.R., Panda, S.K., 2023. Role of long non coding RNA in plants under abiotic and biotic stresses. Plant Physiol. Biochem. 194, 96–110. https://doi.org/10.1016/J.PLAPHY.2022.10.030.

Prasad, P.V.V., Bheemanahalli, R., Jagadish, S.V.K., 2017. Field crops and the fear of heat stress—opportunities, challenges and future directions. Field Crops Res. 200, 114–121. https://doi.org/10.1016/J.FCR.2016.09.024.

Pu, M., Chen, J., Tao, Z., Miao, L., Qi, X., Wang, Y., Ren, J., 2019. Regulatory network of miRNA on its target: coordination between transcriptional and post-transcriptional regulation of gene expression. Cell. Mol. Life Sci. 76 (3), 441–451. https://doi.org/10.1007/S00018-018-2940-7.

Qiu, C.W., Liu, L., Feng, X., Hao, P.F., He, X., Cao, F., Wu, F., 2020. Genome-wide identification and characterization of drought stress responsive microRNAs in Tibetan wild barley. Int. J. Mol. Sci. 21 (8). https://doi.org/10.3390/IJMS21082795.

Rajwanshi, R., Devi, K.J., Sharma, G.R., Lal, B., 2019. Role of miRNAs in plant-microbe interaction. In: Vitro Plant Breeding towards Novel Agronomic Traits: Biotic and Abiotic Stress Tolerance, pp. 167–195. https://doi.org/10.1007/978-981-32-9824-8_10.

Rani, V., Sengar, R.S., 2022. Biogenesis and mechanisms of microRNA-mediated gene regulation. Biotechnol. Bioeng. 119 (3), 685–692. https://doi.org/10.1002/BIT.28029.

Rascio, N., Navari-Izzo, F., 2011. Heavy metal hyperaccumulating plants: how and why do they do it? And what makes them so interesting? Plant Sci. 180 (2), 169–181. https://doi.org/10.1016/J.PLANTSCI.2010.08.016.

Rogers, K., Chen, X., 2013. Biogenesis, turnover, and mode of action of plant microRNAs. Plant Cell 25 (7), 2383–2399. https://doi.org/10.1105/TPC.113.113159.

Rose, L.E., Overdijk, E.J.R., van Damme, M., 2019. Small RNA molecules and their role in plant disease. Eur. J. Plant Pathol. 153 (1). https://doi.org/10.1007/S10658-018-01614-W.

Roy, J., Blervacq, A.S., Créach, A., Huss, B., Hawkins, S., Neutelings, G., 2017. Spatial regulation of monolignol biosynthesis and laccase genes control developmental and stress-related lignin in flax. BMC Plant Biol. 17 (1). https://doi.org/10.1186/S12870-017-1072-9.

Saidi, A., Hajibarat, Z., 2020. Application of Next Generation Sequencing, GWAS, RNA seq, WGRS, for genetic improvement of potato (*Solanum tuberosum* L.) under drought stress. Biocatal. Agric. Biotechnol. 29, 101801. https://doi.org/10.1016/J.BCAB.2020.101801.

Sailaja, B., Voleti, S.R., Subrahmanyam, D., Sarla, N., Prasanth, V.V., Bhadana, V.P., Mangrauthia, S.K., 2014. Prediction and expression analysis of miRNAs associated with heat stress in oryza sativa. Rice Sci. 21 (1), 3–12. https://doi.org/10.1016/S1672-6308(13)60164-X.

Sanz-Carbonell, A., Marques, M.C., Bustamante, A., Fares, M.A., Rodrigo, G., Gomez, G., 2019. Inferring the regulatory network of the miRNA-mediated response to biotic and abiotic stress in melon. BMC Plant Biol. 19 (1). https://doi.org/10.1186/S12870-019-1679-0.

Sanz-Carbonell, A., Marques, M.C., Martinez, G., Gomez, G., 2020. Dynamic architecture and regulatory implications of the miRNA network underlying the response to stress in melon. RNA Biol. 17 (2), 292–308. https://doi.org/10.1080/15476286.2019.1697487.

Schaeffer, A., Wijntjes, C., 2022. Changed degradation behavior of pesticides when present in mixtures. Eco-Environment & Health 1 (1), 23–30. https://doi.org/10.1016/J.EEHL.2022.02.002.

Šečić, E., Kogel, K.H., Ladera-Carmona, M.J., 2021. Biotic stress-associated microRNA families in plants. J. Plant Physiol. 263. https://doi.org/10.1016/J.JPLPH.2021.153451.

Seki, M., Narusaka, M., Abe, H., Kasuga, M., Yamaguchi-Shinozaki, K., Carninci, P., Hayashizaki, Y., Shinozaki, K., 2001. Monitoring the expression pattern of 1300 Arabidopsis genes under drought and cold stresses by using a full-length cDNA microarray. Plant Cell 13 (1), 61–72. https://doi.org/10.1105/TPC.13.1.61.

Sharma, S., Chandra, D., Sharma, A.K., 2021. Rhizosphere Plant–Microbe Interactions Under Abiotic Stress, pp. 195–216. https://doi.org/10.1007/978-981-15-6125-2_10.

Shriram, V., Kumar, V., Devarumath, R.M., Khare, T.S., Wani, S.H., 2016. MicroRNAs as potential targets for abiotic stress tolerance in plants. Front. Plant Sci. 7. https://doi.org/10.3389/FPLS.2016.00817.

Silwal, P., Kim, J.K., Kim, Y.J., Jo, E.K., 2020. Mitochondrial reactive oxygen species: double-edged weapon in host defense and pathological inflammation during infection. Front. Immunol. 11. https://doi.org/10.3389/FIMMU.2020.01649.

Singh, A., Sengar, R.S., Singh, A.K., Singh, S., Kumar, A., Ahmad, K., 2019. Impact of genetic engineering on agriculture food crops. Biotech Today: Int. J. Biol. Sci. 9 (1), 17–24.

Singh, P., Dutta, P., Chakrabarty, D., 2021. miRNAs play critical roles in response to abiotic stress by modulating cross-talk of phytohormone signaling. Plant Cell Rep. 40 (9), 1617–1630. https://doi.org/10.1007/S00299-021-02736-Y.

Singh, S., Fatima, A., Tiwari, S., Prasad, S.M., 2020. Plant responses to radiation stress and its adaptive mechanisms. In: Plant Life under Changing Environment: Responses and Management, pp. 105–122. https://doi.org/10.1016/B978-0-12-818204-8.00006-0.

Singh, S., Parihar, P., Singh, R., Singh, V.P., Prasad, S.M., 2016. Heavy metal tolerance in plants: role of transcriptomics, proteomics, metabolomics, and ionomics. Front. Plant Sci. 6. https://doi.org/10.3389/FPLS.2015.01143.

Song, X., Li, Y., Cao, X., Qi, Y., 2019. MicroRNAs and their regulatory roles in plant-environment interactions. Annu. Rev. Plant Biol. 70, 489–525. https://doi.org/10.1146/ANNUREV-ARPLANT-050718-100334.

Stief, A., Altmann, S., Hoffmann, K., Pant, B.D., Scheible, W.R., Bäurle, I., 2014. Arabidopsis miR156 regulates tolerance to recurring environmental stress through SPL transcription factors. Plant Cell 26 (4), 1792–1807. https://doi.org/10.1105/TPC.114.123851.

Subburaj, 2017. Identification, molecular characterization, and differential expression profiles under γ-radiation stress. J. Plant Biochem. Biotechnol. 26, 330–345.

Sunkar, R., Kapoor, A., Zhu, J.K., 2006. Posttranscriptional induction of two Cu/Zn superoxide dismutase genes in Arabidopsis is mediated by downregulation of miR398 and important for oxidative stress tolerance. Plant Cell 18 (8), 2051–2065. https://doi.org/10.1105/TPC.106.041673.

Sunny, J.S., Snijesh, V.P., Singh, S., 2016. Potential role of mirnas in rheumatoid arthritis. Austin Arthr. 1 (2), 1009.

Tarfeen, N., Nisa, Q., Khair-Ul-Nisa, Kahlief, K., 2022. Antioxidant defense system in plants against biotic stress. In: Antioxidant Defense in Plants, pp. 383–395. https://doi.org/10.1007/978-981-16-7981-0_17.

Tungare, K., Shahu, R., Zambare, V., Agarwal, P., Jobby, R., Nisar, N., Jha, P., 2022. Toxicity mitigation of textile dye reactive blue 4 by hairy roots of Helianthus annuus and testing its effect in in vivo model systems. BioMed Res. Int. 2022.

Tyagi, S., Sharma, S., Ganie, S.A., Tahir, M., Mir, R.R., Pandey, R., 2019. Plant microRNAs: biogenesis, gene silencing, web-based analysis tools and their use as molecular markers. 3 Biotech 9 (11). https://doi.org/10.1007/S13205-019-1942-Y.

Verma, N., Sao, P., Srivastava, A., Singh, S., 2021. Physiological and molecular responses to drought, submergence and excessive watering in plants. In: Harsh environment and plant resilience: Molecular and Functional Aspects, pp. 305–321.

Vora, Z., Pandya, J., Vaikuntapu, P.R., Singh, S., Asha, S., Purama, R.K., 2023. Genome-wide association of defense hormone crosstalk in plants. In: Hormonal Cross-Talk, Plant Defense and Development. Academic Press, pp. 353–371.

Wang, B., Sun, Y.F., Song, N., Wang, X.J., Feng, H., Huang, L.L., Kang, Z.S., 2013. Identification of UV-B-induced microRNAs in wheat. Genet. Mol. Res. : GMR 12 (4), 4213–4221. https://doi.org/10.4238/2013.OCTOBER.7.7.

Wang, Y., Zhu, X., Jiang, X.M., Guo, J., Fu, Z., Zhou, Z., Yang, P., Guo, H., Guo, X., Liang, G., Zeng, P., Xiao, G., Ma, J., Yin, X., Zhang, L.K., Yan, C., Zhang, C.Y., 2021. Decreased inhibition of exosomal miRNAs on SARS-CoV-2 replication underlies poor outcomes in elderly people and diabetic patients. Signal Transduct. Targeted Ther. 6 (1). https://doi.org/10.1038/S41392-021-00716-Y.

Wani, S.H., Kumar, V., Khare, T., Tripathi, P., Shah, T., Ramakrishna, C., Aglawe, S., Mangrauthia, S.K., 2020. miRNA applications for engineering abiotic stress tolerance in plants. Biologia 75 (7), 1063–1081. https://doi.org/10.2478/S11756-019-00397-7.

Wu, L., Liu, S., Qi, H., Cai, H., Xu, M., 2020. Research progress on plant long non-coding RNA. Plants (Basel, Switzerland) 9 (4). https://doi.org/10.3390/PLANTS9040408.

Xing, H., Jiang, Y., Zou, Y., Long, X., Wu, X., Ren, Y., Li, Y., Li, H.L., 2021. Genome-wide investigation of the AP2/ERF gene family in ginger: evolution and expression profiling during development and abiotic stresses. BMC Plant Biol. 21 (1). https://doi.org/10.1186/S12870-021-03329-3.

Yadav, V., Arif, N., Singh, V.P., Deshmukh, R., Sahi, S., Shivaraj, S.M., Tripathi, D.K., Chauhan, D.K., 2020. Roles of microRNAs in plant development and stress tolerance.

In: Plant Life under Changing Environment: Responses and Management, pp. 719–733. https://doi.org/10.1016/B978-0-12-818204-8.00031-X.

Yang, X., Zhang, L., Yang, Y., Schmid, M., Wang, Y., 2021. miRNA mediated regulation and interaction between plants and pathogens. Int. J. Mol. Sci. 22 (6), 1–13. https://doi.org/10.3390/IJMS22062913.

Yang, Y., Guo, J., Cheng, J., Jiang, Z., Xu, N., An, X., Chen, Z., Hao, J., Yang, S., Xu, Z., Shen, C., Xu, M., 2020. Identification of UV-B radiation responsive microRNAs and their target genes in chrysanthemum (Chrysanthemum morifolium Ramat) using high-throughput sequencing. Ind. Crop. Prod. 151, 112484. https://doi.org/10.1016/J.INDCROP.2020.112484.

Zafar, S.A., Zaidi, S.S.E.A., Gaba, Y., Singla-Pareek, S.L., Dhankher, O.P., Li, X., Mansoor, S., Pareek, A., 2020. Engineering abiotic stress tolerance via CRISPR/Cas-mediated genome editing. J. Exp. Bot. 71 (2), 470–479. https://doi.org/10.1093/JXB/ERZ476.

Zaichkina, S.I., Rozanova, O.M., Aptikaeva, G.F., Achmadieva, A.C., Klokov, D.Y., 2004. Low doses of gamma-radiation induce nonlinear dose responses in Mammalian and plant cells. Nonlinearity Biol. Toxicol. Med. 2 (3). https://doi.org/10.1080/15401420490519861.

Zhan, X., Wang, B., Li, H., Liu, R., Kalia, R.K., Zhu, J.K., Chinnusamy, V., 2012. Arabidopsis proline-rich protein important for development and abiotic stress tolerance is involved in microRNA biogenesis. Proc. Natl. Acad. Sci. U.S.A. 109 (44), 18198–18203. https://doi.org/10.1073/PNAS.1216199109.

Zhang, F., Yang, J., Zhang, N., Wu, J., Si, H., 2022. Roles of microRNAs in abiotic stress response and characteristics regulation of plant. Front. Plant Sci. 13. https://doi.org/10.3389/FPLS.2022.919243.

Zhang, L., Xiang, Y., Chen, S., Shi, M., Jiang, X., He, Z., Gao, S., 2022. Mechanisms of MicroRNA biogenesis and stability control in plants. Front. Plant Sci. 13. https://doi.org/10.3389/FPLS.2022.844149.

Zhou, R., Wang, Q., Jiang, F., Cao, X., Sun, M., Liu, M., Wu, Z., 2016. Identification of miRNAs and their targets in wild tomato at moderately and acutely elevated temperatures by high-throughput sequencing and degradome analysis. Sci. Rep. 6. https://doi.org/10.1038/SREP33777.

Zhou, Z.S., Huang, S.Q., Yang, Z.M., 2008. Bioinformatic identification and expression analysis of new microRNAs from Medicago truncatula. Biochem. Biophys. Res. Commun. 374 (3), 538–542. https://doi.org/10.1016/J.BBRC.2008.07.083.

Zhou, Z.S., Song, J.B., Yang, Z.M., 2012. Genome-wide identification of Brassica napus microRNAs and their targets in response to cadmium. J. Exp. Bot. 63 (12), 4597–4613. https://doi.org/10.1093/JXB/ERS136.

CHAPTER

7 Regulatory role of microRNA in plants during development with respect to root and shoot architecture and secondary growth

Manju Pandey, Shikha Srivastava, Anupam K. Pal, Apurwa Singh, Shraddha Pandey
Institute of Pharmacy, Shri Ramswaroop Memorial University, Barabanki, Uttar Pradesh, India

1. Introduction

Plant roots, shoots, and secondary growth all play a crucial role in helping plants establish their entire system to absorb water, minerals, and nutrients from the milieu for their proper growth. Therefore, the basis of a plant's appropriate growth and development depends on optimal root shape, shoot propagation, and secondary growth. The modification of plant at level of root and shoot should be such that it intermittently helps plants to grow well in environments with little water and nutrients. Under such unfavorable environmental circumstances, faster and broader root and shoot growth and development are vital for plant growth. Numerous plant microRNAs (miRNAs) have been retained throughout evolution and are crucial for controlling a wide range of crucial evolving processes, including the regulation of the shoot apical meristem (SAM), the leaf and root system, organ development, and plant floral transit. Plants are the source of siRNAs, which are engaged in the transcriptional and posttranscriptional gene silencing pathways in these plants (Sijen et al., 2001). The creation of the root system architecture is an intricate process that necessitates precise regulation of gene expression related to the hormonal regulation associated with division and cell differentiation. The highly preserved radial tissue shape in the root is regulated by a central vascular cylinder that is influenced by crosstalk between the vascular cylinder and the surrounding endodermis, which is brought about by transcription factors (TFs) and microRNAs that migrate between cells. As example, miRNAs, the significant regulatory inhibitors, form a feedback loop with auxin regulatory growth factors (ARFs), namely miR160-ARF10/16/17, miR390-ARF2/3/4, and miR167-ARF12, to govern the architecture of the root sys-

MicroRNA Advances and Application in Plant Biology. https://doi.org/10.1016/B978-0-443-21682-4.00009-9

tem (Sunny et al., 2016). Plants have the ability to directly synthesis certain miRNAs under abiotic stress, while also inducing low or excessive expression of other miRNAs (Verma et al., 2021). Plants are able to react swiftly to changes in their environment because these miRNAs work on target genes, stress tolerance protein genes, Plant Growth Regulator 9 response protein genes, and TFs linked to stress resistance. MicroRNAs, such as miR393, miR169, miR402, miR397, miR165, miR166, and miR172, help plants to overcome environmental stress of low temperature and help plants to withstand this and grow healthy (Zhao et al., 2019). Similarly in dry environment, miR169 is downregulated and plant that overexpresses miR169a are more susceptible to drought and leaf water loss than wild-type plants. MiR398 in grapevines is majorly involved in the response to many abiotic stressors such as UV radiation, drought, heavy metals, excessive salt, and plant biotic stress by specifically controlling Cu/Zn superoxide dismutase (Leng et al., 2017). Thus, vivid role is played by micro-RNA to promote overall growth of plant as represented in (Table 7.1).

1.1 Involvement of candidate genes in root and shoot development

The intricate processes of root and shoot development in plants are controlled by a multitude of genes different channels (Singh et al., 2019). Numerous essential genes that are essential to the formation of roots and shoots have been found in numerous species of plants. The following list of potential genes related to these processes for root development genes are as follows as: ARFs, AUX/IAAs, and PINs (auxin-responsive genes) are important for the growth of roots since they play a crucial role in auxin signaling and its transport. Auxin receptive gene expression is structured by AUX/IAAs and AUXIN RESPONSE FACTORS (ARFs), else auxin efflux is controlled by PIN-FORMED genes. Similarly, genes SCR (SCARECROW), SHORT ROOT, and AUX1 are involved in asymmetric division of cell and radial patterning of root (Powers et al., 2020). Cytokinin-associated genes and receptors, such as AHK2, AHK3, AHK4, and Type-B ARABIDOPSIS RESPONSE REGULATORs (ARRs), are involved in signaling, influencing root growth and expansion. PLT (PLETHORA) genes, specifically PLT1 and PLT2, are important for specifying root stem cells and maintaining root meristem activity (Santuari et al., 2016). WOX (WUSCHEL-RELATED HOMEOBOX) genes such as WOX5 are vital for maintaining root stem cell functions and amends root growth (Li et al., 2020; Halder et al., 2021).

Candidate genes involved in shoot development include CLV1, CLV2, and CLV3 (CLAVATA) genes that are crucial for shoot meristem upkeep and regulation of stem cells population in the SAM. WUSCHEL (WUS) are genes which are critical for stem cell maintenance under the SAM (Lopes et al., 2021). KNOTTED-LIKE HOMEOBOX (KNOX) genes, including shoot meristem less, play roles in maintaining identity and preventing premature differentiation (Ori et al., 2000). The genes LATERAL ORGAN BOUNDARIES (LOB) regulate lateral organ development in shoots. Growth regulators genes involved in abscisic acid

Table 7.1 Role of miRNAs, the targets, and their roles in plant development.

miRNA	Target	Species	Target function	References
miR156	SPL	*Zea mays* and *Arabidopsis*	Assisting flowering, leaf progression, root maturation, secondary metabolism and abiotic strain and corn growth in *Zea mays*	(Dong et al., 2022; Aukerman and Sakai et al., 2003)
miR159	GAMYB/ GAMYB-like gene	*Arabidopsis*	Progression in reproductive growth, seed development, and expansion of vegetative tissues	(Abdel-Ghany and Pilon, 2008; Ali et al., 2020; Allen et al., 2007; Aukerman and Sakai, 2003)
miR160	ARFs	*Zea mays*, *Arabidopsis*, and *Medicago truncatula*	Embryo, root and leaf progression, and hypocotyl extension	(Sanmamed et al., 2013)
miR164	NAC	*Arabidopsis*, *Oryza*, and *Zea mays*	Meristem periphery singularity, auxiliary development, leaf progression, and lateral root initiation	(Laufs et al., 2004)
miR165/ 166	HD-ZIP III	*Arabidopsis*	Sustaining meristematic cells, adaxial singularity of leaves, lateral root expansion, and procambium distinctiveness	(Williams et al., 2005)
miR167	ARFs	*Arabidopsis* and *Oryza*	Development of male organ, leaves roots, stems, and flowers, embryonic growth, seed expansion, and defense against pathogens	(Martin et al., 2009)
miR169	NF-YA family NF-YA family and CBF	*Zea mays*, *Arabidopsis*, and *Antirrhinum majus*	Supplement of C homeotic gene transcription and root structure	(Cartolano et al., 2007)
miR171	SCL	*Arabidopsis*	Biosynthesis of chlorophyll and floral meristem factor	(Curaba et al., 2013)
miR172	AP2	*H. vulgare*, *Z. mays*, *S. tuberosum*, *Arabidopsis*, and *Oryza*	Vegetative phase modification, replicating carpel, flower meristem distinctiveness and stamen progression, tuberization, and salt forbearance	(Chuck et al., 2007)

Continued

Table 7.1 Role of miRNAs, the targets, and their roles in plant development.—*cont'd*

miRNA	Target	Species	Target function	References
miR319	TCP	*Arabidopsis* and *Solanum lycopersicum*	Leaf development and decay, organ coiling, and hormone biosynthesis and signaling	(Ori et al., 2007)
miR390	TAS3	*Arabidopsis*	Biogenesis of ta-siRNA for ARF suppression, lateral root expansion, and leaf conniving	(Dong et al., 2022; Fahlgren et al., 2006)
miR393	TIR1 and AFB	*Arabidopsis* and *Oryza*	Homeostasis of auxin, leaf shape and number, and lateral root expansion	(Parry et al., 2009)
miR394	LCR	*Arabidopsis*	Meristematic individuality overthrow and leaf tilting	(Baumann, 2013)
miR396	GRF	*Arabidopsis*, *Medicago*, and *Oryza*	Cell proliferation in leaves. Stimulates somatic embryogenesis, panicle implication, and disease-resistance	(Debernardi et al., 2012)
mir397	OsLAC	*Oryza*	Endorses grain yield and panicle branches	(Zhang et al., 2013)
miR824	AGL16	*Arabidopsis*	Stomatous designing	(Bergmann and Sack, 2007)
miR828 and miR858	MYBs	Grapes and cotton	Fiber and flavonol anthocyanin	(Guan et al., 2014)
miR847	IAA28	*Arabidopsis*	Lateral expansion of root	(Wang and Guo, 2015)
miR857	LACCASE7	*Arabidopsis*	Secondary progress	(Abdel-Ghany and Pilon, 2008)
TAS3	ARF3/4 and (only in mosses) AP2-like	All land plants	Vasculature evolution, leaf polarity, and phase changeover	(Dong et al., 2022; Fahlgren et al., 2006)

(ABA), brassinosteroid (BR), and gibberellin (GA) signaling pathways impact elongation, shoot growth, and branching. Genes such as TEOSINTE BRANCHED/CYCLOIDEA/PCF (TCP) play important roles in leaf expansion and affect shoot architecture (Nicolos et al., 2016).

1.2 Importance of microRNA in plant growth

Small RNAs are noncoding that have a length of 20–24 nucleotides. Among these, miRNAs are particularly important for plant development (Dong et al., 2022; Naqvi, 2012). miRNAs either directly cleave target genes or block translation to regulate target expression of gene at the posttranscriptional stage (Tungare et al., 2022). miRNAs remain involved in almost every stage of plant development, including the production of floral organs, the transition from juvenile to adulthood, apical shoot meristem growth, leaf morphogenesis, and flowering period determination. It has been shown that miRNAs are essential for plant development. RNA polymerase II (Pol II) often transcribes them into pri-miRNAs. RNA-induced silencing complexes (RISCs) are formed when these pri-miRNAs are combined with proteins from the ARGONAUTE (AGO) family, following their cleavage by a category of RNase-III nucleases known as Dicer-like proteins (Wilson et al., 2013). Target gene expression and regulation are subsequently mediated by RISCs. miRNAs regulate a variety of processes, including flowering patterns, leaf polarity, and meristem features. miRNAs are essential to coordinating plant development, as evidenced by the numerous impacts that transcription mutations in miRNA or processing complexes typically have on plant form and function (Fig. 7.1). According to Llave et al., miR160, which regulates the expression of ARF10 to control stem cell differentiation at the end of the root meristematic area and dictates the direction of root growth, is associated to the creation of root cap cells during Arabidopsis root growth (Llave et al., 2002). Endogenous miRNAs target associated genes and regulate gene expression during plant development.

Furthermore, miR164 and miR390 have a significant impact on the expansion of plant root organs, such as adventitious root (AR) production, lateral root (LR) growth, and root cap formation. miR164 controls the process of *Arabidopsis thaliana*'s LR development. miR164 mediates NAC1 expression that has been stimulated by auxin, which influences auxin transmission and controls the growth of LRs (Guo et al., 2005). miRNAs control the structure and size of leaves as well. The ratio of miR396 to GRFs (growth regulatory factors) ultimately determines cells number, leaves, and meristem size. Numerous miRNAs work by interacting with hormones as many of these are represented in (Table 7.2). MiRNAs target numerous components of hormone signaling, and plants can quickly and efficiently control their growth, development, and differentiation through the interactions between these components and the miRNAs. In order to influence hormone responses, miRNAs are chosen as intermediates in this signaling process. Alternatively, hormones can be used to regulate particular miRNA levels (Maddala et al., 2023). Thus, majority of miRNA targets are TFs, which play a crucial role in controlling the growth and

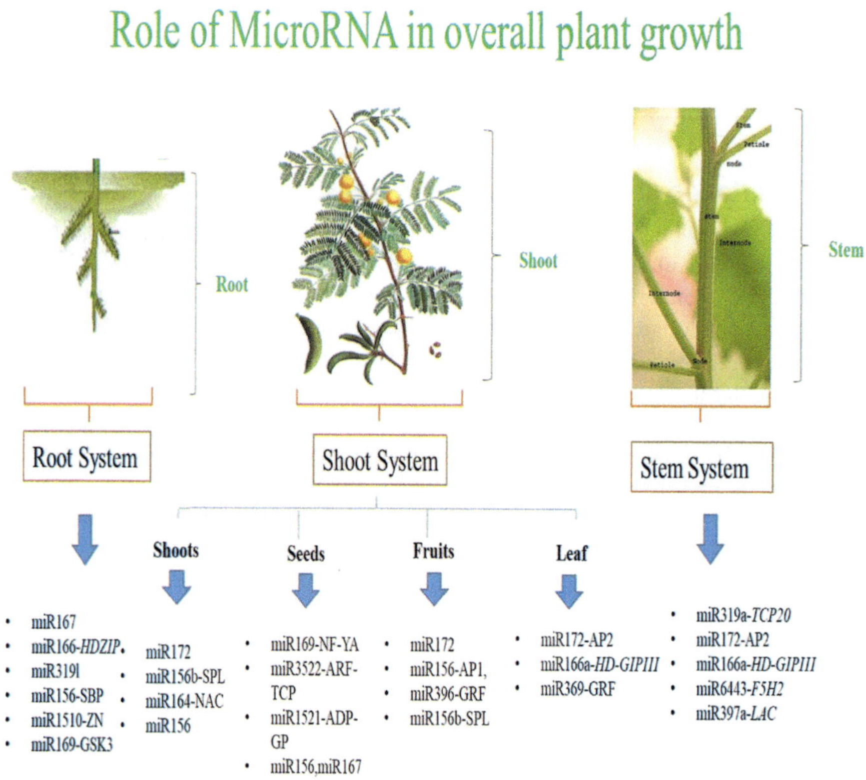

FIGURE 7.1

Role of MicroRNA in overall plant growth.

development of plants. Plants may benefit from the assistance of different families of TFs, which have measured a variety of regulatory networks, in both normal and stressful environmental situations.

2. Role of microRNA in root expansion

The intake of water and minerals, as well as the annexation of land by plants, is significantly facilitated by the root system (Patani et al., 2023). Therefore, the basis of a plant's appropriate growth and development is optimal root shape (Yan et al., 2022). While the AR and LR are generated postembryonically, the primary root (PR) and seminal root (SR) are often consequential from the embryonic radicle (Péret et al., 2009). Though, controlling the manifestation of genes involved in cell differentiation, elongation, and patterning, miRNAs help to control root architecture (Yan et al., 2022). They affect the root system's overall structure as well as the expansion of LRs and root hairs. Throughout root expansion, miRNAs decide

Table 7.2 miRNAs in the regulation of different biological processes in root System.

miRNAs	Species	Regulation of different biological processes	Target gene	References
miR171	*Arabidopsis thaliana*	Root and leaf expansion, meristem enlargement, trichome spreading, diurnal cycle, chlorophyll biosynthesis, flower growth, phytohormone	SCL6-II/III/IV SUVH8 SCL27	(Wang et al., 2010; Liu et al., 2008)
	Oryza sativa	Somatic embryogenesis, phase changeover, root expansion, chlorophyll build-up	SCL	(Luo et al., 2006; Fan et al., 2015)
	Citrus	Somatic embryogenesis	SCL	(Shi et al., 2022)
	Larix leptolepis	Somatic embryogenesis	SCL	(Zhang et al., 2010)
	Lilium	Somatic embryogenesis	*SCL6*	(Li et al., 2017; Yan et al., 2022)
	Raphanus sativus	Somatic embryogenesis	*SCL6*	(Zhai et al., 2014)
	Larix kaempferi	Somatic embryogenesis	*SCL6*	(Zang et al., 2021)
	Solanum lycopersicum	Anther expansion, shoot splitting, leaf manner, plant elevation, flowering time, root expansion, fruit set and growth, phytohormone, phase changeover	*HAM, NSP2L, GRAS24*	(Kravchik et al., 2019; Huang et al., 2017; Hou et al., 2019)

Continued

Table 7.2 miRNAs in the regulation of different biological processes in root System.—*cont'd*

miRNAs	Species	Regulation of different biological processes	Target gene	References
	Hordeum vulgare	Plant tallness, leaf sums, phase changeover, floral meristem determinacy	*SCL*	(Curaba et al., 2013)
	Saccharum officinarum	Stem	*SCL6*	(Sternes and Moyle, 2015)
	Physcomitrium patens	Meristem establishment	*GRAS12*	(Beheshti et al., 2021)
	Medicago truncatula	Root expansion	*NSP2*	(Lauressergues et al., 2015; Branscheid et al., 2011)
	Populus trichocarpa	Phytohormone	*GRAS* *MYB*	(Liu et al., 2014)
	Pyrus	Phytohormone shoot progression	*SCL6/22*	(Jiang et al., 2018)
	Pinus densata	Flower, needle, root, leaf, stem	*SCL6*	(Hai et al., 2018)
	Dendrobium officinale		*SCL*	(Yang et al., 2015)
	Litchi chinensis		*GRAS8/9/24/27*	(Chen et al., 2021)
miR156	*Citrus*	Somatic embryogenesis	*CsSPL (inhibition)*	(Long et al., 2018)
miR857	*Arabidopsis*	Secondary xylem disparity	*LACCASE7*	(Couzigou et al., 2016)
miR165/166	*Arabidopsis*	Vasculature discrepancy/root elongation/meristem dimension	*Class III HP ZIP*	(Couzigou et al., 2016)
miR169	*Arabidopsis*	Root elongation/lateral roots	*NFYA2/10*	(Couzigou et al., 2016)
miR396	*Arabidopsis*	Root elongation/root cap formation	*bHLH/GRFs*	(Couzigou et al., 2016)
miR160	*Arabidopsis*	Root elongation/root cap formation	*ARF10/ARF16*	(Couzigou et al., 2016)

the fate of individual cells (Table 7.2). As miR167 targets the genes ARF6 and ARF8 that controls root growth, encompasses initiation and elongation of LRs as found in the case of soybean nodulation and root growth (Wang et al., 2015a,b). In the root meristem, they control the ratio of cell division to differentiation. These are involved in preserving the function of the root apical meristem (RAM), an area that is essential for root growth and development. miR164 targets TFs that have its place to the NAC (NAM, ATAF, and CUC) domain that assists in controlling the growth of LRs and the upkeep of the root meristem (Feng et al., 2014). Similarly, TCP (TEOSINTE BRANCHED/CYCLOIDEA/PCF) TFs are targeted by miR319 that participates in the regulation of architecture and root growth (Martín et al., 2018). MicroRNA too interacts with hormones and regulates root architect. As auxins, cytokinins, and GAs are perfect examples of plant hormones that networks with miRNAs and regulates root development (Gangwar et al., 2023). For example, AFB (Auxin Signaling F-Box) and TIR1 (Transport Inhibitor Response 1) genes are the targets of miR393 that regulates auxin signaling and effects on root development and progression (Chen et al., 2011). Through influencing the development of root nodules, miRNAs perform directive of symbiotic nitrogen fixation in leguminous plants. MiRNAs have a role in directive of gene expression throughout the symbiotic bacterial relationship-building process. MiRNAs assist plants in countering to a series of environmental signals, such as the availability of nutrients and stress levels (Meddaya et al., 2023). For roots to acclimatize the environmental conditions, this is crucial. For instance, certain miRNAs might influence nutrient uptake and root architecture in response to nutritional deficits (Zhang et al., 2022). For example, miR827 targets NITROGEN LIMITATION ADAPTATION (NLA) gene that controls intake and transportation of phosphate, which affects root development under phosphate shortage (Lin et al., 2013). Deciphering the complex regulatory networks controlling plant root growth requires an understanding of the particular miRNA-target interactions and their subsequent effects on gene expression. New understandings of the molecular processes involving miRNAs in plant growth and adaptation are being uncovered by ongoing study (Samad et al., 2017). Similarly, the role of micro-RNA has been involved in shoot development and discussed further.

2.1 miRNA involved in embryonic root elongation and primary root growth

Here we will discuss about the critical roles of different miRNAs in controlling root length by targeting various genes. In *Arabidopsis*, a miR156-SPL10 module modifies root meristem activity, primarily by adjusting auxin and cytokines (CKs) responses, which in turn controls PR progression and length (Gao et al., 2018; Parveen et al., 2023a,b). miR160 and its intended recipients through influencing cell division and disparity in the root tip, ARF10/16 controls the elongation of roots in *Arabidopsis*. According to (Wang et al., 2005), ARF10ARF16 double mutants and plants over expressing miR160c exhibit unrestrained cell division and inhibited cell differentiation, indicating a critical function for the miR160-ARF10/16 axis in root elongation and development (Vora et al., 2023). Apart from Arabidopsis, several miRNAs have a role in root elongation in crops like rice and barley as a reaction to abiotic stress and nutrient intake. Through the auxin signaling pathway, the miR156-SPL regulatory module plays a significant character in rice root formation (Curaba et al., 2014).

The *WT Lotus japonicus* plants with large SLs modifications have longer roots than the miR156 overexpression plants, although it is unclear if SLs are affected by miR156 and the SPL regulatory mechanism (Wang et al., 2015a,b). miR159 plays a significant role in inhibiting PR progression via its target gene MYB65. It also controls the action of the meristem in root tips by regulating auxin and CK reactions (Xu et al., 2017).

Two controllers of the auxin signaling system are the miRNAs (miR160 and miR167). Many ARF genes that operate as activators or repressors of key auxin reactive genes in plants are modulated in their expression (Di et al., 2015; Dubey et al., 2020). A regulatory network consisting of AtARF6/8/17 and miR160/167 regulates adventitious rooting. MiR167/ARF6/8 is a constructive controller of ARs in this governing system, whereas miR160/ARF17 is a destructive manager. The miR160/ARF10/16/ABSCISIC ACID INSENSITIVE 3 (ABI3) monitoring loop is necessary for a number of opposing processes, such as seed germination, seed latency, and somatic embryogenesis, in addition to controlling root growth (Gutierrez et al., 2009). Secondly, AtABI3 reduces the accretion of miR160 by blocking the transcript of the miRNA encoding gene AtMIR160b. Thus, reduced miR160 post-transcriptionally affects AtARF10/16, and AtARF10/16 limits auxin level by activating AtYUC1/4/10 through AtLEC2-interceded AtYUC1/4/10, which ultimately reins somatic embryogenesis (Chen et al., 2021). Thirdly, under various ABA concentrations, the miR160/ARF10 module also controls seed incubation (Liu et al., 2007).

AtARF6/8 in *Arabidopsis*, miR160 unswervingly chops the mRNA of AtARF10/16/17 (Gutierrez et al., 2009). The pleiotropic developmental abnormalities seen in plants harboring miRNA-resistant forms of these ARFs highlight the critical functions that miR160 and miR167 play in *Arabidopsis* progression and expansion (Liu et al., 2007).

2.2 miRNAs in root radial modeling and tissue differentiation

Plant roots are outwardly arranged in several cell layers round a central vascular cylinder in order to deliver nutrients to the shoot and acclimatize functionally to the soil environment (Khare et al., 2017). The GRAS transcription factor SHORTROOT (SHR), which is articulated in the central vascular tissue, is crucial for the development of root cell layers. Indeed, the endodermis is not specified in SHR mutants (Levesque et al., 2006). It has been found that SHR mutants are unable to identify the endodermis, viz concentric cell layer that is directly adjacent. From cell to cell, the stele produces the SHR protein, which travels to the endodermis, there it stimulates the synthesis of a different TF partner Scarecrow (SCR) (Gallagher et al., 2004). SHR then uses a miRNA-dependent mechanism to return the communication to steal. This specific cell-to-cell communication event involves miR165/166. SHR and SCR work together to induce the transcription of a particular miR165/166 gene in the endodermis. Mature miRNA retreats from endodermis into the stele, where it inhibits PHB mRNA accretion and permits appropriate xylem development (Carlsbecker et al., 2010). These two instances of cell-to-cell contact create a governing loop in *Arabidopsis* that governs radial discrepancy patterns (Boualem et al., 2008).

2.3 miRNA involved in postembryonic lateral root development

The root cap has two primary purposes when positioned in a rootward direction, namely providing protection to the root meristem and stem cells niches (SCNs) during soil foraging and facilitating the sense of gravitropism. miR160 targets the redundant AUXIN RESPONSE FACTORS 10 (ARF10), ARF16, and ARF17 (Mallory et al., 2005). When oriented rootward, the root cap has two main utilities: it protects the root meristem and SCN during soil foraging and enhances the sensation of gravitropism. ARF16, ARF17, and laid off ARF10 are the targets of miR160. ARF17 functions differently from ARF10 and ARF16 in terms of root elongation and is not engaged in miR160-dependent root cap development (Wang et al., 2005).

2.4 miRNA involved in adventitious root development

Several miRNA families have been found to coordinate adventitious and LR architecture. Auxin signaling F-Box in LR progression is regulated by miR393, while ARF10, 16, and 17 are regulated by miR160; NAC1 (NAM-ATAF-CUC transcription factor 1) is regulated by miR167; ARF2, 3, and 4 are controlled by miR1670; and ARF6 and 8 are regulated by miR1670 (Khan et al., 2011; Meng et al., 2010; Rubio-Somoza and Weigel, 2011; Guo et al., 2005).

Research has also been done on the multiplicity of miRNAs in *Japanese apricot* (*Prunus mume*) with regard to floral growth, together with the expression of miR160 and miR167 (Gao et al., 2012). Additionally, miRNA genome outlining has been carried out in *Eucalyptus* species, such as *Eucalyptus globulus*, where miR167 has been found.

Several investigations have recognized monitoring genes common to AR progression and expansion (Hochholdinger and Zimmermann, 2008; Gonin et al., 2019). As a result, miRNAs perform a precarious role in controlling AR production. In many species, the miR156-mediated changeover from the juvenile to the mature phase requires adventitious rooting ability. AR production in the hypocotyl is inhibited by the lowering of miR156 activity, suggesting that its marks SPL proteins also suppress AR production (Curaba et al., 2014; Xu et al., 2016).

According to (Gutierrez et al., 2009), after 7 days of light after 48 h of etiolation, *Arabidopsis* plants with the mutations ARF6-KO, ARF8-KO, and ARF17-OX generated noticeably less ARs in contrast to the wild type. Moreover, ARF17-OX plants and the ARF6 and ARF8 double KOs generated fewer ARs than did the single KOs, suggesting that all three genes influence the phenotypic result. A class of non-coding short RNAs (sRNA) acknowledged as miRNAs are essential for controlling a variety of biological processes, including the adaptation to waterlogging.

2.5 miRNA involved in regulatory network during root development

Several studies on the development of Arabidopsis LR reveal the interaction between auxin and miRNAs. The design of root systems is largely determined by LRs, and the effectiveness of the root function is influenced by the degree of root branch off (Péret et al., 2009). The formation of an LR primordium, which is controlled by the local auxin concentration, is the first step in the expansion of LR (Péret et al., 2009, Dastidar et al., 2019). Auxins and miRNAs interact throughout LR growth to influence how the cell reacts to external stimuli. For instance, miR167 uses its ARF8 target to modify LR architecture in a cell-explicit way in reaction to nitrogen (Gifford et al., 2008). In tallying to their part in the formation and progression of the numerous root types, miRNAs also contribute in definite facets of root growth. For instance, miR160 limits the expression of both ARF10 and ARF16, which regulates the growth of the Arabidopsis root cap and gravity detecting (Wang et al., 2005). The distal stem cells give rise to the root cap cells (columella and LR cap cells). Therefore, the existence of the stem cells is necessary for the creation of the root cap.

Auxin assimilates discrete miRNA governing centers during the formation and progression of LRs. For example, auxin, miR156 levels, and LR number are positively correlated. Apart from creating the RM, the miR156 governing center interacts with auxins to control the quantity and appearance of LRs (Yu et al., 2015).

SPL9, SPL10, and SPL3 that are miR156-targeted suppress LR growth, with SPL10 having the most effect. Crucially, exogenous auxin treatment can activate SPL9 and SPL10, particularly during LR growth. Furthermore, a multifaceted feedback tool of the auxin−miR156−SPL path can be shown, where auxins, SPL9, and SPL10 boost miR156 manifestation, while miR156 destructively controls SPL9 and SPL10 in the roots. This is because auxins also constructively control the production of miR156 in the roots. This response loop may have developed as a means for plants to sustain proper and ongoing root growth, which would guarantee the soil nutrient-seeking activity. According to recent research, the miR156 governing center partially interacts with auxin to govern the growth of AR. In comparison to wild-type plants, transgenic *Malus xiaojinensis* and *Nicotiana benthamiana* plants over stating *M. xiaojinensis* miR156 showed a greater AR number, while MIM156 plants showed a lesser AR number. Indole-3-acetic acid (IAA) conduct of *M. xiaojinensis* stem cuttings in vitro raised the AR quantity in both wild-type and *M. xiaojinensis* MIR156-overexpressing plants; however, IAA treatment could not reverse the MIM156 plants; harm of adventitious rooting potential (Xu et al., 2017).

3. Role of microRNA in shoot development

Plants have the capability to constantly grow different organs all over their life cycle, in contrast to animals. The creation of adjacent organs at the shoot apex in consistent spatial and temporal patterns is the definition of shoot development. Plants need extremely powerful and complicated genetic networks to fully integrate such development plans across many cell and tissue types. The functions of micro-RNAs in these networks have become clearer over the past 20 years, not the least of which is in vegetative shoot progression. In this investigation, we detail ongoing experimental limitations in the field and discuss recent developments in our knowledge of the part of micro-RNA in the directive of vegetative shoot progression. The formation of flowers, the upkeep of the SAM, and the start of leaves are all critical components of shoot expansion in plants that are regulated by microRNAs, or miRNAs. Their apical meristem develops throughout embryonic development and contains a population of stem cells with the capacity for both self-replication and multidirectional differentiation (Fouracre et al., 2016).

miRNA is critically involved in embryonic shoot elongation, shoot growth, shoot tissue differentiation, postembryonic (primary and secondary shoot branches development) and mediated regulatory network during shoot development. The details have been discussed below.

3.1 miRNAs in shoot tissue differentiation

miRNAs such as miR156 and miR157 target SPL (Squamosa promoter-binding protein-like) genes (TF) that regulate shoot development and phase transition. These too affect shoot architecture and regulate changes from vegetative to reproductive phase. Plant development possesses auxin as an essential hormone to differentiate the shoot growth. Thus, ARFs are crucial in auxin signaling that are aims and target of miR160 and miR167 genes (Yang et al., 2006). Targeted by miRNA319 are TFs involved in leaf development and shoot morphogenesis, namely members of the TCP (TEOSINTE BRANCHED/CYCLOIDEA/PCF) family (Koyama et al., 2017). Assisting in the maintenance of the SAM and the formation of leaf polarity, miR390 and TAS3-derived tasiRNAs regulate the expression of ARF3 and ARF4 genes (Ali et al., 2020). miR164 targets genes in the NAC (NAM, ATAF1/2, CUC2) domain that control the formation of lateral organs, especially shoots splitting (Fang et al., 2014). Similarly, TFs APETALA2 (AP2) are the target of miR172. By regulating the time of floral organ development and shoot architecture, it governs the shift to the reproductive phase (Zhu et al., 2011). From the start of leaves and flowers to the upkeep of the SAM and the control of diverse developmental processes, these examples highlight the range of miRNAs involved in various aspects of shoot development.

3.2 miRNA involved in embryonic shoot elongation and shoot growth

Plant embryonic shoot elongation and shoot growth have been found to be significantly influenced by miRNAs, which are important regulators of several processes. miRNAs can modify the expression of genes by either translationally inhibiting the target mRNA or causing mRNA destruction. The expression of genes involved in growth and developmental processes is impacted by this control. miRNA too controls the expression of genes related to differentiation, patterning, and cell division, which affects the shoot's overall development and shape. The growth and development of a plant's above ground organs is largely dependent on the SAM. The pathway known as STM (shoot meristem less)-WUS (Wuschel)-CLV (Clavata) is essential for meristem activity maintenance. Through its ability to target and regulate many genes in the STM-WUS-CLV signaling pathway, miRNA plays a crucial role in the regulation of gene expression networks, directing the development and maintenance of the SAM (Gregory et al., 2018). MiR156 is among these miRNAs. After a great deal of research, it was discovered that miR156 controls many aspects of plant development, including growth and shoot elongation. The SQUAMOSA PROMOTER BINDING PROTEIN-LIKE (SPL) gene family is the target of this action. The control of SPL genes in *Arabidopsis thaliana* by miR156 affects a number of embryonic events, such as the timing of blossoming, the change from the juveniles to the adult phase, and leaf morphology. Shoot construction and growth are controlled by SPL genes, which are regulated by miR156. The gene regulatory networks governing shoot development can be fine-tuned by miRNAs. They frequently work in tandem with other regulatory factors to guarantee accurate regulation of the expression of genes throughout the lengthening and growth of shoots. MiR159 has an impact on the regulation of SAM activity as well as floral development (Gao et al., 2020). Interaction of hormones and genes to regulate shoot apex growth, for example, auxin is an essential hormone that regulates many aspects of the development of plants, including the lengthening of shoots. Through the control of ARF gene expression, miR167 has the ability to indirectly affect the development and growth of shoots. Similarly, MiR172's targets APETALA2 (AP2) that regulates genes' expression and impacts the expansion and lengthening of shoots during plant development and regulates meristem inflorescence (Sang et al., 2022).

3.3 miRNA-mediated regulatory network during shoot development

miRNAs play a role in controlling target gene expression during shoot development. These genes govern distinct aspects of shoot growth, patterning, and differentiation. During shoot development, these miRNA-mediated regulatory networks can be intricate, involving several miRNAs that target several genes. miRNAs (miR165/miR166) control the development of SAMs, leaf polarity, and vascular development by targeting class III HD-ZIP TFs. miR159 precisely targets MYB (Proto-Oncogene protein) TFs that are involved in a number of developmental processes, such as the growth of shoots and reactions to environmental stimuli. The TFs HD-ZIP III are the target of miR165 and miR166. Their roles include vascular tissue development in the shoot and the establishment of leaf polarity (Du et al., 2015). Efficient shoot development and environmental cue adaption are guaranteed by the precise regulation imposed by miRNAs. miR-319 targets the TFs TCP (TEOSINTE BRANCHED1/CYCLOIDEA/PCF), which are important in leaf development and morphogenesis. miR394 arises in the SAM's L1 layer and diffuses all the way down to the organizing center. The SAM-specific gene WUS is downregulated in the OC as a direct result of decreased leaf curling responsiveness (LCR) expression. The inhibition of miR394 on LCR only happens in the OC layer, despite the fact that the concentration of miR394 in the L1 layer is larger than in the OC layer. This suggests that the precise concentration of miR394 is crucial to its activity in *A. thaliana* (Knauer et al., 2013; Song et al., 2012). Through its ability to target and regulate many genes in the STM-WUS-CLV signaling pathway, miRNA plays a crucial role in the regulation of gene expression networks, directing the development and maintenance of the SAM.

3.4 Role of miRNA involved in postembryonic primary and secondary shoot branches development

Several species of plants may differ in the particular miRNAs implicated in the establishment of postembryonic primary as well as secondary shoot branches. Many miRNAs that have been associated with roles in the growth of shoots have been found in *Arabidopsis thaliana*, a model plant that has been thoroughly researched for its genetic and molecular pathways. SQUAMOSA PROMOTER BINDING PROTEIN-LIKE (SPL) TFs include miR156 and miR157. They affect the architecture and branching of shoots and have a role in controlling the change from the vegetal to the reproductive phase. miR165/166 target TFs in the Class III HD-ZIP family. They control multiple aspects that influence plant growth, such as the establishment and upkeep of the apical meristem of the shoot, which has an indirect impact on shoots branching. Targeting TFs such as TEOSINTE BRANCHED/CYCLOIDEA/PCF (TCP), miR319 may have an impact on the branching of shoots and the growth of leaves. miR171 targets gene TFs called SCARECROW-LIKE (SCL) that are important in the upkeep of meristems and shoots development. Research is still ongoing to identify more miRNAs and their distinct functions within the complex regulatory networks that control the growth of plant shoots.

4. Role of microRNA in secondary growth of plants

The development of secondary tissues in plants, including the vascular cambium, secondary xylem, and secondary phloem, is facilitated by the regulation of secondary growth, which is largely dependent on miRNAs (Fig. 7.2). The formation of woody tissues and a rise in stem girth are both facilitated by secondary growth.

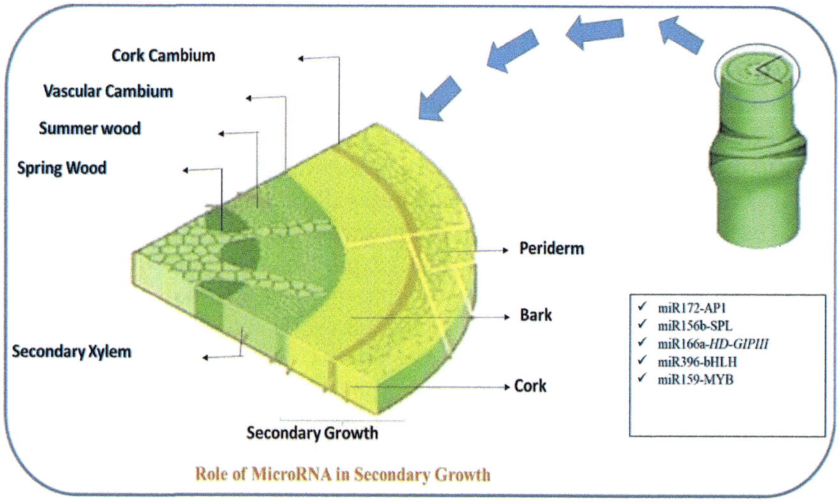

FIGURE 7.2

Role of MicroRNA in secondary growth of plant.

During secondary growth, miRNAs play a role in the creation and upkeep of vascular patterning. They are involved in maintaining the integrity and appropriate differentiation of vascular components, such as arteries and fibers. MiRNAs help to alter gene expression to assist plants in adjusting to shifting environmental conditions external stress including pathogen invasions and drought that can have an impact on secondary growth (Dong et al., 2022). MiRNAs regulate the expression of genes involved in secondary growth processes, which helps to maintain vascular homeostasis (Landskroner et al., 2013). MicroRNA helps in maintaining cell coordination, helps in homeostasis, helps in coordinating hormone signaling, generation of vascular bundles, and overall maintaining balance and improving plant growth.

The following are some important facets of miRNAs' function in secondary growth. As the generation of secondary vascular tissues is brought about by the control of cambium activity, which is a function of miRNAs. Genes involved in the regulation of cell division and differentiation inside the cambial meristem may be targeted by miRNAs. miR164 targets TFs in the NAC domain. Secondary growth is one of the characteristics of plant development that NAC TFs have been linked to. MiR164 may be involved in secondary xylem development and cambium activity regulation. It is established that these miRNAs play a role in controlling auxin signaling. Auxin receptor TIR1 is the target of miR393, whereas GRFs are the target of miR396 to indirectly affect auxin signaling. Auxin is a major regulator of secondary growth and cambial activity. It is possible that activities related to secondary growth are controlled by miRNAs that alter auxin signaling (Dong et al., 2022). MiR319 targets TFs associated with TCP (TEOSINTE BRANCHED/CYCLOIDEA/PCF). The control of cell division and proliferation is mediated by TCP

TFs. By regulating TCP gene expression, miR319 may have an impact on secondary growth (Koyama et al., 2017). Growth-regulating factors (GRFs) are crucial in cell expansion and proliferation, and miR396 targets these factors. It has been connected to the regulation of shoot growth and leaf size. It might affect secondary growth because of its function in cell growth, particularly in connection to cambial activity. It is known that MYB TFs are targeted by miR828 that plays a role in controlling the synthesis of lignin. Since lignin makes up a significant portion of secondary cell walls, miR828 were probably involved in the regulation of secondary growth (Tirumalai et al., 2019). Plants undergo a sequence of developmental transitions, including a change from embryonic to postembryonic growth, as well as a vegetative phase change as they go through the juvenile and adult phases of development before beginning reproductive growth. Both miR172 and miR156 control the transition from the vegetative phase to the flowering stage in all angiosperms (Raihan et al., 2021). To untangle the intricate regulatory networks controlling secondary growth in plants, it is essential to comprehend the particular miRNA-target interactions and their downstream consequences on gene expression. New information regarding the functions of miRNAs in this crucial stage of plant development is still being discovered by ongoing study.

5. Future prospective

It is necessary to investigate miRNA function modes further. miRNAs can enhance plant development by safely changing the gene expression to some extent. Additionally, miRNAs function as variables that respond to the environment, giving plants corresponding phenotypes and encouraging plant evolution and adaptation. More research is required to realize the role of miRNAs and associated unique processes. The precise mechanism by which miRNAs control biological process in particular temporal and geographical patterns remains unclear. However, it is now simpler to modify miRNA expression to boost crop output because of the advancements in genome editing tools. Rice grain yield is evaluated by the total sum number of branches, along with tiller and inflorescence branches. By downregulating target genes OsLAC, OsmiR397 grain size can be enhanced along with panicle branching, and improve grain production in the regulation of kernel size and yield (Zhang et al., 2013). Squamosa promoter binding protein like/miRNA156/miR529 and miR172/ Activator Protein-2 precisely controls the growth of rice tiller (Wang et al., 2015a,b). Zhao et al. demonstrated the significance of miR1432/acyl-CoA thioesterase in hormone synthesis and metabolism of fatty acids that is essential for rice (Zahoa et al., 2019). The particular mechanism by which miRNAs control a biotic process in particular temporal and geographical pattern is still uncertain. Plant hormones and cis-elements of the stress response are found in many miRNA gene promoters, suggesting that controlling the transcription of miRNA genes may be a means of responding to stressors and plant hormones. Researchers studying plants have recently placed a great deal of emphasis on the mobility of sRNA inside and between cells, tissues, and organisms. Thus far, research suggests that miRNA has the ability to migrate and generate gradient distributions across many tissues. Further study is needed to recognize the role of intermediate steps affecting miRNA mobility and its noncellular autonomous function. It is also worthwhile to conduct additional research to clarify the connection between hormone concentration and miRNA flexibility in order to improved understanding. Furthermore, plants are able to synthesize new sRNA in response to environmental stress caused by both biotic along with abiotic factors. For instance, *A. thaliana* generate a significant amount of 22 nt siRNAs in response to shortage of nitrogen that are dependent on DCL2 and RNA-dependent RNA polymerase 6. Though still many doubts still exist about its functions, synthesis, and regulations. As a result, sequencing technology and miRNA research experimentation should be in-depth and more specified. Thus, action mechanism of miRNAs and their directing pathways in prototypical plants could be specified more by utilizing nanopore sequencing and single-cell omics techniques. This will give a solid speculative foundation for comprehending how miRNA controls plant progress and development and can then be applied to plants to get an agricultural benefit.

6. Conclusion

It is crucial to understand and remember the precise roles of miRNAs in root, shoot, and secondary growth development. These whole process and biochemical cooperation are still being researched. Furthermore, investigation is required to understand the precise processes and interactions involved in miRNAs that control the secondary development in plants. The more research at biomolecular level could unravel more realistic picture of interactions, processing, and overall growth of plants. This study at genetic level will definitely bring boon to agroeconomics and provide benefit to whole mankind.

Acknowledgments

The authors are highly thankful to Institute of Pharmacy, Shri Ramswaroop Memorial University to providing opportunity and support to complete this chapter.

References

Abdel-Ghany, S.E., Pilon, M., 2008. MicroRNA-mediated systemic downregulation of copper protein expression in response to low copper availability in Arabidopsis. J. Biol. Chem. 283, 15932–15945.

Ali, S., Khan, N., Xie, L., 2020. Molecular and hormonal regulation of leaf morphogenesis in Arabidopsis. Int. J. Mol. Sci. 21 (14), 5132.

Allen, R.S., Li, J., Stahle, M.I., Dubroue, A., Gubler, F., 2007. Genetic analysis reveals functional redundancy and the major target genes of the Arabidopsis miR159 family. Proc. Natl. Acad. Sci. U.S.A. 104, 16371–16376.

Aukerman, M.J., Sakai, H., 2003. Regulation of flowering time and floral organ identity by a microRNA and its APETALA2-like target genes. Plant Cell 15 (11), 2730–2741.

Baumann, K., 2013. Plant cell biology: mobile miRNAs for stem cell maintenance. Nat. Rev. Mol. Cell Biol. 14, 128.

Beheshti, H., Strotbek, C., Arif, M.A., Klingl, A., Top, O., Frank, W., 2021. PpGRAS12 acts as a positive regulator of meristem formation in Physcomitrium patens. Plant Mol. Biol. 1–13.

Bergmann, D.C., Sack, F.D., 2007. Stomatal development. Annu. Rev. Plant Biol. 58, 163–181.

Boualem, A., Laporte, P., Jovanovic, M., Laffont, C., Plet, J., Combier, J.P., Niebel, A., Crespi, M., Frugier, F., 2008. MicroRNA166 controls root and nodule development in Medicago truncatula. Plant J.: For Cell and Molecul. Biol. 54 (5), 876–887.

Branscheid, A., Devers, E.A., May, P., Krajinski, F., 2011. Distribution pattern of small RNA and degradome reads provides information on miRNA gene structure and regulation. Plant Signal. Behav. 6 (10), 1609–1611.

Bustos-Sanmamed, P., Mao, G., Deng, Y., Elouet, M., Khan, G.A., 2013. Overexpression of miR160 affects root growth and nitrogen-fixing nodule number in Medicago truncatula. Funct. Plant Biol. 40, 1208–1220.

Carlsbecker, A., Lee, J.Y., Roberts, C.J., Dettmer, J., Lehesranta, S., Zhou, J., Lindgren, O., Moreno-Risueno, M.A., Vatén, A., Thitamadee, S., Campilho, A., Sebastian, J., Bowman, J.L., Helariutta, Y., Benfey, P.N., 2010. Cell signalling by microRNA165/6 directs gene dose-dependent root cell fate. Nature 465 (7296), 316–321.

Cartolano, M., Castillo, R., Efremova, N., Kuckenberg, M., Zethof, J., 2007. A conserved microRNA module exerts homeotic control over Petunia hybrida and Antirrhinum majus floral organ identity. Nat. Genet. 39, 901–905.

Chen, Z.H., Bao, M.L., Sun, Y.Z., Yang, Y.J., Xu, X.H., Wang, J.H., Zhu, M.Y., 2011. Regulation of auxin response by miR393-targeted transport inhibitor response protein 1 is involved in normal development in Arabidopsis. Plant Mol. Biol. 77, 619–629.

Chen, B., Fiers, M., Dekkers, B.J., Maas, L., van Esse, G.W., Angenent, G.C., Boutilier, K., 2021. ABA signalling promotes cell totipotency in the shoot apex of germinating embryos. J. Exp. Bot. 72 (18), 6418–6436.

Chen, J., Yan, Q., Li, J., Feng, L., Zhang, Y., Xu, J., Liu, Y., 2021. The GRAS gene family and its roles in seed development in litchi (Litchi chinensis Sonn). BMC Plant Biol. 21, 1–15.

Chuck, G., Cigan, A.M., Saeteurn, K., Hake, S., 2007. The heterochronic maize mutant Corngrass1 results from overexpression of a tandem microRNA. Nat. Genet. 39, 544–549.

Couzigou, J.M., Combier, J.P., 2016. Plant microRNAs: key regulators of root architecture and biotic interactions. New Phytol. 212 (1), 22–35.

Curaba, J., Talbot, M., Li, Z., Helliwell, C., 2013a. Over-expression of microRNA171 affects phase transitions and floral meristem determinancy in barley. BMC Plant Biol. 13 (1), 1–10.

Curaba, J., Talbot, M., Li, Z., Helliwell, C., 2013b. Over-expression of microRNA171 affects phase transitions and floral meristem determinancy in barley. BMC Plant Biol. 13, 6.

Curaba, J., Singh, M.B., Bhalla, P.L., 2014. miRNAs in the crosstalk between phytohormone signalling pathways. J. Exp. Bot. 65 (6), 1425–1438.

Dastidar, M.G., Scarpa, A., Mägele, I., Ruiz-Duarte, P., von Born, P., Bald, L., Maizel, A., 2019. ARF5/MONOPTEROS directly regulates miR390 expression in the *Arabidopsis thaliana* primary root meristem. Plant Direct 3 (2), e00116.

Debernardi, J.M., Rodriguez, R.E., Mecchia, M.A., Palatnik, J.F., 2012. Functional specialization of the plant miR396 regulatory network through distinct microRNA-target interactions. PLoS Genet. 8, e1002419.

Di, D.W., Zhang, C., Guo, G.Q., 2015. Involvement of secondary messengers and small organic molecules in auxin perception and signaling. Plant Cell Rep. 34, 895–904.

Dong, Q., Hu, B., Zhang, C., 2022. microRNAs and their roles in plant development. Front. Plant Sci. 13, 824240.

Du, Q., Wang, H., 2015. The role of HD-ZIP III transcription factors and miR165/166 in vascular development and secondary cell wall formation. Plant Signal. Behav. 10 (10), e1078955.

Dubey, S., Saxena, S., Chauhan, A.S., Mathur, P., Rani, V., Chakrabaroty, D., 2020. Identification and expression analysis of conserved microRNAs during short and prolonged chromium stress in rice (Oryza sativa). Environ. Sci. Pollut. Control Ser. 27, 380–390.

Fahlgren, N., Montgomery, T.A., Howell, M.D., Allen, E., Dvorak, S.K., 2006. Regulation of AUXIN RESPONSE FACTOR3 by TAS3 ta-siRNA affects developmental timing and patterning in Arabidopsis. Curr. Biol. 16, 939–944.

Fan, T., Li, X., Yang, W., Xia, K., Ouyang, J., Zhang, M., 2015. Rice osa-miR171c mediates phase change from vegetative to reproductive development and shoot apical meristem maintenance by repressing four OsHAM transcription factors. PLoS One 10 (5), e0125833.

Fang, Y., Xie, K., Xiong, L., 2014. Conserved miR164-targeted NAC genes negatively regulate drought resistance in rice. J. Exp. Bot. 65 (8), 2119–2135.

Feng, H., Duan, X., Zhang, Q., Li, X., Wang, B., Huang, L., Kang, Z., 2014. The target gene of tae-miR164, a novel NAC transcription factor from the NAM subfamily, negatively regulates resistance of wheat to stripe rust. Mol. Plant Pathol. 15 (3), 284–296.

Fouracre, J.P., Poethig, R.S., 2016. The role of small RNAs in vegetative shoot development. Curr. Opin. Plant Biol. 29, 64–72.

Gallagher, K.L., Paquette, A.J., Nakajima, K., Benfey, P.N., 2004. Mechanisms regulating short-root intercellular movement. Curr. Biol. : Cailiao Baohu 14 (20), 1847–1851.

Gangwar, P., Johri, P., Datta, R., Singh, S., Trivedi, M., 2023. Overview on 1 phytopharmaceuticals and biotechnology of herbal plants. In: Phytopharmaceuticals and Biotechnology of Herbal Plants. CRC Press, pp. 1–17.

Gao, Z., Shi, T., Luo, X., Zhang, Z., Zhuang, W., Wang, L., 2012. High-throughput sequencing of small RNAs and analysis of differentially expressed microRNAs associated with pistil development in Japanese apricot. BMC Genom. 13, 1–14.

Gao, R., Wang, Y., Gruber, M.Y., Hannoufa, A., 2018. miR156/SPL10 modulates lateral root development, branching and leaf morphology in Arabidopsis by silencing AGAMOUS-LIKE 79. Front. Plant Sci. 8, 2226.

Gao, X., Zhang, Q., Zhao, Y.Q., Yang, J., He, H.B., Jia, G.X., 2020. The lre-miR159a-LrGAMYB pathway mediates resistance to grey mould infection in Lilium regale. Mol. Plant Pathol. 21 (6), 749–760.

Gifford, M.L., Dean, A., Gutierrez, R.A., Coruzzi, G.M., Birnbaum, K.D., 2008. Cell-specific nitrogen responses mediate developmental plasticity. Proc. Natl. Acad. Sci. USA 105 (2), 803–808.

Gonin, M., Bergougnoux, V., Nguyen, T.D., Gantet, P., Champion, A., 2019. What makes adventitious roots? Plants 8 (7), 240.

Gregory, E.F., Dao, T.Q., Alexander, M.A., Miller, M.J., Fletcher, J.C., 2018. The signaling peptide-encoding genes CLE16, CLE17 and CLE27 are dispensable for Arabidopsis shoot apical meristem activity. PLoS One 13 (8), e0202595.

Guan, X., Pang, M., Nah, G., Shi, X., Ye, W., 2014. MiR828 and miR858 regulate homoeologous MYB2 gene functions in Arabidopsis trichome and cotton fibre development. Nat. Commun. 5, 3050.

Guo, H.S., Xie, Q., Fei, J.F., Chua, N.H., 2005. MicroRNA directs mRNA cleavage of the transcription factor NAC1 to down regulate auxin signals for Arabidopsis lateral root development. Plant Cell 17 (5), 1376–1386.

Gutierrez, L., Bussell, J.D., Pacurar, D.I., Schwambach, J., Pacurar, M., Bellini, C., 2009. Phenotypic plasticity of adventitious rooting in Arabidopsis is controlled by complex regulation of AUXIN RESPONSE FACTOR transcripts and microRNA abundance. Plant Cell 21 (10), 3119–3132.

Hai, B.Z., Qiu, Z.B., He, Y.Y., Yuan, M.M., Li, Y.F., 2018. Characterization and primary functional analysis of Pinus densata miR171. Biol. Plantarum 62 (2), 318–324.

Halder, T., Liu, H., Chen, Y., Yan, G., Siddique, K.H., 2021. Identification of candidate genes for root traits using genotype–phenotype association analysis of near-isogenic lines in hexaploid wheat (*Triticum aestivum* L.). Int. J. Mol. Sci. 22 (7), 3579.

Hochholdinger, F., Zimmermann, R., 2008. Conserved and diverse mechanisms in root development. Curr. Opin. Plant Biol. 11 (1), 70–74.

Hou, Y., Jiang, F., Zheng, X., Wu, Z., 2019. Identification and analysis of oxygen responsive microRNAs in the root of wild tomato (S. habrochaites). BMC Plant Biol. 19, 1–13.

Huang, W., Peng, S., Xian, Z., Lin, D., Hu, G., Yang, L., Li, Z., 2017. Overexpression of a tomato miR171 target gene Sl GRAS 24 impacts multiple agronomical traits via regulating gibberellin and auxin homeostasis. Plant Biotechnol. J. 15 (4), 472–488.

Jiang, N., Meng, J., Cui, J., Sun, G., Luan, Y., 2018. Function identification of miR482b, a negative regulator during tomato resistance to Phytophthora infestans. Horticulture Research 5.

Khan, G.A., Declerck, M., Sorin, C., Hartmann, C., Crespi, M., Lelandais-Brière, C., 2011. MicroRNAs as regulators of root development and architecture. Plant Mol. Biol. 77, 47–58.

Khare, P., Kishore, K., Sharma, D.K., 2017. A study on the standardization parameters of Cassia angustifolia. Asian J. Pharmaceut. Clin. Res. 10 (7), 329–332.

Knauer, S., Holt, A.L., Rubio-Somoza, I., Tucker, E.J., Hinze, A., Pisch, M., Laux, T., 2013. A protodermal miR394 signal defines a region of stem cell competence in the Arabidopsis shoot meristem. Dev. Cell 24 (2), 125–132.

Koyama, T., Sato, F., Ohme-Takagi, M., 2017. Roles of miR319 and TCP transcription factors in leaf development. Plant Physiol. 175 (2), 874–885.

Kravchik, M., Stav, R., Belausov, E., Arazi, T., 2019. Functional characterization of microRNA171 family in tomato. Plants 8 (1), 10.

Landskroner-Eiger, S., Moneke, I., Sessa, W.C., 2013. miRNAs as modulators of angiogenesis. Cold Spring Harb. Perspect. Med. 3 (2).

Laufs, P., Peaucelle, A., Morin, H., Traas, J., 2004. MicroRNA regulation of the CUC genes is required for boundary size control in Arabidopsis meristems. Development 131, 4311–4322.

Lauressergues, D., Couzigou, J.M., Clemente, H.S., Martinez, Y., Dunand, C., Bécard, G., Combier, J.P., 2015. Primary transcripts of microRNAs encode regulatory peptides. Nature 520 (7545), 90–93.

Leng, X., Wang, P., Zhu, X., Li, X., Zheng, T., 2017. Ectopic expression of CSD1 and CSD2 targeting genes of miR398 in grapevine is associated with oxidative stress tolerance. Funct. Integr. Genomics 17, 697–710.

Levesque, M.P., Vernoux, T., Busch, W., Cui, H., Wang, J.Y., Blilou, I., Benfey, P.N., 2006. Whole-genome analysis of the SHORT-ROOT developmental pathway in Arabidopsis. PLoS Biol. 4 (5), e143.

Li, H., Zhang, J., Yang, Y., Jia, N., Wang, C., Sun, H., 2017. miR171 and its target gene SCL6 contribute to embryogenic callus induction and torpedo-shaped embryo formation during somatic embryogenesis in two lily species. Plant Cell Tissue Organ Cult. 130, 591–600.

Li, Z., Liu, D., Xia, Y., Li, Z., Jing, D., Du, J., Zhang, G., 2020. Identification of the WUSCHEL-related homeobox (WOX) gene family, and interaction and functional analysis of TaWOX9 and TaWUS in wheat. Int. J. Mol. Sci. 21 (5), 1581.

Lin, W.Y., Huang, T.K., Chiou, T.J., 2013. NITROGEN LIMITATION ADAPTATION, a target of microRNA827, mediates degradation of plasma membrane–localized phosphate transporters to maintain phosphate homeostasis in Arabidopsis. Plant Cell 25 (10), 4061–4074.

Liu, P.P., Montgomery, T.A., Fahlgren, N., Kasschau, K.D., Nonogaki, H., Carrington, J.C., 2007. Repression of AUXIN RESPONSE FACTOR10 by microRNA160 is critical for seed germination and post-germination stages. Plant J. 52 (1), 133–146.

Liu, H.H., Tian, X., Li, Y.J., Wu, C.A., Zheng, C.C., 2008. Microarray-based analysis of stress-regulated microRNAs in *Arabidopsis thaliana*. RNA 14 (5), 836–843.

Liu, Z.X., Zeng, C.Z., Zeng, W.X., Xu, G.B., Tan, X.F., 2014. Evolutionary and functional diversity of the poplar MIR171 genes. J. Plant Genet. Resour. 15 (2), 313–319.

Llave, C., Xie, Z., Kasschau, K.D., Carrington, J.C., 2002. Cleavage of Scarecrow-like mRNA targets directed by a class of Arabidopsis miRNA. Science 297 (5589), 2053–2056.

Long, J.M., Liu, C.Y., Feng, M.Q., Liu, Y., Wu, X.M., Guo, W.W., 2018. miR156-SPL modules regulate induction of somatic embryogenesis in citrus callus. J. Exp. Bot. 69 (12), 2979–2993.

Lopes, F.L., Galvan-Ampudia, C., Landrein, B., 2021. WUSCHEL in the shoot apical meristem: old player, new tricks. J. Exp. Bot. 72 (5), 1527–1535.

Luo, Y.C., Zhou, H., Li, Y., Chen, J.Y., Yang, J.H., Chen, Y.Q., Qu, L.H., 2006. Rice embryogenic calli express a unique set of microRNAs, suggesting regulatory roles of microRNAs in plant post-embryogenic development. FEBS Lett. 580 (21), 5111–5116.

Maddala, V.K.S., Singh, S., 2023. Essential oils from plants and their role in nanomaterial synthesis characterization and applications. In: Secondary Metabolites Based Green Synthesis of Nanomaterials and Their Applications. Springer Nature Singapore, Singapore, pp. 191–207.

Mallory, A.C., Bartel, D.P., Bartel, B., 2005. MicroRNA-directed regulation of Arabidopsis AUXIN RESPONSE FACTOR17 is essential for proper development and modulates expression of early auxin response genes. Plant Cell 17 (5), 1360–1375.

Martin, A., Adam, H., Diaz-Mendoza, M., Zurczak, M., Gonzalez-Schain, N.D., 2009. Graft-transmissible induction of potato tuberization by the microRNA miR172. Development 136, 2873–2881.

Martín-Rodríguez, J.Á., Leija, A., Formey, D., Hernández, G., 2018. The MicroRNA319d/TCP10 node regulates the common bean–rhizobia nitrogen-fixing symbiosis. Front. Plant Sci. 9, 1175.

Meddya, S., Meshram, S., Sarkar, D., Datta, R., Singh, S., Avinash, G., Thulasinathan, T., 2023. Plant stomata: an unrealized possibility in plant defense against invading pathogens and stress tolerance. Plants 12 (19), 3380.

Meng, Y., Chen, D., Ma, X., Mao, C., Cao, J., Wu, P., Chen, M., 2010. Mechanisms of microRNA-mediated auxin signaling inferred from the rice mutant osaxr. Plant Signal. Behav. 5 (3), 252–254.

Naqvi, A.R., Sarwat, M., Hasan, S., Roychodhury, N., 2012. Biogenesis, functions and fate of plant microRNAs. J. Cell. Physiol. 227, 3163–3168.

Nicolas, M., Cubas, P., 2016. The role of TCP transcription factors in shaping flower structure, leaf morphology, and plant architecture. In: Plant Transcription Factors. Academic Press, pp. 249–267.

Ori, N., Eshed, Y., Chuck, G., Bowman, J.L., Hake, S., 2000. Mechanisms that control knox gene expression in the Arabidopsis shoot. Development 127 (24), 5523–5532.

Ori, N., Cohen, A.R., Etzioni, A., Brand, A., Yanai, O., 2007. Regulation of LANCEOLATE by miR319 is required for compound-leaf development in tomato. Nat. Genet. 39, 787–791.

Parry, G., Calderon-Villalobos, L.I., Prigge, M., Peret, B., Dharmasiri, S., 2009. Complex regulation of the TIR1/AFB family of auxin receptors. Proc. Natl. Acad. Sci. U.S.A. 106, 22540–22545.

Parveen, R., Jain, T., Parveen, R., Singh, S., 2023a. Genome-wide transcriptome profiling of crop plants. In: Genomics, Transcriptomics, Proteomics and Metabolomics of Crop Plants. Academic Press, pp. 145–171.

Parveen, R., Khare, N., Singh, S., Johri, P., 2023b. An in silico outlook for the detection and surveillance of evolving and persistent plant pathogens. In: Detection, Diagnosis and Management of Soil-Borne Phytopathogens. Springer Nature Singapore, Singapore, pp. 17–40.

Patani, A., Patel, A., Prajapati, D., Khare, N., Singh, S., 2023. Various agriculture crop plant-based bioactive compounds and their use in nanomaterial synthesis and applications. In: Nanomaterials from Agricultural and Horticultural Products. Springer Nature Singapore, Singapore, pp. 223–241.

Péret, B., De Rybel, B., Casimiro, I., Benková, E., Swarup, R., Laplaze, L., Bennett, M.J., 2009. Arabidopsis lateral root development: an emerging story. Trends Plant Sci. 14 (7), 399–408.

Powers, S.K., Strader, L.C., 2020. Regulation of auxin transcriptional responses. Dev. Dynam. 249 (4), 483–495.

Raihan, T., Geneve, R.L., Perry, S.E., Rodriguez Lopez, C.M., 2021. The regulation of plant vegetative phase transition and rejuvenation: miRNAs, a key regulator. Epigenomes 5 (4), 24.

Rubio-Somoza, I., Weigel, D., 2011. MicroRNA networks and developmental plasticity in plants. Trends Plant Sci. 16 (5), 258–264.

Samad, A.F., Sajad, M., Nazaruddin, N., Fauzi, I.A., Murad, A.M., Zainal, Z., Ismail, I., 2017. MicroRNA and transcription factor: key players in plant regulatory network. Front. Plant Sci. 8, 565.

Sang, Q., Vayssières, A., Ó'Maoiléidigh, D.S., Yang, X., Vincent, C., Bertran Garcia de Olalla, E., Coupland, G., 2022. MicroRNA172 controls inflorescence meristem size through regulation of APETALA2 in Arabidopsis. New Phytol. 235 (1), 356–371.

Santuari, L., Sanchez-Perez, G.F., Luijten, M., Rutjens, B., Terpstra, I., Berke, L., Heidstra, R., 2016. The PLETHORA gene regulatory network guides growth and cell differentiation in Arabidopsis roots. Plant Cell 28 (12), 2937–2951.

Shi, Q.F., Long, J.M., Yin, Z.P., Jiang, N., Feng, M.Q., Zheng, B., Wu, X.M., 2022. miR171 modulates induction of somatic embryogenesis in citrus callus. Plant Cell Rep. 41 (6), 1403–1415.

Sijen, T., Vijn, I., Rebocho, A., van Blokland, R., Roelofs, D., 2001. Transcriptional and post-transcriptional gene silencing are mechanistically related. Curr. Biol. 11, 436–440.

Singh, A., Sengar, R.S., Singh, A.K., Singh, S., Kumar, A., Ahmad, K., 2019. Impact of genetic engineering on agriculture food crops. Biotech Today: Int. J. Biol. Sci. 9 (1), 17–24.

Song, J.B., Huang, S.Q., Dalmay, T., Yang, Z.M., 2012. Regulation of leaf morphology by microRNA394 and its target LEAF CURLING RESPONSIVENESS. Plant Cell Physiol. 53 (7), 1283–1294.

Sunny, J.S., Snijesh, V.P., Singh, S., 2016. Potential role of mirnas in rheumatoid arthritis. Austin Arthritis 1 (2), 1009.

Sternes, P.R., Moyle, R.L., 2015. Deep sequencing reveals divergent expression patterns within the small RNA transcriptomes of cultured and vegetative tissues of sugarcane. Plant Mol. Biol. Rep. 33, 931–951.

Tirumalai, V., Swetha, C., Nair, A., Pandit, A., Shivaprasad, P.V., 2019. miR828 and miR858 regulate VvMYB114 to promote anthocyanin and flavonol accumulation in grapes. J. Exp. Bot. 70 (18), 4775–4792.

Tungare, K., Shahu, R., Zambare, V., Agarwal, P., Jobby, R., Nisar, N., Jha, P., 2022. Toxicity mitigation of textile dye reactive blue 4 by hairy roots of Helianthus annuus and testing its effect in in vivo model systems. BioMed Res. Int. 2022.

Verma, N., Sao, P., Srivastava, A., Singh, S., 2021. Physiological and molecular responses to drought, submergence and excessive watering in plants. In: Harsh environment and plant resilience: molecular and functional aspects, pp. 305–321.

Vora, Z., Pandya, J., Vaikuntapu, P.R., Singh, S., Asha, S., Purama, R.K., 2023. Genome-wide association of defense hormone crosstalk in plants. In: Hormonal Cross-Talk, Plant Defense and Development. Academic Press, pp. 353–371.

Wang, J.W., Wang, L.J., Mao, Y.B., Cai, W.J., Xue, H.W., Chen, X.Y., 2005. Control of root cap formation by microRNA-targeted auxin response factors in Arabidopsis. Plant Cell 17 (8), 2204–2216.

Wang, L., Mai, Y.X., Zhang, Y.C., Luo, Q., Yang, H.Q., 2010. MicroRNA171c-targeted SCL6-II, SCL6-III, and SCL6-IV genes regulate shoot branching in Arabidopsis. Mol. Plant 3 (5), 794–806.

Wang, Y., Li, K., Chen, L., Zou, Y., Liu, H., Tian, Y., Li, X., 2015a. MicroRNA167-directed regulation of the auxin response factors GmARF8a and GmARF8b is required for soybean nodulation and lateral root development. Plant Physiol. 168 (3), 984–999.

Wang, Y., Wang, Z., Amyot, L., Tian, L., Xu, Z., Gruber, M.Y., Hannoufa, A., 2015b. Ectopic expression of miR156 represses nodulation and causes morphological and developmental changes in Lotus japonicus. Mol. Genet. Genom. 290, 471–484.

Williams, L., Grigg, S.P., Xie, M., Christensen, S., Fletcher, J.C., 2005. Regulation of Arabidopsis shoot apical meristem and lateral organ formation by microRNA miR166g and its AtHD-ZIP target genes. Development 132, 3657–3668.

Wilson, R.C., Doudna, J.A., 2013. Molecular mechanisms of RNA interference. Annu. Rev. Biophys. 42, 217–239.

Xu, M., Hu, T., Zhao, J., Park, M.Y., Earley, K.W., Wu, G., Poethig, R.S., 2016. Developmental functions of miR156-regulated SQUAMOSA PROMOTER BINDING PROTEIN-LIKE (SPL) genes in *Arabidopsis thaliana*. PLoS Genet. 12 (8), e1006263.

Xu, X., Li, X., Hu, X., Wu, T., Wang, Y., Xu, X., Han, Z., 2017. High miR156 expression is required for auxin-induced adventitious root formation via MxSPL26 independent of PINs and ARFs in Malus Xiaojinensis. Front. Plant Sci. 8, 1059.

Yan, X.X., Liu, X.Y., Hong, C.U.I., Zhao, M.Q., 2022. The roles of microRNAs in regulating root formation and growth in plants. J. Integr. Agric. 21 (4), 901–916.

Yan, R., Song, S., Li, H., Sun, H., 2022. Functional analysis of the eTM-miR171-SCL6 module regulating somatic embryogenesis in Lilium pumilum DC. Fisch. Horticulture Research 9, uhac045.

Yang, J.H., Han, S.J., Yoon, E.K., Lee, W.S., 2006. Evidence of an auxin signal pathway, microRNA167-ARF8-GH3, and its response to exogenous auxin in cultured rice cells. Nucleic Acids Res. 34 (6), 1892–1899.

Yang, Z., Yang, D., Ding, X., Gao, Y., Li, D., Xu, T., 2015. MicroRNA expression profiles in conventional and micropropagated Dendrobium officinale. Genes & Genomics 37, 315–325.

Yu, N., Niu, Q.W., Ng, K.H., Chua, N.H., 2015. The role of miR156/SPL s modules in Arabidopsis lateral root development. Plant J. 83 (4), 673–685.

Zang, Q.L., Zhang, Y., Han, S.Y., Li, W.F., Qi, L.W., 2021. Transcriptional and post-transcriptional regulation of the miR171-LaSCL6 module during somatic embryogenesis in Larix kaempferi. Trees (Berl.) 35, 145–154.

Zhai, L., Xu, L., Wang, Y., Huang, D., Yu, R., Limera, C., Liu, L., 2014. Genome-wide identification of embryogenesis-associated microRNAs in radish (Raphanus sativus L.) by high-throughput sequencing. Plant Mol. Biol. Rep. 32, 900–915.

Zhang, S., Zhou, J., Han, S., Yang, W., Li, W., Wei, H., Qi, L., 2010. Four abiotic stress-induced miRNA families differentially regulated in the embryogenic and non-embryogenic callus tissues of Larix leptolepis. Biochem. Biophys. Res. Commun. 398 (3), 355–360.

Zhang, Y.C., Yu, Y., Wang, C.Y., Li, Z.Y., Liu, Q., 2013. Overexpression of microRNA OsmiR397 improves rice yield by increasing grain size and promoting panicle branching. Nat. Biotechnol. 31, 848–852.

Zhang, F., Yang, J., Zhang, N., Wu, J., Si, H., 2022. Roles of microRNAs in abiotic stress response and characteristics regulation of plant. Front. Plant Sci. 13, 919243.

Zhao, Y.F., Peng, T., Sun, H.Z., Teotia, S., Wen, H.L., 2019. MiR1432-OsACOT (acyl-CoA thioesterase) module determines grain yield via enhancing grain filling rate in rice. Plant Biotechnol. J. 17, 712–723.

CHAPTER

Homeostasis in plant microRNAs

8

Ankita Acharya[1], Rituparna Chatterjee[1], Ritam Dutta[1], Srabani Karmakar[2]
[1]Department of Pharmaceutical Technology, JIS University, Agarpara, Kolkata, India; [2]Kingston College of Science, Barasat, Kolkata, India

1. Introduction

In this era of microtechnology, microRNA or miRNA becomes an interesting fact to the scientists. In cell biology, after microRNA invention there was a new beginning. Recently, there has been significant and steadily increasing interest among researchers in the study of microRNAs and their critical roles in cellular functions and investigations into various diseases. It has been founded that in bloodstream in all over the body and cells, the microRNA is found. Mature microRNAs (miRNAs) are a group of naturally existing, small noncoding RNA molecules that typically range from 21 to 25 nucleotides in length (1). It has a major role in expression of gene in human and gene silencing. Gene expression in all complex organisms' cells and tissues is finely controlled and heavily influenced by many situations (such as development, changes in the environment, diseases, or drugs). MiRNA can be found generally in flora, fauna and in some cases in virus also. The major pathway of miRNA production is generally through central dogma, from DNA strand to RNA to miRNA and then to matured microRNAs finally. Typically, miRNAs interact with target mRNAs primarily at the 3′ untranslated region (3′ UTR), leading to mRNA degradation and translational repression. MiRNAs can also interact with other regions such as the 5′ UTR, coding sequence, and gene promoters in certain cases (2). But the interesting data is that miRNA, which is produced from Transcription process, can be the cause of activation and synchronization of transcription process by itself. The gene expression profile index of miRNAs has accepted nowadays as an answer of the cancer. For instance, research has demonstrated that specific levels of circulating miRNAs are implicated in processes such as inflammation, angiogenesis, and cardiac muscle contractility. The intensity and duration of exercise directly influence these miRNA levels, suggesting a potential role in mediating the physiological adaptation of the heart to exercise (3), (4, 5). In 1993, the Ambros and Ruvkun research teams identified the initial microRNA (miRNA), *lin-4*, in *Caenorhabditis elegans* (Lee et al., 1993; Wightman et al., 1993) sparking a revolution in molecular biology. Simultaneously the miRNA target gene is locked by Gary Ravkun. The two incident creates an enlightened pathway of discovery of novel mechanism of posttranscriptional gene regulation. Ambros and

MicroRNA Advances and Application in Plant Biology. https://doi.org/10.1016/B978-0-443-21682-4.00015-4

Ruvkun further investigated *lin-4* and *lin-14*, confirming that *lin-4* is a small noncoding RNA rather than a protein-coding RNA (Almeida et al., 2011; Lee et al., 2004a). In 2000, the Ruvkun laboratory identified a second short RNA in *C. elegans* originating from the let-7 gene (Reinhart et al., 2000). Discoverment of more than thousands of miRNAs through artificial intelligence and random gene cloning and sequencing.

In typical physiological circumstances, miRNAs function as feedback mechanisms that safeguard essential biological processes such as cell proliferation, differentiation, and apoptosis (Esquela-Kerscher and Slack, 2006; Bruce et al., 2015). Dysregulation of a single miRNA or a small group of miRNAs can substantially alter the expression patterns of hundreds of mRNAs, potentially leading cells towards transformation (Sotiropoulou et al., 2009; Ben-Hamo and Efroni, 2015). Lymphocytic leukemia was the first human disease associated with chronic microRNA dysregulation. Several investigations using blood samples and postmortem brain tissue from individuals with mental health disorders have revealed changes in gene expression (Kumarasinghe et al., 2012). MicroRNA is having a major role in cardiomyopathies in case of cardiovascular disorder and stroke. miRNAs can be described as a regulation center of diverse physiological and biological development with more proof of modern studies. The study and analysis of miRNA can depict various information of any dysregulation of function of human body system that cause to a disease such as hepatic, cardiovascular, metabolic, cancer, or immunity disorder. Two enzymes, ST6GAL1 and ST6GAL2, are responsible for miRNA modification. ST6GAL1 is distributed throughout the body, while ST6GAL2 is mostly found in the brain, breast, and colon (Lundberg et al., 2010; Uhlén et al., 2015).

2. Biogenesis and assembly of miRNA

The biogenesis of microRNAs occurs in two main steps involving cleavage at both nuclear and cytoplasmic levels. MicroRNAs are initially transcribed as a precursor molecule called pri-miRNA within the nucleus, either from their own promoters or by sharing promoters with their host genes (Lee et al., 2002). The process of miRNA biogenesis starts with the post- or cotranscriptional processing of RNA polymerase II/III transcripts (Ha and Kim, 2014). RNA polymerase II transcribes microRNAs from DNA to pri-miRNA, and the big main RNA transcript has a 5′ cap and a 3′ polyadenylation (Lee et al., 2004b). Only microRNAs that possess a 5′ cap and 3′ poly-A tail, totaling approximately 85 nucleotides in length, proceed to the subsequent step in the nuclear process (Du and Zamore, 2005). The initial stage in this process involves nuclear cleavage and maturation facilitated by a protein complex referred to as a "microprocessor." This microprocessor comprises two components: Drosha, which is an RNase III endonuclease, and DGCR8, which is an RNA-binding double-stranded protein (Lee et al., 2003; Han et al., 2004; Gregory et al., 2004). A recent study revealed that DGCR8 can enhance the stability of Drosha through protein-protein interactions as a result of cross-regulation. Conversely, Drosha protein negatively regulates DGCR8 levels by cleaving at the 5′ end. Sometimes microprocessor, which associated with protein complex, produce special and particular pre-miRNA (Michlewski et al., 2008; Han et al., 2009). RNA splicing helps to remove introns and join with exons to continue the translation process and through direct splicing of mRNA intron in alternative pathway produce mirtrons. This is the end of nuclear event and now it is the time to transport the complex from nucleus to cytoplasm and starting of cytoplasmic event. Pre-miRNAs, and potentially mirtrons if they are generated in the nucleus, undergo further processing in the cytoplasm by forming a complex with the nucleocytoplasmic transporter exportin-5 and RanGTP. This complexation event helps to prevent the degradation of nucleus and transitive in translocation in cytoplasm (Zeng and Cullen, 2004; Lund and Dahlberg, 2006; Bohnsack et al., 2004). In plant system this transportation is done by HASTY (Cambiagno et al., 2021). Then the further matured pre-miRNA accomplished by RNase III endonuclease Dicer presents also in all eukaryotes except yeast. Dicer is located within RER or cytoplasm in human having multiple domains e.g., PAZ domain (Piwi/Argonaute/Zwilli), which is mainly involved in cleavage (Zhang et al., 2002; Provost et al., 2002). MicroRNA-miRNA complex or miRNA duplex can be produced either by single cleavage or by double cleavage. In a single cleavage step, a complex involving Dicer, the human immunodeficiency virus transactivating response RNA-binding protein (TRBP), and the Protein Kinase R-activating protein (PACT) translocates miRNA (Lee et al., 2006; Chendrimada et al., 2005; Perron and Provost, 2008). Intermolecular dimerization of Dicer's two RNaseIII domains creates a single processing center with two catalytic sites that cleaves the single strand opposite the one previously cleaved by Drosha in the pre-miRNA. The finished product is a double-stranded miRNA duplex with 3′ overhanging ends (Zhang et al., 2004). The specific cleavage site of the pre-microRNA is determined by measuring the distance between the PAZ and RNase domains as if using a ruler (Perron and Provost, 2008). The alternative approach is called twofold cleavage. Diedrichs and Haber suggest a new Ago2-dependent pre-miRNA model to explain this mechanism. They propose that TRBP and Ago2 form a protein complex that recognizes and binds to the pre-miRNA through the PAZ domain of Dicer and/or Ago2. Ago2 then cleaves a single strand of the pre-miRNA about 11−12 nucleotides from the 3′ end, resulting in a nicked hairpin structure termed "Ago2-cleaved precursor miRNA" or "ac-pre-miRNA". The Dicer enzyme subsequently processes the ac-pre-miRNA to generate the double-stranded miRNA duplex (MacFarlane and Murphy, 2010). This duplex consists of two strands: the guide strand, also known as the mature miRNA strand, and the passenger strand (miRNA) (Kim, 2005; Bartel, 2004).

In the final stage of miRNA biogenesis, the RNA duplex unwinds, producing a mature single-stranded miRNA. This single-stranded miRNA then interacts with the RNA-induced silencing complex (RISC), which binds to the target mRNA and initiates gene silencing (Bartel, 2004). It is thought that the RISC creation process in humans is ATP independent (Gregory et al., 2005). Other miRNA regulate the miRNA biogenesis pathway as well. For instance, in mouse, miR-709 binds to specific complementary regions of the primary miR-15a/16-1, thereby inhibiting the processing of Drosha (Tang et al., 2012).

3. Relationship of miRNA homeostasis with NPK macronutrients

One critical aspect of a plant's life cycle essential for its growth and development is macronutrients. They can be the main growth limiting factor in modern agriculture, though fast recycling solves this problem in noncultivated system. Nitrogen (N), phosphorus (P), and potassium (K) are the three primary essential nutrients that often limit crop management, and they are commonly supplied as fertilizers. The nutrients effects in cellular permeability, osmotic regulation, and act as essential metabolite. In this sketch, miRNA becomes an extra regulatory key in the terms of NPK nutrients cope upping. Various research have been done on this factor. Some studies are able to explain the intrinsic relation between the target and miRNA and the how they help in uptake of nutrients and the method of modulation in plant for assimilation with the NPK level (Kulcheski et al., 2015). By physiological and molecular adaptive response, lack of nutrients is deal and effects on plant breeding. Nitrogen is a particularly important nutrient because it is the fundamental structural building component for nucleic acids and amino acids, as well as the regulator of protein synthesis and plays a pivotal role for amino acid and carbon metabolism (Frink et al., 1999). Plants fulfill their nitrogen needs by absorbing nitrogen from the soil in the form of nitrate (NO_3^-), ammonia/ammonium (NH_3/NH_4^+), or urea [CO $(NH_2)_2$], as well as from free amino acids or organic nitrogen through microbial symbiosis in legumes (Williams and Miller, 2001). So nitrogenous fertilizer are added in a large amount in every year but this can emission of nitrous oxide and cause global warming if conversion is not complete (Montzka et al., 2011). Several microRNA have been introduced in recent years to design the N stress and many of those are capable to regulation positively or negatively. Researchers studied nitrogen-responsive miRNAs in repressed phloem sap of *Brassica napus* and identified that miR2111, miR169, and miR827-like sequences were highly influenced by nitrogen status using deep sequencing technology. They also observed that members of the same miRNA families can show varying expression patterns in response to nitrogen deficiency in *Arabidopsis* (Pant et al., 2009). Rice is also a successful crop in associated with miRNA regulation under N stress (Nischal et al., 2012). There are a pattern and the causing microRNAs in negative or positive regulation of N signaling in roots and shoots. miR156 and Squamosa promoter binding protein like (SPL) play crucial roles, with miR156 notably targeting 11 out of the 17 SPL genes (Rhoades et al., 2002). SPL protein clay has an impact on plant growth. Another example is miR160 and auxin response factor (ARF) 16, which are preserved and negatively regulate in N deficient condition. It binds with auxin promoter elements, control root cap formation. DNA binding proteins, ARFs, only present in plants can control the transcription, which is regulated by auxin (Wang et al., 2005). miR167 and ARF8 play roles in regulating adventitious rooting, and in *Arabidopsis*, their expression is influenced by nitrogen stimulus. Studying the differential gene expression across various cell types spanning from inner to outer layers of root tissues is essential to analyze tissue-specific responses and understand how these tissues coordinate their responses to nitrogen signals. In a study, it has been noticed that if the atmosphere is N rich, then lateral root formation is started under starvation indicates cell specific adjustment in lateral root regulation (Gifford et al., 2008). miRNA 169 and Nuclear Factor Y Subunit A (NFYA), also known as heme-activated protein (HAP) or CCAAT-box binding-factor (CBF), are subject to regulation under various conditions. Plant NFYA has been linked to nodule differentiation and drought resistance. During N deprivation, three members of this family, NFYA3, NFYA5, and NFYA8, are substantially activated in shoots and roots, while their repressor, miR169, is inhibited. Nutrient deficiency, such as N, can activate oxidative stresses signals, and NFYA5 expression can modulate downstream genes involved in oxidative stress responses, such as those encoding a subunit of the cytochrome *b*6-f complex, glutathione S-transferases (GST), peroxidases, and an oxidoreductase family protein (Zhao et al., 2011; Combier et al., 2006). The connection between miR171 and Scarecrow-Like 6 (SCL6) effects primary root elongation. N-starvation promotes miR171 induction, which inhibits their targets (SCL6-II, SCL6-III, and SCL6-IV) and so suppresses primary roots extension during this stress situation (Liang et al., 2012). miR292 and AFB3 have a special relation and they express in roots in terms of nitrate. AFB3 expression regulation by nitrate and metabolites produced downstream of nitrate reduction may constitute a mechanism for promptly and precisely adjusting root development based on the availability of external and internal nitrate. AFB3's fast downregulation by miR393 provides a fine-tuned mechanism for the root system to adapt dynamically to N in real time (Vidal et al., 2010). According to one study, miR444a regulates NO_3^- signaling in rice root development and nitrate buildup. Plants overexpressing this miRNA showed decreased expression of the four MADS-box genes as well as reduced nitrate-induced root lateral expansion (Yan et al., 2014). Plants use a particular mechanism to boost nutrients intake with the help of microorganisms found in soil known as N-fixation and microorganism care called N-fixing bacteria. Biological N-fixers reduce dinitrogen into NH_4^+. This type of symbioses occurs between plant legumes and microorganisms known as rhizobia. There is a signaling method between legumes and rhizobium called nod signals. miR159 and miR393 were consistently increased, but miR160 and miR169 were downregulated in response to rhizobia. Two of these miRNAs, miR393 and miR160, have already been found in plant-pathogen interactions and may be implicated in the enhancement of plant basal immunity (Subramanian et al., 2008). Many recent studies are ongoing on soybean nodulation, genetic legumes studies with miRNA. The next, the phosphorus is another important macronutrient and it is the basics for growth, development and propagation of a plant. Plant uptakes P directly as orthophosphate or Pi is done by a special kind of tissue and protein. Two proteins are there, which are present for uptake named-phosphate transporter (PHT1) and phosphate 1 (PHO1). In *Arabidopsis*, nine genes are there for encoding the PHT1 (Yang et al., 2020; Mudge et al., 2002). There are 900–3000 genes are involved in about assimilation and allocation of Pi. The miRNAs muR399 and miR827 plays an essential role in PHT1 and PHO1 in repression and accumulation during crisis period. In *Arabidopsis*, Phosphate Starvation 1 (IPS1) induces long noncoding gene that control miR399 family members. IPS1 has a sequence that is similar to miR399, but it has unpaired nucleotides at the anticipated cleavage site. It has been shown that IPS1 is not cleaved, but instead sequesters miR399 and blocks its function, adding another level of control in PSR (Franco-Zorrilla et al., 2007). Various miRNAs are present in *Arabidopsis* and express their profile in different tissues such as miR163, miR399 in stem, miR19, mi miR395 in root, miR398 in seedling (Kulcheski et al., 2015). In signaling pathways potassium (K^+) are an essential macroconstituents and maintain metabolism during growth and reproduction and response at cold, hyper light condition, hormones. Maximum plants have their physiological mechanism for uptake of K ion and this ion cause biotic and abiotic stresses (Zörb et al., 2014). Potassium regulated genes, which cause homeostasis belong to family ATK1 and KT/KUP/HAK types and it works in root to shoot (Adams and Shin, 2014). Potassium has a remarkable regulatory ability of fine tuning against various stresses. Even in humans, genes of potassium channel works on heart and lung physiology (Vidal et al., 2010). Till now the miRNA, which have the relation with potassium ion channel signaling is miR444a. Involvement of miRNA in N and Pi accumulation has been deeply evaluated. The expression profile of these miRNAs and their corresponding targets (MADS-23, MADS-27a, MADS-27b, and MADS-57) under K^+ deficiency in rice roots have also been investigated. Because of this situation, miR444a levels have decreased slightly. The MADS-23 target was highly stimulated in comparison to the control scenario (Yan et al., 2014). Another study suggested that NH_4^+ and K^+ competition for protein allows for miRNA regulation in barley and *Arabidopsis*. So, it can be depicted that miRNA level and their homeostasis can be affected by any environmental changes. Because miRNA is directly associated to its target function, it can be observed that the majority of miRNAs are engaged in NPK deficiency and are also related to the adaptation process to stress circumstances (Kulcheski et al., 2015).

4. Phosphate regulation in *Arabidopsis* by miRNA (why only *Arabidopsis*?)

After considering NPK constituents, undoubtedly phosphorus is a crucial macroconstituents in plant body for metabolism, enzyme reaction and energy transfer. By a mechanism in which inorganic phosphate (Pi) homeostasis is regulated for adaptation in any changes of environment. Because any change of environment can change the Pi level in plant body. This mechanism is done by ubiquitin conjugating E2 enzyme (Chiou et al., 2006). Ubiquitination is a very particular and highly maintained, ATP dependent biological process, have an essential role in homeostasis (Guo et al., 2023). This equilibrium is regulated by a specific microRNA, miR399. In *Arabidopsis thaliana*, when Pi is depleted, miR399 is increased while its target gene, an ubiquitin-conjugating E2 enzyme, is downregulated (Chiou et al., 2006). Now, why do we exclusively employ *Arabidopsis* rather than other species? To any botanist or plant researcher some parameters are checked. Some basics reason to choose *Arabidopsis* is its small genome size, increase in production in very short time, can be manipulated easily, genetically tractable and the genes present in *Arabidopsis* gives an overall representation of plant kingdom. Many biochemists, biologists, geneticists have chosen *Arabidopsis* foe the genome changing study. After conducting different studies on phosphate regulation, many criteria, such as the process of plant accession and allocation in response to Pi famine, are examined in terms of Pi homeostasis. The *Arabidopsis* Affymetrix ATH1 GeneChip analysis indicated that the expression of At2g33770, a putative miR399 target gene, was fivefold downregulated in Pi-starved roots. miR399 was projected to target various places on the 59 untranslated region (UTR) of At2g33770 mRNA, which encodes an E2 enzyme coupled with ubiquitin, a protein degradation pathway component that is ubiquitin-dependent. And in this process, miR399 regulates Pi starvation negatively, and miR399 expression is elevated in high order in Pi starvation, resulting in miR399 overexpression due to Pi starvation. Overexpression of miR399 inhibits the accumulation of E2 transcripts in the presence of elevated Pi levels. But the production more than normal level has a correlation with E2 mRNA decreased level. If miR399 accumulate in a high level plant body indicated good level of RNA expression (Sunkar and Zhu, 2004). A study indicated that miR399-overexpressing plants had leaves with P levels ranging from 1.8% to 2.0%. Plants with Pi toxicity have been shown to exhibit necrosis or chlorosis on the margins of mature leaves (Pearse et al., 2006). miR399 overexpression impairs Pi mobility again. Inorganic phosphate is up taken by roots through xylem and finally to the leaves. During Pi starvation, old leaves are initially more effected and results purple in color due to anthocyanin. Micro RNA over expression results the leaves remain green. On these leaves, high 33P radioactivity buildup was linked to a chlorosis or necrosis phenotype (Mimura et al., 1996). But the real Pi starvation toxicity is unknown still now. A working hypothesis that is depicted by research conclude that as E2 is the key for protein regulation and high E2 regulation in roots in Pi cause Pi downregulation by direct or indirect mechanism. The Pi toxicity is assumed to done by miR399 overexpression. During starvation of Pi, E2 expression is repressed by miR399 overexpression. Pi allocation in plant by Pi homeostasis is still under research. There are different approaches to analyze and investigate the homeostasis such as plant materials and growth conditions, transgene production and transformation of *Arabidopsis*, P content test, and Pi uptake assay (Chiou et al., 2006).

5. Copper regulation in *Arabidopsis* by miRNA

Cytoplasm of plant cells possess a significant amount of copper proteins, which include plastocyanin, copper/zinc superoxide dismutase, and cytochrome *c* oxidase (Abdel-Ghany, 2009). Copper conditions in limit cause a decrease in the expression of copper/zinc superoxide dismutase, resulting in an increase in the expression of iron superoxide dismutase, present in chloroplasts. We present evidence that, when copper is limited, a micro-RNA called *miR398* directs the degradation of copper/zinc superoxide dismutase mRNA, which mediates the regulation in *Arabidopsis thaliana*. The target of *miR398* is a subunit of the mitochondrial cytochrome *c* oxidase and analysis of sequence showed that the encoded transcripts of cytosolic copper/zinc superoxide dismutase and COX5b-1 (Abdel-Ghany et al., 2005a). Under a low range of copper levels (0.2−0.5 μm), through *miR398* regulation takes place, which concludes that *miR398* is involved in low-level copper regulation. On the other hand, another major copper protein, plastocyanin, was not regulated through *miR398*, which is essential for photosynthetic electron flow in higher plants. In plants, *miR398* is a key factor in copper homeostasis under copper-limited conditions. It regulates the stability of mRNAs of major copper proteins (Abdel-Ghany et al., 2005b).

In plants delivery of copper to various subcellular locations is very important for the activation of enzymes. Plastocyanin (PC) is one of the most well-known copper proteins in tissues, which perform photosynthesis (Abdel-Ghany and Pilon, 2008). Electron transportation is facilitated between cytochrome b_6f complex and photosystem I by this proteins placement in chloroplasts thylakoidal cells lumen.

In some algae a heme-containing cytochrome c_6 is acting as an alternative to PC under low copper conditions, in higher plants PC function replacement protein is not found (Andres-Colas et al., 2006).

In plants another major copper enzyme is copper/zinc superoxide dismutase (Cu, Zn-SOD). The conversion of O_2 to H_2O_2 was done by scavenging of reactive oxygen species (ROS) with the help of superoxide dismutase (SOD) (Arnon, 1949).

In green tissues of *Arabidopsis*, the major isomeric forms of Cu, Zn-SOD are localized to the cytoplasm (CSD1) and to the stroma of chloroplast (CSD2), whereas a minor activity has shown by the peroxisomal isoform (CSD3) (Bernal et al., 2012).

A SOD is called FSD1, found in chloroplasts, which uses an iron cofactor. Manganese requiring MnSOD localizes to mitochondria (Bernal et al., 2007).

Coppers involvement in mitochondria as a cofactor in respiration in electron transport chain and other processes goes beyond its roles in photosynthesis and superoxide scavenging in chloroplasts, in the cytochrome *c* oxidase (COX) complex (Blaby-Haas and Merchant, 2014).

Keeping an optimal energy budget in a cell may require for the allocation of available copper between these two organelles.

In plant cells the delivery system of copper to PC and Cu, Zn-SOD has been studied in some detail (copper enters the cell by means of the CopT family of transporters a cytosolic form of the copper chaperone for SOD (CCS) delivers copper to CSD1).

In *Arabidopsis* for efficient transport of copper across plastid envelop and thylakoid membrane there are two P-type ATPases, PAA1 and PAA2, are required for efficient copper delivery across the plastid envelope and the thylakoid membrane, respectively (Blaby-Haas et al., 2014).

Copper can be transferred into the stromal form of CCS to target stromal CSD2 after transporting it into the stroma by PAA1.

On the other hand, stromal copper can also be transferred to PC through PAA2.

The generation of superoxide occurs when one electron is reduced from oxygen at PSI in illuminated chloroplasts.

Stromal Fe-SOD combined action or Cu, Zn-SOD enzymes, has been used to reduce potential damage caused by superoxide or superoxide-derived ROS, which convert superoxide to oxygen and peroxide, and by stromal ascorbate peroxidases, which convert the formed peroxide into water. In chloroplast where plants encounter a high risk of generating damaging ROS, scavenging systems of ROS is also present to neutralize the effect (Boutigny et al., 2014). Copper delivery is crucial for energy metabolism and superoxide scavenging, but excess copper can be toxic and plants growth on tissue culture media containing copper above 20 μm causes visible damage to plants (Burkhead et al., 2009).

The Fenton reaction may involve free copper ions, which can create highly reactive and damaging hydroxyl radicals. Strict regulation of copper delivery in cells can minimize the risk of free copper ions (Carr and Winge, 2003). An estimation was given for free copper concentration in a yeast cell to be less than 10^{-18} m, but the total amount of copper concentration was estimated to be around 70 μm (Catty et al., 2011).

In plants, the chloroplast is particularly sensitive to copper toxicity, which can be understand by the phenotypes of mutants in genes that encode chloroplast copper transporters. The *Arabidopsis paa1 mutant* exhibits decreased sensitivity to excess copper, which greatly hinders copper delivery to the stroma (Chen et al., 2011).

The thylakoid lumen does not receive copper from *paa2*, which causes hypersensitivity to high copper concentrations. When there is an excess of copper, major copper proteins such as PC and Cu, Zn-SOD may act as a means of removing excess copper. Regulating the abundance of these copper sinks could play a role in the process of copper homeostasis. Three copper proteins increase their levels when copper concentrations increase, as previously demonstrated, CSD1, CSD2, and to a lesser extent PC increase.

Copper availability at the mRNA abundance level influences the abundance of CSD1 and CSD2 in regulation. When copper is scarce, the expression of CSD1 and CSD2 ceases, and FSD1, which requires iron, is increased. In contrast, apo-PC without copper is unstable the condition where PC level is likely to be a consequence of copper delivery to the thylakoid lumen (Chu et al., 2005).

Maintaining the activity of plastid SOD and efficiently delivering limited copper to the PC would be possible through regulation of SOD expression (Chu et al., 2010).

In plants the small 21–22-long RNA molecules, which are known as microRNAs contribute to regulate the expression of specific genes by directing an endo-ribonuclease machinery to degrade target mRNAs.

MicroRNAs are involved in many developmental processes, but they can also respond to stress and also in controlling nutritional status of plant including sulfur and phosphate (Bernal et al., 2007). The *CSD1*, *CSD2*, and *COX5b-1* mRNAs targeted by *miR398*, based on computational analysis of microRNA family, were identified and predicted (Constabel et al., 2000). In response to environmental stress, it was found in a recent report that the *miR398* family is responsible for controlling the expression of *CSD1* and *CSD2*. Regulation of *miR398* is predominantly in response to copper. Even when stress conditions occur, there is no accumulation of *CSD1* and *CSD2* transcripts and proteins when copper is below a critical threshold. When copper is sufficient, the expression of *miR398* can be influenced by copper and stress in a slightly additive way (Divol et al., 2013).

6. Iron regulation in *Arabidopsis* by miRNA

MicroRNAs (miRNAs) are a class of short RNAs that are endogenous and do not code for themselves, and they bind to their target mRNAs to inhibit their translation or destroy them. Several miRNAs regulate how plants react in conditions of sulfate and phosphate destitution. A crucial component required for numerous biological activities, including respiration and photosynthesis, is iron (Fe). Although having abundant on arable land, iron is rarely available in its oxidized state, Fe(III), because of its poor solubility. Contrarily, toxicity in plants can develop as a result of Fe overload, which can happen under anaerobic circumstances in acidic soils. Therefore, it is important to closely monitor the management of Fe levels in plants. Fe is primarily controlled by the absorption, translocation, and recycling procedures. In the Strategy I plant species *Arabidopsis*, Fe is absorbed through a process in which Fe(II) is transported across the root plasma membrane by Fe-regulated transporters (IRTs) after Fe(III) has been decreased to the more soluble Fe(II) form on the root surface by a ferric-chelate reductase (FRO). In contrast, grasses (Plants of Strategy II) exude high-affinity phytosiderophores that bind Fe(III) and mobilize it from the soil. The plasma membrane of root cells has particular transporters that are in charge of absorbing Fe(III)-phytosiderophore complexes (Varotto et al., 2002).

Due to Fe deficit associated to Fe acquisition and translocation, multiple genes have been activated (Connolly et al., 2003). FRO2 expression, which codes for ferric-chelate reductase, was upregulated in *Arabidopsis* when Fe levels were low. In addition, the expression of the primary pathway for entering Fe transporter, which is encoded by AtIRT1, is elevated in cells. The important gene AtFIT1 (Fer-like iron deficiency-induced transcription factor 1), which codes for a basic helix-loop-helix (bHLH) transcription factor, is required for the transcription of FRO2 and IRT1 with reference to the expression of Fe acquisition genes under Fe deficit (Eide et al., 1996). Fe shortage has been shown to cause up-regulation of phytosiderophore expression in Strategy II plants. However, it is unclear how these genes are upregulated and how they manifest themselves. In transgenic tobacco plants, Kobayashi et al. identified the novel Fe deficiency-responsive cis-acting elements IDE1 and IDE2, which cooperatively drive Fe deficiency-specific gene expression in roots. The promoter region of several Fe deficiency-inducible graminaceous genes included homologous sequences of IDE1/IDE2, which is intriguing because it indicates that conserved processes involving IDE have coordinated the regulation of these genes in response to Fe shortage.

Interaction between transcription factors (TFs) and cis-regulatory DNA sequences, also known as cis-elements, in the upstream region of genes is necessary for gene regulation to occur at the transcriptional level because this regulates several target genes responsible for different physiological responses downstream. Many computer methods have been developed to predict promoters or other cis-elements upstream of target genes. Because some of these cis-elements are known to be directly involved in the regulation of gene transcription under Fe scarcity, we will be able to use them to identify potential genes of interest. In a recent study, 249 promoters for 212 rice pre-miRNA sequences were found along with genomic analysis of cis-elements (such as the TATA-box) surrounding the promoter regions of rice microRNA (miRNA) genes (Henriques et al., 2002).

Thus, the computational technique makes it easier to find the target genes across the entire genome and to analyze their functional annotation.

A subclass of endogenous noncoding short RNAs known as microRNAs is crucial for translation suppression or posttranscriptional control of gene expression (Romheld, 1987). Other small interfering RNAs (siRNA), such as trans-acting siRNAs, siRNAs derived from naturally occurring antisense transcripts, and siRNAs linked to repeats, are also members of the family of endogenous small RNAs to which miRNAs belong. In plants, miRNAs regulate several genes involved in development (Marschner et al., 1986), as well as those responding to abiotic stresses (Ivanov et al., 2012; Bauer et al., 2007). The regulation of nutrient deficit in plants is also influenced by a large number of miRNA families (Romheld and Marschner, 1986; Romheld, 1987; Robinson et al., 1999). Although these miRNAs were found to respond to environmental stresses, no plant miRNAs linked to Fe homeostasis have been identified to date. For this study, we generated a library of size-fractionated RNAs from *Arabidopsis* cultivated in a Fe-deficient medium. Eight conserved members of five families were among the cloned miRNAs. The transcriptional research reveals that these miRNAs exhibit differential expression in response to Fe deficiency. In addition, we looked for IDE1/IDE2 motifs upstream of every miRNA gene throughout the whole *Arabidopsis* genome. It was predicted that homologous IDE1 and IDE2 sequences would encircle the promoter regions of 24 miRNAs. A transcriptional research revealed that the majority of miRNA genes were expressed more when Fe was deficient.

Iron is a necessary component for plants, but too much iron can be hazardous, hence iron homeostasis is closely maintained in plants. So, the list below includes various specialized uptake techniques.

Plant that takes up reduced iron.

Fe^{3+} is first reduced to Fe^{2+} by proton extrusion, which aids in its solubilization, before being taken up by a particular transporter into the root epidermis (Marschner et al., 1986; Romheld and Marschner, 1986; Romheld, 1987). The P-type H+-ATPase AHA2 of the Strategy I plant *Arabidopsis thaliana* extrudes protons into the rhizosphere (Santi and Schmidt, 2009). Iron shortage in the root epidermis is induced through ferric iron reduction by the ferric chelate reductase FRO2 (FERRIC REDUCTION OXIDASE) (Robinson et al., 1999; Connolly et al., 2003). Finally, IRT1 (IRON-REGULATED TRANSPORTER (Eide et al., 1996; Henriques et al., 2002; Varotto et al., 2002; Vert et al., 2002) is responsible for ferrous iron translocation in root cells. The transcription factor FIT (FER-LIKE IRON DEFICIENCY-INDUCED TRANSCRIPTION FACTOR) controls AHA2, FRO2, and IRT1 expression (Colangelo and Guerinot, 2004; Jakoby et al., 2004; Ivanov et al., 2012; Bauer et al., 2007). IRT1 and FRO2 are exclusively induced under iron insufficient conditions (Colangelo and Guerinot, 2004) and only in leaves under iron deficient ectopic expression of IRT1 and FRO2. This is true even under robust constitutive FIT expression. FIT protein-protein interactions have been discovered that can explain this behavior, which indicates that posttranslational regulation happened in the case of FIT activity (Jakoby et al., 2004). Chlorosis, stunted development, and mortality are signs of iron deprivation in loss-of-function mutants of fit (Colangelo and Guerinot, 2004; Jakoby et al., 2004).

The "transportome", which among other things contains genes involved in transition metal detoxification, is activated by iron shortage, activating many functional modules. FIT, IRT1, NAS1, NAS2, FRD3, and the gene for a protein belonging to the 2OG-Fe(II) oxygenase family are examples of iron-responsive genes whose expression is ethylene-dependent. EIN3/EIL1 appear to be necessary for the photosystems to reorganize in order to minimize photooxidative damage, and this might also be accomplished in the presence of an iron deficiency by increasing iron uptake through EIN3/EIL1-mediated mechanisms. In *Arabidopsis*, a copper deficiency results in secondary iron deficiency and increases IRT1 and FRO2 expression. There is interaction between copper, iron absorption, and phosphate deprivation, and there are cues for various copper functions under these conditions. In A. thaliana rosette leaves, enhanced CuSOD expression was shown to functionally substitute FeSODs (iron-containing superoxide dismutases) when iron levels were low. This was made possible by the discovery that microRNAs negatively control CuSOD (copper-containing superoxide dismutase) genes. Following the creation of iron shortage, different sets of genes are up- and down-regulated at various time periods, according to time course transcriptomic analysis (Bauer et al., 2007).

In Fe-deficient conditions, there was an up-regulation of roughly 50 miRNAs and a down-regulation of about 31 miRNAs. The differently expressed miRNAs may have a significant impact on citrus plants ability to withstand Fe shortages. Citrus plants may be more resilient to stress because of decreased expression of miR172; reduced expression of miR395 may regulate S homeostasis; increased expression of miR319 and miR477 may inhibit plant growth; reduced expression of miR398 and miR408 may regulate Cu homeostasis and activate Cu/Zn superoxide dismutase activity; and reduced expression of miR395 may regulate lignin accumulation. The miRNAs identified in this study offered a framework for comprehending the ways citrus plants have adapted to iron scarcity.

7. Sulfur regulation by miRNA

Since cysteine and methionine are not synthesized by our bodies, sulfur-containing amino acids must be consumed through food (Allen et al., 2005). Microbes and plants convert inorganic sulfur from the environment into amino acids. In addition to amino acids, secondary metabolites of plants also contain sulfur-like compounds, such as glucosinolates, which are beneficial to humans as phytochemicals that prevent cancer in the diet (Liang et al., 2015; Maruyama-Nakashita et al., 2006).

Due to variations in the sulfur content of the soil, plants constantly adapt to their environment. Low sulfur levels caused the activation of high-affinity sulfate transporters (the SULTR1 family), which results in increased root sulfate absorption efficiency (Talalay and Fahey, 2001). The recently discovered *Arabidopsis* transcription factor SULFUR LIMITATION1 (SLIM1) is one of the primary regulators of the S response. SULTR1, which is essential for sulfate absorption, is expressed as a result of SLIM1, a high-affinity sulfate transporter. SLIM1 coordinatedly regulated genes involved in sulfur metabolism (both for assimilation and reuse of sulfate) in S conditions, according to transcriptome profiles of slim1 mutants (Jones-Rhoades et al., 2006).

Noncoding short RNAs regulate expression of gene posttranscriptionally in addition to transcriptionally. MicroRNA (miRNA), a subclass of these RNA types, is produced from precursor molecules (pre-miRNA) having a hairpin structure (Huang et al., 2010). RNA-induced silencing complexes (RISCs), which are composed of ribonucleoproteins and contain mature miRNAs, are 21 nucleotides long and short. The miRNA directs the RISC to mRNAs that have a target site that is complementary to the miRNA. Although RISC cleaves the majority of miRNA-targeted plant mRNAs, it has recently been shown that RISC also inhibits translation, just like in animals (Brodersen et al., 2008). A portion of plant miRNAs target genes involved in metabolism, while the majority of plant miRNAs control transcription factors expression. One of them, microRNA-395 (miR395), is increased by S conditions and has been demonstrated to modulate two ATP sulfurylases (APS1 and APS4) and a low-affinity sulfate transporter (SULTR2;1). However, sulfur deficiency results in the upregulation of several miR395 target genes, which raises concerns regarding the mechanism governing miRNA regulation.

Several previously known genes were required for miRNA biogenesis (Huang et al., 2010). Little is known about the regulation of miRNAs during transcription.

MYB, a transcription factor is associated with the phosphate starvation response 1 (PHR1), controls the expression of miR399. In turn, activation of miR399 expression also depends on inorganic phosphate deficiency (Bari et al., 2006). Phr1 mutants significantly reduced miR399 expression. Because mature miR395 was not induced in slim1 mutants, it was determined in the current investigation that the transcription factor SLIM1 also controls protein-coding genes expression involved in sulfur metabolism in addition to controlling the accumulation of miR395. However, we found that SLIM1 was not required for the induction of SULTR2;1, suggesting that the induction of SULTR2;1 mRNA is likely the result of another sulfur-sensitive component. We further show that MIR395 loci are expressed in root, leaf, and stem vasculature. Both show increased accumulation in the root vasculature in the S environment, although miR395 targets SULTR2;1. The target gene and precise expression patterns of miRNAs—SULTR2;1 mRNA is deposited in xylem parenchyma cells while miR395 is expressed in partner phloem cells—explain this puzzle.

Many studies have been conducted to determine and confirm the role of S-sensitive miRNAs. These S-sensitive miRNAs control sulfate uptake, transport, and assimilation in plants by generally targeting multiple transcription factors involved in the auxin signaling cascade and stress response (Huang et al., 2010). Deep sequencing revealed five novel miRNAs and 27 conserved miRNAs expressed in *Arabidopsis* during SO2 stress. In SO2-treated samples, novel miRNAs, miR66 and miR67 increased more than six-fold, while miR14, miR20 and miR43 decreased seven-fold compared with control. The targets of differentially produced miRNAs were connected to cellular and metabolic activities, signal transmission, and nutritional homeostasis, according to comparative deep sequencing of *Arabidopsis* sRNAs treated with various nutrient deficiencies, including C, N, and S. Under conditions of C, N, and/or S deprivation, respectively, miR169b/c, miR826, and miR395 were all particularly upregulated. In contrast, under conditions of C, N, and S deficiency, many miRNAs, including miR167, miR172, miR397, miR398, miR399, miR408, miR775, miR827, miR841, miR857, and miR2111, were suppressed. According to (Liang et al., 2015), sequencing of Brassica napus sRNA population revealed novel conserved miRNAs sensitive to S deficiency and Cd stress. This revealed the target genes of miRNAs and their involvement in both stressors in concert. After thorough sequencing analysis, *Chlamydomonas reinhardtii* miRNAs showed differential expression under S- and sulfur-deficient conditions. Furthermore, studies suggest that prohibitin (PHB), Squamosa-promoter binding (SPB), and Sulfur Limitation 1 (SLIM1) are sulfur-responsive transcription factors that bind to the promoter of diet-responsive miRNAs (Sun et al., 2012).

8. Homeostasis associated with nutrients by miRNA regulation

Small regulatory RNAs are increasingly being recognized as one of the most ubiquitous signaling molecules that regulate posttranscriptional gene expression. These small RNAs are typically less than 30 nucleotides in length and interact with their target mRNAs through perfect complementary base pairing. The binding of small regulatory RNAs to their target mRNAs attenuates gene expression through various mechanisms, such as mRNA degradation, translational repression and modulation of the transcriptional machinery. Small regulatory RNAs, in some cases, may also modify the stability or function of transcription factors that further modify gene expression. Small regulatory RNAs play a critical role in a variety of biological processes, including developmental, differential, apoptotic and homeostatic processes. They are emerging as important regulators of cell processes and are becoming an invaluable tools in posttranscriptional gene regulation research (Bartel, 2004). In plants, two main types of small RNAs, small interfering RNAs (siRNAs) and microRNAs (miRNAs), are likely to be involved in the silencing of gene expression. Small interfering RNA (siRNA) is a type of RNA that is used in RNA interference (RNAi) mediated gene silencing technologies. siRNAs are 20–24 nucleotide (n) regulatory RNAs, small in size, that are derived from endogenous genome regions or exogenously derived nucleic acid duplexes (Jamalkandi and Masoudi-Nejad, 2009; Khraiwesh et al., 2012). Dicer-like enzymes (in *Arabidopsis*, DCL) break the double-stranded duplex with a hairpin loop into short, ideal duplexes. These duplexes are then loaded onto the RISC (RNA induced silencing complex). The RISC recognizes the target gene(s) or transcript(s) and binds to suitable complementary nucleotide sequence(s) to control gene expression at posttranscriptional or transcriptional levels (Schauer et al., 2002). However, miRNAs (20–24 nucleotides lengths) are extracted from endogenous primary transcript(s) after processing. When mature miRNA is processed, they are incorporated into RISC complex(s) and directed towards degradation/translational inhibition (Lee et al., 2004b; Vazquez et al., 2010).

In plants, miRNAs are mainly involved in several physiological processes, including growth, development and organogenesis, as well as responses to various biotic and abiotic stresses (Sunkar and Zhu, 2004; Chen, 2004; Kidner and Martienssen, 2005; Kruszka et al., 2012; Lu et al., 2008; Zhang et al., 2008). Many novel miRNAs are involved in the uptake and transport of important nutrients in plants (Fischer et al., 2013; Jones-Rhoades and Bartel, 2004; Kehr, 2013). We have reviewed the current state of the art of nutrient-responsive miRNAs and their roles in nutrient homeostasis in plants, which may help us to understand miRNAs as key regulators of nutrient load in plants. In the future, miRNAs may uncover a new paradigm in crop biofortification or mineral bioavailability in cereal grains.

8.1 miRNAs in nutrient homeostasis

8.1.1 miRNA and phosphorus nutrition

Pi is an inorganic micronutrient that plants need to grow and develop, is frequently a limiting element in plant nutrition. Fertilized land contains acceptable amount of P, but plants are limited in their ability to absorb very small amounts of P due to poor availability. In order to solubilize exogenous Pi or release Pi from organic substances, and to increase gene expression of particular exporter/importer transport proteins in order to adapt to phosphate starvation, plants synthesis many nucleases and phosphatases (Abel, 2011; Poirier et al., 2002). Small RNAs (miRNAs) in particular, have been associated with differential regulation of phosphate-related gene expression in a variety of plants, including *Arabidopsis*, Rice, Wheat, Barley, Maize, Soybean, White Lupin, Tomato, and other plants based on recent studies (Pant et al., 2009; Gu et al., 2010; Hackenberg et al., 2013; Lundmark et al., 2010; Pei et al., 2013; Xu et al., 2013; Zhao et al., 2013a; Zhu et al., 2010). In *Arabidopsis*, the lack of pi results in the synthesis of miR399 a-f families and is rapidly transferred into the shoots. In the UTR (5′ Untranslated region), 200−400 kb upstream of the phosphate over accumulator 2 (PHO2) gene, miR399 attaches to the five homologous sites and causes mRNA destruction (Pant et al., 2008). PHO2, a major transporter for phosphate mobilization, interacts with E2, an enzyme that conjugates with ubiquitin. High levels of pi accumulate in the shoots due to decreased PHO2. The function and regulation of miR399 generation have been well demonstrated. Myeloblastosis (MYB) is a myelodysplastic disorder. The myeloblastosis (MYB) transcription factors such as phosphate starvation response 1 (PHR1) binding sites, such as *P1BS*, *PHO*, and *PHO*-like, are present upstream of miR399 promoter region (Bari et al., 2006; Xu et al., 2013; Zeng et al., 2010). In addition, miR399f expression is regulated by MYB2 transcription factor binding to the *cis*-acting regions of MIR399f promoter as shown in *Arabidopsis* (Baek et al., 2013). Long noncoding (NSC) RNA is another signaling molecule induced by phosphate starvation 1 (IPS1) that protects the *PHO2* transcript and disrupts miR399-mediated *PHO2* gene regulation (Franco-Zorrilla et al., 2007; Hu et al., 2011; Huang et al., 2011).

Phosphate limitation also increases the abundance of three additional miRNAs in *Arabidopsis*: miR778, miR827, and miR2111. Interestingly, within 3 h of Pi reintroduction, the abundance of both miR2111 and miR778 rapidly decreases by approximately 2-fold. There are two copies of the gene miR2111 in the genome of *Arabidopsis* and both loci exhibit similar phosphate response activities (Pant et al., 2009). In recent studies, the dual reactions of miR827, miR399, miR2111 and miR827 have been further elucidated. miR2111 activity reverses under nitrogen (N)-starvation as opposed to phosphate (P)-starvation (Liang et al., 2012). The miR827 and miR399 miRNAs are also known to function in N-starvation by targeting the *nitrogen limitation adaptation* (*NLA*) gene and are thought to enhance the expression of the PHO2 transporter. When nitrate levels are low, the *NLA* mutation plays a significant role in the phosphorus homeostasis of the plant by excessive Pi accumulation. A similar condition is seen in the *Arabidopsis* plants that is overexpressed to miR827 and supports the gene regulation of the phosphate transporters mediated by miR827 (Kant et al., 2011). In rice, Pi accumulation is followed by a disturbance of phosphorus mobilization in old leaves by miR827 manifestation (Wang et al., 2012). Notably, the expression of miR398a is regulated by P, carbon (C) and N limitation, indicating a more general role in mineral homeostasis. Repression due to the C limitation correlates with the induction of the sucrose (Dugas and Bartel, 2008). miR159a is up regulated in soybean roots when there is a P deficiency while miR319a, miR396a, miR398b, miR1507a are down regulated. The differential regulation of miRNAs during P-starvation depends on the frequency of phosphorus responsive motifs (P responsive motifs) in the promoter of the miRNA genes. The number of P-responsive motifs in the *cis*-acting region of miRNA genes is reported to be higher than those of nonresponding miRNA genes (Zeng et al., 2010).

8.1.2 miRNAs in nitrogen uptake and transport

Nitrogen, a critical component of protein, chlorophyll, nucleic acids, and other substances, is required for plant growth and development. Except for the mechanism of biological N fixation in legumes, N transport is dependent on the external acquisition of N by roots from the soil. By up- or down-regulating a particular collection of exporter or importer proteins, plants can adjust to N-limited soil conditions. Several tiny miRNA families are specifically in control of the differential regulation of transporters. Solexa high-throughput sequencing can divide the miRNA-responsive populations in *Arabidopsis* into two groups based on the prevalence of N-starvation. MiR156, MiR169, MiR171, MiR160, MiR319, MiR826, MiR829, MiR839, and MiR846 are among the N-starvation-induced (NSI) miRNA families, but MiR167, MiR172, MiR399, MiR395, MiR850, MiR857, MiR863, and MiR827 are not (Liang et al., 2012). In addition, it has been discovered that 15 and 14 miRNA families, respectively, are responsive in N-limiting conditions in rice and maize (Nischal et al., 2012; Xu et al., 2011). The miR156 family has been discovered in the highest abundance in *Arabidopsis*, and miR156h is considered to be the most significant of the three miR156 family members. MiR156, on the other hand, adversely regulates miR172, which prevents the reproductive phase by extending the juvenile stage. By targeting auxin response factor (ARF16/17) and SCL6 regulatory proteins, respectively, miR170 promotes primary roots growth under N-starvation, but miR160 inhibits lateral roots development (Liang et al., 2012). In contrast, miR167 biogenesis is disturbed in N-limiting conditions, which reduces ARF6/8 expression (Gifford et al., 2008; Jones-Rhoades and Bartel, 2004).

In Medicago truncatula, miR169 and miR172 control HAP2 and AP2 genes expression, which is crucial for nodule growth. By sustaining low N in the roots, the reduced abundances of miR169 during N-limitation increase HAP2 gene expression and the subsequent differentiation of nodule formation (Pant et al., 2009). However, diverse *Arabidopsis*, soybean, maize, and other plant species have been found to express various miR169 family members, including miR169a, miR169bc, miR169d-g, and miR169h-n. During N-starvation, they are directly linked to the up-regulation of nitrate transporters (Zhao et al., 2011; Liang et al., 2012; Wang et al., 2013; Xu et al., 2011; Zhao et al., 2012; Zhao et al., 2013b). The expression of miR169d-g has been found to increase in conditions of low N availability.

8.1.3 miRNAs in sulfur homeostasis

Sulfur (S), an essential inorganic mineral, is mostly absorbed by roots from the soil as sulfate. When the S is exposed to high levels of stress, it is absorbed into a large number of Fe−S proteins, cofactors and lipoic acids, which are involved in the primary and secondary metabolism (Rausch and Wachter, 2005), as well as cysteine, methionine, glutathione, glucosinolate compounds, and various Fe−S proteins. Sulfate is transported through the xylem or phloem in *Arabidopsis* by a number of cell specific transporters. These transporters include Sulfate Transporters 1;1, 2;1 and 2; 2. According to S abundance or famine, miR395 primarily controls the transporters expression (Kawashima et al., 2009). It's interesting to note that, in contrast to sulfate lack, sulfate limitation promotes miR395 development and the low affinity sulfate transporter SULTR2;1. The xylem parenchyma is where SULTR2;1 is mainly present. When there is a sulfate deficit, miR395 plays an important role in remobilizing sulfate ions between leaves and is very abundant in phloem parenchyma (Liang et al., 2010). This makes it more easier for sulfate ions to move from roots to shoots because miR395 limits the expression of SULTR 2;1 in xylem parenchyma. Additionally, a lack of S causes the roots to produce more of the protein SULFUR LIMITATION 1 (SLIM1), which in turn triggers the activation of other sulfate transporters and increases the uptake of S (Kawashima et al., 2011; Liang and Yu, 2010). By blocking the expression of ATP sulfurylase genes such as APS1, APS3, and APS4, which catalyze the initial step of S assimilation, miR395's role in S assimilation has also been clarified (Matthewman et al., 2012). As seen in Brassica napus (Huang et al., 2010), the expression levels of miR156, miR160, miR164, miR167, miR168, and miR394 are also affected by S deficiency.

8.1.4 miRNAs in copper homeostasis

An essential micronutrient known as copper (Cu) is predominantly used in the electron transport chain as a cofactor of metabolic enzymes and protein complexes. It is a crucial component of the compound plastocyanin, which actively takes part in the movement of electrons through the chloroplast granules during photosynthesis. By serving as a cofactor of Copper/Zinc superoxide dismutase (CSD), Cu plays a significant part in the defense against oxidative stress reactions. Superoxide dismutase SOD1 (CCS1) gene expression is downregulated under Cu restriction by the activation of miR398 (Beauclair et al., 2010; Sunkar et al., 2006). The CSD1 and CSD2 apoprotein receive the Cu ions from CCS1, a chaperone protein. Because of the lack of Cu, iron (Fe) superoxide dismutase replaces the function of Cu/Zn superoxide in Cu-deprived conditions. In *Arabidopsis*, miR398 expression suppresses the cytochrome *c* oxidase subunit, a mitochondrial internal membrane protein encoded by cyclo-mono-oxygenase (Abdel-Ghany and Pilon, 2008). Cu starvation has been shown to upregulate other miRNA families such as miR397, miR408 and miR857, which in turn reduce the expression of the laccase and plastocyanin genes. MiR408 inhibits the three laccase genes (LAC), including LAC3, LAC12, and LAC13, whereas miR397 degrades the mRNA of LAC2, LAC4, and LAC17. LAC7 transcripts are largely the target of miR857. Notably, miRNAs involved in maintaining Cu homeostasis promote the production of plastocyanin by inhibiting the manufacture of nonessential Cu enzymes and maintain Cu homeostasis by changing the availability of Cu among different protein subgroups (Gifford et al., 2008).

8.1.5 miRNA in other mineral homeostasis

Phaseolus has reduced levels of expression for a large number of genes, including several transcription factors such as TEOSINTE-LIKE 1, CYCLOIDEA, PROLIFERATING CELL FACTOR 1 (TCP), HAPLESS (HAP2), SCARECROW-LIKE, NO APICAL MERISTEM (NAC), *Arabidopsis* transcription activation factor, CUP SHAPED COTYLEDON, and serine threonine protein kinase (Valdes-Lopez et al., 2010).

Because they are the primary cofactors for a number of important metabolic enzymes, such as Fe–S cluster proteins and ferredoxin molecules, iron and zinc are essential micronutrients for plants (Couturier et al., 2013; Forieri et al., 2013). The small amounts of Fe and zinc found in cereal grains in the food we eat have a major impact on human nutrition. The absorption, transport and loading of mineral ions in cereal grains directly influence their bioavailability of minerals (Aung et al., 2013). Recently, it was discovered that in *Arabidopsis*, miRNA-mediated control of Fe-related transporters and storage proteins occurs. A different way that Fe deficiency regulates miR398 is in the opposite direction from how Cu deficiency regulates it. Fe deficiency lowers the expression of the following miR397, miR398a, miR398b, miR398c, miR398s, miR399, miR408 and miR2111, whereas Cu deficiency increases the expression of these genes. Cu deficiency controls *CSD1*, *CSD2* expression (Buhtz et al., 2010; Waters et al., 2012). With regard to the investigation of gene expression during Fe homeostasis, the Cu–Fe interaction is thus another novel discovery. Also, eight miRNAs from five families (miR159; miR164; miR172; miR173; miR394) were previously identified as Fe-responsive families in a short RNA library population of *Arabidopsis*. It is intriguing that the Fe deficiency responsive cis-acting elements 1 and 2 (IDE1/IDE2), which mimic the Fe-responsive gene families that are controlled during Fe deficiencies, were discovered in the promoters of 24 miRNA genes in *Arabidopsis*. However, the roles of other miRNAs in Fe transport and storage have not been clearly established.

The overexpression of several miRNA families such as miR166, miR171, miR172, miR398, miR399, miR319, etc. targets several members of the gene families such as transporters, leads to the Zn deficiency in Sorghum bicolor (Li et al., 2013). Interestingly, two members of miRNA family have been found to play opposite roles in regulating expression of CSD gene family. Reduced miR528 increased the number of CSD transcripts in seeds and leaves while increased miR398 decreased gene expression of CSD in roots. miRNAs also play a role in the transport of micro and macro nutrients by acting as endogenous signals (Kehr, 2013; Liu et al., 2009; Marín-González et al., 2012). Other metal stresses, such as Al, Cd, Hg and Cd stress have also been shown to increase or decrease several miRNAs. It is thought that these miRNAs play a role in plants adaptive mechanisms by regulating the expression of several stress-related genes (Zeng et al., 2012; Zhang et al., 2013; Zhou et al., 2012).

9. Biotic stress associated with miRNA

Globally, nematode parasites, insect pests, bacteria, fungi, and viruses affect crop plant quality and yield. Agrochemicals have been widely used to manage infections and pests, which has resulted in biotic factors becoming resistant to agrochemicals. To satisfy the demands of the expanding population, research has turned to developing crop plants that are resistant to pathogens and pests while also increasing yields. Expression of regulatory genes, including miRNAs, which regulate the expression of defense-related genes in plants, happens in response to biotic stress. Due to biotic stress, miRNAs may be overexpressed because they have been targeted by the inhibitors of the genes that are resistant to biotic stress and are therefore overexpressed. The miRNAs aid in the downregulation of resistance genes as a result of biotic stress through negative regulation. The host plant's genes involved in defense are interfered with by the diseases or pests' production of miRNAs. In contrast, in resistant plants, biotic stress causes an overexpression of the miRNAs that positively control the resistant genes. Since viruses and pests can be controlled to improve crop quality, miRNAs can be used for this purpose.

The first miRNA, miR393, has been characterized as biotic stress dependent (Navarro et al., 2006). The miRNA (Windels and Vazquez, 2011), which targets the Transport Inhibitor Response 1 (TIR1)/Auxin-signaling F-Box (AFB) auxin co-receptors (Sunkar and Zhu, 2004; Jones-Rhoades and Bartel, 2004; Naseem et al., 2015; Pruss et al., 2008; Robert-Seilaniantz et al., 2011), is found in 15 different plant species. A modified RACE test (Navarro et al., 2006) has demonstrated that miR393 is cleaving the transcripts of TIR1, AFB2, and AFB3 (Navarro et al., 2006). When there is no auxin, the auxin/indole acetic acid (AUX/IAA) proteins attached to the auxin response factor (ARF) proteins are inactivated. TIR1 facilitates the ubiquitination of the AUX/IAA protein during the sensing of auxin. TIR1 releases the ARFs, which in turn activate (or suppress, depending on the member of the ARF family) the transcription of the auxin responsive genes (Dharmasiri et al., 2005a,b). TIR1 is involved in this process together with AFB1, AFB2, and AFB3. Flagellin 22 (flg22), a bacterial Microbe-Associated Molecular Pattern (MAMP), induces the production of miR393, which in turn causes RNA-mediated suppression of TIR1/AFB proteins and regulation of auxin signaling (Lee et al., 1993; Navarro et al., 2006). Salicylic acid (SA) but not jasmonic acid (JA) signaling is repressed by auxin, therefore increasing miR393 and thereby suppressing auxin signaling raises SA levels, which accounts for the increased resistance to biotrophic bacteria (Naseem et al., 2015). For instance, miR393 overexpression in *Arabidopsis* confers resistance to hemibiotrophic bacteria *Pseudomonas syringae* (Windels and Vazquez, 2011), *Phytophthora capsici* (Hou et al., 2019), and biotrophic oomycetes *Hyaloperonospora arabidopsidis* (Robert-Seilaniantz et al., 2011), but renders these plants susceptible to necrotrophic pathogens such as *Alternaria brassicicola* (Robert-Seilaniantz et al., 2011). The same results were obtained for soybean with the help of the *Phytophthora sojae* (Wong et al.,

2014) and for *Nicotiana tabacum* with the help of the gall-forming Agrobacterium tumefaciens (Pruss et al., 2008), but not by the avirulent strain. This miRNA is sensitive to MAMPs, as demonstrated by the increased miR393 observed in the roots of susceptible and resistant cultivars, as well as in roots treated with heat-inactivated P. sojae cells. The expression of miR393 was also increased in maize due to the presence of the hemibiotrophic mold *Colletotrichum graminicola* (Balmer et al., 2017).

In contrast, miR393 down-regulation was discovered in aubergine challenged with *Verticillium dahliae* (Yang et al., 2013), supporting the notion that this is a plant defense mechanism. This necrotrophic fungus showed increased susceptibility to infection when the transport of auxin was blocked or the seedlings were treated with SA or IAA. Taken together, these results suggest that miR393 is increased by MAMPs, bacterial infection, oomycetes and biotrophic, hemibiotrophic fungi, and decreased by necrotrophic fungi. We illustrated its function when plants were exposed to *Pseudomonas syringae* (Lee et al., 1993; Fahlgren et al., 2010; Gonzá et al., 2012; Hou et al., 2019) and *Verticillium dahliae* (Yang et al., 2013), two of the few miRNAs that respond to more than one biotic stressor and play a role in modulating the auxin signaling pathways.

The alteration of miR393-induced secondary metabolite formation that results from tryptophan, including IAA, camalexin (CL), and indole glucosinolates (IG). Both CL and IG exhibit antimicrobial activities, with CL being more effective in killing necrotrophic fungi by cell membrane damage and IG being more effective in killing insects, biotrophic pathogens, and bacteria (Robert-Seilaniantz et al., 2011). Since miR393 exerts direct antibacterial activity through the production of isoflavonoid, it also functions as a positive regulator, whose activation increases the defensive responses (Wong et al., 2014). Our results demonstrate that miRNA393 functions as a tool for precise regulation of hormone balance and secondary metabolites formation, essential for a well-functioning immune system against pathogens and pests.

153 miRNAs in all were shown to be sensitive to the biotrophic powdery mildew fungus *Blumeria graminis*, and miR393 was one of them. There is a difference in miR393 expression between the almost isogenic wheat lines JD8 and JD8-Pm30, the latter of which contains the powdery mildew R gene Pm30. This indicates that this miRNA plays a role in the plant defense responses (González-Lamothe et al., 2012).

Notably, during AGO2 loading, the complementary miR393b contributes to resistance to bacterial infection (Zhang et al., 2011). AGO2- miR393 module induction was observed in *Arabidopsis* postinoculation of *P. syringae*. This module targets and blocks the MEMB12, which is an SNARE gene located in the Golgi region, resulting in increased secretion of PR1, a pathogenesis-related protein. This miRNA/miRNA duplex is a demonstration of multilayered immune regulation using several AGO family members in *Arabidopsis*.

10. Conclusion

The genetic control of P and N transporters under the corresponding nutritional stress situations has been the focus of significant research efforts. The involvement of miRNAs in phosphate metabolism, including inositol phosphate or phytic acid production, is an interesting area for future investigation because phytate is a significant source of inorganic phosphorus. The cereals biomass is influenced by nitrate metabolism and the N to C ratio. There is a need for further research on the functions of miRNA in the nitrate transporters and the control of metabolism, in which many enzymes as aspartate amino transferase, glutamine synthase, and glutamate dehydrogenase are involved. Future studies on differentially expressed miRNAs and their regulatory functions in different Fe and Zn transporters related to metal homeostasis could help to establish a new platform for Fe and Zn loading. The edible parts of wheat grains contain only small amount of the two essential dietary minerals Fe and Zn. Several biotechnological techniques have been used to improve the concentration of these metals in milled grain, but few studies have been conducted on the function of miRNAs in controlling individual transporters or transcription factors in the Fe supply. The regulatory mechanism of Fe-related gene is essential for understanding dietary Fe requirements, and can provide a clear picture of how genes are regulated during nutrient homeostasis.

A total of 153 miRNAs were found to respond to the biotrophic powdery mildew fungus *Blumeria graminis*, and miR393 was one of them. Interestingly, miR393 showed differential expression in the powdery mildew R gene-containing JD8-Pm30 wheat line, demonstrating the key role that this miRNA plays in plant defense responses.

References

Abdel-Ghany, S.E., 2009. Contribution of plastocyanin isoforms to photosynthesis and copper homeostasis in *Arabidopsis thaliana* grown at different copper regimes. Planta 229, 767–779. https://doi.org/10.1007/s00425-008-0869-z.

Abdel-Ghany, S.E., Pilon, M., 2008. MicroRNA-mediated systemic down-regulation of copper protein expression in response to low copper availability in *Arabidopsis*. J. Biol. Chem. 283, 15932–15945. https://doi.org/10.1074/jbc.M801406200.

Abdel-Ghany, S.E., Muller-Moule, P., Niyogi, K.K., Pilon, M., Shikanai, T., 2005a. Two P-type ATPases are required for copper delivery in *Arabidopsis thaliana* chloroplasts. Plant Cell 17, 1233–1251. https://doi.org/10.1105/tpc.104.030452.

Abdel-Ghany, S.E., Burkhead, J.L., Gogolin, K.A., Andres-Colas, N., Bodecker, J.R., Puig, S., et al., 2005b. AtCCS is a functional homolog of the yeast copper chaperone Ccs1/Lys7. FEBS Lett. 579, 2307–2312.

Abel, S., 2011. Phosphate sensing in root development. Curr. Opin. Plant Biol. 14, 303–309. https://doi.org/10.1016/j.pbi.2011.04.007.

Adams, E., Shin, R., 2014. Transport, signaling, and homeostasis of potassium and sodium in plants. J. Integr. Plant Biol. 56 (3), 231–249.

Allen, E., Xie, Z., Gustafson, A.M., Carrington, J.C., 2005. microRNA-directed phasing during trans-acting siRNA biogenesis in plants. Cell 121, 207–221.

Almeida, M.I., Reis, R.M., Calin, G.A., December 2011. MicroRNA history: discovery, recent applications, and next frontiers. Mutat. Res. 717 (1–2), 1–8.

Andres-Colas, N., Sancenon, V., Rodriguez-Navarro, S., Mayo, S., Thiele, D.J., Ecker, J.R., et al., 2006. The *Arabidopsis* heavy metal P-type ATPase HMA5 interacts with metallochaperones and functions in copper detoxification of roots. Plant J. 45, 225–236. https://doi.org/10.1111/j.1365-313X.2005.02601.x.

Arnon, D.I., 1949. Copper enzymes in isolated chloroplasts. Polyphenoloxidase in *Beta vulgaris*. Plant Physiol. 24, 1–15.

Aung, M.S., Masuda, H., Kobayashi, T., Nakanishi, H., Yamakawa, T., Nishizawa, N.K., 2013. Iron biofortification of Myanmar rice. Front. Plant Sci. 4, 158. https://doi.org/10.3389/fpls.2013.00158.

Baek, D., Kim, M.C., Chun, H.J., Kang, S., Park, H.C., Shin, G., et al., 2013. Regulation of miR399f transcription by AtMYB2 affects phosphate starvation responses in *Arabidopsis*. Plant Physiol. 161, 362–373. https://doi.org/10.1104/pp.112.205922.

Balmer, A., De Paoli, E., Si-Ammour, A., Mauch-Mani, B., Balmer, D., 2017. Signs of Silence: Small RNAs and Antifungal Responses in *Arabidopsis thaliana* and *Zea mays* Plant Engineering. InTech. https://doi.org/10.5772/intechopen.69795.

Bari, R., Pant, B.D., Stitt, M., Scheible, W.-R., 2006. PHO2, microRNA399, and PHR1 define a phosphate-signaling pathway in plants. Plant Physiol. 141, 988–999.

Bartel, D.P., January 2004. MicroRNAs: genomics, biogenesis, mechanism, and function. Cell 116 (2), 281–297.

Bauer, P., Ling, H.Q., Guerinot, M.L., 2007. FIT, the FER-like iron deficiency induced transcription factor in *Arabidopsis*. Plant Physiol. Biochem. 45 (5), 260–261.

Beauclair, L., Yu, A., Bouche, N., 2010. microRNA directed cleavage and translational repression of the copper chaperone for superoxide dismutase mRNA in *Arabidopsis*. Plant J. 62, 454–462. https://doi.org/10.1111/j.1365-313X.2010.04162.x.

Ben-Hamo, R., Efroni, S., 2015. MicroRNA regulation of molecular pathways as a generic mechanism and as a core disease phenotype. Oncotarget 6 (3), 1594–1604.

Bernal, M., Testillano, P.S., Alfonso, M., del Carmen Risueño, M., Picorel, R., Yruela, I., 2007. Identification and subcellular localization of the soybean copper P1B-ATPase GmHMA8 transporter. J. Struct. Biol. 158, 46–58. https://doi.org/10.1016/j.jsb.2006.10.016.

Bernal, M., Casero, D., Singh, V., Wilson, G.T., Grande, A., Yang, H., et al., 2012. Transcriptome sequencing identifies SPL7-regulated copper acquisition genes FRO4/FRO5 and copper dependence of iron homeostasis in *Arabidopsis*. Plant Cell 24, 738–761. https://doi.org/10.1105/tpc.111.090431.

Blaby-Haas, C.E., Merchant, S.S., 2014. Lysosome-related organelles as mediators of metal homeostasis. J. Biol. Chem. 289, 28129–28136. https://doi.org/10.1074/jbc.R114.592618.

Blaby-Haas, C.E., Padilla-Benavides, T., Stübe, R., Argüello, J.M., Merchant, S.S., 2014. Evolution of a plant-specific copper chaperone family for chloroplast copper homeostasis. Proc. Natl. Acad. Sci. U.S.A. 111, E5480–E5487. https://doi.org/10.1073/pnas.1421545111.

Bohnsack, M.T., Czaplinski, K., Görlich, D., February 2004. Exportin 5 is a RanGTP-dependent dsRNA-binding protein that mediates nuclear export of pre-miRNAs. RNA 10 (2), 185.

Boutigny, S., Sautron, E., Finazzi, G., Rivasseau, C., Frelet-Barrand, A., Pilon, M., et al., 2014. HMA1 and PAA1, two chloroplast-envelope PIB-ATPases, play distinct roles in chloroplast copper homeostasis. J. Exp. Bot. 65, 1529–1540. https://doi.org/10.1093/jxb/eru020.

Brodersen, P., Sakvarelidze-Achard, L., Bruun-Rasmussen, M., Dunoyer, P., Yamamoto, Y.Y., Sieburth, L., Voinnet, O., 2008. Widespread translational inhibition by plant miRNAs and siRNAs. Science 320 (5880), 1185–1190.

Bruce, J.P., Hui, A.B.Y., Shi, W., Perez-Ordonez, B., Weinreb, I., Xu, W., et al., 2015. Identification of a microRNA signature associated with risk of distant metastasis in nasopharyngeal carcinoma. Oncotarget 6 (6), 4537–4550.

Buhtz, A., Pieritz, J., Springer, F., Kehr, J., 2010. Phloem small RNAs, nutrient stress responses, and systemic mobility. BMC Plant Biol. 10, 64. https://doi.org/10.1186/1471-2229-10-64.

Burkhead, J.L., Reynolds, K.A., Abdel-Ghany, S.E., Cohu, C.M., Pilon, M., 2009. Copper homeostasis. New Phytol. 182, 799–816. https://doi.org/10.1111/j.14698137.2009.02846.x.

Cambiagno, D.A., Giudicatti, A.J., Arce, A.L., Gagliardi, D., Li, L., Yuan, W., et al., March 2021. HASTY modulates miRNA biogenesis by linking pri-miRNA transcription and processing. Mol. Plant 14 (3), 426–439.

Carr, H.S., Winge, D.R., 2003. Assembly of cytochrome c oxidase within the mitochondrion. Acc. Chem. Res. 36, 309–316. https://doi.org/10.1021/ar0200807.

Catty, P., Boutigny, S., Miras, R., Joyard, J., Rolland, N., Seigneurin-Berny, D., 2011. Biochemical characterization of AtHMA6/PAA1, a chloroplast envelope Cu(I)-ATPase. J. Biol. Chem. 286, 36188–36197. https://doi.org/10.1074/jbc.M111.241034.

Chen, X., 2004. A microRNA as a translational repressor of APETALA2 in *Arabidopsis* flower development. Science 303, 2022–2025. https://doi.org/10.1126/science.1088060.

Chen, C.C., Chen, Y.Y., Tang, I.C., Liang, H.M., Lai, C.C., Chiou, J.M., et al., 2011. *Arabidopsis* SUMO E3 ligase SIZ1 is involved in excess copper tolerance. Plant Physiol. 156, 2225–2234. https://doi.org/10.1104/pp.111.178996.

Chendrimada, T.P., Gregory, R.I., Kumaraswamy, E., Norman, J., Cooch, N., Nishikura, K., et al., August 2005. TRBP recruits the Dicer complex to Ago2 for microRNA processing and gene silencing. Nature 436 (7051), 740–744.

Chiou, T.J., Aung, K., Lin, S.I., Wu, C.C., Chiang, S.F., Su, C.L., 2006. Regulation of phosphate homeostasis by MicroRNA in *Arabidopsis*. Plant Cell 18 (2), 412.

Chu, C.C., Lee, W.C., Guo, W.Y., Pan, S.M., Chen, L.J., Li, H.M., et al., 2005. A copper chaperone for superoxide dismutase that confers three types of copper/zinc superoxide dismutase activity in *Arabidopsis*. Plant Physiol. 139, 425–436. https://doi.org/10.1104/pp.105.065284.

Chu, H.H., Chiecko, J., Punshon, T., Lanzirotti, A., Lahner, B., Salt, D.E., et al., 2010. Successful reproduction requires the function of *Arabidopsis* Yellow Stripe-Like1 and Yellow Stripe-Like3 metal-nicotianamine transporters in both vegetative and reproductive structures. Plant Physiol. 154, 197–210. https://doi.org/10.1104/pp.110.159103.

Colangelo, E.P., Guerinot, M.L., 2004. The essential basic helix-loop-helix protein FIT1 is required for the iron deficiency response. Plant Cell 16 (12), 3400–3412.

Combier, J.P., Frugier, F., De Billy, F., Boualem, A., El-Yahyaoui, F., Moreau, S., et al., November 2006. MtHAP2-1 is a key transcriptional regulator of symbiotic nodule development regulated by microRNA169 in Medicago truncatula. Genes Dev. 20 (22), 3084.

Connolly, E.L., Campbell, N.H., Grotz, N., Prichard, C.L., Guerinot, M.L., 2003. Overexpression of the FRO2 ferric chelate reductase confers tolerance to growth on low iron and uncovers posttranscriptional control. Plant Physiol. 133 (3), 1102–1110.

Constabel, C.P., Yip, L., Patton, J.J., Christopher, M.E., 2000. Polyphenol oxidase from hybrid poplar: cloning and expression in response to wounding and herbivory. Plant Physiol. 124, 285–295. https://doi.org/10.1104/pp.124.1.285.

Couturier, J., Touraine, B., Briat, J.-F., Gaymard, F., Rouhier, N., 2013. The iron-sulfur cluster assembly machineries in plants: current knowledge and open questions. Front. Plant Sci. 4, 259. https://doi.org/10.3389/fpls.2013.00259.

Dharmasiri, N., Dharmasiri, S., Estelle, M., 2005a. The F-box protein TIR1 is an auxin receptor. Nature 435, 441–445. https://doi.org/10.1038/nature03543.

Dharmasiri, N., Dharmasiri, S., Weijers, D., Lechner, E., Yamada, M., Hobbie, L., Ehrismann, J.S., Jürgens, G., Estelle, M., 2005b. Plant development is regulated by a family of auxin receptor F box proteins. Dev. Cell 9, 109–119. https://doi.org/10.1016/j.devcel.2005.05.014.

Divol, F., Couch, D., Conéjéro, G., Roschzttardtz, H., Mari, S., Curie, C., 2013. The *Arabidopsis* YELLOW STRIPE LIKE4 and 6 transporters control iron release from the chloroplast. Plant Cell 25, 1040–1055. https://doi.org/10.1105/tpc.112.107672.

Du, T., Zamore, P.D., November 2005. microPrimer: the biogenesis and function of microRNA. Development 132 (21), 4645–4652.

Dugas, D.V., Bartel, B., 2008. Sucrose induction of *Arabidopsis* miR398 represses two Cu/Zn superoxide dismutases. Plant Mol. Biol. 67, 403–417. https://doi.org/10.1007/s11103-008-9329-1.

Eide, D., Broderius, M., Fett, J., Guerinot, M.L., 1996. A novel iron-regulated metal transporter from plants identified by functional expression in yeast. Proc. Natl. Acad. Sci. U.S.A. 93 (11), 5624–5628.

Esquela-Kerscher, A., Slack, F.J., April 2006. Oncomirs - microRNAs with a role in cancer. Nat. Rev. Cancer 6 (4), 259–269.

Fahlgren, N., Jogdeo, S., Kasschau, K.D., Sullivan, C.M., Chapman, E.J., Laubinger, S., Smith, L.M., Dasenko, M., Givan, S.A., Weigel, D., Carrington, J.C., 2010. MicroRNA gene evolution in *Arabidopsis lyrata* and *Arabidopsis thaliana*. Plant Cell 22, 1074–1089. https://doi.org/10.1105/tpc.110.073999.

Fischer, J.J., Beatty, P.H., Good, A.G., Muench, D.G., 2013. Manipulation of microRNA expression to improve nitrogen use efficiency. Plant Sci. 210, 70–81. https://doi.org/10.1016/j.plantsci.2013.05.009.

Forieri, I., Wirtz, M., Hell, R., 2013. Toward new perspectives on the interaction of iron and sulfur metabolism in plants. Front. Plant Sci. 4, 357. https://doi.org/10.3389/fpls.2013.00357.

Franco-Zorrilla, J.M., Valli, A., Todesco, M., Mateos, I., Puga, M.I., Rubio-Somoza, I., et al., August 2007. Target mimicry provides a new mechanism for regulation of microRNA activity. Nat. Genet. 39 (8), 1033–1037.

Frink, C.R., Waggoner, P.E., Ausubel, J.H., February 1999. Nitrogen fertilizer: retrospect and prospect. Proc. Natl. Acad. Sci. U.S.A. 96 (4), 1175–1180.

Gifford, M.L., Dean, A., Gutierrez, R.A., Coruzzi, G.M., Birnbaum, K.D., January 2008. Cell-specific nitrogen responses mediate developmental plasticity. Proc. Natl. Acad. Sci. U.S.A. 105 (2), 803–808.

González-Lamothe, R., El Oirdi, M., Brisson, N., Bouarab, K., 2012. The conjugated auxin indole-3-acetic acid-aspartic acid promotes plant disease development. Plant Cell 24, 672–777. https://doi.org/10.1105/tpc.111.095190.

Gregory, R.I., Yan, K.P., Amuthan, G., Chendrimada, T., Doratotaj, B., Cooch, N., et al., November 2004. The Microprocessor complex mediates the genesis of microRNAs. Nature 432 (7014), 235–240.

Gregory, R.I., Chendrimada, T.P., Cooch, N., Shiekhattar, R., November 2005. Human RISC couples microRNA biogenesis and posttranscriptional gene silencing. Cell 123 (4), 631–640.

Gu, M., Xu, K., Chen, A., Zhu, Y., Tang, G., Xu, G., et al., 2010. Expression analysis suggests potential roles of microRNAs for phosphate and arbuscular mycorrhizal signaling in *Solanum lycopersicum*. Physiol. Plant. 138, 226–237. https://doi.org/10.1111/j.1399-3054.2009.01320.

Guo, H.J., Rahimi, N.R., Tadi, P., March 2023. Biochemistry, ubiquitination. StatPearls. https://www.ncbi.nlm.nih.gov/books/NBK556052/.

Ha, M., Kim, V.N., July 2014. Regulation of microRNA biogenesis. Nat. Rev. Mol. Cell Biol. 15 (8), 509–524.

Hackenberg, M., Huang, P.J., Huang, C.Y., Shi, B.J., Gustafson, P., Langridge, P.A., 2013. Comprehensive expression profile of microRNAs and other classes of non-coding small RNAs in barley under phosphorous-deficient and -sufficient conditions. DNA Res. 20, 109–125. https://doi.org/10.1093/dnares/dss037.

Han, J., Lee, Y., Yeom, K.H., Kim, Y.K., Jin, H., Kim, V.N., December 2004. The Drosha-DGCR8 complex in primary microRNA processing. Genes Dev. 18 (24), 3016–3027.

Han, J., Pedersen, J.S., Kwon, S.C., Belair, C.D., Kim, Y.K., Yeom, K.H., et al., January 2009. Posttranscriptional crossregulation between Drosha and DGCR8. Cell 136 (1), 75–84.

Henriques, R., Jasik, J., Klein, M., Martinoia, E., Feller, U., Schell, J., Pais, M.S., Koncz, C., 2002. Knock-out of *Arabidopsis* metal transporter gene IRT1 results in iron deficiency accompanied by cell differentiation defects. Plant Mol. Biol. 50 (4–5), 587–597.

Hou, Y., Zhai, Y., Feng, L., Karimi, H.Z., Rutter, B.D., Zeng, L., Choi, D.S., Zhang, B., Gu, W., Chen, X., Ye, W., Innes, R.W., Zhai, J., Ma, W., 2019. A Phytophthora effector suppresses trans-kingdom RNAi to promote disease susceptibility. Cell Host Microbe 25, 153–165. https://doi.org/10.1016/j.chom.2018.11.007.

Hu, B., Zhu, C., Li, F., Tang, J., Wang, Y., Lin, A., et al., 2011. LEAF TIP NECROSIS1 plays a pivotal role in the regulation of multiple phosphate starvation responses in rice. Plant Physiol. 156, 1101–1115. https://doi.org/10.1104/pp.110.170209.

Huang, S.Q., Xiang, A.L., Che, L.L., Chen, S., Li, H., Song, J.B., et al., 2010. A set of miRNAs from Brassica napus in response to sulphate deficiency and cadmium stress. Plant Biotechnol. J. 8, 887–899. https://doi.org/10.1111/j.1467-7652.2010.00517.x.

Huang, C.Y., Shirley, N., Genc, Y., Shi, B., Langridge, P., 2011. Phosphate utilization efficiency correlates with expression of low-affinity phosphate transporters and noncoding RNA, IPS1, in barley. Plant Physiol. 156, 1217–1229. https://doi.org/10.1104/pp.111.178459.

Ivanov, R., Brumbarova, T., Bauer, P., 2012. Fitting into the harsh reality: regulation of iron-deficiency responses in dicotyledonous plants. Mol. Plant 5 (1), 27–42.

Jakoby, M., Wang, H.Y., Reidt, W., Weisshaar, B., Bauer, P., 2004. FRU (BHLH029) is required for induction of iron mobilization genes in *Arabidopsis thaliana*. FEBS Lett. 577 (3), 528–534.

Jamalkandi, S.A., Masoudi-Nejad, A., 2009. Reconstruction of *Arabidopsis thaliana* fully integrated small RNA pathway. Funct. Integr. Genomics 9, 419–432. https://doi.org/10.1007/s10142-009-0141-z.

Jones-Rhoades, M.W., Bartel, D.P., 2004. Computational identification of plant microRNAs and their targets, including a stress-induced miRNA. Mol. Cell. 14, 787–799. https://doi.org/10.1016/j.molcel.2004.05.027.

Jones-Rhoades, M.W., Bartel, D.P., Bartel, B., 2006. MicroRNAs and their regulatory roles in plants. Annu. Rev. Plant Biol. 57, 19–53.

Kant, S., Peng, M., Rothstein, S.J., 2011. Genetic regulation by NLA and microRNA827 for maintaining nitrate-dependent phosphate homeostasis in *Arabidopsis*. PLoS Genet. 7. https://doi.org/10.1371/journal.pgen.1002021 e1002021.

Kawashima, C.G., Yoshimoto, N., Maruyama-Nakashita, A., Tsuchiya, Y.N., Saito, K., Takahashi, H., 2009. Sulphur starvation induces the expression of microRNA-395 and one of its target genes but in different cell types. Plant J. 57, 313–321. https://doi.org/10.1111/j.1365-313X.2008.03690.x.

Kawashima, C.G., Matthewman, C.A., Huang, S., Lee, B.R., Yoshimoto, N., Koprivova, A., et al., 2011. Interplay of SLIM1 and miR395 in the regulation of sulfate assimilation *Arabidopsis*. Plant J. 66, 863–876. https://doi.org/10.1111/j.1365-313X.2011.04547.x.

Kehr, J., 2013. Systemic regulation of mineral homeostasis by micro RNAs. Front. Plant Sci. 4, 145. https://doi.org/10.3389/fpls.2013.00145.

Khraiwesh, B., Zhu, J.K., Zhu, J., 2012. Role of miRNAs and siRNAs in biotic and abiotic stress responses of plants. Biochim. Biophys. Acta 1819, 137–148. https://doi.org/10.1016/j.bbagrm.2011.05.001.

Kidner, C.A., Martienssen, R.A., 2005. The developmental role of microRNA in plants. Curr. Opin. Plant Biol. 8, 38–44. https://doi.org/10.1016/j.pbi.2004.11.008.

Kim, V.N., May 2005. MicroRNA biogenesis: coordinated cropping and dicing. Nat. Rev. Mol. Cell Biol. 6 (5), 376–385.

Kruszka, K., Pieczynski, M., Windels, D., Bielewicz, D., Jarmolowski, A., Kulinska-Szweykowska, Z., et al., 2012. Role of micro RNAs and other sRNAs of plants in their changing environments. J. Plant Physiol. 169, 1664–1672. https://doi.org/10.1016/j.jplph.2012.03.009.

Kulcheski, F.R., Córrea, R., Gomes, I.A., De Lima, J.C., Margis, R., June 2015. NPK macro-nutrients and microRNA homeostasis. Front. Plant Sci. 6 (June).

Kumarasinghe, N., Tooney, P.A., Schall, U., July 2012. Finding the needle in the haystack: a review of microarray gene expression research into schizophrenia. Aust. N. Z. J. Psychiatr. 46 (7), 598–610.

Lee, R.C., Feinbaum, R.L., Ambros, V., December 1993. The *C. elegans* heterochronic gene lin-4 encodes small RNAs with antisense complementarity to lin-14. Cell 75 (5), 843–854.

Lee, Y., Jeon, K., Lee, J.T., Kim, S., Kim, V.N., September 2002. MicroRNA maturation: stepwise processing and subcellular localization. EMBO J. 21 (17), 4663–4670.

Lee, Y., Ahn, C., Han, J., Choi, H., Kim, J., Yim, J., et al., September 2003. The nuclear RNase III Drosha initiates microRNA processing. Nature 425 (6956), 415–419.

Lee, R., Feinbaum, R., Ambros, V., 2004a. A short history of a short RNA. Cell 116 (2 Suppl. l).

Lee, Y., Kim, M., Han, J., Yeom, K.H., Lee, S., Baek, S.H., et al., 2004b. MicroRNA genes are transcribed by RNA polymerase II. EMBO J. 23 (20), 4051–4060.

Lee, Y., Hur, I., Park, S.Y., Kim, Y.K., Mi, R.S., Kim, V.N., February 2006. The role of PACT in the RNA silencing pathway. EMBO J. 25 (3), 522–532.

Li, Y., Zhang, Y., Shi, D., Liu, X., Qin, J., Ge, Q., et al., 2013. Spatial-temporal analysis of zinc homeostasis reveals the response mechanisms to acute zinc deficiency in *Sorghum bicolor*. New Phytol. 200, 1102–1115. https://doi.org/10.1111/nph.12434.

Liang, G., Yu, D., 2010. Reciprocal regulation among and miR395 posttranscriptionally APS and SULTR2; 1 *Arabidopsis thaliana*. Plant Signal. Behav. 5, 1257–1259. https://doi.org/10.1111/j.1365-313X.

Liang, G., Yang, F., Yu, D., 2010. MicroRNA395 mediates regulation of sulfate accumulation and allocation in *Arabidopsis thaliana*. Plant J. 62, 1046–1057. https://doi.org/10.1111/j.1365-313X.2010.04216.x.

Liang, G., He, H., Yu, D., November 2012. Identification of nitrogen starvation-responsive microRNAs in *Arabidopsis thaliana*. PLoS One 7 (11).

Liang, G., Ai, Q., Yu, D., 2015. Uncovering miRNAs involved in crosstalk between nutrient deficiencies in *Arabidopsis*. Sci. Rep. 5, e11813. https://doi.org/10.1038/srep11813.

Liu, T.Y., Chang, C.Y., Chiou, T.J., 2009. The long-distance signaling of mineral macronutrients. Curr. Opin. Plant Biol. 12, 312–319. https://doi.org/10.1016/j.pbi.2009.04.004.

Lu, S., Sun, Y.H., Chiang, V.L., 2008. Stress-responsive microRNAs in populus. Plant J. 55, 131–151. https://doi.org/10.1111/j.1365-313X.2008.03497.x.

Lund, E., Dahlberg, J.E., 2006. Substrate selectivity of exportin 5 and Dicer in the biogenesis of microRNAs. Cold Spring Harbor Symp. Quant. Biol. 71, 59–66.

Lundberg, E., Fagerberg, L., Klevebring, D., Matic, I., Geiger, T., Cox, J., et al., 2010. Defining the transcriptome and proteome in three functionally different human cell lines. Mol. Syst. Biol. 6.

Lundmark, M., Korner, C.J., Nielsen, T.H., 2010. Global analysis of microRNA in *Arabidopsis* in response to phosphate starvation as studied by locked nucleic acid-based microarrays. Physiol. Plant. 140, 57–68. https://doi.org/10.1111/j.1399-3054.2010.01384.x.151.

MacFarlane, L.-A., Murphy, P.R., December 2010. MicroRNA: biogenesis, function and role in cancer. Curr. Genom. 11 (7), 537.

Marín-González, E., Suárez-López, P., 2012. "And yet it moves": cell-to-cell and long-distance signaling by plant microRNAs. Plant Sci. 196, 18–30. https://doi.org/10.1016/j.plantsci.2012.07.009.

Marschner, H., Romheld, V., Kissel, M., 1986. Different strategies in higher-plants in mobilization and uptake of iron. J. Plant Nutr. 9 (3–7), 695–713.

Maruyama-Nakashita, A., Nakamura, Y., Tohge, T., Saito, K., Takahashi, H., 2006. *Arabidopsis* SLIM1 is a central transcriptional regulator of plant sulphur response and metabolism. Plant Cell 18, 3235–3251.

Matthewman, C.A., Kawashima, C.G., Huska, D., Csorba, T., Dalmay, T., Kopriva, S., 2012. miR395 is a general component of the sulfate assimilation regulatory network in *Arabidopsis*. FEBS Lett. 586, 3242–3248. https://doi.org/10.1016/j.febslet.2012.06.044.

Michlewski, G., Guil, S., Semple, C.A., Cáceres, J.F., November 2008. Posttranscriptional regulation of miRNAs harboring conserved terminal loops. Mol. Cell 32 (3), 383–393.

Mimura, T., Sakano, K., Shimmen, T., 1996. Studies on the distribution, re-translocation and homeostasis of inorganic phosphate in barley leaves. Plant. Cell Environ. 19 (3), 311–320.

Montzka, S.A., Dlugokencky, E.J., Butler, J.H., August 2011. Non-CO2 greenhouse gases and climate change. Nature 476 (7358), 43–50.

Mudge, S.R., Rae, A.L., Diatloff, E., Smith, F.W., 2002. Expression analysis suggests novel roles for members of the Pht1 family of phosphate transporters in *Arabidopsis*. Plant J. 31 (3), 341–353.

Naseem, M., Srivastava, M., Tehseen, M., Ahmed, N., 2015. Auxin crosstalk to plant immune networks: a plant-pathogen interaction perspective. Curr. Protein Pept. Sci. 16, 389–394. https://doi.org/10.2174/1389203716666150330124911.

Navarro, L., Dunoyer, P., Jay, F., Arnold, B., Dharmasiri, N., Estelle, M., Voinnet, O., Jones, J.D.G., 2006. A plant miRNA contributes to antibacterial resistance by repressing auxin signaling. Science 312, 436–439. https://doi.org/10.1126/science.1126088.

Nischal, L., Mohsin, M., Khan, I., Kardam, H., Wadhwa, A., Abrol, Y.P., et al., December 2012. Identification and comparative analysis of microRNAs associated with low-N tolerance in rice genotypes. PLoS One 7 (12).

Pant, B.D., Buhtz, A., Kehr, J., Scheible, W.R., 2008. MicroRNA399 is a long-distance signal for the regulation of plant phosphate homeostasis. Plant J. 53, 731–738. https://doi.org/10.1111/j.1365-313X.2007.03363.x.

Pant, B.D., Musialak-Lange, M., Nuc, P., May, P., Buhtz, A., Kehr, J., et al., 2009. Identification of nutrient-responsive *Arabidopsis* and rapeseed MicroRNAs by comprehensive real-time polymerase chain reaction profiling and small RNA sequencing. Plant Physiol. 150 (3), 1541.

Pearse, S.J., Veneklaas, E.J., Cawthray, G., Bolland, M.D.A., Lambers, H., February 2006. *Triticum aestivum* shows a greater biomass response to a supply of aluminium phosphate than Lupinus albus, despite releasing fewer carboxylates into the rhizosphere. New Phytol. 169 (3), 515–524.

Pei, L., Jin, Z., Li, K., Yin, H., Wang, J., Yang, A., 2013. Identification and comparative analysis of low phosphate tolerance associated microRNAs in two maize genotypes. Plant Physiol. Biochem. 70, 221–234. https://doi.org/10.1016/j.plaphy.2013.05.043.

Perron, M.P., Provost, P., January 2008. Protein interactions and complexes in human microRNA biogenesis and function. Front. Biosci. 13 (7), 2537.

Poirier, Y., Bucher, M., 2002. Phosphate transport and homeostasis in *Arabidopsis*. In: Somerville, R., Meyerowitz, E.M., Rockvilli, M.D. (Eds.), The *Arabidopsis*, vol. 1. American Society of Plant Biologists).

Provost, P., Dishart, D., Doucet, J., Frendewey, D., Samuelsson, B., Rådmark, O., November 2002. Ribonuclease activity and RNA binding of recombinant human Dicer. EMBO J. 21 (21), 5864.

Pruss, G.J., Nester, E.W., Vance, V., 2008. Infiltration with Agrobacterium tumefaciens induces host defense and development-dependent responses in the infiltrated zone. Mol. Plant Microbe Interact. 21, 1528–1538. https://doi.org/10.1094/MPMI-21-12-1528.

Rausch, T., Wachter, A., 2005. Sulfur metabolism: a versatile platform for launching defence operations. Trends Plant Sci. 10, 503–509. https://doi.org/10.1016/j.tplants.2005.08.006.

Reinhart, B.J., Slack, F.J., Basson, M., Pasquienelll, A.E., Bettlnger, J.C., Rougvle, A.E., et al., February 2000. The 21-nucleotide let-7 RNA regulates developmental timing in *Caenorhabditis elegans*. Nature 403 (6772), 901–906.

Rhoades, M.W., Reinhart, B.J., Lim, L.P., Burge, C.B., Bartel, B., Bartel, D.P., August 2002. Prediction of plant microRNA targets. Cell 110 (4), 513–520.

Robert-Seilaniantz, A., MacLean, D., Jikumaru, Y., Hill, L., Yamaguchi, S., Kamiya, Y., Jones, J.D.G., 2011. The microRNA miR393 re-directs secondary metabolite biosynthesis away from camalexin and towards glucosinolates. Plant J. 67, 218–231. https://doi.org/10.1111/j.1365-313X.2011.04591.x.

Robinson, N.J., Procter, C.M., Connolly, E.L., Guerinot, M.L., 1999. A ferric-chelate reductase for iron uptake from soils. Nature 397 (6721), 694–697.
Romheld, V., 1987. Different strategies for iron acquisition in higher-plants. Physiol Plant 70 (2), 231–234.
Romheld, V., Marschner, H., 1986. Evidence for a specific uptake system for iron phytosiderophores in roots of grasses. Plant Physiol. 80 (1), 175–180.
Santi, S., Schmidt, W., 2009. Dissecting iron deficiency-induced proton extrusion in *Arabidopsis* roots. New Phytol. 183 (4), 1072–1084.
Schauer, S.E., Jacobson, S.E., Meinke, D.W., Ray, A., 2002. DICER-LIKE1: blind men and elephants in *Arabidopsis* development. Trends Plant Sci. 7, 487–491. https://doi.org/10.1016/S1360-1385(02)02355-5.
Sotiropoulou, G., Pampalakis, G., Lianidou, E., Mourelatos, Z., August 2009. Emerging roles of microRNAs as molecular switches in the integrated circuit of the cancer cell. RNA 15 (8), 1443.
Subramanian, S., Fu, Y., Sunkar, R., Barbazuk, W.B., Zhu, J.K., Yu, O., April 2008. Novel and nodulation-regulated microRNAs in soybean roots. BMC Genom. 9.
Sun, S.B., Gu, M., Cao, Y., Huang, X.P., Zhang, X., Ai, P.H., et al., 2012. A constitutive expressed phosphate transporter, OsPht1;1, modulates phosphate uptake and translocation in phosphate-replete rice. Plant Physiol. 159, 1571–1581. https://doi.org/10.1104/pp.112.196345.
Sunkar, R., Zhu, J.K., 2004. Novel and stress-regulated microRNAs and other small RNAs from *Arabidopsis*. Plant Cell 16 (8), 2001–2019.
Sunkar, R., Kapoor, A., Zhu, J.K., 2006. Posttranscriptional induction of two Cu/Zn superoxide dismutase genes in *Arabidopsis* is mediated by downregulation of miR398 and important for oxidative stress tolerance. Plant Cell 18, 2051–2065. https://doi.org/10.1105/tpc.106.041673.
Talalay, P., Fahey, J.W., 2001. Phytochemicals from cruciferous plants protect against cancer by modulating carcinogen metabolism. J. Nutr. 131, 3027S–3033S.
Tang, R., Li, L., Zhu, D., Hou, D., Cao, T., Gu, H., et al., March 2012. Mouse miRNA-709 directly regulates miRNA-15a/16-1 biogenesis at the posttranscriptional level in the nucleus: evidence for a microRNA hierarchy system. Cell Res. 22 (3), 504.
Uhlén, M., Fagerberg, L., Hallström, B.M., Lindskog, C., Oksvold, P., Mardinoglu, A., et al., January 2015. Proteomics. Tissue-based map of the human proteome. Science 347 (6220).
Valdes-Lopez, O., Yang, S.S., Aparicio-Fabre, R., Graham, P.H., Reyes, J.L., Vance, C.P., et al., 2010. MicroRNA expression profile in common bean (*Phaseolus vulgaris*) under nutrient deficiency stresses and manganese toxicity. New Phytol. 187, 805–818. https://doi.org/10.1111/j.1469-8137.2010.03320.x.
Varotto, C., Maiwald, D., Pesaresi, P., Jahns, P., Salamini, F., Leister, D., 2002. The metal ion transporter IRT1 is necessary for iron homeostasis and efficient photosynthesis in *Arabidopsis thaliana*. Plant J. 31 (5), 589–599.
Vazquez, F., Legrand, S., Windels, D., 2010. The biosynthetic pathways and biological scopes of plant small RNAs. Trends Plant Sci. 15, 337–345. https://doi.org/10.1016/j.tplants.2010.04.001.
Vert, G., Grotz, N., Dedaldechamp, F., Gaymard, F., Guerinot, M.L., Briat, J.F., Curie, C., 2002. IRT1, an *Arabidopsis* transporter essential for iron uptake from the soil and for plant growth. Plant Cell 14 (6), 1223–1233.
Vidal, E.A., Araus, V., Lu, C., Parry, G., Green, P.J., Coruzzi, G.M., et al., March 2010. Nitrate-responsive miR393/AFB3 regulatory module controls root system architecture in *Arabidopsis thaliana*. Proc. Natl. Acad. Sci. U.S.A. 107 (9), 4477–4482.

Wang, J.W., Wang, L.J., Mao, Y.B., Cai, W.J., Xue, H.W., Chen, X.Y., 2005. Control of root cap formation by MicroRNA-targeted auxin response factors in *Arabidopsis*. Plant Cell 17 (8), 2204–2216.

Wang, C., Huang, W., Ying, Y., Li, S., Secco, D., Tyerman, S., et al., 2012. Functional characterization of the rice SPXMFS family reveals a key role of OsSPX-MFS1 in controlling phosphate homeostasis in leaves. New Phytol. 196, 139–148. https://doi.org/10.1111/j.1469-8137.2012.04227.x.

Wang, Y., Zhang, C., Hao, Q., Sha, A., Zhou, R., Zhou, X., et al., 2013. Elucidation of miRNAs-mediated responses to low nitrogen stress by deep sequencing of two soybean genotypes. PLoS One 8, e67423. https://doi.org/10.1371/journal.pone.0067423.

Waters, B.M., Mcinturf, S.A., Stein, R.J., 2012. Rosette iron deficiency transcript and microRNA profiling reveals links between copper and iron homeostasis in *Arabidopsis thaliana*. J. Exp. Bot. 63, 5903–5918. https://doi.org/10.1093/jxb/ers239.

Wightman, B., Ha, I., Ruvkun, G., December 1993. Posttranscriptional regulation of the heterochronic gene lin-14 by lin-4 mediates temporal pattern formation in *C. elegans*. Cell 75 (5), 855–862.

Williams, L.E., Miller, A.J., 2001. Transporters responsible for the uptake and partitioning of nitrogenous solutes. Annu. Rev. Plant Physiol. Plant Mol. Biol. 52, 659–688.

Windels, D., Vazquez, F., 2011. Mir393: integrator of environmental cues in auxin signaling? Plant Signal. Behav. 6, 1672–1675. https://doi.org/10.4161/psb.6.11.17900.

Wong, J., Gao, L., Yang, Y., Zhai, J., Arikit, S., Yu, Y., Duan, S., Chan, V., Xiong, Q., Yan, J., Li, S., Liu, R., Wang, Y., Tang, G., Meyers, B.C., Chen, X., Ma, W., 2014. Roles of small RNAs in soybean defense against Phytophthora sojae. Infection Plant J. 79, 928–940. https://doi.org/10.1111/tpj.12590.

Xu, Z., Zhong, S., Li, X., Li, W., Rothstein, S.J., Zhang, S., et al., 2011. Genome-wide identification of microRNAs in response to low nitrate availability in maize leaves and roots. PLoS One 6, e28009. https://doi.org/10.1371/journal.pone.0028009.

Xu, F., Liu, Q., Chen, L., Kuang, J., Walk, T., Wang, J., et al., 2013. Genome-wide identification of soybean microRNAs andtheir targets reveals their organ-specificity and responses to phosphate starvation. BMC Genom. 14, 66. https://doi.org/10.1186/1471-2164-14-66.

Yan, Y., Wang, H., Hamera, S., Chen, X., Fang, R., 2014. miR444a has multiple functions in the rice nitrate-signaling pathway. Plant J. 78 (1), 44–55.

Yang, C., Li, D., Mao, D., Liu, X., Ji, C., Li, X., Zhao, X., Cheng, Z., Chen, C., Zhu, L., 2013. Overexpression of microRNA319 impacts leaf morphogenesis and leads to enhanced cold tolerance in rice (*Oryza sativa* L.). Plant Cell Environ. 36, 2207–2218. https://doi.org/10.1111/pce.12130.

Yang, J., Zhou, J., Zhou, H.J., Wang, M.M., Liu, M.M., Ke, Y.Z., et al., March 2020. Global survey and expressions of the phosphate transporter gene families in Brassica napus and their roles in phosphorus response. Int. J. Mol. Sci. 21 (5).

Zeng, Y., Cullen, B.R., 2004. Structural requirements for pre-microRNA binding and nuclear export by Exportin 5. Nucleic Acids Res. 32 (16), 4776.

Zeng, H.Q., Zhu, Y.Y., Huang, S.Q., Yang, Z.M., 2010. Analysis of phosphorus-deficient responsive miRNAs and cis elements from soybean (*Glycine max* L.). J. Plant Physiol. 167, 1289–1297. https://doi.org/10.1016/j.jplph.2010.04.017.

Zeng, Q.Y., Yang, C.Y., Ma, Q.B., Li, X.P., Dong, W.W., Nian, H., 2012. Identification of wild soybean miRNAs and their target genes responsive to aluminum stress. BMC Plant Biol. 12, 182. https://doi.org/10.1186/1471-2229-12-182.

Zhang, H., Kolb, F.A., Brondani, V., Billy, E., Filipowicz, W., November 2002. Human Dicer preferentially cleaves dsRNAs at their termini without a requirement for ATP. EMBO J. 21 (21), 5875–5885.

Zhang, H., Kolb, F.A., Jaskiewicz, L., Westhof, E., Filipowicz, W., July 2004. Single processing center models for human Dicer and bacterial RNase III. Cell 118 (1), 57–68.

Zhang, B., Pan, X., Stellwag, E.J., 2008. Identification of soybean microRNAs and their targets. Planta 229, 161–182. https://doi.org/10.1007/s00425-008-0818-x.

Zhang, X., Zhao, H., Gao, S., Wang, W.C., KatiyarAgarwal, S., DaHuang, H., Raikhel, N., Jin, H., 2011. *Arabidopsis* argonaute 2 regulates innate immunity via miRNA393*-Mediated silencing of a Golgi-localized SNARE gene, MEMB12. Mol. Cell 42, 356–366. https://doi.org/10.1016/j.molcel.2011.04.010.

Zhang, L.W., Song, J.B., Shu, X.X., Zhang, Y., Yang, Z.M., 2013. miR395 is involved in detoxification of cadmium in *Brassica napus*. J. Hazard Mater. 250–251, 204–211. https://doi.org/10.1016/j.jhazmat.2013.01.053.

Zhao, M., Ding, H., Zhu, J.K., Zhang, F., Li, W.X., June 2011. Involvement of miR169 in the nitrogen-starvation responses in *Arabidopsis*. New Phytol. 190 (4), 906–915.

Zhao, M., Tai, H., Sun, S., Zhang, F., Xu, Y., Li, W.X., 2012. Cloning and characterization of maize miRNAs involved in responses to nitrogen deficiency. PLoS One 7, e29669. https://doi.org/10.1371/journal.pone.0029669.

Zhao, X., Liu, X., Guo, C., Gu, J., Xiao, K., 2013a. Identification and characterization of microRNAs from wheat (*Triticum aestivum* L.) under phosphorus deprivation. J. Plant Biochem. Biotechnol. 22, 113–123. https://doi.org/10.1007/s13562-012-0117-2.

Zhao, Y., Xu, Z., Mo, Q., Zou, C., Li, W., Xu, Y., et al., 2013b. Combined small RNA and degradome sequencing reveals novel miRNAs and their targets in response to low nitrate availability in maize. Ann. Bot. 112, 633–642. https://doi.org/10.1093/aob/mct133.

Zhou, Z.S., Song, J.B., Yang, Z.M., 2012. Genome-wide identification of *Brassica napus* microRNA and their targets in response to cadmium. J. Exp. Bot. 63, 4597–4613. https://doi.org/10.1093/jxb/ers136.

Zhu, Y.Y., Zeng, H.Q., Dong, C.X., Yin, X.M., Shen, Q.R., Yang, Z.M., 2010. MicroRNA expression profiles associated with phosphorus deficiency in white lupin (*Lupinus albus* L.). Plant Sci. 178, 23–29. https://doi.org/10.1016/j.plantsci.2009.09.011.

Zörb, C., Senbayram, M., Peiter, E., May 2014. Potassium in agriculture–status and perspectives. J. Plant Physiol. 171 (9), 656–669.

Further readings

Baggish, A.L., Hale, A., Weiner, R.B., Lewis, G.D., Systrom, D., Wang, F., et al., August 2011. Dynamic regulation of circulating microRNA during acute exhaustive exercise and sustained aerobic exercise training. J Physiol. 589 (Pt 16), 3983–3994.

Clauss, S., Wakili, R., Hildebrand, B., Kääb, S., Hoster, E., Klier, I., et al., 2016. MicroRNAs as biomarkers for acute atrial remodeling in marathon runners (the miRathon study – a sub-study of the Munich marathon study). PLoS One 11 (2).

Cohu, C.M., Abdel-Ghany, S.E., Reynolds, K.A.G., Onofrio, A.M., Bodecker, J.R., Kimbrel, J.A., et al., 2009. Copper delivery by the copper chaperone for chloroplast and cytosolic copper/zincsuperoxide dismutases: regulation and unexpected phenotypes in an *Arabidopsis* mutant. Mol. Plant 2, 1336–1350. https://doi.org/10.1093/mp/ssp084.

Cui, S., Sun, B., Yin, X., Guo, X., Chao, D., Zhang, C., et al., December 2017. Time-course responses of circulating microRNAs to three resistance training protocols in healthy young men. Sci. Rep. 7 (1).

O'Brien, J., Hayder, H., Zayed, Y., Peng, C., August 2018. Overview of microRNA biogenesis, mechanisms of actions, and circulation. Front. Endocrinol. 9 (AUG), 402.

CHAPTER

MicroRNA tools for gene silencing

9

Arathi Radhakrishnan[1], Mehak Manchanda[1], Ranjanee Chattopadhayay[1], Rajpal Srivastav[1,2]

[1]*Amity Institute of Biotechnology, Amity University, Noida, Uttar Pradesh, India;* [2]*Department of Science & Technology, Ministry of Science & Technology, New Delhi, India*

1. Introduction

MicroRNA is an important factor of gene silencing in living organisms working under different cellular conditions. Gene silencing, mediator of gene regulation, controls biological processes such cellular responses, immunity, growth, and development. Though the discussion on gene silencing is of interest in the recent years but the first paper published on RNA silencing was in 1928. Wingard's study revealed tobacco plants with necrotic, sick tobacco ringspot virus, with higher leaves becoming asymptomatic and resistant to subsequent infection. Gene silencing is a molecular biological process that involves the downregulation or inhibition of the expression of a gene (Dana et al., 2017; Hood et al., 2004). It may happen at the translational or transcriptional levels. It is a natural mechanism to control diseases by suppressing or silencing specific genes that are associated with a disease or induced by pathogens (Mocellin and Provenzano, 2004; Redberry and Grace, 2006). Two categories—transcriptional gene silencing (TGS) and posttranscriptional gene silencing (PTGS) in plants—are used to describe the regulation of gene silencing (Fig. 9.1). TGS methylates the 5′ untranslated region, which obstructs transcription, whereas PTGS methylates a coding region, which results in transcript deterioration. It has been demonstrated that a number of species possess VIGS (Virus induced gene silencing), cosuppression, and position variation as gene silencing mechanisms (Ahmed et al., 2021). The virus VIGS, which causes the albino phenotype, has been connected to petunia plants (Sunny et al., 2016).

The role of miRNAs is very important in determining the process of gene silencing through different ways (Verma et al., 2021). MicroRNA-based methods are used to enhance posttranscriptional quality features in a variety of plants and crops. The impact of abiotic factors can reduce plant growth and crop yields such as salinity significantly impacts crop productivity. The tolerance to cellular stress is mediated by various molecular determinants (Nag et al., 2010; Srivastav et al., 2019). Plant stress tolerance is largely enhanced by natural resistance genes and associated miRNAs. The microRNAs that improve plant tolerance to salt, heat, and drought include miR159, miR160, miR167, miR169, miR393, and miR397 (Nag et al., 2010).

MicroRNA Advances and Application in Plant Biology. https://doi.org/10.1016/B978-0-443-21682-4.00014-2

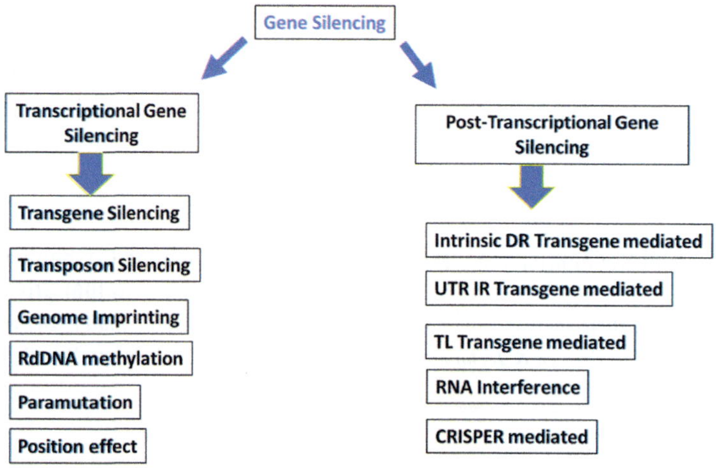

FIGURE 9.1

Various ways of gene silencing including transcriptional and posttranscriptional control.

Moreover, gene silencing methods such as RNA interference and CRISPR/Cas9 can be used to enhance the nutritional value, shape, color, and shelf life of plants and crops. In *Caenorhabditis elegans*, it has been demonstrated that RNAi causes the albino phenotype (Conte et al., 2015). First discovered in prokaryotes, CRISPR is a potent gene-silencing tool used in eukaryotes for genetic engineering. Recent study has been centered on using RNA interference (RNAi), and CRISPR to improve plants tolerance to stress, infections, and drought (Singh et al., 2019).

Transposons cause mutations within a genome and PTGS avoids their effects. Moreover, methylation silences genes, which lowers transposon activity (Tungare et al., 2022). Mutations in MET1 and DDM1 increase transposon activity in *Arabidopsis* and *C. elegans*. These methods are essential for plant growth and agricultural production (Xiaoguo et al., 2017). Further, the development of stress tolerant crops and plants with increased resistance to biotic and abiotic stressors and higher yields is a promising application of gene silencing. The PTGS pathway induces gene silencing in homologous transgenes and viral genes. Plants employ the TGS and PTGS pathways in conjunction with siRNAs to inhibit transposon activity and prevent genomic instability (El-Sappah et al., 2021). The focus of the article is to emphasize the function of miRNA, the implications of miRNA as tools for gene silencing, the molecular mechanisms involved in gene silencing, and their critical roles in plant growth and development. Gene silencing can occur through various mechanisms as discussed in the following sections.

2. Transcriptional gene silencing (TGS)

Transcriptional gene silencing hampers transcriptional machinery to block the expression of certain genes, hence silencing the genes. It occurs in the nucleus by blocking a promoter region and eventually blocking transcription. RNA-directed DNA methylation, paramutation, genomic imprinting, transposon silencing, and position effect are some of the mechanisms that can lead to TGS (Patani et al., 2023).

RNA-Directed DNA Methylation (RdDM) is a phenomenon using noncoding RNA to add methyl to specific DNA sequences. This pathway incorporates siRNA along with RISC (ribonuclease induced silencing complex: responsible for binding and degrading translation of that RNA). RdDM was first discovered in *Arabidopsis thaliana*, which is accomplished by biosynthesis of siRNAs and followed methylation. RNA-DNA sequence homology nucleotide combinations such as CG, CHG, and CHH (H is A, C, or T) are the most common sites of DNA methylation (Peñuela, 2022). Chromomethylase 3 (CMT3), methyltransferase 1 (MET1), and DNMTs methylate the symmetric sequences CG and CHG, while DNMTs, chromomethylase 2 (CMT2), and domain-rearranged methyltransferase 2 methylate the asymmetric sequence CHH (DRM2). The process of DNA methylation involves proteins including Polymerase V, AGO4, DNMT, DRM2, and SRA-domain-containing proteins such as SUVH1, SUVH2, and SU(VAR)3—9. Through H3K9me, these proteins oversee creating a chromatic domain in heterochromatin. The SUVH gene family in *Arabidopsis* has 10 members (Groth, 2016). The effects on methylation of SUVH1-4 are distinct. SUV39H1 is involved in H3K9me heterochromatin and regulates facultative heterochromatin. It binds to DNMT to regulate a range of epigenetic changes, including methylation of DNA and histones (Gangwar et al., 2023).

Genome Imprinting is an epigenetic process that influences gene expression, and it can activate or silence target genes as per the requirements. The complete X chromosome is silenced in marsupials (Deakin et al., 2009). Only 2% of plants and animals exhibit differential imprinting, a form based on the dominance of one parental group. Most imprinted genes express themselves in the CZE endosperm. The Pr1 reciprocal kernel-color allele, for example, causes colorless or spotted kernels in maize, but the same allele inherited from the mother produces colored kernels. Due to chromatin remodeling, methylated paternal alleles stay dormant and demethylated maternal alleles remain active. In *Arabidopsis thaliana*, DNA glycosylase differentially demethylates the DEMETER gene to activate imprinting. The Zea mays gene (Zm575) controls endosperm development in maize, whereas OsGL1 controls grain size and shape in Oryza sativa (Schoft et al., 2011).

In plants, notably in maize, a process known as *paramutation* occurs when one allele can result in heritable alterations in another, changing the expression of the gene. To explain paramutation, three models are used: the RNA model, physical interaction, and RNA-physical interaction. According to the RNA model, paramutation is a process in which paramutagenic repeats are translated into mRNA, transformed into dsRNA, and then cut into siRNA by a Dicer-like protein. A paramutable allele caused by RdDM is silenced using methylated free-state siRNAs. In the physical model, paramutagenic and paramutable alleles come into physical touch with one another. In the combo-model, RdDM interacts with both paramutable and paramutagenic alleles (Hale et al., 2007).

Transposon Silencing is also an important part of transcriptional gene silencing. When target sequences are found inside or close to genes, DNA methylation by RdDM is essential for shutting down genes, transposons, repetitive elements in the genome. Epigenetic regulation is essential for preserving genomic integrity and controlling gene expression in plants. The target DNA sequences of the 24-nt siRNAs and 21-nt miRNAs are recognized through complementary base pairing. They are incorporated into the RNA-Induced Transcriptional Silencing complex (RITS), which is leased to GS and comprises AGO4 and DRD1 (Lippman et al., 2003; Ito, 2012).

There are several methods for *transgenes silencing*, including VIGS, IR, and hairpin RNAs. In these procedures, the transgene is inserted into heterochromatin or junk DNA, and complex DNA structures are created at single locus. The original sense ssRNA is converted into a dsRNA by the actions of RDR6, SGS3, SDE3, and RISC, and the RISC subsequently cuts the dsRNA. The sense transgene's ssRNA and the antisense ssRNA either directly or indirectly mix to form the dsRNA and silence the gene expression. Further, in another method of gene silencing called *position effect*, the genomic location where a transgene is introduced affects the expression of that gene; transgenes located near or inside heterochromatin are more likely to be silenced. Transgenes in open, active regions of chromatin are less likely to be silenced. Epigenetic theories explain molecular mechanisms such as cis and trans interactions. A complete model of interactions between heterochromatin patches and chromosomal rearrangements that lead to aberrant gene activation is known as metabolic interaction or nuclear compartmentalization.

3. RNA, biogenesis and gene silencing

There are two types of biogenesis pathways: canonical and noncanonical. The polymerase, RDR2, and Dicer homolog 3 (DCL3) determine which pathway is used. The conventional process results in lysine 9 (K9) being.

methylated by DNA transcription factor 1 interacting to unmethylated histone 3 (H3). Classy 1(CLSY 1) is a chromatin-remodeling protein that helps polymerase IV and RDR2 initiate this transcription. Endoribonuclease DCL3 breaks down dsRNA into 23–24 nt siRNAs that intercalate with AGO4 or AGO6 to methylate the target molecules (Fig. 9.2). Exogenous siRNAs are often introduced into cells from external sources. They are also referred to as synthetic siRNA molecules, viral RNA, or dsRNA of foreign origin. Long dsRNA molecules produced by the cell are used to make endogenous siRNAs. An enzyme known as Dicer is responsible for processing both exogenous and endogenous siRNAs. Long dsRNA is cut into siRNA duplexes by the enzyme Dicer, which are generally 21–23 nucleotides long. Polymerase II transcribes the satellite gene into RNA, which then binds to the AGO protein through a noncanonical pathway. RdRp converts one TAS transcript into dsRNA, while miRNAs cleave other TAS transcripts. Then, using this dsRNA, the Dicer-like protein produces transacting siRNAs, which activate the

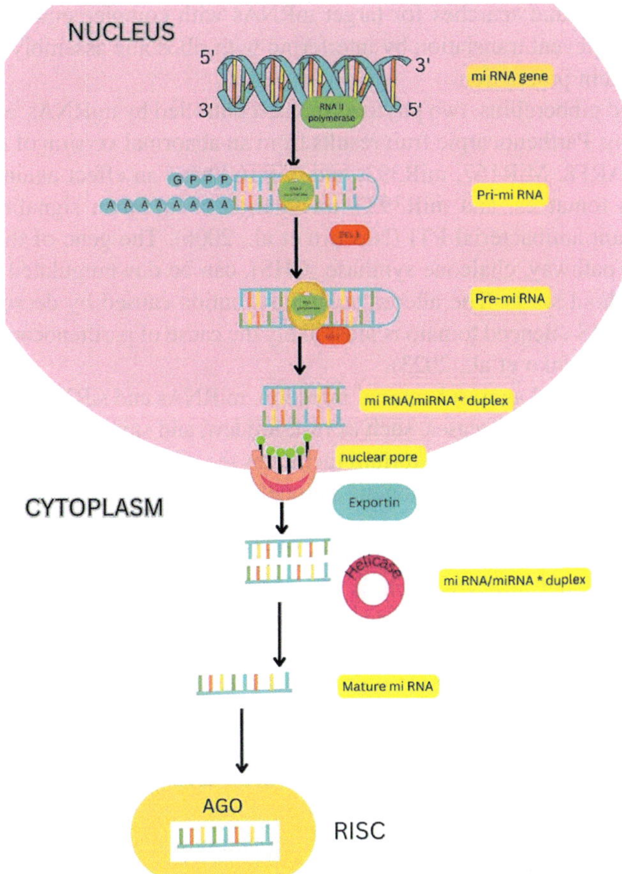

FIGURE 9.2

The process of microRNA biogenesis.

corresponding gene silencing pathways (Lippman et al., 2003). Next, the siRNA duplex that is produced is loaded in with the RNA-induced silencing complex (RISC). To direct RISC to the target mRNA, just one strand of the siRNA is selected. After being positioned onto RISC, the guide siRNA strand helps RISC recognize and bind to complementary target mRNA. The binding often causes the target mRNA to be cleaved, halting protein translation. The synthesis of 22–24 nt-long siRNAs is indirectly triggered by siRNA and responds to viral infection (Vishwakarma et al., 2024).

miRNAs are short RNA molecules produced by genes and undergo processing and change to become mature. Primary miRNA transcripts are processed from precursor miRNAs, which are converted into mature miRNAs by enzymes such as Drosha and Dicer. Mature miRNAs are placed onto RISC during synthesis, where the guide miRNA strand searches for target mRNAs with complementary sequences. MiRNAs can prevent translation by interfering with ribosome assembly or stability, reducing protein production.

Auxin and gibberellins, two phytohormones controlled by miRNAs, are linked to parthenocarpy. Parthenocarpic fruit results from an abnormal version of the miR167 target gene ARF8. MiR167, miR393, and miR160 have an effect against bacterial infections in tomatoes, and miR393's modulation of an auxin signaling pathway improved plant antibacterial PTI (Navarro et al., 2006). The gene of the flavonoid biosynthetic pathway, chalcone synthase (CHS), can be downregulated to produce tomatoes without seeds. The altered auxin distribution caused by decreased flavonoids in the CHS-silenced tomato is most likely the cause of parthenocarpy (Navarro et al., 2006; Rudenko et al., 2023).

In order to defend against bacterial infection, miRNAs and siRNAs are essential. Note that 70% of plant diseases, such as rust, mildew, and smut, are caused by fungi (Rabuma et al., 2022). The haustorium facilitates the exchange of nutrients and signals between the host and pathogen. Through diverse processes, gene silencing strengthens a plant's resistance to a range of pathogens. In wheat and barley, Avra10 limits fungus *Blumeria graminis*, miR1138 in wheat counters *Puccinia graminis* infection, and miRNAs such as Md-miRln20 negatively regulate resistance against *Glomerella* leaf spot respectively (Nowara et al., 2010; Rabuma et al., 2022; Zhang et al., 2019). MicroRNAs, such as miR156 and miR397, impact grain quantity and quality in a range of crops and are crucial for improving posttranscriptional quality features. miR159 controls stem elongation and floral development, miR160 is essential for rice growth, development, and immunity. Several elements of plant growth, including leaf polarity, anthers development, barley blooming, vegetative and reproductive growth, and tomato flowering, are regulated by miR164, miR166, miR159, miR172, miR156, and miR4376. Resistance to grapevine virus A was also created in tobacco by synthetic miRNAs (Liu et al., 2017).

4. Posttranscriptional gene silencing

There is another important gene regulation mechanism, which works after the transcription process. Posttranscriptional gene silencing is a technique that specifically targets and eliminates some genes' mRNA transcripts. Posttranscriptional gene silencing (PTGS) is mediated by a variety of mechanisms, such as RNA interference and clustered regularly interspaced short palindromic repeats (Patani et al., 2024).

4.1 RNA interference

First discovered in the year 1998 by American scientists, Andrew Z. As per Fire and Craig C. Mello, RNA interference (RNAi) is a defense mechanism in plants and worms (Agrawal et al., 2003) that helps them resist infections or pathogen growth (Muhammad et al., 2019). It is an evolutionary conserved mechanism, and it regulates gene expressions by inhibiting the translation of specific mRNAs (Maddala et al., 2024). RNAi prevents the messenger RNA (mRNA) of genes from being translated into proteins, hence suppressing the gene expression. In growth, development, immune functioning, and defense against viral infections, RNAi is an important method for regulating gene expression. The dsRNA-binding domain (dsRBP), AGO, RDR, and DCL make up the entire RNAi gene silencing system . For instance, plant resistance to a variety of pathogens can be introduced using RNA interference.

The two main RNA interference (RNAi) pathways in eukaryotic cells are the short interfering RNA (siRNA) pathway and the microRNA (miRNA) pathway (Meddya et al., 2023). Although both routes use small RNA molecules to silence target genes, they differ in their origins and modes of operation. Gene silencing is a useful tool for suppressing bacterial infections that lead to cankers, leaf spots, and scabs in the targeted plant parts. These infections can cause severe illnesses. It uses double-strand break (dsRNA) to enable sequence-specific suppression of gene expression (Saurabh et al., 2014). The central molecules of RNAi are siRNAs and miRNAs. The RNA-Induced Silencing Complex (RISC), a protein complex, is directed by these small RNAs to find or target mRNAs that have complementary sequences. RISC leads to cleaving of mRNA to suppress its translation and therefore completing the procedure of gene silencing .

A long dsRNA is cleaved with the help of enzyme, Dicer. It is a ribonuclease enzyme and plays a central role in RNAi. It leads to the formation of these small RNA molecules. The small molecule, siRNA, or miRNA contains sequences, which are complementary to the target mRNA. One strand of small RNA gets incorporated into the RISC leading to the formation of either RISC-siRNA complex or RISC-miRNA complex (Matranga et al., 2005). Then, RISC-siRNA complex is formed, siRNA duplex unwinds, the passenger strand degrades, and the other strand becomes the guiding strand for RNAi and gets bound to RISC. This complex so formed is known as RISC assembly. The RISC assembly now binds to the targeted mRNA. Protein Ago, found within this complex, catalyzes the cleavage of the target mRNA, and proceeds to degrade it. Several RISC-siRNA/RISC-miRNA complexes frequently target the same mRNA to increase the gene silencing effect (Parveen et al., 2023). This is an effective way to express genes as well.

RNAi is used to develop crops with enhanced properties such as higher yield, higher resistance to pests, diseases, and abiotic stress. It agrees with the prospect of sustainable growth & development without harmful environmental effects. RNAi has been used to develop virus resistant papaya by targeting the viral genes responsible for papaya ringspot virus (Azad et al., 2024) and to accentuate the nutritional content of fruits and crops. Further, with this fruit ripening can also be altered. RNAi has been successfully employed to reduce the allergenic proteins in apples, thus making them much safer. RNAi has been utilized in Bt maize to accentuate the resistance against rootworm pests (Baum et al., 2007). Low-expression of the RACK1 gene has led to the development of rice that can withstand drought (Dahong et al., 2009). miRNA regulates the levels of-carotene and lutein in potatoes, with transgenic lines holding higher levels of-carotene than those under the control of the Cauliflower mosaic virus (Lu et al., 2006). RNAi is being used to design genetically modified cotton, by targeting and silencing essential genes in cotton bollworms, thus creating Cotton bollworm resistance. *Puccinia striiformis f.* sp. *tritici* (Pst), the fungus that causes Wheat Stripe Rust disease, can be effectively and permanently controlled with the use of RNAi (Ni et al., 2017).

4.2 CRISPR mediated gene silencing

Bacteria and archaea have prokaryotic genomes that include a set of short, repetitive DNA sequences (El-Sappah et al., 2021). Prokaryotic genomes contain small segments of unique and distinct DNA sequences known as spacers and are found between the repetitive DNA sequences (Dana et al., 2017). These clusters of repeats designated as CRISPR. During successive infections, CRISPR sequences are employed to find and remove DNA from related bacteriophages. As a result, the spacers serve as molecular memory. Therefore, these sequences play a crucial part in the prokaryotic organisms' antiviral (or antiphage) defensive acquired immunity (Hille et al., 2018). CRISPR-based method is being used for precise modification of DNA sequences and modification & regulation of genes in plants. It is also a negative feedback mechanism, which knocks down gene expressions in plants and takes part in plant defense mechanisms (Dangquan et al., 2021).

It has been found that the CRISPR scaffold RNAs (scRNAs) prove out to be the founding models for synthetic locus-specific transcriptional programming in eukaryotes (Zalatan et al., 2017). The CRISPR-Cas9 system can introduce preplanned and tightly regulated changes in the genome. It relies on a few key molecules, i.e., Guide RNA (gRNA), which is approximately 20 base-pair long, these are predesigned or synthetic RNA, and can be located within a long RNA scaffold. These scaffolds are produced to be complementary to the target DNA. This aids Cas9 to move to the specific site for gene slicing. Protein Cas9, is an endonuclease enzyme, acting like a molecular scissors, to cut dsDNA at specific locations in the genome guided by gRNA (Figs. 9.3 and 9.4). Cas9 follows gRNA to the exact, precise location in the DNA sequence and cuts this dsDNA across both its strands (Gan et al., 2022).

The CRISPR cluster is processed by the enzyme Cas9 after being translated into a precursor RNA. They together are called CRISPR RNAs (crRNAs). To simply this complex process, the guide RNA (gRNA) combines with crRNA and hence, they combine their functions in a single RNA molecule. It is now known as the Cas9 complex. The Cas9 protein consists of two key parts - HNH and RuvC, which catalyze cleavage of the strands. These two domains aid in the cleaving process. A brief sequence that is next to the target DNA and acts as a recognition site is called the protospacer adjacent motif (PAM) (Gleditzsch et al., 2019). After finding the target DNA, the Cas9 complex identifies the PAM. The DNA then experiences a double-strand break (DSB) as a result of the strands being cleaved by the HNH and RuvC domains.

Here, utilizing the cell's innate DNA repair mechanism, the CRISPR-Cas9 process copies DNA. The mechanism for DNA repair can proceed in two ways. Gene

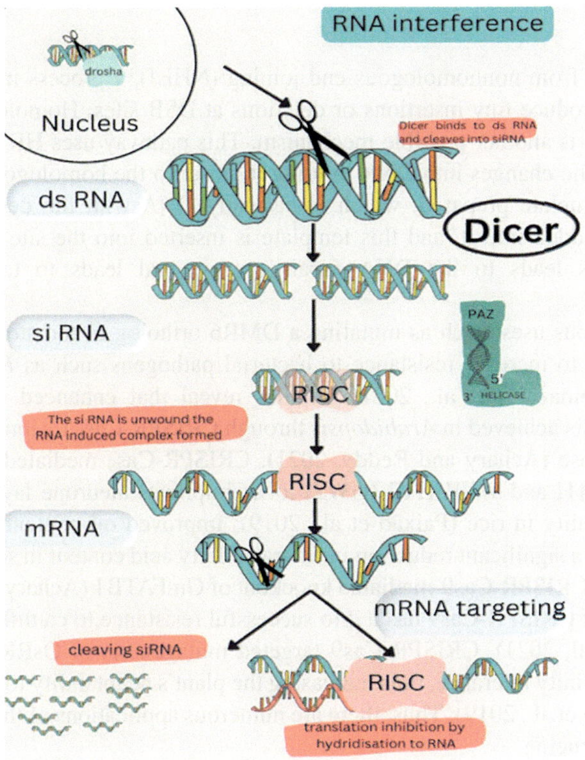

FIGURE 9.3

Schematic representation of the RNAi mediated gene silencing.

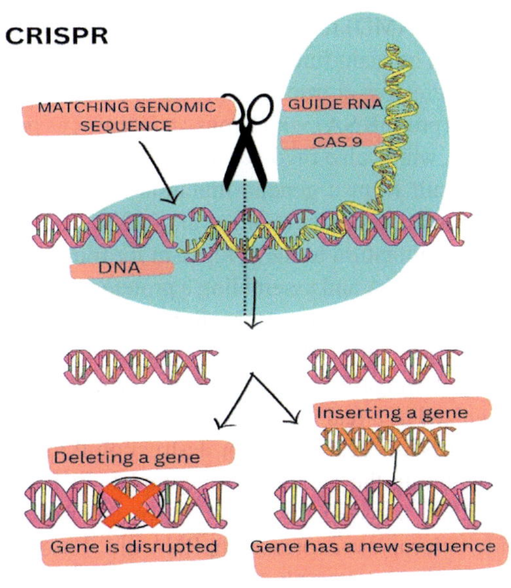

FIGURE 9.4

Representation of CRISPER-Cas mediated gene silencing.

slicing results from nonhomologous end joining (NHEJ), a process in which cells frequently introduce tiny insertions or deletions at DSB sites. Homology-Directed Repair (HDR) is another possible mechanism. This pathway uses HR to insert targeted or specific changes into DNA to bring it closer to the homologous sequence. The repair template prepared, which is basically DNA with the desired genetic changes embedded into it, and this template is inserted into the site of the DSB. These process leads to the DNA repair process and leads to target genetic modifications.

It has various uses, such as mutating a DMR6 ortholog in tomato plants using CRISPR-Cas9 to increase resistance to bacterial pathogens such as *Pseudomonas syringae* (Thomazella et al., 2021). Studies reveal that enhanced resistance to drought stress is achieved in *Arabidopsis* through CRISPR-Cas9 fusion with histone acetyltransferase (Achary and Reddy, 2021). CRISPR-Cas9 mediated mutation in GRAIN WIDTH and WEIGHT2 (GW2) locus improves aleurone layer and grain nutritional quality in rice (Paixão et al., 2019). Improved oil content in soybeans is the result of a significant reduction in saturated fatty acid content in soybean seeds caused by the CRISPR-Cas9-mediated knockout of GmFATB1 (Achary et al., 2021). Application of CRISPR-Cas9 has led to successful resistance to cauliflower mosaic virus (Ma et al, 2021). CRISPR-Cas9 targeted mutagenesis of OsRR22 gene enhances rice salinity tolerance, thus increasing the plant's adaptability to salinity conditions (Zhang et al., 2019). Thus, there are numerous applications of the CRISPR in plant gene silencing.

4.3 Nonsense-mediated decay

To control gene expression, the essential cellular quality control process known as nonsense-mediated decay (NMD) recognizes and target mRNA that contain premature termination codons. The main function of this mechanism in eukaryotic cells is to stop the synthesis of defective or truncated proteins. The target gene is transcribed into a precursor mRNA (pre-mRNA) with exons and introns as part of the nonsense-mediated decay process, which also involves transcription and splicing. A nonsense mutation, which can result from a genetic mutation or a splicing mistake, might create a premature termination codon (Vora et al., 2023). The nucleus exports this mRNA with the nonsense mutation to the cytoplasm for translation, where the ribosome starts decoding the mRNA and assembling a protein (Wang et al., 2019). When the ribosome comes across the premature termination codon, several protein complexes, including Upf1, Upf2, and Upf3, are recruited to start the nonsense-mediated decay process (Arciga-Reyes et al., 2006). Whether the premature termination codon is recognized as a real stop codon, or a nonsense mutation depends on NMD variables. The mRNA is designated for destruction if the ribosome stops and NMD is triggered, inhibiting the creation of incomplete or aberrant proteins. The nucleus-localized hpRNA transgene was ineffective at protecting against insect gene silencing, while chloroplast-expressed hpRNA did. Based on these results, it appears that cross-kingdom gene silencing in insects cannot be induced by siRNAs that have been processed inside plant cells. Rather, the target organism must process the dsRNA. There are important aspects of NMD pathway in plants to provide resistance to pathogens. The NMD mutants in plants have been shown to have association with pathogen infections (Rayson et al., 2012).

4.4 Other tools of gene silencing

There are other ways also in Posttranscriptional Gene Silencing. Plants use PTGS-based gene silencing mechanisms, such as Direct Repeat, 3′ UTR Inverted-Repeat, and Terminator-less Transgenes, to regulate gene expression under various circumstances. *Intrinsic direct-repeat transgenes* induce potent gene silencing with almost complete efficiency in both *Arabidopsis* and maize. The accumulation of siRNAs, which are associated with silencing, is suggested to be part of PTGS. When a sequence lacking homology to the direct repeats is transcriptionally fused at the 3′ end, these transgenes have the ability to cause transitive silencing (Ma and Mitra, 2002). *UTR Inverted repeat (IR) Transgene mediated GS-* In tomato and *Arabidopsis*, siRNAs form of transgenic structure that causes gene silencing. The IR sequence is followed by a terminator in a sense target gene sequence. There is uncertainty regarding the mechanisms. It is possible to identify siRNAs linked to gene silencing by looking upstream of the IR sequence in the target gene. *Terminator-less Transgene mediated GS*—here, the siRNAs induced by 3′ UTR IR transgenes in tomato and *Arabidopsis* can induce efficient gene silencing. With this structure, a target gene sequence that is not homologous to the terminator sequence in front of the IR is cloned (Douglas et al., 1994). For example, it is extensively used in Flavr Savr tomatoes by introduction of antisense gene that interferes with the production of polygalacturonase enzyme that degrades pectin present in the cell wall, therefore making it resistant to rotting (Kanno et al., 2013).

5. Applications and future perspectives

Gene silencing can be used for epigenetic modification where the gene expression is regulated and altered without changing the DNA sequence in the target plant/crop or animal. Gene silencing is thus becoming a powerful tool in genetic engineering. Scientists have been utilizing the gene slicing technique in drug development. The target gene can be silenced using various techniques such as RNAi, CRISPR-Cas9, TGS or PTGS. It is being used extensively utilized to generate pest resistant crops, which target and kill only the targeted insects or pests. It is also used to design commercially important crops having higher nutritional content, higher yield, size, aroma, taste, texture, and other desirable traits. Gene silencing can reform the health and pharmaceutical industry with effective medicines having reduced side effects (Khare et al., 2017). It could also prove to effectively suppress disease causing genes (Radhakrishnan et al., 2024). Even though the low probability, gene silencing might unintentionally affect nontarget genes. The investigations are required in these directions. Further, the application of NGS technologies has accelerated genomic analysis at rapid pace (Srivastav and Suneja, 2019). Targeted mutagenesis in plants has led to the development of CRISPR technology, which can alter or modify the nucleotide sequences of specific genes or genomic loci (Gleditzsch et al., 2019; Li et al., 2020). This strategy may be used to produce genetically stable gene knock-out lines. RNA silencing methods continue to be helpful for analyzing gene activity and improving crops quality. RNA silencing methods can recover mutants for gene function studies since knocking out critical genes can be detrimental to plants. In animals, gene silencing may trigger immune responses, which in turn might lead to unexpected and adverse reactions in the body. It also raises ethical concerns as it leads to silencing of genes. Long-term safety of gene silencing techniques, limited applicability and regulatory challenges are still under area of investigation, which also needs to be addressed.

CRISPR/Cas9-like technique enables tissue-specific gene silencing utilizing a transgene produced in the tissue. RNAi technique can also be applied for gene mapping of various plants and to understand the traits of commercially important crops including tomato development (Shinozaki et al., 2018; Liu et al., 2021). It can increase agricultural qualities, create crops that are resistant to climate change, boost nutritional value, and combine with precision agriculture techniques. Research on RNA silencing in plants will provide insights into the epigenetic regulation of gene expression. These findings can be combined with genome editing technologies to enhance comprehension and facilitate novel applications. The research showed that CRISPR-Cas9 -mediated editing of BADH2 can trigger revolution of rice fragrance (Liu et al., 2018). Thus, functional genomics, the manufacturing of biopharmaceuticals, and environmental preservation are further applications of RNA silencing. However, it is necessary to address ethical issues, public acceptance, and regulatory considerations. Regulatory organizations will need to create standards and safety evaluations for genetically modified crops as RNA silencing technology develops. Plants that are virus-resistant, like those in papaya and potatoes, can be controlled by RNA silencing processes. A mechanism based on RNA silencing was shown in papaya plants producing viral coat proteins by a substantial drop in coat protein mRNA. The improved tolerance to the oomycete disease in transgenic potatoes that expressed hpRNA constructs against syntaxin-related 1 (SYR1) suggests that syntaxins play a role in secretory defense responses in potatoes. *Helicoverpa armigera*, damages agricultural yields by feeding on cotton and other crops. Insect resistance to gossypol, a poisonous cotton metabolite, is caused by the CYP6AE14 gene. It is possible to suppress gene expression in the midguts of insects by using transgenic tobacco plants. The ability of transgenic tobacco plants to effectively inhibit the development of root-knot worms provided evidence of their usefulness as a delivery system for dsRNA-mediated gene silencing.

6. Conclusive remarks

Gene silencing tools employing miRNA has revolutionized genetic engineering in agriculture in recent years. It is a precise and powerful technique, which regulates the expression of target gene. It is utilized extensively in the plant and crop industry to enhance yields, nutritional content, taste, size, aroma, reduce allergic properties, and other characteristics. The microRNAs also play significant roles in modulating biotic and abiotic stress responses. The impact of the biotic and abiotic stress leads to huge agricultural loss in the current scenario. To map the methods to address these stresses and enhance plant growth and yield, it is therefore necessary to comprehend the gene regulatory mechanisms and gene silencing linked to stress tolerance.

Acknowledgments

AR and RS acknowledges INSPIRE, Department of Science and Technology, Ministry of Science and Technology, Government of India, New Delhi, India for fellowship.

References

Achary, V.M.M., Reddy, M.K., 2021. CRISPR-Cas9 mediated mutation in GRAIN WIDTH and WEIGHT2 (GW2) locus improves aleurone layer and grain nutritional quality in rice. Sci. Rep. 11, 21941. https://doi.org/10.1038/s41598-021-00828-z.

Agrawal, N., Dasaradhi, P.V., Mohmmed, A., Malhotra, P., Bhatnagar, R.K., Mukherjee, S.K., December 2003. RNA interference: biology, mechanism, and applications. Microbiol. Mol. Biol. Rev. 67 (4), 657–685. https://doi.org/10.1128/MMBR.67.4.657-685.2003.

Ahmed, F.F., Hossen, M.I., Sarkar, M.A., Konak, J.N., Zohra, F.T., Shoyeb, M., Mondal, S., September 2, 2021. Genome-wide identification of DCL, AGO and RDR gene families and their associated functional regulatory elements analyses in banana (Musa acuminata). PLoS One 16 (9) e0256873.

Arciga-Reyes, L., Wootton, L., Kieffer, M., Davies, B., 2006. UPF1 is required for nonsense-mediated mRNA decay (NMD) and RNAi in Arabidopsis. Plant J. 47, 480–489.

Azad, M.A., Amin, L., Sidik, N.M., March 17, 2014. Gene technology for papaya ringspot virus disease management. Sci. World J. 2014, 768038. https://doi.org/10.1155/2014/768038.

Baum, J., Bogaert, T., Clinton, W., et al., 2007. Control of coleopteran insect pests through RNA interference. Nat. Biotechnol. 25, 1322–1326. https://doi.org/10.1038/nbt1359.

Conte, D.J., MacNeil, L.T., Walhout, A.J.M., Mello, C.C., 2015. RNA interference in *Caenorhabditis elegans*. Curr. Protoc. Mol. Biol. 109. https://doi.org/10.1002/0471142727.mb2603s109, 26.3.1–26.3.30.

Da-hong, L., Hui, L., Yang, Y.-L., Zhen, P.-P., Liang, J.-S., 2009. Down-regulated expression of RACK1 gene by RNA interference enhances drought tolerance in rice. Rice Sci. 16 (1), 14–20. https://doi.org/10.1016/S1672-6308(08)60051-7.

Dana, H., Chalbatani, G.M., Mahmoodzadeh, H., Karimloo, R., Rezaiean, O., Moradzadeh, A., Mehmandoost, N., Moazzen, F., Mazraeh, A., Marmari, V., Ebrahimi, M., Rashno, M.M., Abadi, S.J., Gharagouzlo, E., 2017. Molecular mechanisms and biological functions of siRNA. IJBS 13 (2), 48–57.

Deakin, J.E., Chaumeil, J., Hore, T.A., et al., 2009. Unravelling the evolutionary origins of X chromosome inactivation in mammals: insights from marsupials and monotremes. Chromosome Res. 17, 671–685. https://doi.org/10.1007/s10577-009-9058-6.

Dorer, D.R., Henikoff, S., 1994. Expansions of transgene repeats cause heterochromatin formation and gene silencing in Drosophila. Cell 77 (7), 993–1002. https://doi.org/10.1016/0092-8674(94)90439-1.

El-Sappah, A.H., Yan, K., Huang, Q., Islam, M.M., Li, Q., Wang, Y., Khan, M.S., Zhao, X., Mir, R.R., Li, J., El-Tarabily, K.A., Abbas, M., September 7, 2021. Comprehensive mechanism of gene silencing and its role in plant growth and development. Front. Plant Sci. 12, 705249. https://doi.org/10.3389/fpls.2021.705249.

Gan, W.C., Ling, A.P.K., March 24, 2022. CRISPR/Cas9 in plant biotechnology: applications and challenges. Biotechnologia 103 (1), 81–93. https://doi.org/10.5114/bta.2022.113919.

Gangwar, P., Johri, P., Datta, R., Singh, S., Trivedi, M., 2023. Overview on 1 phytopharmaceuticals and biotechnology of herbal plants. In: In Phytopharmaceuticals and Biotechnology of Herbal Plants, pp. 1–17 (CRC Press).

Gleditzsch, D., Pausch, P., Müller-Esparza, H., Özcan, A., Guo, X., Bange, G., Randau, L., April 2019. PAM identification by CRISPR-Cas effector complexes: diversified mechanisms and structures. RNA Biol. 16 (4), 504–517. https://doi.org/10.1080/15476286.2018.1504546.

Groth, M., Moissiard, G., Wirtz, M., Wang, H., García-Salinas, C., Ramos-Parra, P.A., Jacobsen, S.E., 2016. Mthfd1 controls dna methylation in arabidopsis. Nat. Commun. 7 (1). https://doi.org/10.1038/ncomms11640.

Hale, C.J., Stonaker, J.L., Gross, S.M., Hollick, J.B., October 16, 2007. A novel Snf2 protein maintains trans-generational regulatory states established by paramutation in maize. PLoS Biol. 5 (10), e275. https://doi.org/10.1371/journal.pbio.0050275.

Hille, F., Richter, H., Wong, S.P., Bratovič, M., Ressel, S., Charpentier, E., March 8, 2018. The biology of CRISPR-Cas: backward and forward. Cell 172 (6), 1239–1259. https://doi.org/10.1016/j.cell.2017.11.032.

Hood, E., March 2004. RNAi: what's all the noise about gene silencing? Environ. Health Perspect. 112 (4), A224–A229. https://doi.org/10.1289/ehp.112-a224.

Ito, H., January 2012. Small RNAs and transposon silencing in plants. Dev. Growth Differ. 54 (1), 100–107. https://doi.org/10.1111/j.1440-169X.2011.01309.x.

Kanno, T., Yoshikawa, M., Habu, Y., 2013. Locus-specific requirements of DDR complexes for gene-body methylation of TAS genes in *Arabidopsis thaliana*. Plant Mol. Biol. Rep. 31, 1048–1052. https://doi.org/10.1007/s11105-012-0554-z.

Khare, P., Kishore, K., Sharma, D.K., July 1, 2017. A study on the standardization parameters of *Cassia angustifolia*. Asian J. Pharmaceut. Clin. Res. 10 (7), 329–332.

Li, Y., Li, J., Chen, Z., Wei, Y., Qi, Y., Wu, C., 2020. OsmiR167a-targeted auxin response factors modulate tiller angle via fine-tuning auxin distribution in rice. Plant Biotechnol. J. 18, 2015–2026. https://doi.org/10.1111/pbi.13360.

Lippman, Z., May, B., Yordan, C., Singer, T., Martienssen, R., December 2003. Distinct mechanisms determine transposon inheritance and methylation via small interfering RNA and histone modification. PLoS Biol. 1 (3), E67. https://doi.org/10.1371/journal.pbio.0000067.

Liu, S.R., Zhou, J.J., Hu, C.G., Wei, C.L., Zhang, J.Z., 2017. MicroRNA-mediated gene silencing in plant defense and viral counter-defense. Front. Microbiol. 8, 1801. https://doi.org/10.3389/fmicb.2017.01801.

Liu, H., Soyars, C.L., Li, J., Fei, Q., He, G., Peterson, B.A., Meyers, B.C., Nimchuk, Z.L., Wang, X., March 7, 2018. CRISPR/Cas9-mediated resistance to cauliflower mosaic virus. Plant Direct 2 (3), e00047. https://doi.org/10.1002/pld3.47.

Liu, Q., Yang, F., Zhang, J., Liu, H., Rahman, S., Islam, S., Ma, W., She, M., 2021. Application of CRISPR/Cas9 in crop quality improvement. Int. J. Mol. Sci. 22 (8), 4206. https://doi.org/10.3390/ijms22084206.

Lu, S., Van Eck, J., Zhou, X., Lopez, A.B., O'Halloran, D.M., Cosman, K.M., Conlin, B.J., Paolillo, D.J., Garvin, D.F., Vrebalov, J., Kochian, L.V., Küpper, H., Earle, E.D., Cao, J., Li, L., December 2006. The cauliflower or gene encodes a DnaJ cysteine-rich domain-containing protein that mediates high levels of beta-carotene accumulation. Plant Cell 18 (12), 3594–3605. https://doi.org/10.1105/tpc.106.046417.

Ma, C., Mitra, A., July 2002. Intrinsic direct repeats generate consistent post-transcriptional gene silencing in tobacco. Plant J. 31 (1), 37–49. https://doi.org/10.1046/j.1365-313x.2002.01332.x.

Ma, J., Sun, S., Whelan, J., Shou, H., 2021. CRISPR/Cas9-Mediated knockout of GmFATB1 significantly reduced the amount of saturated fatty acids in soybean seeds. Int. J. Mol. Sci. 22, 3877. https://doi.org/10.3390/ijms22083877.

Maddala, V.K., Singh, S., Bandi, J., January 2024. Soil properties influencing nutrient signaling in plant soil systems. In: In Essential Minerals in Plant-Soil Systems, pp. 373–393 (Elsevier).

Matranga, C., Tomari, Y., Shin, C., Bartel, D.P., Zamore, P.D., November 2005. Passenger-strand cleavage facilitates assembly of siRNA into Ago2-containing RNAi enzyme complexes. Cell 123 (4), 607–620. https://doi.org/10.1016/j.cell.2005.08.044.

Meddya, S., Meshram, S., Sarkar, D., Datta, R., Singh, S., Avinash, G., Kumar Kondeti, A., Savani, A.K., Thulasinathan, T., September 25, 2023. Plant stomata: an unrealized possibility in plant defense against invading pathogens and stress tolerance. Plants 12 (19), 3380.

Mocellin, S., Provenzano, M., November 2004. RNA interference: learning gene knock-down from cell physiology. J. Transl. Med. 2 (1), 39. https://doi.org/10.1186/1479-5876-2-39.

Muhammad, T., Zhang, F., Zhang, Y., Liang, Y., January 10, 2019. RNA interference: a natural immune system of plants to counteract biotic stressors. Cells 8 (1), 38. https://doi.org/10.3390/cells8010038.

Nag, A., Jack, T., 2010. Sculpting the flower; the role of microRNAs in flower development. Curr. Top. Dev. Biol. 91, 349–378. https://doi.org/10.1016/S0070-2153(10)91012-0.

Navarro, L., Dunoyer, P., Jay, F., Arnold, B., Dharmasiri, N., Estelle, M., Voinnet, O., Jones, J.D.G., 2006. A plant miRNA contributes to antibacterial resistance by repressing auxin signaling. Science 312, 436–439. https://doi.org/10.1126/science.1126088.

Ni, M., Ma, W., Wang, X., Gao, M., Dai, Y., Wei, X., Zhu, Z., 2017. Next-generation transgenic cotton: pyramiding rnai and bt counters insect resistance. Plant Biotechnol. J. 15 (9), 1204–1213. https://doi.org/10.1111/pbi.12709.

Nowara, D., Gay, A., Lacomme, C., Shaw, J., Ridout, C., Douchkov, D., Hensel, G., Kumlehn, J., Schweizer, P., September 2010. HIGS: host-induced gene silencing in the obligate biotrophic fungal pathogen *Blumeria graminis*. Plant Cell 22 (9), 3130–3141. https://doi.org/10.1105/tpc.110.077040.

Parveen, R., Khare, N., Singh, S., Johri, P., March 29, 2023. An in silico outlook for the detection and surveillance of evolving and persistent plant pathogens. In: In Detection, Diagnosis and Management of Soil-borne Phytopathogens, pp. 17–40 (Singapore: Springer Nature Singapore).

Patani, A., Patel, A., Prajapati, D., Khare, N., Singh, S., August 1, 2023. Various agriculture crop plant-based bioactive compounds and their use in nanomaterial synthesis and applications. In: In Nanomaterials from Agricultural and Horticultural Products, pp. 223–241 (Singapore: Springer Nature Singapore).

Patani, A., Prajapati, D., Shukla, K., Patel, M., Patani, P., Patel, A., Singh, S., January 1, 2024. Environmental stress–induced alterations in the micro-and macronutrients status of plant. In: In Essential Minerals in Plant-Soil Systems, pp. 181–195 (Elsevier).

Peñuela, M., Gallo-Franco, J.J., Finke, J., Rocha, C., Gkanogiannis, A., Ghneim-Herrera, T., Lorieux, M., 2022. Methylation in the CHH context allows to predict recombination in rice. Int. J. Mol. Sci. 23 (20), 12505. https://doi.org/10.3390/ijms232012505.

Rabuma, T., Gupta, O.P., Chhokar, V., 2022. Recent advances and potential applications of cross-kingdom movement of miRNAs in modulating plant's disease response. RNA Biol. 19 (1), 519–532. https://doi.org/10.1080/15476286.2022.2062172.

Radhakrishnan, A., Gangopadhyay, R., Sharma, C., Kapardar, R.K., Sharma, N.K., Srivastav, R., 2024. Unwinding helicase MCM functionality for diagnosis and therapeutics of replication abnormalities associated with cancer: a review. Mol. Diagn. Ther. 28, 249–264. https://doi.org/10.1007/s40291-024-00701-5.

Rayson, S., Arciga-Reyes, L., Wootton, L., De Torres Zabala, M., Truman, W., Graham, N., Grant, M., Davies, B., 2012. A role for nonsense-mediated mRNA decay in plants: pathogen responses are induced in *Arabidopsis thaliana* NMD mutants. PLoS One 7 (2), e31917. https://doi.org/10.1371/journal.pone.0031917.

Redberry, G., 2006. Gene Silencing: New Research. Nova Science Publishers, New York. ISBN 9781594548321.

Roca Paixão, J.F., Gillet, F.X., Ribeiro, T.P., et al., 2019. Improved drought stress tolerance in Arabidopsis by CRISPR/dCas9 fusion with a Histone AcetylTransferase. Sci. Rep. 9, 8080. https://doi.org/10.1038/s41598-019-44571-y.

Rudenko, N.N., Vetoshkina, D.V., Marenkova, T.V., Borisova-Mubarakshina, M.M., 2023. Antioxidants of non-enzymatic nature: their function in higher plant cells and the ways of boosting their biosynthesis. Antioxidants 12 (11), 2014. https://doi.org/10.3390/antiox12112014.

Saurabh, S., Vidyarthi, A.S., Prasad, D., March 2014. RNA interference: concept to reality in crop improvement. Planta 239 (3), 543–564. https://doi.org/10.1007/s00425-013-2019-5.

Schoft, V.K., Chumak, N., Choi, Y., Hannon, M., Garcia-Aguilar, M., Machlicova, A., Slusarz, L., Mosiolek, M., Park, J.S., Park, G.T., Fischer, R.L., Tamaru, H., 2011. Function of the DEMETER DNA glycosylase in the *Arabidopsis thaliana* male gametophyte. Proc. Natl. Acad. Sci. U.S.A. 108 (19), 8042–8047. https://doi.org/10.1073/pnas.1105117108.

Shinozaki, Y., Nicolas, P., Fernandez-Pozo, N., et al., 2018. High-resolution spatiotemporal transcriptome mapping of tomato fruit development and ripening. Nat. Commun. 9, 364. https://doi.org/10.1038/s41467-017-02782-9.

Singh, A., Sengar, R.S., Singh, A.K., Singh, S., Kumar, A., Ahmad, K., 2019. Impact of genetic engineering on agriculture food crops. Biotech Today: Int. J. Biol. Sci. 9 (1), 17–24.

Srivastav, R., Suneja, G., 2019. Recent advances in microbial genome sequencing. In: Tripathi, V., Kumar, P., Tripathi, P., Kishore, A., Kamle, M. (Eds.), Microbial Genomics in Sustainable Agroecosystems, first ed., vol. 2. Springer. ISBN 9789813298606.

Srivastav, R., Sharma, R., Tandon, S., Tandon, C., 2019. Role of DHH superfamily proteins in nucleic acids metabolism and stress tolerance in prokaryotes and eukaryotes. Int. J. Biol. Macromol. 127, 66–75. https://doi.org/10.1016/j.ijbiomac.2018.12.123.

Sunny, J.S., Snijesh, V.P., Singh, S., 2016. Potential role of mirnas in rheumatoid arthritis. Austin Arthritis 1 (2), 1009.

Thomazella, D.P.T., Seong, K., Mackelprang, R., Dahlbeck, D., Geng, Y., Gill, U.S., Qi, T., Pham, J., Giuseppe, P., Lee, C.Y., Ortega, A., Cho, M.J., Hutton, S.F., Staskawicz, B., July 6, 2021. Loss of function of a DMR6 ortholog in tomato confers broad-spectrum disease resistance. Proc. Natl. Acad. Sci. U.S.A. 118 (27). https://doi.org/10.1073/pnas.2026152118 e2026152118.

Tungare, K., Shahu, R., Zambare, V., Agarwal, P., Jobby, R., Nisar, N., Alabdallah, N.M., Al-Saeed, F.A., Johri, P., Singh, S., Saeed, M., July 2022. Toxicity mitigation of textile dye reactive blue 4 by hairy roots of Helianthus annuus and testing its effect in in vivo model systems. BioMed Res. Int. 25, 2022.

Verma, N., Sao, P., Srivastava, A., Singh, S., 2021. Physiological and molecular responses to drought, submergence and excessive watering in plants. Harsh Environ. Plant Resil. 305–321.

Vishwakarma, S.K., Ilyas, T., Shahid, M., Malviya, D., Kumar, S., Singh, S., Johri, P., Singh, U.B., Singh, H.V., January 11, 2024. Bacillus spp.: nature's gift to agriculture and humankind. In: In Applications of Bacillus and Bacillus Derived Genera in Agriculture, Biotechnology and Beyond, pp. 1–36 (Singapore: Springer Nature Singapore).

Vora, Z., Pandya, J., Vaikuntapu, P.R., Singh, S., Asha, S., Purama, R.K., January 2023. Genome-wide association of defense hormone crosstalk in plants. In: In Hormonal Cross-Talk, Plant Defense and Development, 1, pp. 353–371 (Academic Press).

Wang, J., Mei, J., Ren, G., 2019. Plant microRNAs: biogenesis, homeostasis, and degradation. Front Plant Sci. 10, 360. https://doi.org/10.3389/fpls.2019.00360.

Zalatan, J.G., 2017. CRISPR-cas RNA scaffolds for transcriptional programming in yeast. Methods Mol. Biol. 1632, 341–357. https://doi.org/10.1007/978-1-4939-7138-1_22.

Zhang, Y., Zhang, Q., Hao, L., Wang, S., Wang, S., Zhang, W., Xu, C., Yu, Y., Li, T., 2019. A novel miRNA negatively regulates resistance to Glomerella leaf spot by suppressing expression of an NBS gene in apple. Hortic. Res. 6, 93. https://doi.org/10.1038/s41438-019-0175-x.

Zhang, A., Liu, Y., Wang, F., et al., 2019. Enhanced rice salinity tolerance via CRISPR/Cas9-targeted mutagenesis of the OsRR22 gene. Mol. Breeding 39, 47. https://doi.org/10.1007/s11032-019-0954-y.

Zhang, D., Zhang, Z., Unver, T., Zhang, B., 2021. CRISPR/Cas: a powerful tool for gene function study and crop improvement. J. Adv. Res. 29, 207–221. https://doi.org/10.1016/j.jare.2020.10.003.

Zhu, X., Qi, T., Yang, Q., He, F., Tan, C., Ma, W., Voegele, R.T., Kang, Z., Guo, J., December 2017. Host-induced gene silencing of the MAPKK gene PsFUZ7 confers stable resistance to wheat Stripe rust. Plant Physiol. 175 (4), 1853–1863. https://doi.org/10.1104/pp.17.01223.

Further readings

Abdel-Ghany, S.E., Pilon, M., 2008. MicroRNA-mediated systemic downregulation of copper protein expression in response to low copper availabilityin Arabidopsis. J. Biol. Chem. 283, 15932–15945. https://doi.org/10.1074/jbc.M801406200.

Allen, E., Xie, Z., Gustafson, A.M., Carrington, J.C., 2005. microRNAdirected phasing during trans-acting siRNA biogenesis in plants. Cell 121, 207–221. https://doi.org/10.1016/j.cell.2005.04.004.

Ambesajir, A., Kaushik, A., Kaushik, J.J., Petros, S.T., 2012. RNA interference: a futuristic tool and its therapeutic applications. Saudi J. Biol. Sci. 19 (4), 395–403. https://doi.org/10.1016/j.sjbs.2012.08.001.

Axtell, M.J., 2013. Classification and comparison of small RNAs from plants. Ann.Rev. Plant Biol. 64, 137–159. https://doi.org/10.1146/annurev-arplant-050312-120043.

Baulcombe, D., 2004. RNA silencing in plants. Nature 431, 356–363.

Du, J., Johnson, L.M., Groth, M., Feng, S., Hale, C.J., Li, S., et al., 2014. Mechanism of DNA methylation-directed histone methylation by KRYPTONITE. Mol. Cell 55, 495–504. https://doi.org/10.1016/j.molcel.2014.06.009.

Islam, M.M., Li, Q., Wang, Y., Khan, M.S., Zhao, X., Mir, R.R., Li, J., El-Tarabily, K.A., Abbas, M., 2021. Comprehensive mechanism of gene silencing and its role in plant growth and development. Front. Plant Sci. 12, 705249. https://doi.org/10.3389/fpls.2021.705249.

Kurihara, Y., Watanabe, Y., 2004. Arabidopsis micro-RNA biogenesis through Dicer-like 1 protein functions. Proc. Natl. Acad. Sci. U.S.A. 101, 12753–12758.

Li, S., Xia, L., 2019. Precise gene replacement in plants through CRISPR/Cas genome editing technology: current status and future perspectives. aBIOTECH 1 (1), 58–73. https://doi.org/10.1007/s42994-019-00009-7.

Meister, G., Tuschl, T., 2004. Mechanisms of gene silencing by doublestranded RNA. Nature 431, 343–3.

Mette, M.F., Aufsatz, W., van der Winden, J., Matzke, M.A., Matzke, A.J., 2000. Transcriptional silencing and promoter methylation triggered by double-stranded RNA. EMBO J. 19 (19), 5194–5201.

Molesini, B., Pandolfini, T., Rotino, G.L., Dani, V., Spena, A., 2009. Aucsia gene silencing causes Parthenocarpic fruit development in tomato. Plant Physiol. 149, 534–548.

Yadav, R.K., Chattopadhyay, D., 2011. Enhanced viral intergenic regionspecific short interfering RNA accumulation and DNA methylation correlates with resistance against a geminivirus. Mol. Plant Microbe Interact. 24, 1189–1197.

Ye, R., Wang, W., Iki, T., Liu, C., Wu, Y., Ishikawa, M., Zhou, X., Qi, Y., 2012. Cytoplasmic assembly and selective nuclear import of Arabidopsis Argonaute4/siRNA complexes. Mol. Cell 46, 859–870.

Zhang, H., He, X., Zhu, J.K., 2013. RNA-directed DNA methylation in plants: where to start? RNA Biol. 10 (10), 1593–1596.

CHAPTER

Application of microRNAs in horticultural crops

10

Sanhita Chowdhury[1], Vinod Kumar Dubey[1], Sanjay Kumar Sahoo[1], Amardeep Kour[2], Satnam Singh[2]

[1]*Department of Entomology, Post Graduate College of Agriculture, Dr. Rajendra Prasad Central Agricultural University, Samastipur, Bihar, India;* [2]*Punjab Agricultural University, Regional Research Station, Faridkot, Punjab, India*

1. Introduction

One of the crucial research areas being handled by agriculture experts all over the world is focused to address food as well as nutritional security. While the natural supplies are static, the global population is growing geometrically. Numerous agricultural enhancement methods have been created as a result of recent developments in the fields of plant biotechnology and molecular biology to gear up the global food demand. MicroRNAs (miRNAs) in this direction offer a wider scope to explore the application of these small naturally occurring molecular entities for the improvement of crop plants (Sunny et al., 2016).

Small RNAs can be categorized into four main groups based on their origin, manner of processing, and association with effector proteins: small interfering RNAs (siRNA), microRNAs (miRNA), small RNAs produced from transfer RNA (tsRNAs) and PIWI-interacting RNAs (piRNA, animals only) (Treiber et al., 2019). A Dicer-like (DCL) enzyme generates plant miRNAs from specific stem sections of larger source transcripts, and these miRNAs differ from other types of short RNAs in many ways. Nearly every element of regular plant growth and development, including reactions to light, nutrients, and a variety of biotic and abiotic stresses, is influenced by miRNAs.

miRNAs are the evolutionarily conserved RNAs. These endogenous, small (21–24 nucleotides in length) RNAs are foremost cytoplasmic regulator at posttranscriptional gene expressions by binding to partly complementary sequences in the target messenger RNAs (mRNAs) (Verma et al., 2021). The most fascinating aspect of miRNAs is how DNA sequences are translated into primary miRNAs, which are subsequently converted into precursor miRNAs and mature miRNAs. To cause mRNA degradation and translational repression, miRNAs bind with the 3′ UTR of their target mRNAs. Nevertheless, miRNAs also engage in interactions with gene promoters, coding sequences, and 5′ UTR (O'Brien et al., 2018). Numerous recent studies on miRNAs help to better understand their biogenesis, functional characteristics, and possible use in improving crop quality and yield.

MicroRNA Advances and Application in Plant Biology. https://doi.org/10.1016/B978-0-443-21682-4.00003-8

2. History

MicroRNAs were entirely unheard 2 decades ago, both in context of their existence and significance. The major focus of research up to that point concentrated on genes that code for proteins. The study of all non-protein-coding sequences was set aside by the traditional assumption that DNA is translated into RNA, which is subsequently transcribed into protein. The study of molecular biology has undergone a revolution since the identification of lin-4 as the first microRNA (miRNA) from *Caenorhabditis elegans* in 1993 by Ambros and Ruvkun groups (Lee et al., 1993). *Lin-4* is necessary for the normal temporal control of diverse postembryonic developmental events in *C. elegans*. Since then, miRNAs have been reported in other nematode species as well in plants, *Drosophila*, and mammals. In plants, the target mRNA may be cut by the 22 nt miRNA, and the fragment is then processed by the enzymes RNA-DEPENDENT RNA POLYMERASE 6 (RDR6) and DICER-LIKE four to create secondary 21 nt siRNA. Multicopy miRNA families are produced by the repeated replication of the miRNAs. The de novo emergence of a miRNA family may occasionally contribute to regulatory novelty during plant evolution. For instance, miR824, a recently identified miRNA family in *Brassica* species, helps to regulate flowering time by targeting AGAMOUS-LIKE 16 (Hu et al., 2014).

3. Biogenesis of microRNAs and its conservation

Plant genome can encode hundreds of miRNA genes. MiRNAs are either categorized as "intergenic" (found between two protein-coding genes) or "introgenic" depending on where they are placed in the genome (processed from introns of their host transcripts). In the nucleus, miRNAs undergo their biogenesis and are often transcribed by RNA Polymerase II into long pri-miRNAs (primary transcripts) (Singh et al., 2019). The 5′ end of the majority miRNAs is capped with a specifically modified nucleotide, while the 3′ end is polyadenylated with multiple adenosines (Pantaleo et al., 2010). These pri-miRNAs are folded into hairpin-like structures that have an upper stem, a lower stem, two arms, and a terminal loop (Fig. 10.1).

They are further cleaved by RNase-III nuclease enzymes called DICER-LIKE (DCL1) into miRNAs precursors (pre-miRNAs), which are of 70 nt-350 nt. With the aid of two accessory proteins, HYL1 (Hyponastic Leaves 1) and the zinc-finger protein Serrate (SE), DCL1 catalyzes the production of the majority of miRNAs (Fang and Spector, 2007; Dong et al., 2008). Afterwards, DCL1 breaks down these hairpin-looped structures to produce miRNAs duplexes in the nucleus. These duplexes are subsequently methylated at their 2′-OH position of their 3′ ends by the conserved HUA enhancer 1 (HEN1) and with the help of plant homolog of Exportin 5 (EXPO5), and Hasty (HST) are exported to the cytoplasm (Peláez et al., 2012). The produced duplexes are then loaded into the RISC (RNA-induced silencing complex) in the cytoplasm (Patani et al., 2023). Recent findings suggest the assembly of RISC in the nucleus and its export to the cytoplasm with the help of EXPO1. The

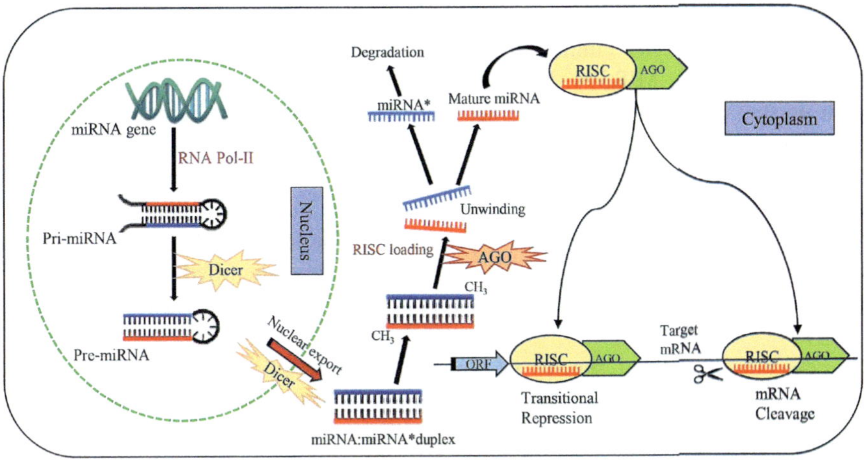

FIGURE 10.1

Biogenesis and mode of action of miRNAs.

ARGONAUTE (AGO) comprises of four major domains such as N-terminal domain (N), the PAZ (PIWI/Argonaute/Zwille) domain, MID domain, and the PIWI (P-element-induced wimpy tested) domain. It forms one of the major structural proteins in the RISC. The PAZ domain can bind to RNA while the PIWI domain exhibits RNAse H activity (Dong et al., 2022). Unwinding of the duplexes is primarily carried out by the AGO1 protein after loading on the RISC (Iki, 2017; Gangwar et al., 2023). One strand of the miRNA duplex is directed to the exosome by a small-RNA degrading nuclease for degradation as the mature miRNA is incorporated into the RISC containing AGO proteins (Sun et al., 2012). In addition, the mature miRNA primarily mediates gene silencing through target cleavage and/or translation inhibition. It does this by directing the RISC to target genes through base pairing (Wang et al., 2019a).

Many miRNAs, including those found in gymnosperms, mosses, monocots, and eudicots, are thought to have undergone evolutionary conservation (Cuperus et al., 2011). MiRNA families may be divided into two classes based on the conservation and diversity of miRNAs during the evolution of the plant kingdom (Patani et al., 2024). Since the young miRNAs typically only occur in a few species and are frequently expressed at low levels or only when certain conditions are met. They generally exist in a small number of species as a result of which they are evolutionarily nonconserved contrary to the evolutionarily conserved and frequently highly expressed ancient miRNAs (Qin et al., 2014).

The plant species show evolutionary conservation of 36 miRNA families, which have been functionally validated (Maddala et al., 2024). Eight miRNA families, viz., MIR156, MIR159/319, MIR160, MIR166, MIR171, MIR408, MIR390/391, and MIR395 are present in the common ancestor of all embryophytes, while MIR396 family is common to all ancestors of tracheophytes. All spermatophytes share

MIR397 and MIR398 family members as common ancestors. Every angiosperm has one of the 10 miRNA families (MIR162, MIR164, MIR167, MIR168, MIR169, MIR172, MIR393, MIR394, MIR399, and MIR827). MIR529 and MIR536 were both present in the eudicots and tracheophytes, but not in the common ancestors of the embryophytes (Cuperus et al., 2011). The conserved miRNAs have conserved targets and thus better suited for use in trait enhancement because of their vital role in plant growth and development. Recent research has shown that the miR164-targeted NAC genes *OMTN1–OMTN6*, which inhibit drought tolerance in rice during reproductive stage. *TaNAC21/22*, a novel NAC gene identified in wheat confers stripe rust resistance and is negatively regulated by miR164 in wheat (Feng et al., 2014).

Few closely related species or some specific species have nonconserved or lineage-specific miRNAs. These are regarded as transient products and energy wasters due to their inexplicit processing, poor expression, and lack of functional targets (Meddaya et al., 2023). Nonetheless, they occasionally expressed in large amounts in particular tissues to perform physiological roles in unique environmental adaptations. MiR824 was initially discovered to be a nonconserved miRNA produced by the duplication of the founder gene *Agamous-like 16* (*AGL16*). The no BLAST hits confirmed the absence of miR824 orthologs in any of the plant species (Rajagopalan et al., 2006). MiR824 and its target *AGL16* is identified in *A. thaliana*, but highly conserved in the Brassicaceae family. The development of the stomatal complex is impeded by *AGL16* expression. miRNA resistant *AGL16* overexpression appreciably boosts the stomatal density in *Arabidopsis* compared with wild-type *AGL16* (Kutter et al., 2007).

4. Molecular techniques for isolation and characterization

The first and most important step in understanding the roles of miRNAs is to identify them. Genetic screening, direct cloning and sequencing, and predictions based on bioinformatics tools have been employed for the miRNAs discovery (Schommer et al., 2012). By isolating, cloning, and sequencing short RNA populations in *A. thaliana*, the miRNAs, such as miR171, miR164, miR159, and miR156, were discovered for first time in plant species (Reinhart et al., 2002; Llave et al., 2002). Owing to sequence redundancy with other miRNA-coding genes that have comparable or identical variations, or simply because tiny RNAs are less in size and relatively few miRNAs have been discovered through genetic screening. Nonetheless, it was demonstrated that the *Arabidopsis* approach of miRNAs discovery using activation-tagging and genetic screening has proved to be an efficient methodology for isolating major miRNAs mutants such as miR172a-2 (Aukerman and Sakai, 2003). Genetic screening helped in the discovery of the JAW (jagged and wavy) locus, which generates a microRNA (miR-JAW) that may regulate the cleavage of the mRNAs of TCP genes involved in leaf growth (TFs) (Palatnik et al., 2003). MiRNAs binding sites can gain or lose throughout the coevolution of target genes and their associated miRNAs due to nucleotide changes, insertions, and deletions (Guo et al., 2008). MiRNAs identification can be aided by screening for particular developmental defects brought on by such mutations. Thousands of miRNAs have been found in plants over the past decade using both computational and experimental approaches. The conserved miRNAs were discovered in cotton through BLAST (Basic Local Alignment Search Tool) of known miRNA sequences against various nucleic acid sequence databases such as ESTs and genomic sequences. This was possible because high levels of sequence complementarity are present between plant miRNAs and their target mRNAs (Parveen et al., 2023). The early experimental methods for finding and evaluating the function of the plant miRNAs comprised direct cloning and methods based on genetic screening (Reinhart et al., 2002). The introduction of next-generation deep sequencing (NGS) technology after direct cloning and Sanger sequencing proved to be the robust tool both for miRNA discovery and identification of respective target genes in plants (Peláez et al., 2012; Djami-Tchatchou and Dubery, 2015). However, sometimes it was necessary to confirm the expression of the discovered miRNAs through northern blotting and highly sensitive methods such as semiquantitative PCR or quantitative PCR (Zhang and Wang, 2015).

By employing SILAC (stable isotope labelling with amino acids in cell culture) and microarray data processing, a quantitative mass spectrometry-based technique can be used to examine the downstream effects of miRNA regulation on various proteins level associated with their respective target genes (Vora et al., 2023). Besides this protein foot printing, western blotting etc., can also be employed for miRNA target investigations (Baek et al., 2008). miRNAs operate as rheostats to regulate the production of proteins in the majority of interactions. After miRNAs were discovered, searchable miRNA databases such as miRBase (http://www.mirbase.org) have been created and these are serving as helpful tool for searching miRNA with keywords such as miRNA name, keyword, references, and annotation (Djami-Tchatchou et al., 2017).

5. Mechanisms of microRNAs mediated gene regulation

The majority of studies to far have demonstrated that miRNAs bind to a particular region in the 3′ UTR of their target mRNAs to inhibit translation and induce mRNA deadenylation and capping (Ipsaro et al., 2015). In addition, the coding sequence, promoter regions, 5′ UTR, and other mRNA sections have all been identified to have miRNA binding sites. miRNA binding with the promoter region has been demonstrated to boost the transcription, also attaching to the 5′ UTR, and coding regions silences gene expression. More research is required to fully understand the functional importance of this interaction.

The miRNA in plants are highly complementary to their target gene mRNA because of the cleavage-type mode of action. Plant translation inhibition mechanisms have only recently been discovered. Depending on the complementarity between miRNAs and their targets, cleavage and inhibitory mechanisms are often coordinated (Yu et al., 2017; O'Brien et al., 2018). These small RNAs have enough variability to match practically RNA encoded in a genome with a particular short RNA, controlling essentially every aspect of plant life and frequently impacting regulation through transcription factors (TFs) (Khare et al., 2017). miRNAs are significant regulators of gene expression, and their mechanisms of action are being studied extensively. Regardless of their origin, miRNAs are 20−40 nt long, and has been demonstrated that due to their high degree of complementarity, they mostly influence gene expression at the posttranscriptional stage (Sun, 2012). Later research established that translational repression occurs when an incomplete miRNA pairing results in the target mRNA's degradation whereas a perfect miRNA pairing results in the opposite (Sun, 2012). Cleavage starts the removal of the poly (A) tail, which causes the target mRNAs to become unstable and degrade. MiRNAs can reinforce transcriptional programs and decrease random variations in the transcript copy number by leveraging transcriptional silence to offer biological processes robustness as well act as posttranscriptional repressors of gene expression.

6. Functional role of microRNAs

MiRNAs, which are tiny RNAs, have been identified as important epigenetic regulators in plants since they were discovered in *Arabidopsis* in 2002. They mainly modulate coding or noncoding RNA expression after transcription by modulating posttranscriptional regulation. MiRNAs are thought to be engaged in a range of biological activities under developmental and environmental cues, biotic and abiotic stresses as per details presented in (Table 10.1). The majority of functional research has been conducted on model horticultural crops such as vegetables, fruits, ornamental, and medicinal plants have also been studied, but intriguing and encouraging findings have shown that miRNAs are crucial in controlling the development of quality traits such as color, size, shape, flavor, texture, growth, development, ripening, and senescence, as well as nutrition.

MiRNAs are used by plants to control a wide range of biological functions, including genomic integrity, development, metabolism, and resiliency to environmental challenges (Li et al., 2017). In recent years, scientists have discovered that miRNAs are involved in the regulation of horticultural traits (Chen et al., 2018; Ma et al., 2020; Wang et al., 2020a). Since the first report of miRNA in tomato in 2007, over 100 horticultural crops have been identified as having miRNA from ˃10,000 miRNAs with most of them deposited in the sRNAanno database (Chen et al., 2021).

7. Targets of microRNAs

miRNA is targeted for cleavage, translational repression, or decay by miRNAs to regulate gene expression. In order to understand how miRNAs influence on plant

Table 10.1 Role of miRNAs in major horticultural crops against various biotic and abiotic environmental factors.

miRNAs	Plants/ species	Targets/genes	Targets functions	Key references
miR160, and miR167	Grapevine, *Vitis vinifera*	Auxin response factor	Development, stress response	Wang et al. (2014)
miR169a, miR160e, and miR167b	Apple, *Malus domestica*	Auxin response factor transcription factors	Bacterial disease (Fire blight resistance)	Kaja et al. (2015)
miR408	Orange, *Citrus sinensis*	Cu homeostasis, superoxide dismutase	Tolerance to Boron deficiency	Lu et al. (2015)
miR169	Tomato, *Solanum lycopersicum*	NF-YA transcription factors	Drought tolerance	Zhang et al. (2011c)
miR164	Cassava, *Manihot esculenta*	NAC transcription factors	Drought tolerance	Patanun et al. (2013)
miR393	Coffee, *Coffea arabica* and *Coffea canephora*	Transport inhibitor-like protein, DNA-binding proteins, GRR1-like protein	Chitin, cold, salt stress, and water deprivation	Akter et al. (2014)
miR156	Tea, *Camellia sinensis*	SBP/SPL transcription factors	Cold stress	Zhu and Luo (2013)
miRNA169, miRNA159, miRNA396, and miRNA393	All land plants	Suppressing ARF-mediated expression of auxin responsive gene	Response to drought, salinity, heat, and cold stress	Asefpour Vakilian (2020)
miRNA156a, and miRNA167a/c	Chinese cabbage, *Brassica* spp.	APS	Exposure to Cd in root and their up regulation	Huang et al. (2010)
mi159a	Tomato, *Solanum lycopersicum*	Nucleoprotein (N)	Resistance against Tospoviruses	Milter et al. (2016)
miR531	Tomato, *Solanum lycopersicum*	Mitogen-activated protein kinase	Resistance against whitefly, *Bemisia tabaci*	Ketao et al. (2018)
miR482	Apple, *Malus domestica*	Nucleotide-binding site	Regulate resistance against leaf spot disease, *Glomerella cingulata*	Zhang et al. (2019)

Continued

Table 10.1 Role of miRNAs in major horticultural crops against various biotic and abiotic environmental factors.—*cont'd*

miRNAs	Plants/ species	Targets/genes	Targets functions	Key references
miR159a	Lily, *Lilium regale*	*LrGAMYB*	Resistance against gray mold, *Botrytis elliptica*	Gao et al. (2020)
miR7997a, and miR7997c	Chilli, *Capsicum* spp.	Transcriptome	Drought tolerance and UV-B damage	Zhou et al. (2022)

development, it is imperative to identify their targets. This has been accomplished using a variety of approaches over the past few years. Genetic approaches are used to identify miRNA targets, based on abnormal expression of targeted mRNAs in miRNA loss-of-function mutants. Numerous studies have demonstrated that plants' growth and development, quality, flavor and texture regulations are controlled by miRNAs and the targets of those miRNAs (Table 10.2).

8. Application of microRNAs in horticultural plants

Several applications of microRNAs in horticultural crops are briefly described below (Fig. 10.2).

8.1 MicroRNAs regulate organ development

miRNAs play a fundamental role in plant growth and development. The final products of horticultural plants are usually leaves and shoot (vegetables), flowers (ornamental plants), and fruit (fruit trees). According to the major biological functions miRNAs pathways can be divided into four main categories: flower development, fruit development, leaf development, and shoot development. These key biological roles of miRNAs have been described in *Arabidopsis* as well as many other crop plants and classified as flower, fruit, leaf, and shoot development.

8.1.1 Flower development

A crucial part of a plant, the flower signifies the changes from vegetative to reproductive growth. Fruit trees must blossom in order to produce fruit and generate flowers. The miR172 plays an important role in flower development as the majority of its targets are AP2 transcription factors, which negatively regulate FT (Flowering Locus T) involved in flower development (Yamaguchi and Abe, 2012). SPL transcription factors are abundant in the target genes of several transition phase miRNAs including miR156 (Rhoades et al., 2002). Early flowering is encouraged by the

Table 10.2 miRNAs and their targeted role involve in regulating growth and development in different plants.

miRNAs	Plants/species	Targets/ genes	Targets functions	Key references
miR156	Tomato, *Lycopersicum esculentum*	SPL	Ovary, and fruit development	Silva et al. (2014)
miR160	Dragon eyes, *Dimocarpus longan*	ARF10/ 16/17	Embryo development, and hormone signalling transduction	Lin et al. (2015)
miR397a	Pear, *Pyrus communis*	LAC	Lignin synthesis, fruit development, and formation of fruit quality	Wu et al. (2014)
miR172	Tomato, *Lycopersicum esculentum*	AP2a	Fruit ripening, and development	Gao et al. (2015)
miR164/ 169	Cucumber, *Cucumis sativus*	NAC	Fruit-spine development	Liu et al. (2018)
miR167a	Tomato, *Lycopersicum esculentum*	ARF6, ARF8	Floral development defects, and female sterility	Liu et al. (2014)
miR172	Potato, *Solanum tuberosum*	AP2	Tuberization, and *salt tolerance*	Lian et al. (2021)
miR858	Grape, *Vitis vinifera*	MYBs	Promote anthocyanin, and flavonol accumulation	Tirumalai et al. (2019)
miR319	*Arabidopsis* and Tomato *Solanum lycopersicum*	TCP family TCP	Leaf development and senescence, organ curvature, and hormone biosynthesis and signalling	Bresso et al. (2018)
miR169	Snapdragon, *Antirrhinum majus*	NF-YA	Exerts homeotic control, and root architecture	Cartolano et al. (2007)
miR160/ 172	Sweet potato, *Ipomoea batatas*	ARF	Growth and development, and Starch accumulation	Din et al. (2014)
miR395, miR172, miR319, miR396, and miR397	Cassava, *Manihot esculenta*	TF, GF	Starch biosynthesis/ metabolism	Chen et al. (2015b)
miR156	Apple, *Malus domestica*	SPL2-like/33	Antagonized the light-induced anthocyanin accumulation in the fruit	Yang et al. (2019)
miR172	Apple, *Malus domestica*	AP2	Overexpression of miR172 reduced fruit size or no fruit and altered floral organ development	Yao et al. (2015)

Continued

Table 10.2 miRNAs and their targeted role involve in regulating growth and development in different plants.—*cont'd*

miRNAs	Plants/species	Targets/ genes	Targets functions	Key references
miR858	Kiwifruit, *Actinidia deliciosa*	MYBC1	Fruit transiently overexpressing of miR858 did not present coloration and anthocyanin was hardly detected	Li et al. (2020)
miR393	Musk melon, *Cucumis melo*	AFB2	Overexpressing miR393 delayed fruit ripening	Bai et al. (2020)
miR858	Persimmon, *Diospyros kaki*	MYB19/ 20	Overexpression of dk-miR858b reduced the content of proanthocyanidin (PA) in persimmon fruit discs	Yang et al. (2020c)
miR399	Strawberry, *Fragaria ananassa*	PHO2	Transgenic fruit overexpressing miR399a displayed higher sugar content	Wang et al. (2017)
miR166	Chinese cabbage, *Brassica rapa*	REV1/2	Overexpressing miR166g altered the leaf curvature	Ren et al. (2018)
miR396	Lettuce, *Lactuca sativa*	GRF5	Smaller leaves were observed in miR396a overexpression lines	Zhang et al. (2021)
miR156	Chrysanthemum, *Chrysanthemum* spp.	SPL3/5/9	Virus-based ectopic expression of miR156 restored the early flowering phenotype caused by NF-YB8 silencing	Wei et al. (2017)
miR396	Periwinkle, *Vinca minor*	SVP	Expressing CrMIR396 in arabidopsis caused abnormal flowers with curved sepals/petals, enlarged stigmas, and late-flowering	Yang et al. (2015)
miR156	Petunia, *Petunia* spp.	SPLs	Overexpression of ph-miR157a in arabidopsis and petunia resulted in delayed flowering, dwarf plant stature, increased branches, and reduced organ size	Yu et al. (2015)

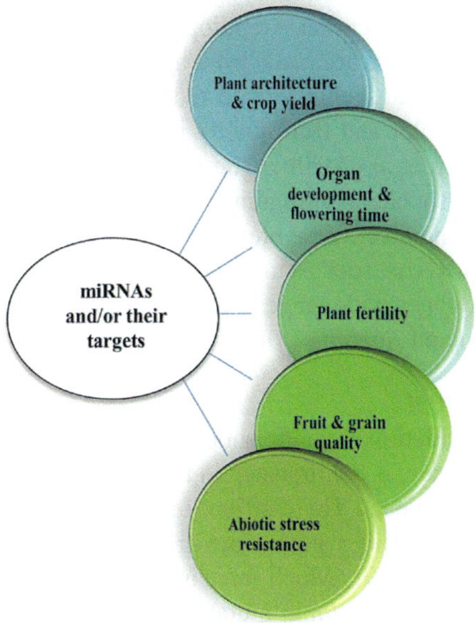

FIGURE 10.2

miRNAs and/or their target genes that participate in all aspects of plants.

direct binding of SPL3/4/5 proteins to the FD-FT complex, which activates genes associated with flowering (Jung et al., 2016). *SPL9* and *SPL10* are specifically targeted by MiR156, which binds to the promoter of MIR172b and controls flowering time through the miR172-TOE-TF pathway. Plant flowering is influenced by the miR156-SPL-miR172-AP2-FT pathway. A phenotype of early flowering was observed in *Arabidopsis* when apple MdmiR172e was overexpressed (Zhao et al., 2015). Moreover, species-specific short RNAs, such as citrus miR3954 targeting NAC TF and two lengthy noncoding genes, regulate timing of flowering. Several miR3954-targeted NAC TFs are cleaved by phasiRNAs produced by lncRNAs (Liu et al., 2017). Flowering-time regulation has also been reported in horticultural plants by miR156 and miR172. Flowering time is regulated by miR156, miR159, miR172, miR393, and many other molecules, such as miR165/166, miR172, and miR319 (Nag et al., 2009; Wang et al., 2017; Wei et al., 2017).

8.1.2 Fruit development

Breeders should focus on fruit size and shape when breeding fleshy fruit. The size of the fruit influences both harvest output and consumer decisions (Ripoll et al., 2015). MiR172 silenced *Arabidopsis* plants have a smaller silique than wild type plants. MiR172 targets AP2 TFs in order to influence the expression of *AG* (*AGAMOUS*) and *FUL* (*FRUITFUL*) in the silique and ovary (Yant et al., 2010). On the contrary, the overexpression of miR172e in apple severely reduces the size of the fruit (Yao et al., 2015). In further research, it was found that miR172 affects fruit size according to the type of fruit. Evidences suggest that miR172 plays a different role in various tissues based on the fact that Arabidopsis siliques are made from ovary tissues whereas fleshy apple fruits are made from hypanthium tissues. Further the overexpression of miR172 in transgenic tomato reduced the fruit size by 4–5 folds in addition to texture and color and fruit quality in terms of sugar content and aroma (Yao et al., 2016). Wang et al (2020b). The miR164-NAC combination influences not just ethylene synthesis but also fragrance generation (through terpenes synthase (AaTPS1) activity) and hemicellulose breakdown since NAC TFs have various roles in fruit (AdMAN1). MiRNAs, such as miR160, miR167, miR172, miR390, miR393, miR828, and miR858, play a wide range of roles in fruit development, and from fruit initiation to coloration and ripening (Yao et al., 2015).

8.1.3 Leaf development

Leaf primordial differentiation from the shoot apical meristem (SAM), followed by the growth of the leaf blade, is the first step in the formation of a leaf and involves a number of regulatory elements. The distribution of cells determines SAM organogenesis (Veit, 2009). MiR160 regulate auxin response factors (ARFs) such as ARF10, ARF16, and ARF17, which are involved in leaf development. The miR160 cleavage-resistant *A. thaliana* mutant *arf10* and *arf17* exhibit an excessive number of cotyledons and a serrated, curled-up edge to their leaves (Liu et al., 2007). The size and structure of leaves are also regulated by miRNAs. Ultimately, miR396 and GRFs balance to determine the cell number in leaves and meristem size (Debernardi et al., 2014). MiR319-TCP4 is responsible for controlling the senescence of leaves (Sun et al., 2017). The high sequence similarity of miR159 and miR319 and curled leaves of miR159a and miR159b double mutant suggested involvement of miR159 in leaf development (Allen et al., 2007). TAS3-ARF, TCP, and GRF, as well as miR390, miR319, and miR396 regulate their target genes to affect morphogenesis and growth polarity in leaves (Rodriguez et al., 2010). Further variable expression of miR164, miR169, and miR396 in leaves with curvatures suggests that the role of these miRNAs in leaf shape (Ren et al., 2018).

8.1.4 Shoot development

In contrast to animals, plants continue to grow new organs across their life period. A group of stem cells capable of multidirectional differentiation and self-replication emerge at the embryo's apical meristem. SAM plays a crucial role in the development of a plant's above-ground organs. STM (Shoot Meristem Less)-WUS (Wuschel)-CLV (Clavata) path ways maintain meristem activity (Somssich et al., 2016). In the meantime, miR394-LCR plays a variety of roles in stem cell regulation (Kumar et al., 2019). Moreover, four types of microRNA/target modules were identified and characterized for their functions in regulating the development of tea plant shoot tips and leaf shapes: miR319b/CsTCP3-4, miR164b/CsCUC, miR396/CsGRF-GIF, and miR165b/HD-ZIPIII (Yu et al., 2021). Over-expressing sly-miR156 significantly reduced the expression of these target genes, as well as tomato *FLOWERING LOCUS T* (*FT*) ortholog *SFT*. A novel phenotype is revealed by these studies that are regulated by miR156 (Zhang et al., 2011a). As petunia plants ages, the level of PhmiR156/157 is reduced in shoot apical buds and leaves; moreover, most miR156/157-targeted PhSPLs display complementary expression patterns (Zhou et al., 2021).

8.2 MicroRNAs regulate quality change

Quality, in general, refers to all the positive attributes that meet a given standard, including color, size, form, flavor, texture, nutrition, etc. Horticultural crops are valued according to their market and commercial potential and their growth, development, ripening, and senescence are all intimately tied to quality development. Recent research has revealed how small RNAs, such as miRNAs, regulate horticultural traits (Ma et al., 2020). Here we have discussed one by one on the major quality development traits.

8.2.1 Color

In horticultural crops, anthocyanin accumulation is mainly attributed to MYB transcription factors, which are responsible for coloration. Silencing of MiR858 caused increase in the level of its target mRNA, *MYB7-like* resulting in tomato anthocyanin buildup (Jia et al., 2015a). In tomato, miR828 also inhibits anthocyanin biosynthesis when phosphate is deficient (Jia et al., 2015b). MiR828 may inhibit anthocyanin metabolism since it down regulates the expression of its target gene *MYB113* and causes lighter-colored leaves when *V. vinifera* miR828 is overexpressed in *Arabidopsis* (Chen et al., 2019). It was also recently demonstrated in an apple, *Malus domestica* that miR828-TAS4-MYB1 modulate the accumulation of anthocyanin in apple cells (Zhang et al., 2020). Anthocyanin biosynthesis in red kiwifruit, *Actinidia arguta* is negatively regulated by miR858, which targets *MYBC1*, an R2R3-type MYB. Furthermore, miR858 overexpression and knockdown of *MYBC1* results in inhibition of anthocyanin synthesis (Li et al., 2020). The ripe fruit of a transgenic apple plant expressing miR172 exhibits a severe russeting and a loss in red color. Inhibiting the biosynthesis of flavonoids in apples by upregulating the anthocyanin activator *MYB10*, miR172 inhibits the expression of AP2. Light-induced coloration interacts with one another is revealed in an interesting study suggesting miR7125 mediates the crosstalk between lignin metabolism and light-induced coloration in apple (Hu et al., 2021).

8.2.2 Size

Consumer preferences are often influenced by the size of horticultural crops. Gene *AP2* plays a positive role in the development of apple fruit, which is primarily composed of sepal tissue. When miR172 is overexpressed, *AP2* is silenced and fruit size is severely reduced. Under insertional transposon allele of MIR172, larger fruits with up regulated *AP2* expression exhibit lower levels of miR172 (Hu et al., 2021). It has been shown that miR168 plays an important role in tomato in the initiation and growth of fruits by interacting with AGO1s (Xian et al., 2014). Overexpressing miR159 causes silencing of target *GAMYB1/2*, which results in precocious fruit initiation and obligatory parthenocarpy in tomato plants (da Silva et al., 2017). Short tandem target mimic (STTM) approaches lead to down regulation of miR396a/b and up regulation of their target GRFs, leading to bigger tomatoes (Cao et al., 2016). A knock-down mutant of miR1917 in tomato has also been reported to produce bigger fruit, as well as up regulated expression of *CTR4* gene and alteration in ethylene response (Yang et al., 2020a).

8.2.3 Shape

Quality in horticultural crops is also determined by the type of crop being under consideration. Fruit shape is one of the important characteristics in terms of commercial value and in tomato miRNAs governing the shape have been functionally characterized. The overexpressed MIR166b in tomatoes leads to fruit growing out of fruit. Additionally, ectopic carpels and fruit fusion are produced by overexpressing *REV* (35S::REVRis) in a miR166-resistant mutant. Single overexpression of *REV* does not affect phenotype, indicating that REV's function depends primarily on miR166 post transcriptionally (He et al., 2014). The Aux/IAA family genes control fruit expansion, and in tomato, inhibiting miR160 dramatically increases the expression of the miR160 target gene *ARF10A* (Damodharan et al., 2016). MiR159 regulates fruit morphology through its target *GAMYB2*, which is involved in GA biosynthesis. Blocking of miR159 resulted in flattened fruits with more locules (Zhao et al., 2022). The quality of leafy vegetables is closely related to the shape of the leaves (including petioles). Overexpression of MIR319a in Chinese cabbage results in straightened leaf tips, wavy leaf margins, and changed leaves (Mao et al., 2014). He et al. (2020). Moreover, it has been demonstrated that broccoli miR390a overexpression in *Arabidopsis* can enhance the growth of lateral organs and increase the biomass of transgenic plants (Fig. 10.3).

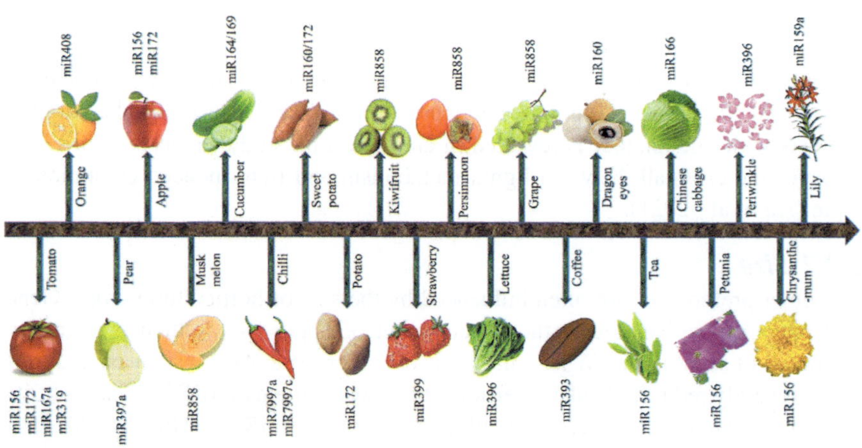

FIGURE 10.3

Main miRNAs involved in representative horticultural plants species.

8.2.4 Ripening

Fruit crops go through a distinct ripening process, which not only signals about the quality of fruit but indicates the start of irrevocable senescence. The tomato, a fruit by botanical definition but a vegetable by horticultural definition, is the first fruit to demonstrate how miRNAs influence fruit ripening. Fruit ripening depends on transcription factors encoded by these two target genes (Karlova et al., 2013). Overexpression of miR157 delays fruit ripening in tomato by targeting SPL-CNR (Chen et al., 2015a). A reduction in miR172 expression accelerates fruit ripening, thereby increasing ethylene biosynthesis, while repressing only *AP2a* and no other ripening-related genes (Chung et al., 2020). The spliced versions of *CTR4* that allows MiR1917 target for the overexpression of *CTR4* to improve the ethylene response, and hasten fruit ripening (Wang et al., 2018). In grapes, strawberries, sweet cherries, *Prunus avium*, bananas, *Musa acuminate*, kiwifruit, and *Litchi chinensis*, miRNAs linked to fruit ripening have been identified. Fruits with widely varied ripening traits share miR172-AP2, miR156/157-SPL and miR164-NAC as common miRNA-target modules indicating evolutionary conservation of miRNAs (Wang et al., 2020b). Since miR393 suppresses the expression of the auxin receptor gene *AFB2*, transgenic melon, *Cucumis melo* plants overexpressing miR393 have shown delay in fruit ripening (Bai et al., 2020).

8.2.5 Flavor and texture

Fruit flavor and texture can also be controlled by miRNAs, these two factors greatly affect the edible quality of fruit. Strawberry, *Fragaria ananassa* may express miR399 in response to sugar content (Li et al., 2013). Furthermore, improvement in fruit quality through enhancement of glucose, fructose and soluble solid content has been associated with overexpression of miR399a in strawberry. It has also been found that overexpressed miR399a is a potential candidate that can drastically enhance the quality of strawberries by increasing levels of glucose, fructose, and soluble (Wang et al., 2017). As tomatoes ripen, glutamate accumulates, which contributes to their umami flavor (Mounet et al., 2007). The poor flavor in pears, *Pyrus bretschneideri*, is known to be associated with increased lignification of stone cells. *LAC* gene is silenced by transient overexpression of miR397a in pear fruit, resulting in reduced lignin content, and stone cell number (Xue et al., 2019). Recent findings suggest that miR7125 regulates lignin-anthocyanin balance in apples by targeting CCR implicated in lignin production and being regulated by MYB transcription factors (Hu et al., 2021).

8.3 MicroRNAs regulate plant responses to environmental stresses

Several RNA silencing techniques, such as hairpin (hpRNA) transgene and artificial miRNA (amiRNA) techniques have been employed to investigate the function of genes. Manipulation of pathways gene's metabolic expression in crop plants can generate pathogen and pest resistance. RNA silencing has been successfully employed in horticultural plants to increase crop yields and improve disease and insect resistance.

8.3.1 Increasing resistance to biotic stresses

Horticultural crops are severely affected by biotic stresses caused by viruses, bacteria, and fungi, as well as insects, and nematodes. Globally, crop losses attributed to biotic stress tune to 30% (Flood, 2010). Many unique and conserved miRNAs that control the genes associated with biotic stresses have been discovered in tomatoes by high-throughput sequencing research. The miRNAs studied and characterized in response to various biotic stresses are described below.

8.3.1.1 Virus

MiRNAs appear to play a key role in plant antiviral defenses according to several studies conducted previously (Bester et al., 2017). As viruses multiply and infect through a variety of strategies, they are particularly difficult to control. One of the earliest instances of total immunity to potato virus Y (PVY) was demonstrated in virus-resistant potato plants expressing viral *HCPro* (helper-component proteinase) gene transcripts both in sense and antisense orientations (Waterhouse et al., 1998), Potato varieties have been developed that are highly resistant to three strains of PVY (Missiou et al., 2004). Recently, tomato plants expressing hpRNA from potato spindle tuber viroid (PSTVd) were found to be resistant to the virus (Schwind et al., 2009). Double infections with Potato virus X (PVX) and PVY or Plum pox virus (PPV) in *N. benthamiana* caused severe symptoms as well as a higher shift in miRNA156, miRNA171, miRNA398, and miRNA168 accumulation as compared with single infections (Pachec et al., 2012). Virus-resistant plants produced by the production of viral coat proteins exhibited RNA silencing mechanism with virus resistance in papaya plant (Kertbundit et al., 2007). Three tomato miRNAs (miR156, miR159, and miR403) have been shown to target the tomato leaf curl virus (ToLCV) (Tousi et al., 2017). The plant immune response involves the miRNA targets NB-LRR, NAC, MYB, zinc finger transcription factor, fructose two to six bisphosphate, and HDZIP protein. Other oxidative stress-responsive genes include *Auxin response factor* (*ARF*), *superoxide dismutase* (*SOD*), and *basic blue copper protein* (Kundu et al., 2017).

8.3.1.2 Bacteria

In tomato plants, the expression of miRNA167, miRNA160, miRNA390, miRNA393, and miRNA846 has been discovered to decrease auxin signaling after *Pseudomonas syringae* or *Xanthomonas axonopodis* infection by negatively regulating the mRNAs of F-box auxin receptors *AFB2*, *AFB3*, and *TIR1* (Jodder et al., 2017). MiRNAs that regulate plant hormone signaling pathways, such as ABA, jasmonic acid (JA) and auxin signaling, are also involved in defense against bacterial pathogens (Zhang et al., 2011b). *P. syringae* was suppressed by *Bacillus cereus* as a result of up regulation of their target genes followed by pretreatment with *B. cereus* (Niu et al., 2016). A study was conducted on cassava, *Manihot esculenta* showed that *X. axonopodis* infected plants downregulated several miRNA families, including miRNA397, miRNA398, and miRNA408 (Pérez-Quintero et al., 2012).

8.3.1.3 Insect pests

The lepidopteran polyphagous pests, *Helicoverpa armigera* feeds on cotton and many horticultural crops causing sizable yield losses. The *CYP6AE14* cytochrome P450 gene is believed to be responsible for insect tolerance to toxin gossypol even at otherwise inhibitory concentrations. It was shown that hpRNA against *CYP6AE14* expressed by transgenic tobacco plants was capable of inhibiting gene expression in the mid guts of insects (Mao et al., 2007). Insects may have been controlled using this strategy by feeding them on engineered plants producing dsRNAs targeting vital genes. Transgenes targeting the β-actin gene of the colorado potato beetle, *Leptinotarsa decemlineata* were developed recently to express in the chloroplast of potato, which provided strong resistance to herbivore. Whole length hpRNA can build up at high levels because chloroplasts lack DCLs, which may help to explain why chloroplast-expressed hpRNA is just as effective as DCL-expressed hpRNA. Similarly, a transgene expressing hpRNA in the nucleus failed to provide effective protection. These results suggest that siRNAs generated in plant cells are unable to produce cross-kingdom gene silencing and dsRNA processing in the target organism is the prerequisite to induce gene silencing. *Chrysanthemum morifolium* was shown to produce miRNA159a, which is predicted to be a GAMYB-like 2 when aphids feed on its leaves (Xia et al., 2015). Gibberellin (GA) signaling pathway is affected by the product of this enzyme (Woodger et al., 2003). The tea plant *Camellia sinensis* Linn showed differential expression of 139 unique miRNAs and 36 recognized miRNAs in response to geometrid infestation stress (Jeyaraj et al., 2017). In addition, some of those miRNAs encoded transcription factors that are involved in stress response, including ethylene-responsive transcription factors and squamosa promoter-binding-like proteins (Jeyaraj et al., 2017).

8.3.1.4 Nematode

Root knot nematode (*Meloidogyne* spp.) infects numerous plant species and has caused significant economic losses all over the world. The expression of putative zinc finger transcription factor *MjTis11* of *M. javanica* was drastically decreased in transgenic tobacco plants expressing hpRNA construct (Fairbairn et al., 2007). The host mediated dsRNA delivery can be utilized to silence vital genes of parasites such as root-knot nematodes, *M. incognita*, infection leads to root galls because of genetic reprogramming of vascular cells, which transforms them into massive feeding cells. *Arabidopsis* miR159abc triple mutants were less susceptible to infection by *M. incognita*, according to a study, suggesting that miRNA159 suppresses the expression of MYB33, which plays a crucial role in the early stages of infection, thereby influencing plant response to root knot nematodes (Medina et al., 2017). Studies with roots infected with *M. incognita* suggest differential expression of 16 miRNAs in cotton crop (Pan et al., 2019).

8.3.1.5 Fungus

Eschen-Lippold et al. (2012) reported that to reduce the risk of oomycete infection of late blight, *Phytophthora infestans*, transgenic potato plants expressed hpRNAs targeting Syntaxin-related 1 (SYR1). Transgenic potatoes accumulate salicylic acid and PR1 transcripts after infection with *P. infestans*. A cytological examination revealed abnormal callose deposition and reduced papilla formation in potato as a result of infection with *P. infestans*. Fusiform rust disease, affecting loblolly pine, *Pinus taeda* is characterized by thread-like galls on branches or stems that are brought on by *Cronartium quercuum* f. sp. *fusiforme* an indigenous rust fungus (Lu et al., 2007). Studies with *Verticillium* wilt in cotton suggest 24 of the 37 novel miRNAs were associated with genes involved in endocytosis, plant-pathogen interactions, MAPK (mitogen activated protein kinase) signaling, the synthesis of terpenoid backbones, isoquinoline alkaloid, primary bile acids and secondary metabolites (He et al., 2014). The low expression of the miRNAs such as miRNA8163, miRNA8165, miRNA8170, miRNA8175, and miRNA8178 was also anticipated to mediate the plant defensive response (He et al., 2014). Garlic (*Allium sativum* L.) infected with basal plate rot fungus *Fusarium oxysporum* f. sp. *cepae* (FOC) exhibited upregulation of miRNA164a, miRNA168a (which targets the protein AGO1) and miRNA393. Besides overexpression of miRNAs, the transgenic garlic also showed weaker disease phenotype, poor growth of the fungus in infected tissues, and up regulation of defense-responsive genes, and increased resistance to FOC (Chand et al., 2017). An apple cultivar Golden delicious exhibited overexpression of Md-miRNA156ab and Md-miRNA395 and higher susceptibility followed by infection with apple leaf spot fungus *Alternaria alternaria* f. sp. *mali* (ALT1) (Zhang et al., 2017a).

8.3.2 Tolerance to abiotic stress

Abiotic stress refers to detrimental effects of nonliving elements on a plant under certain environmental conditions. Plant growth is negatively affected by a number of physical and chemical factors. Low or high temperatures, high salinity, a lack of or an abundance of water, heavy metals, and UV light are a few examples of such factors. Plants must adjust to such stresses in order to survive, hence they need strong adaptive strategies to prevent or withstand the negative impacts (Wang et al., 2019). The miRNAs play an important role in several important and model plants when exposed to different abiotic stress conditions.

8.3.2.1 Salinity and drought tolerance

Abiotic stresses such as drought and salinity adversely affect crop yields, and have been extensively studied. To cope with such stress, plants adapt several physiological, biochemical, and molecular strategies (Kumar, 2014; Wani et al., 2020). Recent machine learning models have demonstrated that the miRNAs 169, 159, 396, and 393 play a significant role in plants in response to salinity, drought, heat, and cold stress (Asefpour Vakilian, 2020). Drought and salinity stress up regulate several miRNAs, including miRNA167, miRNA393, and miRNA160. Suppressing ARF-mediated expression of auxin sensitive genes may act as a mediator of stress tolerance (Sunkar et al., 2012). Shorter petals, stamens, and styles were the result of overexpression of miRNA167 in tomato plants (Liu et al., 2014). Changes in leaf shape and stomatal closure were correlated with expression of the cassava miRNA-167 target genes (*MesARF6/8*) during drought stress (Phookaew et al., 2014). Down regulation of *ARF10* is caused by drought stress and under such conditions MiR160 activates its target and implicates auxin as a stress mediator (Bouzroud et al., 2018).

8.3.2.2 Temperature stress tolerance

Seasonal variations and different geographic regions often expose plants to extreme temperatures. Several miRNAs have been shown to respond to heat and cold stress. MiRNAs that respond to heat stress, such as spi-miRNA6300-gma, spi-miRNA166c-3p, and spi-miRNA166g 3p-osa, regulate the induction of HSPs (heat shock proteins) (Ding et al., 2020). The "Yingshuang" cultivar of cold-resistant tea also showed down regulation of 43 miRNAs, and up regulation of 31 miRNAs. LEA proteins, zinc ion binding and transmembrane amino acid transporter proteins, and WBP (WW-domain-binding protein) are all regulated by some differentially expressed miRNAs in tea. LEA proteins assist in preserving an organism's biological structures and viability during cold stress. Zinc ion binding not only contributes to healthy respiratory function but also makes abscisic acid stress ripening proteins more structurally complicated. WPB binds to WW-domains, which are tiny protein modules involved in the control of cell signaling (Zhang et al., 2014). MiRNA169 is crucial for plants' ability to withstand heat stress (Pegler et al., 2019).

8.3.3 Heavy metal stress tolerance

Abiotic stress caused by heavy metals can negatively affect plant growth, development and productivity by altering physiological and metabolic processes (Gupta et al., 2014). Heavy metal contamination in soil is caused by anthropogenic and human activities. Xu et al. (2013) constructed small RNA libraries from radish, *Raphanus sativus* with or without cadmium (Cd) stress in order to unravel Cd-responsive miRNAs and their regulatory networks. A study with bean species, *Phaseolus vulgaris* revealed that the miRNAs miRNA1515, miRNA1510/2110, miRNA1508, and miRNA1532 respond to Mn and have heat shock proteins, proteins with nucleotide-binding sites with leucine rich repeats, calcium-dependent protein kinases and receptor kinases as their targets (Valdés-López et al., 2010).

8.3.3.1 Ultra violet radiation

Plant growth and development are negatively affected by elevated ultra violet (UV−B) radiation caused by the depletion of the ozone layer, primarily through reactive oxygen species (ROS)-mediated oxidative bursts (Kruszka et al., 2012; Gill et al., 2015). Yang et al. (2020b) studied with *Chrysanthemum morifolium* identified 137 UV-B responsive miRNAs suggesting that UV-B exposure influences transcriptional regulation. Five genes associated to chlorogenic acid and flavonoids (*ANS*, *F3H*, *4CL*, *PAL*, and *UGT88*) and three glycolysis-related genes (*GAPDH*, *PK*, and *GEG*) were also found to be targets of UV-B responsive miRNAs, indicating that they contribute to the buildup of bioactive components.

9. Conclusion and future perspectives

For the development of high quality, high yield, and stress-tolerant crops, a thorough understanding of miRNA function is of great importance. Bioinformatics and molecular biology tools have been employed to predict and confirm an increasing number of miRNAs and their target genes in plants. Particularly when transcription factors are directly paired with traditional cis-regulatory promoters, miRNAs regulate at the posttranscriptional level. The miRNA involved in plant stress response has a more degree of complicated mechanism, though, as the majority of these target genes are transcription factors. As they may control complex agronomic properties, plant miRNAs have emerged as interesting targets for crop development, leading to increased yield, quality, and stress tolerance. MiRNAs may play an important role in plants, based on what was outlined above. Transcription factors are a target gene for miRNAs, and depending on their subunits, they can perform a variety of transcriptional activation actions. These factors play a key role in controlling plant development and acclimation along with the gene targets that encode proteins or enzymes involved in plant metabolism. Overexpression or repression of miRNAs may have an impact on plant phenotypes by up- or down-regulating the expression of their downstream target genes. For a complete understanding, synchronized investigations of miRNAs and their target mRNAs are essential. MiRNA candidates can be predicted and analyzed using degradome sequencing technology, which is helpful for comprehending their posttranscriptional function. Artificial miRNAs (amiRNAs), potent methods for silencing endogenous genes, will be the effective way to suppress the expression of target genes for identifying miRNA activities in both developmental control, and abiotic stress. AmiRNA-mediated gene silencing has also been reported to explain the functional importance of many genes.

Short tandem target mimics (STTM), which can be utilized for functional dissection of miRNAs in research as well as for examining relationships and functions of miRNA regulatory networks, are a very efficient and specific method for inducing miRNA degradation. A variety of biotechnological tools, such as synthetic and endogenous miRNAs, CRISPR/Cas-mediated miRNA gene editing, and stress tolerance markers based on miRNAs, are available for modifying miRNA expression. Plants are generally affected by the combination of two or more stresses, such as high temperature, and drought. Crop plants may be able to improve their tolerance to abiotic stress by regulating miRNAs. Based on a preliminary statistical analysis, several miRNAs (miR156, miR159, miR160, miR164, and miR172) play diverse role in plants under diverse stresses. Plant protection strategies currently do not fully utilize miRNA-based techniques, even though the biological roles of target genes and miRNA-based regulatory mechanisms are still being investigated. Plant miRNA research faces many challenges due to adversity. Conservative miRNAs play a significant role in growth and development, and nonconservative (specific) miRNAs play an important role in stress resistance. In plants, most miRNAs are highly conserved. As previously shown, miRNAs regulate gene expression in a specific manner across time and space, and in different species at the same time. In addition to regulating root, stem, leaf, flower organ, and fruit growth and development, miRNAs also respond to biotic and abiotic stresses to regulate growth and development. Therefore, miRNA-mediated regulatory networks need to be explored for better and thorough understanding of their target genes revealed. Under current scenario for both sustainable horticulture development and genetic enhancement of horticultural crops are being supported to conduct studies in these directions.

References

Akter, A., Islam, M.M., Mondal, S.I., Mahmud, Z., Jewel, N.A., Ferdous, S., et al., 2014. Computational identification of miRNA and targets from expressed sequence tags of coffee (*Coffea arabica*). Saudi J. Biol. Sci. 21, 3–12.

Allen, R.S., Li, J., Stahle, M.I., Dubroue, A., Gubler, F., 2007. Genetic analysis reveals functional redundancy and the major target genes of the Arabidopsis miR159 family. Proc. Natl. Acad. Sci. U.S.A. 104, 16371–16376.

Asefpour Vakilian, K., 2020. Machine learning improves our knowledge about miRNA functions towards plant abiotic stresses. Sci. Rep. 10, 3041.

Aukerman, M.J., Sakai, H., 2003. Regulation of flowering time and floral organ identity by a microRNA and its APETALA2-like target genes. Plant Cell 15, 2730–2741.

Baek, D., Villén, J., Shin, C., Camargo, F.D., Gygi, S.P., Bartel, D.P., 2008. The impact of microRNAs on protein output. Nature 455, 64–71.

Bai, S., Tian, Y., Tan, C., Bai, S., Hao, J., Hasi, A., 2020. Genome-wide identification of microRNAs involved in the regulation of fruit ripening and climacteric stages in melon (*Cucumis melo*). Hortic. Res. 7, 106.

Bester, R., Burger, J.T., Maree, H.J., 2017. Differential expression of miRNAs and associated gene targets in grapevine leaf roll-associated virus 3-infected plants. Arch. Virol. 162, 987–996.

Bouzroud, S., Gouiaa, S., Hu, N., Bernadac, A., Mila, I., Bendaou, N., et al., 2018. Auxin response factors (ARFs) are potential mediators of auxin action in tomato response to biotic and abiotic stress (*Solanum lycopersicum*). PLoS One 13 e0193517.

Bresso, E.G., Chorostecki, U., Rodriguez, R.E., Palatnik, J.F., Schommer, C., 2018. Spatial control of gene expression by miR319-regulated TCP transcription factors in leaf development. Plant Physiol. 176, 1694–1708.

Cao, D., Wang, J., Ju, Z., Liu, Q., Li, S., Tian, H., et al., 2016. Regulations on growth and development in tomato cotyledon, flower and fruit via destruction of miR396 with short tandem target mimic. Plant Sci. 247, 1–12.

Cartolano, M., Castillo, R., Efremova, N., Kuckenberg, M., Zethof, J., 2007. A conserved microRNA module exerts homeotic control over Petunia hybrida and Antirrhinum majus floral organ identity. Nat. Genet. 39, 901–905.

Chand, S.K., Nanda, S., Mishra, R., Joshi, R.K., 2017. Multiple garlic (*Allium sativum* L.) microRNAs regulate the immunity against the basal rot fungus *Fusarium oxysporum* f. sp. cepae. Plant Sci. 257, 9–21.

Chen, C., Li, J., Feng, J., Liu, B., Feng, L., Yu, X., et al., 2021. sRNAanno-a database repository of uniformly annotated small RNAs in plants. Hortic. Res. 8, 1–8.

Chen, C., Zeng, Z., Liu, Z., Xia, R., 2018. Small RNAs, emerging regulators critical for the development of horticultural traits. Hortic. Res. 5, 1–14.

Chen, Q., Deng, B., Gao, J., Zhao, Z., Chen, Z., Song, S., et al., 2019. Comparative analysis of miRNA abundance revealed the function of Vvi-miR828 in fruit coloring in root restriction cultivation grapevine (*Vitis vinifera* L.). Int. J. Mol. Sci. 20, 4058.

Chen, W., Kong, J., Lai, T., Manning, K., Wu, C., Wang, Y., et al., 2015a. Tuning LeSPL-CNR expression by SlymiR157 affects tomato fruit ripening. Sci. Rep. 5, 7852.

Chen, X., Xia, J., Xia, Z., Zhang, H., Zeng, C., Lu, C., et al., 2015b. Potential functions of microRNAs in starch metabolism and development revealed by miRNA transcriptome profiling of cassava cultivars and their wild progenitor. BMC Plant Biol. 15, 1–11.

Chung, M.Y., Nath, U.K., Vrebalov, J., Gapper, N., Lee, J.M., Lee, D.J., et al., 2020. Ectopic expression of miRNA172 in tomato (*Solanum lycopersicum*) reveals novel function in fruit development through regulation of an AP2 transcription factor. BMC Plant Biol. 20, 1–15.

Cuperus, J.T., Fahlgren, N., Carrington, J.C., 2011. Evolution and functional diversification of miRNA genes. Plant Cell 23, 431–442.

da Silva, E.M., Silva, G.F.F.E., Bidoia, D.B., da Silva Azevedo, M., de Jesus, F.A., Pino, L.E., et al., 2017. Micro RNA 159-targeted Sl GAMYB transcription factors are required for fruit set in tomato. Plant J. 92, 95–109.

Damodharan, S., Zhao, D., Arazi, T., 2016. A common mi RNA 160-based mechanism regulates ovary patterning, floral organ abscission and lamina outgrowth in tomato. Plant J. 86, 458–471.

Debernardi, J.M., Mecchia, M.A., Vercruyssen, L., Smaczniak, C., Kaufmann, K., 2014. Post-transcriptional control of GRF transcription factors by microRNA miR396 and GIF co-activator affects leaf size and longevity. Plant J. 79, 413–426.

Din, M., Barozai, M.Y.K., Baloch, I.A., 2014. Identification and functional analysis of new conserved microRNAs and their targets in potato (*Solanum tuberosum* L.). Turk. J. Bot., Le 38, 1199–1213.

Ding, Y., Huang, L., Jiang, Q., Zhu, C., 2020. MicroRNAs as important regulators of heat stress responses in plants. J. Agric. Food Chem. 68, 11320–11326.

Djami-Tchatchou, A.T., Dubery, I.A., 2015. Lipopolysaccharide perception leads to dynamic alterations in the microtranscriptomes of *Arabidopsis thaliana* cells and leaf tissues. BMC Plant Biol. 15, 79.

Djami-Tchatchou, A.T., Sanan-Mishra, N., Ntushelo, K., Dubery, I.A., 2017. Functional roles of microRNAs in agronomically important plants-potential as targets for crop improvement and protection. Front. Plant Sci. 8, 378.

Dong, Q., Hu, B., Zhang, C., 2022. microRNAs and their roles in plant development. Front. Plant Sci. 13, 824240.

Dong, Z., Han, M.H., Fedoroff, N., 2008. The RNA-binding proteins HYL1 and SE promote accurate in vitro processing of pri-miRNA by DCL1. Proc. Natl. Acad. Sci. U.S.A. 105, 9970–9975.

Eschen-Lippold, L., Landgraf, R., Smolka, U., Schulze, S., Heilmann, M., Heilmann, I., et al., 2012. Activation of defense against Phytophthora infestans in potato by down-regulation of syntaxin gene expression. New Phytol. 193, 985–996.

Fairbairn, D.J., Cavallaro, A.S., Bernard, M., Mahalinga-Iyer, J., Graham, M.W., Botella, J.R., 2007. Host-delivered RNAi: an effective strategy to silence genes in plant parasitic nematodes. Planta 226, 1525–1533.

Fang, Y.D., Spector, D.L., 2007. Identification of nuclear dicing bodies containing proteins for microRNA biogenesis in living Arabidopsis plants. Curr. Biol. 17, 818–823.

Feng, H., Duan, X., Zhang, Q., Li, X., Wang, B., Huang, L., et al., 2014. The target gene of tae-miR164, a novel NAC transcription factor from the NAM subfamily, negatively regulates resistance of wheat to stripe rust. Mol. Plant Pathol. 15, 284–296.

Flood, J., 2010. The importance of plant health to food security. Food Secur. 2, 215–231.

Gangwar, P., Johri, P., Datta, R., Singh, S., Trivedi, M., 2023. Overview on 1 phytopharmaceuticals and biotechnology of herbal plants. In: Phytopharmaceuticals and Biotechnology of Herbal Plants. CRC Press, pp. 1–17.

Gao, C., Ju, Z., Cao, D., Zhai, B., Qin, G., Zhu, H., Zhu, B., 2015. Micro RNA profiling analysis throughout tomato fruit development and ripening reveals potential regulatory role of RIN on micro RNA s accumulation. Plant Biotechnol. J. 13, 370–382.

Gao, X., Zhang, Q., Zhao, Y.Q., Yang, J., He, H.B., Jia, G.X., 2020. The lre-miR159a-LrGA-MYB pathway mediates resistance to grey mould infection in Lilium regale. Mol. Plant Pathol. 21, 749–760.

Gill, S.S., Anjum, N.A., Gill, R., Jha, M., Tuteja, N., 2015. DNA damage and repair in plants under ultraviolet and ionizing radiations. Sci. World J. e250158.

Guo, X., Gui, Y., Wang, Y., Zhu, Q.H., Helliwell, C., Fan, L., 2008. Selection and mutation on microRNA target sequences during rice evolution. BMC Genom. 9, 454.

Gupta, O.P., Sharma, P., Gupta, R.K., Sharma, I., 2014. MicroRNA mediated regulation of metal toxicity in plants: present status and future perspectives. Plant Mol. Biol. 84, 1–18.

He, L., Wang, Y., Jia, Y., Yang, Y., Han, X., Yuan, J., et al., 2020. Ectopic overexpression of bol-miR390a from broccoli (*B. oleracea* L var. italica) increases lateral branches in Arabidopsis. Plant Growth Regul. 92, 547–558.

He, X., Sun, Q., Jiang, H., Zhu, X., Mo, J., Long, L., et al., 2014. Identification of Novel microRNAs in the Verticillium Wilt-Resistant Upland Cotton Variety KV-1 by High-Throughput Sequencing, vol. 3. Springer Plus, pp. 1–9.

Hu, G., Fan, J., Xian, Z., Huang, W., Lin, D., Li, Z., 2014. Overexpression of SlREV alters the development of the flower pedicel abscission zone and fruit formation in tomato. Plant Sci. 229, 86–95.

Hu, J.Y., Zhou, Y., He, F., Dong, X., Liu, L.Y., Coupland, G., et al., 2014. miR824-regulated AGAMOUS-LIKE16 contributes to flowering time repression in Arabidopsis. Plant Cell 26, 2024–2037.

Hu, Y., Cheng, H., Zhang, Y., Zhang, J., Niu, S., Wang, X., et al., 2021. The MdMYB16/MdMYB1-miR7125-MdCCR module regulates the homeostasis between anthocyanin and lignin biosynthesis during light induction in apple. New Phytol. 231, 1105–1122.

Huang, S.Q., Xiang, A.L., Che, L.L., Chen, S., Li, H., Song, J.B., et al., 2010. A set of miRNAs from Brassica napus in response to sulphate deficiency and cadmium stress. Plant Biotechnol. J. 8, 887–899.

Iki, T., 2017. Messages on small RNA duplexes in plants. J. Plant Res. 130, 7–16.

Ipsaro, J.J., Joshua-Tor, L., 2015. From guide to target: molecular insights into eukaryotic RNA-interference machinery. Nat. Struct. Mol. Biol. 22, 20–28.

Jeyaraj, A., Liu, S., Zhang, X., Zhang, R., Shangguan, M., Wei, C., 2017. Genome-wide identification of microRNAs responsive to Ectropis oblique feeding in tea plant (*Camellia sinensis* L.). Sci. Rep. 7, 1–16.

Jia, X., Liu, H., Shen, J., Li, F., Ding, N., Sun, Y., et al., 2015b. Negative regulation of anthocyanin biosynthesis in tomato by microRNA828 under phosphate deficiency. Sci. Agric. Sin. 48, 2911–2924.

Jia, X., Shen, J., Liu, H., Li, F., Ding, N., Gao, C., et al., 2015a. Small tandem target mimic-mediated blockage of microRNA858 induces anthocyanin accumulation in tomato. Planta 242, 283–293.

Jodder, J., Basak, S., Das, R., Kundu, P., 2017. Coherent regulation of miR167a biogenesis and expression of auxin signaling pathway genes during bacterial stress in tomato. Physiol. Mol. Plant Pathol. 100, 97–105.

Jung, J.H., Lee, H.J., Ryu, J.Y., Park, C.M., 2016. SPL3/4/5 integrate developmental aging and photoperiodic signals into the FT-FD module in Arabidopsis flowering. Mol. Plant 9, 1647–1659.

Kaja, E., Szcześniak, M.W., Jensen, P.J., Axtell, M.J., McNellis, T., Makałowska, I., 2015. Identification of apple miRNAs and their potential role in fire blight resistance. Tree Genet. Genomes 11, 812.

Karlova, R., Van Haarst, J.C., Maliepaard, C., Van de Geest, H., Bovy, A.G., Lammers, M., et al., 2013. Identification of microRNA targets in tomato fruit development using high-throughput sequencing and degradome analysis. J. Exp. Bot. 64, 1863–1878.

Kertbundit, S., Pongtanom, N., Ruanjan, P., Chantasingh, D., Tanwanchai, A., Panyim, S., et al., 2007. Resistance of transgenic papaya plants to Papaya ringspot virus. Biol. Planta. 51, 333–339.

Ketao, W., Xiaomei, S.U., Xia, C.U.I., Yongchen, D.U., Zhang, S., Jianchang, G., 2018. Identification and characterization of microRNA during Bemisia tabaci infestations in *Solanum lycopersicum* and *Solanum habrochaites*. Hortic. Plant J. 4, 62–72.

Khare, P., Kishore, K., Sharma, D.K., 2017. A study on the standardization parameters of Cassia angustifolia. Asian J. Pharmaceut. Clin. Res. 10 (7), 329–332.

Kruszka, K., Pieczynski, M., Windels, D., Bielewicz, D., Jarmolowski, A., Szweykowska-Kulinska, Z., et al., 2012. Role of microRNAs and other sRNAs of plants in their changing environments. J. Plant Physiol. 169, 1664–1672.

Kumar, A., Gautam, V., Kumar, P., Mukherjee, S., Verma, S., 2019. Identification and co-evolution pattern of stem cell regulator miR394s and their targets among diverse plant species. BMC Evol. Biol. 19, 55.

Kumar, R., 2014. Role of microRNAs in biotic and abiotic stress responses in crop plants. Appl. Biochem. Biotechnol. 174, 93–115.

Kundu, A., Paul, S., Dey, A., Pal, A., 2017. High throughput sequencing reveals modulation of microRNAs in Vigna mungo upon Mungbean Yellow Mosaic India Virus inoculation highlighting stress regulation. Plant Sci. 257, 96–105.

Kutter, C., Scho, B.H., Stadler, M., Meins, F.J., Si-Ammour, A., 2007. MicroRNA-mediated regulation of stomatal development in Arabidopsis. Plant Cell 19, 2417–2429.

Lee, R.C., Feinbaum, R.L., Ambros, V., 1993. The *C. elegans* heterochronic gene lin-4 encodes small RNAs with antisense complementarity to lin-14. Cell 75, 843–854.

Li, S., Cong, Y., Liu, Y., Wang, T., Shuai, Q., Chen, N., et al., 2017. Optimization of Agrobacterium-mediated transformation in soybean. Front. Plant Sci. 8, 246.

Li, Y., Cui, W., Qi, X., Lin, M., Qiao, C., Zhong, Y., et al., 2020. MicroRNA858 negatively regulates anthocyanin biosynthesis by repressing AaMYBC1 expression in kiwifruit (Actinidia arguta). Plant Sci. 296, 110476.

Li, Z., Peng, J., Wen, X., Guo, H., 2013. Ethylene-insensitive3 is a senescence-associated gene that accelerates age-dependent leaf senescence by directly repressing miR164 transcription in Arabidopsis. Plant Cell 25, 3311–3328.

Lian, H., Wang, L., Ma, N., Zhou, C.M., Han, L., 2021. Redundant and specific roles of individual MIR172 genes in plant development. PLoS Biol. 19 e3001044.

Lin, Y., Lai, Z., Tian, Q., Lin, L., Lai, R., Yang, M., et al., 2015. Endogenous target mimics down-regulate miR160 mediation of ARF10,-16, and-17 cleavage during somatic embryogenesis in Dimocarpus longan Lour. Front. Plant Sci. 6, 956.

Liu, N., Wu, S., Van Houten, J., Wang, Y., Ding, B., Fei, Z., et al., 2014. Down-regulation of AUXIN RESPONSE FACTORS 6 and 8 by microRNA 167 leads to floral development defects and female sterility in tomato. J. Exp. Bot. 65, 2507–2520.

Liu, P.P., Montgomery, T.A., Fahlgren, N., Kasschau, K.D., Nonogaki, H., 2007. Repression of AUXIN RESPONSE FACTOR10 by microRNA160 is critical for seed germination and post-germination stages. Plant J. 52, 133–146.

Liu, X., Wang, T., Bartholomew, E., Black, K., Dong, M., Zhang, Y., et al., 2018. Comprehensive analysis of NAC transcription factors and their expression during fruit spine development in cucumber (*Cucumis sativus* L.). Hortic. Res. 5, 31.

Liu, Y., Ke, L., Wu, G., Xu, Y., Wu, X., Xia, R., et al., 2017. miR3954 is a trigger of phasiRNAs that affects flowering time in citrus. Plant J. 92, 263–275.

Llave, C., Xie, Z., Kasschau, K.D., Carrington, J.C., 2002. Cleavage of scarecrow-like mRNA targets directed by a class of Arabidopsis miRNA. Science 297, 2053–2056.

Lu, S., Sun, Y.H., Amerson, H., Chiang, V.L., 2007. MicroRNAs in loblolly pine (*Pinus taeda* L.) and their association with fusiform rust gall development. Plant J. 51, 1077–1098.

Lu, Y.B., Qi, Y.P., Yang, L.T., Guo, P., Li, Y., Chen, L.S., 2015. Boron deficiency responsive microRNAs and their targets in Citrus sinensis leaves. BMC Plant Biol. 15, 271.

Ma, L., Shi, Y.N., Grierson, D., Chen, K.S., 2020. Research advance in regulation of fruit quality characteristics by microRNAs. Food Qualt. Saf. 4, 1–8.

Maddala, V.K.S., Singh, S., Bandi, J., 2024. Soil properties influencing nutrient signaling in plant soil systems. In: Essential Minerals in Plant-Soil Systems. Elsevier, pp. 373–393.

Mao, Y., Wu, F., Yu, X., Bai, J., Zhong, W., He, Y., 2014. MicroRNA319a-targeted Brassica rapa ssp. pekinensis TCP genes modulate head shape in Chinese cabbage by differential cell division arrest in leaf regions. Plant Physiol. 164, 710–720.

Mao, Y.B., Cai, W.J., Wang, J.W., Hong, G.J., Tao, X.Y., Wang, L.J., et al., 2007. Silencing a cotton bollworm P450 monooxygenase gene by plant-mediated RNAi impairs larval tolerance of gossypol. Nat. Biotechnol. 25, 1307–1313.

Meddya, S., Meshram, S., Sarkar, D., Datta, R., Singh, S., Avinash, G., Thulasinathan, T., 2023. Plant stomata: an unrealized possibility in plant defense against invading pathogens and stress tolerance. Plants 12 (19), 3380.

Medina, C., Da Rocha, M., Magliano, M., Ratpopoulo, A., Revel, B., Marteu, N., et al., 2017. Characterization of microRNAs from Arabidopsis galls highlights a role for miR159 in the plant response to the root-knot nematode *Meloidogyne incognita*. New Phytol. 216, 882–896.

Missiou, A., Kalantidis, K., Boutla, A., Tzortzakaki, S., Tabler, M., Tsagris, M., 2004. Generation of transgenic potato plants highly resistant to potato virus Y (PVY) through RNA silencing. Mol. Breed. 14, 185–197.

Mitter, N., Zhai, Y., Bai, A.X., Chua, K., Eid, S., Constantin, M., et al., 2016. Evaluation and identification of candidate genes for artificial microRNA-mediated resistance to tomato spotted wilt virus. Virus Res. 211, 151–158.

Mounet, F., Lemaire-Chamley, M., Maucourt, M., Cabasson, C., Giraudel, J.L., Deborde, C., et al., 2007. Quantitative metabolic profiles of tomato flesh and seeds during fruit development: complementary analysis with ANN and PCA. Metabolomics 3, 273–288.

Nag, A., King, S., Jack, T., 2009. miR319a targeting of TCP4 is critical for petal growth and development in Arabidopsis. Proc. Natl. Acad. Sci. U.S.A. 106, 22534–22539.

Niu, Q., Li, J., Cai, D., Qian, M., Jia, H., Bai, S., Zheng, X., 2016. Dormancy-associated MADS-box genes and microRNAs jointly control dormancy transition in pear (Pyrus pyrifolia white pear group) flower bud. J. Exp. Bot. 67, 239–257.

O'Brien, J., Hayder, H., Zayed, Y., Peng, C., 2018. Overview of MicroRNA biogenesis, mechanisms of actions, and circulation. Front. Endocrinol. 9, 402.

Pacheco, R., García-Marcos, A., Barajas, D., Martiáñez, J., Tenllado, F., 2012. PVX–potyvirus synergistic infections differentially alter microRNA accumulation in Nicotiana benthamiana. Virus Res. 165, 231–235.

Palatnik, J.F., Allen, E., Wu, X., Schommer, C., Schwab, R., Carrington, J.C., et al., 2003. Control of leaf morphogenesis by microRNAs. Nature 425, 257–263.

Pan, X., Nichols, R.L., Li, C., Zhang, B., 2019. MicroRNA-target gene responses to root knot nematode (*Meloidogyne incognita*) infection in cotton (*Gossypium hirsutum* L.). Genomics 111, 383–390.

Pantaleo, V., Szittya, G., Moxon, S., Miozzi, L., Moulton, V., Dalmay, T., et al., 2010. Identification of grapevine microRNAs and their targets using high-throughput sequencing and degradome analysis. Plant J. 62, 960–976.

Parveen, R., Khare, N., Singh, S., Johri, P., 2023. An in silico outlook for the detection and surveillance of evolving and persistent plant pathogens. In: Detection, Diagnosis and Management of Soil-Borne Phytopathogens. Springer Nature Singapore, Singapore, pp. 17–40.

Patani, A., Patel, A., Prajapati, D., Khare, N., Singh, S., 2023. Various agriculture crop plant-based bioactive compounds and their use in nanomaterial synthesis and applications. In: Nanomaterials from Agricultural and Horticultural Products. Springer Nature Singapore, Singapore, pp. 223–241.

Patani, A., Prajapati, D., Shukla, K., Patel, M., Patani, P., Patel, A., Singh, S., 2024. Environmental stress–induced alterations in the micro-and macronutrients status of plant. In: Essential Minerals in Plant-Soil Systems. Elsevier, pp. 181–195.

Patanun, O., Lertpanyasampatha, M., Sojikul, P., Viboonjun, U., Narangajavana, J., 2013. Computational identification of microRNAs and their targets in Cassava (Manihot esculenta Crantz.). Mol. Biotechnol. 53, 257–269.

Pegler, J.L., Oultram, J.M., Grof, C.P., Eamens, A.L., 2019. Profiling the abiotic stress responsive microRNA landscape of *Arabidopsis thaliana*. Plants 8, 58.

Peláez, P., Trejo, M.S., Iñiguez, L.P., Estrada-Navarrete, G., Covarrubias, A.A., Reyes, J.L., et al., 2012. Identification and characterization of microRNAs in Phaseolus vulgaris by high-throughput sequencing. BMC Genom. 13, 83.

Pérez-Quintero, Á.L., Quintero, A., Urrego, O., Vanegas, P., López, C., 2012. Bioinformatic identification of cassava miRNAs differentially expressed in response to infection by Xanthomonas axonopodis pv. manihotis. BMC Plant Biol. 12, 1–11.

Phookaew, P., Netrphan, S., Sojikul, P., Narangajavana, J., 2014. Involvement of miR164-and miR167-mediated target gene expressions in responses to water deficit in cassava. Biolog. Plant. 58, 469–478.

Qin, Z., Li, C., Mao, L., Wu, L., 2014. Novel insights from non-conserved microRNAs in plants. Front. Plant Sci. 5, 586.

Rajagopalan, R., Vaucheret, H., Trejo, J., Bartel, D.P., 2006. A diverse and evolutionarily fluid set of microRNAs in *Arabidopsis thaliana*. Genes Dev. 20, 3407–3425.

Reinhart, B.J., Weinstein, E.G., Rhoades, M.W., Bartel, B., Bartel, D.P., 2002. MicroRNAs in plants. Genes Dev. 16, 1616–1626.

Ren, W., Wang, H., Bai, J., Wu, F., He, Y., 2018. Association of microRNAs with types of leaf curvature in Brassica rapa. Front. Plant Sci. 9, 73.

Rhoades, M.W., Reinhart, B.J., Lim, L.P., Burge, C.B., Bartel, B., Bartel, D.P., 2002. Prediction of plant microRNA targets. Cell 110, 513–520.

Ripoll, J.J., Bailey, L.J., Mai, Q.A., Wu, S.L., Hon, C.T., Chapman, E.J., et al., 2015. microRNA regulation of fruit growth. Nat. Plants 1, 15036.

Rodriguez, R.E., Mecchia, M.A., Debernardi, J.M., Schommer, C., Weigel, D., Palatnik, J.F., 2010. Control of cell proliferation in *Arabidopsis thaliana* by microRNA miR396. Development 137, 103–112.

Schommer, C., Bresso, E.G., Spinelli, S.V., Palatnik, J.F., 2012. Role of microRNA miR319 in plant development. In: Sunkar, R. (Ed.), MicroRNAs in Plant Development and Stress Responses, Signaling and Communication in Plants, vol 15. Springer-Verlag, Berlin, pp. 29–47.

Schwind, N., Zwiebel, M., Itaya, A., Ding, B., Wang, M.B., Krczal, G., et al., 2009. RNAi-mediated resistance to Potato spindle tuber viroid in transgenic tomato expressing a viroid hairpin RNA construct. Mol. Plant Pathol. 10, 459–469.

Silva, G.F.F.E., Silva, E.M., da Silva Azevedo, M., Guivin, M.A.C., Ramiro, D.A., Figueiredo, C.R., et al., 2014. Micro RNA 156-targeted SPL/SBP box transcription factors regulate tomato ovary and fruit development. Plant J. 78, 604–618.

Singh, A., Sengar, R.S., Singh, A.K., Singh, S., Kumar, A., Ahmad, K., 2019. Impact of genetic engineering on agriculture food crops. Biotech Today: Int. J. Biol. Sci. 9 (1), 17–24.

Somssich, M., Je, B.I., Simon, R., Jackson, D., 2016. CLAVATA-WUSCHEL signaling in the shoot meristem. Development 143, 3238–3248.

Sun, G., 2012. MicroRNAs and their diverse functions in plants. Plant Mol. Biol. 80, 17–36.

Sun, G., Stewart Jr., C.N., Xiao, P., Zhang, B., 2012. MicroRNA expression analysis in the cellulosic biofuel crop Switchgrass (Panicum virgatum) under abiotic stress. PLoS One 7, e32017.

Sun, X., Wang, C., Xiang, N., Li, X., Yang, S., 2017. Activation of secondary cell wall biosynthesis by miR319-targeted TCP4 transcription factor. Plant Biotechnol. J. 15, 1284–1294.

Sunkar, R., Li, Y.F., Jagadeeswaran, G., 2012. Functions of microRNAs in plant stress responses. Trends Plant Sci. 17, 196–203.

Sunny, J.S., Snijesh, V.P., Singh, S., 2016. Potential role of mirnas in rheumatoid arthritis. Austin Arthritis 1 (2), 1009.

Tirumalai, V., Swetha, C., Nair, A., Pandit, A., Shivaprasad, P.V., 2019. MiR828 and miR858 regulate VvMYB114 to promote anthocyanin and flavonol accumulation in grapes. J. Exp. Bot. 70, 4775–4792.

Tousi, N., Eini, O., Ahmadvand, R., Carra, A., Miozzi, L., Noris, E., et al., 2017. In silico prediction of miRNAs targeting ToLCV and their regulation in susceptible and resistant tomato plants. Australas. Plant Pathol. 46, 379–386.

Treiber, T., Treiber, N., Meister, G., 2019. Regulation of microRNA biogenesis and its crosstalk with other cellular pathways. Nat. Rev. Mol. Cell Biol. 20, 5–20.

Valdés-López, O., Yang, S.S., Aparicio-Fabre, R., Graham, P.H., Reyes, J.L., Vance, C.P., et al., 2010. MicroRNA expression profile in common bean (Phaseolus vulgaris) under nutrient deficiency stresses and manganese toxicity. New Phytol. 187, 805–818.

Veit, B., 2009. Hormone mediated regulation of the shoot apical meristem. Plant Mol. Biol. 69, 397–408.

Verma, N., Sao, P., Srivastava, A., Singh, S., 2021. Physiological and molecular responses to drought, submergence and excessive watering in plants. Harsh Environ. Plant Resil. 305–321.

Vora, Z., Pandya, J., Vaikuntapu, P.R., Singh, S., Asha, S., Purama, R.K., 2023. Genome-wide association of defense hormone crosstalk in plants. In: Hormonal Cross-Talk, Plant Defense and Development. Academic Press, pp. 353–371.

Wang, C., Leng, X., Zhang, Y., Kayesh, E., Zhang, Y., Sun, X., et al., 2014. Transcriptome-wide analysis of dynamic variations in regulation modes of grapevine microRNAs on their target genes during grapevine development. Plant Mol. Biol. 84, 269–285.

Wang, J., Mei, J., Ren, G., 2019a. Plant microRNAs: biogenesis, homeostasis, and degradation. Front. Plant Sci. 10, 360.

Wang, K., Ding, Y., Cai, C., Chen, Z., Zhu, C., 2019. The role of C2H2 zinc finger proteins in plant responses to abiotic stresses. Physiol. Plantru. 165, 690–700.

Wang, W., Shi, T., Ni, X., Xu, Y., Qu, S., Gao, Z., 2018. The role of miR319a and its target gene TCP4 in the regulation of pistil development in Prunus mume. Genome 61, 43–48.

Wang, W.Q., Allan, A.C., Yin, X.R., 2020a. Small RNAs with a big impact on horticultural traits. Crit. Rev. Plant Sci. 39, 30–43.

Wang, W.Q., Wang, J., Wu, Y.Y., Li, D.W., Allan, A.C., Yin, X.R., 2020b. Genome-wide analysis of coding and non-coding RNA reveals a conserved miR164-NAC regulatory pathway for fruit ripening. New Phytol. 225, 1618–1634.

Wang, Y., Zhang, J., Cui, W., Guan, C., Mao, W., Zhang, Z., 2017. Improvement in fruit quality by overexpressing miR399a in woodland strawberry. J. Agric. Food Chem. 65, 7361–7370.

Wang, Y., Zou, W., Xiao, Y., Cheng, L., Liu, Y., Gao, S., et al., 2018. MicroRNA1917 targets CTR4 splice variants to regulate ethylene responses in tomato. J. Exp. Bot. 69, 1011–1025.

Wani, S.H., Kumar, V., Khare, T., Tripathi, P., Shah, T., Ramakrishna, C., et al., 2020. miRNA applications for engineering abiotic stress tolerance in plants. Biology 75, 1063–1081.

Waterhouse, P.M., Graham, M.W., Wang, M.B., 1998. Virus resistance and gene silencing in plants can be induced by simultaneous expression of sense and antisense RNA. Proc. Natl. Acad. Sci. U.S.A. 95, 13959–13964.

Wei, Q., Ma, C., Xu, Y., Wang, T., Chen, Y., Lu, J., et al., 2017. Control of chrysanthemum flowering through integration with an aging pathway. Nat. Commun. 8, 829.

Woodger, F.J., Millar, A., Murray, F., Jacobsen, J.V., Gubler, F., 2003. The role of GAMYB transcription factors in GA-regulated gene expression. J. Plant Growth Regul. 22, 176–184.

Wu, J., Wang, D., Liu, Y., Wang, L., Qiao, X., Zhang, S., 2014. Identification of miRNAs involved in pear fruit development and quality. BMC Genom. 15, 1–19.

Xia, X., Shao, Y., Jiang, J., Du, X., Sheng, L., Chen, F., et al., 2015. MicroRNA expression profile during aphid feeding in chrysanthemum (*Chrysanthemum morifolium*). PLoS One 10 e0143720.

Xian, Z., Huang, W., Yang, Y., Tang, N., Zhang, C., Ren, M., et al., 2014. miR168 influences phase transition, leaf epinasty, and fruit development via SlAGO1s in tomato. J. Exp. Bot. 65, 6655–6666.

Xu, L., Wang, Y., Zhai, L., Xu, Y., Wang, L., Zhu, X., et al., 2013. Genome-wide identification and characterization of cadmium-responsive microRNAs and their target genes in radish (*Raphanus sativus* L.) roots. J. Exp. Bot. 64, 4271–4287.

Xue, C., Yao, J.L., Qin, M.F., Zhang, M.Y., Allan, A.C., Wang, D.F., et al., 2019. PbrmiR397a regulates lignification during stone cell development in pear fruit. Plant Biotechnol. J. 17, 103–117.

Yamaguchi, A., Abe, M., 2012. Regulation of reproductive development by non-coding RNA in Arabidopsis: to flower or not to flower. J. Plant Res. 125, 693–704.

Yang, C.Y., Huang, Y.H., Lin, C.P., Lin, Y.Y., Hsu, H.C., Wang, C.N., et al., 2015. MicroRNA396-targeted SHORT VEGETATIVE PHASE is required to repress flowering and is related to the development of abnormal flower symptoms by the phyllody symptoms1 effector. Plant Physiol. 168, 1702–1716.

Yang, S., Zhang, M., Xu, L., Luo, Z., Zhang, Q., 2020c. MiR858b inhibits proanthocyanidin accumulation by the repression of DkMYB19 and DkMYB20 in persimmon. Front. Plant Sci. 11, 576378.

Yang, T., Ma, H., Zhang, J., Wu, T., Song, T., Tian, J., et al., 2019. Systematic identification of long noncoding RNA s expressed during light-induced anthocyanin accumulation in apple fruit. Plant J. 100, 572–590.

Yang, T., Wang, Y., Liu, H., Zhang, W., Chai, M., Tang, G., et al., 2020a. MicroRNA1917-CTR1-LIKE PROTEIN KINASE 4 impacts fruit development via tuning ethylene synthesis and response. Plant Sci. 291, 110334.

Yang, Y., Guo, J., Cheng, J., Jiang, Z., Xu, N., An, X., et al., 2020b. Identification of UV-B radiation responsive microRNAs and their target genes in chrysanthemum (Chrysanthemum morifolium Ramat) using high-throughput sequencing. Ind. Crop. Prod. 151, 112484.

Yant, L., Mathieu, J., Dinh, T.T., Ott, F., Lanz, C., Wollmann, H., et al., 2010. Orchestration of the floral transition and floral development in Arabidopsis by the bifunctional transcription factor APETALA2. Plant Cell 22, 2156–2170.

Yao, J.L., Xu, J., Cornille, A., Tomes, S., Karunairetnam, S., Luo, Z., et al., 2015. A micro RNA allele that emerged prior to apple domestication may underlie fruit size evolution. Plant J. 84, 417–427.

Yao, J.L., Tomes, S., Xu, J., Gleave, A.P., 2016. How microRNA172 affects fruit growth in different species is dependent on fruit type. Plant Signal. Behav. 11 e1156833.

Yu, S., Li, P., Zhao, X., Tan, M., Ahmad, M.Z., Xu, Y., et al., 2021. CsTCPs regulate shoot tip development and catechin biosynthesis in tea plant (*Camellia sinensis*). Hortic. Res. 8, 1–21.

Yu, Y., Jia, T., Chen, X., 2017. The 'how' and 'where' of plant microRNAs. New Phytol. 216, 1002–1017.

Yu, Z.X., Wang, L.J., Zhao, B., Shan, C.M., Zhang, Y.H., Chen, D.F., et al., 2015. Progressive regulation of sesquiterpene biosynthesis in Arabidopsis and Patchouli (Pogostemon cablin) by the miR156-targeted SPL transcription factors. Mol. Plant 8, 98–110.

Zhang, B., Tong, Y., Luo, K., Zhai, Z., Liu, X., Shi, Z., 2021. Identification of GROWTH-REGULATING FACTOR transcription factors in lettuce (Lactuca sativa) genome and functional analysis of LsaGRF5 in leaf size regulation. BMC Plant Biol. 21, 1–13.

Zhang, B., Wang, Q., 2015. MicroRNA-based biotechnology for plant improvement. J. Cell. Physiol. 230, 1–15.

Zhang, B., Yang, H.J., Yang, Y.Z., Zhu, Z.Z., Li, Y.N., Qu, D., et al., 2020. mdm-miR828 participates in the feedback loop to regulate anthocyanin accumulation in apple peel. Front. Plant Sci. 11, 608109.

Zhang, Q., Li, Y., Zhang, Y., Wu, C., Wang, S., Hao, L., et al., 2017. Md-miR156ab and Md-miR395 target WRKY transcription factors to influence apple resistance to leaf spot disease. Front. Plant Sci. 8, 526.

Zhang, W., Shang, G., Xiang, Z., Padmanabhan, C., Zheng, C., Xuefeng, Z., et al., 2011b. Bacteria-responsive microRNAs regulate plant innate immunity by modulating plant hormone networks. Plant Mol. Biol. 75, 93–105.

Zhang, X., Zou, Z., Gong, P.J., Zhang, J.H., Ziaf, K., Li, H.X., et al., 2011c. Overexpression of microRNA169 confers enhanced drought tolerance to tomato. Biotechnol. Lett. 33, 403–409.

Zhang, X., Zou, Z., Zhang, J., Zhang, Y., Han, Q., Hu, T., et al., 2011a. Over-expression of sly-miR156a in tomato results in multiple vegetative and reproductive trait alterations and partial phenocopy of the sft mutant. FEBS Lett. 585, 435–439.

Zhang, Y., Zhang, Q., Hao, L., Wang, S., Wang, S., Zhang, W., et al., 2019. A novel miRNA negatively regulates resistance to Glomerella leaf spot by suppressing expression of an NBS gene in apple. Hortic. Res. 6, 93.

Zhang, Y., Zhu, X., Chen, X., Song, C., Zou, Z., Wang, Y., et al., 2014. Identification and characterization of cold-responsive microRNAs in tea plant (Camellia sinensis) and their targets using high-throughput sequencing and degradome analysis. BMC Plant Biol. 14, 1–18.

Zhao, P., Wang, F., Deng, Y., Zhong, F., Tian, P., Lin, D., Huang, T., 2022. Sly-miR159 regulates fruit morphology by modulating GA biosynthesis in tomato. Plant Biotechnol. J. 20, 833–845.

Zhao, Q., Sun, C., Liu, D.D., Hao, Y.J., You, C.X., 2015. Ectopic expression of the apple Md-miR172e gene alters flowering time and floral organ identity in Arabidopsis. Plant Cell Tissue Organ Cult. 123, 535–546.

Zhou, Q., Shi, J., Li, Z., Zhang, S., Zhang, S., Zhang, J., 2021. miR156/157 targets SPLs to regulate flowering transition, plant architecture and flower organ size in petunia. Plant Cell Physiol. 62, 839–857.

Zhou, Y., Mumtaz, M.A., Zhang, Y., Yang, Z., Hao, Y., Shu, H., et al., 2022. Response of anthocyanin biosynthesis to light by strand-specific transcriptome and miRNA analysis in Capsicum annuum. BMC Plant Biol. 22, 1–13.

Zhu, Q.W., Luo, Y.P., 2013. Identification of miRNAs and their targets in tea (Camellia sinensis). Biomed. Biotechnol. 14, 916–923.

Further reading

Zhang, H., Zhang, J., Yan, J., Gou, F., Mao, Y., Tang, G., et al., 2017. Short tandem target mimic rice lines uncover functions of miRNAs in regulating important agronomic traits. Proc. Nat. Acad. Sci. U.S.A. 114, 5277–5282.

CHAPTER

11

MicroRNAs in monocot plants like rice, wheat, and sugarcane

E. Abhishek[1], Kotte Bhargava[1], P. Bhavana[2]
[1]*ICAR-Indian Agricultural Research Institute, New Delhi, India;* [2]*ICAR-Research Complex for Eastern Region, Farming Systems Research Centre Hill and Plateau Region, Ranchi, Jharkhand, India*

1. Introduction

MicroRNAs, or small noncoding RNA molecules, play a major role in the regulation of gene expression and plant growth. MicroRNAs, or small noncoding RNA molecules, play a major role in the regulation of gene expression and plant growth. According to Chen et al. (2005), these entities possess a length of around 21 nucleotides and exert posttranscriptional control on gene expression by their interaction with target mRNA, resulting in either its degradation or inhibition of translation.

The first identification of miRNAs occurred in the worm *Caenorhabditis elegans* in 1993, subsequently leading to their detection in several species, including plants. It has been determined that microRNAs (miRNAs) play a major role in a number of physiological processes in plants, such as signaling pathways, stress response mechanisms, and developmental processes. Small noncoding RNAs known as microRNAs (miRNAs) are essential for controlling posttranscriptional gene expression. They have undergone thorough examination in several organisms, including plants. Monocots, being one of the most extensive plant taxa, have significant significance in global agriculture and crop cultivation, specifically in relation to rice, maize, and wheat. In recent years, there has been an increased emphasis on understanding the functions of miRNAs in monocots.

This chapter will provide a comprehensive summary of the existing knowledge on monocot miRNAs, including their production, control, and role. This study aims to examine the variety and evolutionary patterns of microRNAs (miRNAs) in monocotyledonous plants, while also investigating their involvement in crucial biological mechanisms such as development, stress responses, and disease resistance.

1.1 Micro RNAs and Si RNAs

Short RNA molecules called microRNAs (miRNAs) and small interfering RNAs (siRNAs) control the expression of certain genes. Yet, there are differences between them in terms of biogenesis and action mechanisms.

MicroRNA Advances and Application in Plant Biology. https://doi.org/10.1016/B978-0-443-21682-4.00012-9

Single-stranded RNAs known as microRNAs are produced from genomic DNA and are broken down by the RNAse III enzymes Drosha and Dicer in the cytoplasm and nucleus, respectively. Mature microRNAs associate with the RNA-induced silencing complex (RISC) when they bind to target messenger RNAs (mRNAs), usually in the 3′ untranslated region (UTR). According to Bartel et al. (2009), this relationship leads to mRNA degradation and/or translational inhibition.

On the other hand, siRNAs are synthetic or exogenous, double-stranded RNAs that can be introduced into cells by transfection or viral delivery. Dicer processes them into 21–25 nucleotide duplexes, which are incorporated into the RISC and guide the complex to target mRNAs, resulting in their demise (Zamore et al., 2000).

Thus, the primary distinctions between miRNAs and siRNAs are their synthesis, origin, and method of activity. MiRNAs are endogenous and influence gene expression posttranscriptionally, whereas siRNAs can be exogenous or synthetic and cause gene silence through mRNA degradation.

1.2 MiRNAs in monocots

MiRNAs have been found and characterized in a number of monocot species, including rice, wheat, barley, and maize.

For instance, Zhang et al. (2006) reported using deep sequencing and computational methods to identify more than 700 microRNAs (miRNAs) in rice (Oryza sativa). These microRNAs, or miRNAs, regulate a number of functions, including the growth of leaves, the onset of blooming, and the body's reaction to stress (Jeyaraj et al., 2014). Li et al. (2010) investigated the involvement of miRNAs in maize kernel development and stress responses. Additionally, Yao et al. (2007) established a connection between miRNAs and disease resistance, drought tolerance, and grain production in wheat. Chen et al. (2013) demonstrated the impact of miRNAs on various facets of plant growth and development, as well as their role in modulating barley responses to biotic and abiotic stressors.

Over all, miRNAs play significant roles in controlling gene expression in monocots, and functional studies of them have shed light on the molecular processes underlying plant growth, stress reactions, and adaptability (Table 11.1).

2. Biogenesis of microRNA in monocots

2.1 Biogenesis of miRNA respect to rice

The identification of rice miRNAs has experienced rapid expansion, primarily facilitated by the utilization of next-generation sequencing methodologies tailored for short RNA detection. Kozomara and Griffiths-Jones (2014) reported a comprehensive inventory of 592 miRNAs in rice, yielding a total of 713 mature miRNA products, as documented in the latest miRNA Sanger database. Within the nucleus, RNA polymerase II synthesizes primary miRNAs (pri-miRNAs) from the majority of plant microRNA genes, forming long hairpin-shaped molecules. The length of these pri-miRNA hairpins spans a range of approximately 70 to several hundred bases (ChatGPT).

Table 11.1 List of major miRNAs in cereals.

Crop	miRNA	Target gene	Functional involvement
Rice	miR393	Auxin receptor gene	Drought tolerance
	miR820	DRM2	Response to salt stress
	miR167	ARF transcription factors	Cold stress
	miR397	L-ascorbate transcriptase	Heat stress
Wheat	miR397	L-ascorbate transcriptase	Development
	miR395	ATP sulfurylase	Abiotic stress
	miR1435		Ion transportation
Maize	miR156	Squamosa binding protein	Development
	miR160	ARF transcription factors	Endosperm development
	miR164	NAC1	Stress response

Following transcription, pri-miRNAs undergo several processing steps within the nucleus, ultimately leading to the formation of a single hairpin-shaped structure known as pre-miRNA. Dicer-like (DCL) enzymes then cleave this pre-miRNA, resulting in the production of mature microRNA and its corresponding star molecule. These mature microRNAs are subsequently transported to the cytoplasm by HASTY (HST) and incorporated into the RNA-induced silencing complex (RISC). By binding extensively or nearly completely to target mRNA, microRNAs facilitate mRNA cleavage or translational repression (Wheeler et al., 2009).
MicroRNA (miRNA) biogenesis and its mode of function.

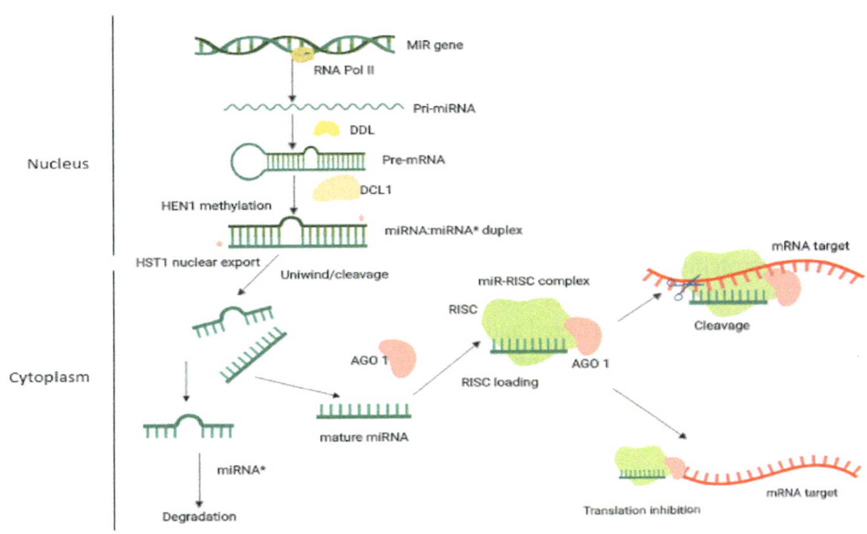

MicroRNAs in plants often induce the suppression of specific target messenger RNAs (mRNAs) by directly cleaving complementary transcripts, with many of these mRNAs originating from critical regulatory genes, including those involved in transcriptional regulation and F-box protein production (Jones-Rhoades et al., 2004). The reciprocal interactions between microRNAs and their target genes govern multiple downstream proteins and signaling pathways, contributing to the establishment of a sophisticated regulatory network (Raghuram et al., 2014).

2.2 Biogenesis of miRNA in wheat

The biogenesis of miRNAs in wheat involves a complex and well-coordinated process that begins with their transcription from miRNA genes, followed by their processing, maturation, and functional incorporation into RNA-induced silencing complexes (RISCs) (Zhang et al., 2015).

The primary transcription of miRNA genes in wheat is carried out by RNA polymerase II, resulting in the production of long, single-stranded RNA molecules known as primary miRNAs (pri-miRNAs). These pri-miRNAs are typically several hundred nucleotides in length and contain one or more stem-loop structures that serve as precursors for miRNA maturation (Li et al., 2016).

The pri-miRNA processing in wheat happens in two steps. Dicer-like 1 (DCL1) first cleaves pri-miRNAs to generate precursor miRNAs (pre-miRNAs), which are typically 70–80 nucleotides in length. Exportin-5, a nuclear export protein, transports pre-miRNAs from the nucleus to the cytoplasm. DCL1 further processes pre-miRNAs in the cytoplasm to produce mature miRNA duplexes that are approximately 21 nucleotides long.

It then loads the mature miRNA duplexes into an Argonaute protein, a crucial part of RISCs. The duplexes are unraveled, and one of the strands, the mature miRNA, is preferentially integrated into the RISC while the other strand, the miRNA* or miRNA passenger strand, is normally destroyed. The mature miRNA functions as a guide molecule to identify and attach to complementary messenger RNAs (mRNAs) in the cytoplasm once it has been incorporated into the RISC. Post-transcriptional gene silencing results from this binding's effects on mRNA degradation or translational repression.

MiRNA synthesis in wheat is a multistep, intricately regulated process involving numerous proteins and enzymes. Through molecular and genetic research, our understanding of this process has substantially improved. In wheat crops, a number of miRNAs have been discovered and characterized, and their functions in a number of biological processes, including as growth, stress responses, and grain quality, have been clarified (Song et al., 2019).

3. Role of miRNA in growth and development

3.1 Rice

According to Wu et al., the transition of rice from vegetative to reproductive phases is governed by miR156, a conserved plant miRNA targeting SPL transcription factors, which regulates the timing of flowering and other aspects of reproductive growth, thus serving as a key regulator of shoot development.

In rice cultivation, maximizing plant productivity is the primary goal, with variables such as tillering, panicle initiation, branching, and grain size also influential. Breeders have introduced the concept of "ideal plant architecture" (IPA), encompassing traits such as reduced tillers, minimal unproductive tillers, increased grain yield per panicle, and sturdy stems. IPA1, a semidominant quantitative trait locus, was identified in 2010 by two research groups. Takeda et al. (2003) identified and characterized IPA1, which encodes OsSPL14, a transcription factor with attributes similar to SQUAMOSA-PROMOTER BINDING PROTEIN-LIKE. OsSPL14 activates significant genes such as OsTB1, known for its inhibitory effect on lateral branching. Jiao et al. (2010) suggested that upregulation of OsSPL14 expression stimulates panicle branching and enhances grain output. Xie et al. (2006) demonstrated that OsSPL14's posttranscriptional suppression by miR156 influences rice traits. Mutation in the miR156-targeted region of OsSPL14 led to improved traits in transgenic rice, as reported by Jiao et al. (2010). Early leaf maturity and rapid leaf initiation in rice are linked to miR156 overexpression (Xie et al., 2012). Additionally, Fu et al. (2012) found that overexpressing Osa-miR156b in switchgrass improved biomass production and fodder digestibility.

Guo et al. (2013) suggested that the miRNA/MADS/TCP/D14 regulatory network, particularly involving OsMIR444a, OsMADS57, and D14, may regulate rice tillering. Inhibitory effects of OsMIR444a on OsMADS57 expression may influence D14 and tillering, potentially benefiting high-grain yield rice breeding programs.

3.2 Wheat

MicroRNAs (miRNAs) serve as regulators of essential processes governing the growth and development of wheat plants, including seed germination, root and shoot development, flowering timing, and responses to various stresses. A significant number of miRNAs linked to wheat's immune response against pathogens underscore their crucial role in the plant's defense mechanisms, crucial for overall growth (Li et al., 2017). Wang et al. (2018) highlight the importance of reproductive growth in determining wheat grain yield and quality, with specific miRNAs expressed during early grain development stages. Abiotic stress, particularly salt stress, mediated by miRNAs, demonstrates their role in enhancing wheat's resilience (Sharma et al., 2018). Recent studies, as per Li et al. (2019), indicate miRNAs' involvement in regulating phosphate signaling and metabolism in wheat, essential for nutrient uptake and utilization, thus influencing plant growth. Drought stress, a significant abiotic stressor, prompts a stress response mediated by miRNAs identified by Hanif et al. (2021), further illustrating their role in wheat's adaptive mechanisms.

3.3 Sugarcane

miRNAs have been found in sugarcane (Saccharum spp.) to have important roles in governing several aspects of plant growth and development, including shoot and root development, flowering, and stress responses.

3.3.1 Role of miRNAs in sugarcane shoot and root development

MiRNAs have been shown to regulate sugarcane shoot and root development. As an example, miR156, miR164, and miR166 have been demonstrated to affect the formation and development of axillary buds by targeting genes that control shoot branching and tillering in sugarcane (Lakhotia et al., 2017; Bally et al., 2018). Furthermore, miR160, miR393, and miR164, which target genes involved in root growth, lateral root formation, and root hair synthesis, have been connected to sugarcane root development (Macedo et al., 2015; Gupta et al., 2017). Zhang et al. (2020) performed high-throughput sequencing of sugarcane short RNAs at several developmental stages to identify miRNAs that affect flowering time. A thorough analysis of small RNA sequencing data revealed numerous miRNAs with varied expression patterns across distinct stages of sugarcane flowering.

3.3.2 Role of miRNAs in sugarcane flowering

MiRNAs have been shown to regulate sugarcane flowering, a critical stage in crop output. For example, miRNAs miR156, miR172, and miR5077 have been shown to affect the timing of flowering in sugarcane by directing their attention to the genes SPLs (SQUAMOSA PROMOTER BINDING PROTEIN-LIKE) and AP2-like transcription factors (Cardoso-Silva et al., 2014; Zhang et al., 2020). MiR159 and miR172 have also been associated with influencing floral growth in sugarcane (Cardoso-Silva et al., 2014).

3.3.3 Role of miRNAs in sugarcane root development

Garsmeur et al. (2018) developed a monoploid reference sequence for the complex sugarcane genome in order to discover and characterize miRNAs involved in sugarcane root growth. The researchers utilized high-throughput sequencing, bioinformatics analysis, and experimental validation to identify and annotate miRNAs expressed throughout various stages of sugarcane root growth. Based on their projected target genes, they investigated the potential regulatory roles of these miRNAs in sugarcane root growth.

This work emphasizes the involvement of miRNAs in sugarcane root formation, which is a key phase in nutrient intake, water absorption, and overall plant growth. Understanding the regulatory activities of miRNAs in sugarcane root growth can provide critical insights into enhancing root-related characteristics in this important bioenergy and sugar crop.

4. Importance of micro-RNA for tolerance to biotic stress in cereals

Rice, one of the world's most important staple crops, relies heavily on microRNAs (miRNAs) to regulate its tolerance to biotic stress. Several studies have found distinct miRNAs that regulate rice's defensive mechanisms against biotic stressors such as infections and pests.

4.1 Rice

Several miRNAs in rice are variably expressed after infection with the blast fungus Magnaporthe oryzae, a devastating rice pathogen. The study indicated that these miRNAs are crucial in controlling rice's immune responses to the pathogen, and their overexpression or silencing dramatically influenced rice plants' tolerance to blast disease (Li et al., 2014). Campo et al. (2017) identified a new miRNA in rice which targets as an alternate spliced transcript of the Nramp6 gene, which is important in pathogen resistance. The study found that this miRNA regulates rice defense responses against pathogens and that overexpressing it in rice plants improved pathogen resistance.

Baldrich and San Segundo offered an overview of the activities of miRNAs in rice innate immunity, including their role in the modulation of defense mechanisms to biotic stimuli (2016). The review underlined miRNAs' involvement in controlling the expression of genes involved in defense mechanisms, as well as their potential as targets for increasing rice resistance to biotic stress.

The function of miRNAs in the development of vascular and root development (Dong et al., 2022).

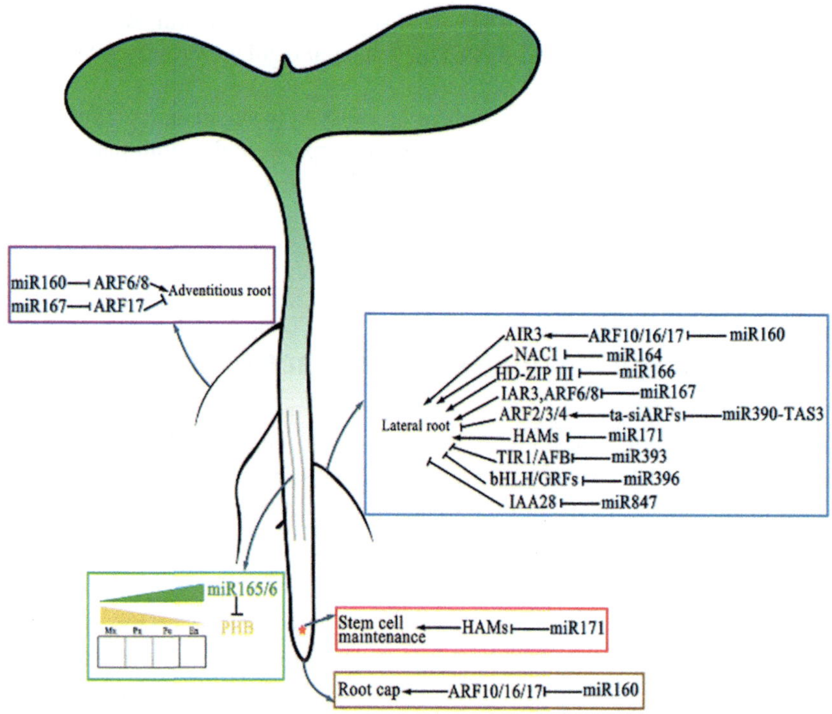

4.2 Wheat

Wheat is a popular cereal crop that is grown extensively throughout the world. MicroRNAs (miRNAs) play an important role in wheat tolerance to biotic stress. According to various research, certain miRNAs regulate several wheat defense systems against biotic stresses such as pest and diseases.

Li et al. (2012) identified miRNAs in wheat that regulate innate immune receptors, crucial components of the plant's defense against pathogens. They observed that these miRNAs influence the expression of immune receptors, impacting the plant's ability to resist infections when either overexpressed or suppressed. Xin et al. (2010) discovered a diverse set of miRNAs in wheat responding to powdery mildew, a common fungal disease, and noted their sensitivity to heat stress, often concurrent with biotic stress in field conditions. This suggests a potential role for these miRNAs in regulating wheat's response to both biotic and abiotic stressors.

Wang et al. (2018) investigated how miRNAs in wheat spikes, a critical stage for reproduction, respond to cold treatment. They identified miRNAs with altered expression patterns under cold stress, suggesting their involvement in regulating wheat's tolerance to cold stress, which can exacerbate vulnerability to biotic stresses.

These studies underscore the significance of miRNAs in regulating wheat's tolerance to biotic stresses such as pathogens. They emphasize miRNAs' role in modulating the expression of defense-related genes and their potential as targets for enhancing wheat's resilience to biotic stress.

4.3 Sugarcane

Recent research has clarified how miRNAs control sugarcane's ability to withstand brown rust disease, a serious fungal disease that reduces sugarcane productivity. The scientists discovered particular miRNAs that are responsible for controlling the production of genes linked to the plant's immune response and defense mechanisms against the pathogen that causes brown rust. This study underlines the regulatory function of miRNAs in this process and offers significant new insights into the molecular mechanisms driving sugarcane's tolerance to brown rust disease.

The Sugarcane Mosaic Virus (SCMV), a common viral disease in sugarcane, was studied in this study to determine how miRNAs regulate sugarcane responses to infection. They discovered a number of miRNAs that were expressed differently between healthy and SCMV-infected sugarcane plants. It was discovered that these miRNAs bind to genes that are involved in a number of defense processes, such as pathogen identification, signal transduction, and defense response. In order to confer tolerance to biotic stresses such as illnesses on sugarcane, the study revealed that miRNAs are vital in the control of the plant's defense responses to viral infections (Santos et al., 2019).Recent research has clarified how miRNAs control sugarcane's ability to withstand brown rust disease, a serious fungal disease that reduces sugarcane productivity. The scientists discovered particular miRNAs that are responsible for controlling the production of genes linked to the plant's immune response and defense mechanisms against the pathogen that causes brown rust. This study underlines the regulatory function of miRNAs in this process and offers significant new insights into the molecular mechanisms driving sugarcane's tolerance to brown rust disease.

The Sugarcane Mosaic Virus (SCMV), a common viral disease in sugarcane, was studied in this study to determine how miRNAs regulate sugarcane responses to infection. They discovered a number of miRNAs that were expressed differently between healthy and SCMV-infected sugarcane plants. It was discovered that these miRNAs bind to genes that are involved in a number of defense processes, such as pathogen identification, signal transduction, and defense response. In order to confer tolerance to biotic stresses such as illnesses on sugarcane, the study revealed that miRNAs are vital in the control of the plant's defense responses to viral infections (Santos et al., 2019).

5. Importance of micro-RNA for tolerance to abiotic stresses in cereals

MicroRNAs (miRNAs) are acknowledged as crucial regulators of abiotic stress tolerance in cereal crops such as rice, wheat, and sugarcane. Drought, salinity, and extreme temperatures are among the various abiotic factors that adversely affect crop growth and development, ultimately leading to decreased yields. Nonetheless, miRNAs play a pivotal role in modulating signaling pathways and gene expression in plants' responses to external stimuli, thereby bolstering their capacity to endure stress.

Numerous studies have provided evidence of miRNAs' involvement in enhancing wheat's ability to withstand abiotic stress. For instance, Xie et al. (2018) demonstrated that miR156 targets OsSPL genes, critical for plant growth and stress responses, positively affecting rice plants' drought tolerance. Similarly, Ding et al. (2018) revealed the role of miR398 in wheat in regulating copper homeostasis, crucial for enhancing resistance to oxidative stress. Gentile et al. (2015) showed that MiR408 modulates sugarcane's response to drought stress by targeting genes involved in scavenging reactive oxygen species (ROS). These findings underscore miRNAs' significance as important regulators of abiotic stress tolerance in cereal crops such as rice, wheat, and sugarcane, despite the adverse effects of various abiotic factors on crop growth and development, ultimately leading to reduced yields.

Numerous studies have contributed evidence regarding the involvement of miRNAs in enhancing wheat's resilience to abiotic stress. For instance, Xie et al. (2018) demonstrated that miR156 targets OsSPL genes, pivotal for plant growth and stress responses, thereby positively affecting rice plants' drought tolerance. Similarly, Ding et al. (2018) revealed the role of miR398 in wheat in regulating copper homeostasis, crucial for enhancing resistance to oxidative stress. Gentile et al. (2015) showed that MiR408 modulates sugarcane's response to drought stress by targeting genes involved in scavenging reactive oxygen species (ROS). These findings underscore miRNAs' significance in bolstering wheat's capacity to withstand abiotic stress.

Role of miRNAs in abiotic stress tolerance like heat and drought tolerance (Zhang et al., 2022).

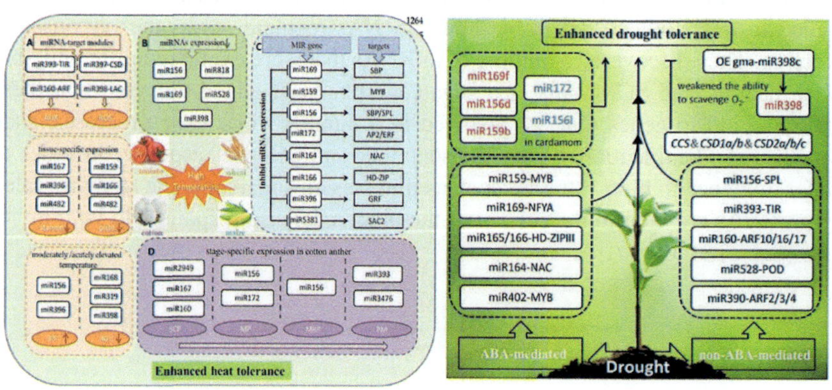

Furthermore, miRNAs play a role in fostering tolerance to various abiotic stressors. Li et al. demonstrated that rice miR319 enhances drought and salt tolerance by targeting stress signaling genes OsTCP21 and OsTCP23. Similarly, Li et al. (2019) found that wheat miR171 regulates heat and drought tolerance by targeting stress-responsive genes.

The utilization of high-throughput sequencing and bioinformatics approaches in identifying stress-responsive miRNAs and their target genes contributes to the growing body of evidence supporting miRNAs' significant involvement in cereals' ability to withstand abiotic stress. Zhang et al. investigated miRNA expression profiling in drought-stressed rice plants, identifying stress-responsive miRNAs such as miR166, miR168, and miR169, which target genes related to plant development and stress responses. In another study by Ding et al., miRNAs such as miR156, miR164, and miR171 were found to be sensitive to drought stress in wheat, targeting genes crucial for stress signaling pathways. The aforementioned results underscore the significant importance of microRNAs (miRNAs) in the regulation of wheat plant responses to abiotic stressors. Altering the expression or function of miRNA might greatly enhance the ability of agricultural plants, such as rice, wheat, and sugarcane, to withstand stress. Further investigation is now underway to elucidate the specific mechanisms behind abiotic stress reactions in cereal crops, with the identification of certain microRNAs (miRNAs) and their potential gene targets for enhancing agricultural productivity.

6. Future prospects of miRNAs in cereals

The prospects of microRNAs (miRNAs) in cereals, including rice, wheat, and sugarcane, hold great promise for crop improvement and sustainable agriculture. MiRNAs have been identified as important regulators of various plant processes, including development, growth, and stress responses. With advancements in molecular biology and genetic engineering, the potential applications of miRNAs in cereal crops are vast and can revolutionize crop breeding and management strategies.

One of the future applications of miRNAs in cereals is their use in crop development projects through genetic engineering. Researchers may be able to improve critical cereal features including stress tolerance, yield, and nutritional content by changing the expression or activity of certain miRNAs. Overexpression of stress-responsive miRNAs, for example, can give increased tolerance to abiotic stresses such as drought, salinity, and temperature extremes, enhancing crop performance in harsh climatic conditions.

Another potential application of miRNAs in cereals is their use as diagnostic and prognostic tools for crop management. MiRNA expression patterns can serve as molecular markers for plant health and stress status, allowing for early detection and monitoring of plant diseases, pest attacks, and environmental stresses. This can facilitate timely interventions, such as targeted pesticide applications or irrigation adjustments, to mitigate crop losses and improve overall crop health.

Furthermore, miRNAs can also be utilized in crop breeding programs to accelerate the development of new cultivars with desired traits. By identifying and selecting miRNAs associated with desirable agronomic traits, breeders can use miRNA-guided breeding strategies to speed up the process of developing improved cereal varieties with enhanced traits such as yield, disease resistance, and stress tolerance. This can help shorten the breeding cycle and lead to the development of more resilient and high-performing crop varieties.

Additionally, the use of miRNAs in crop protection strategies is another future prospect. MiRNAs can be utilized as biopesticides by targeting specific genes in pests or pathogens, thereby providing a sustainable and environmentally friendly approach for pest and disease management in cereals. This can potentially reduce the reliance on chemical pesticides, minimize environmental impacts, and enhance the sustainability of cereal crop production.

Furthermore, advances in high-throughput sequencing and bioinformatics tools have made it easier to identify novel miRNAs and their target genes in cereals. This opens the door to the discovery of new miRNAs with distinct roles, as well as the understanding of complex miRNA-mediated regulatory networks in cereals. This knowledge can be used to create novel tactics for crop improvement and stress tolerance in cereals.

7. Conclusion

In summary, microRNAs (miRNAs) play a pivotal role in regulating various aspects of cereal crops, encompassing development, growth, and responses to stress. Their potential is significant in enhancing cereal crop performance, resilience to abiotic stresses, and promoting sustainable agricultural practices. MiRNAs offer avenues for improvement in crop breeding programs through genetic engineering, diagnostics, breeding techniques, and pest management strategies. They can be manipulated or overexpressed to bolster stress tolerance, increase yield, and improve nutritional content in cereals. Additionally, miRNAs can serve as molecular markers for monitoring plant health and stress levels, enabling early detection and management of diseases, pests, and environmental stresses. Furthermore, the ongoing discovery of novel miRNAs and their target genes through high-throughput sequencing technologies and bioinformatics tools presents further opportunities for innovative strategies in cereal crop enhancement. The future prospects of miRNAs in cereals are promising and have the potential to revolutionize cereal crop production, contributing to the development of sustainable and resilient agricultural systems.

References

Bally, J., Nakasugi, K., Jia, F., Jung, H., Ho, S.Y.W., Wong, M., Paul, C.M., Naim, F., Wood, C.C., Crowhurst, R., et al., 2018. The extremophile *Nicotiana benthamiana* has traded viral defence for early vigour. Nat. Plants 4, 186–194.

Bartel, D.P., January 23, 2009. MicroRNAs: target recognition and regulatory functions. Cell 136 (2), 215–233. https://doi.org/10.1016/j.cell.2009.01.002.

Campo, S., Peris-Peris, C., Siré, C., Moreno, A.B., Donaire, L., Zytnicki, M., Notredame, C., Llave, C., San Segundo, B., 2017. Identification of a novel microRNA (miRNA) from rice that targets an alternatively spliced transcript of the Nramp6 (Natural resistance-associated macrophage protein 6) gene involved in pathogen resistance. New Phytol. 213, 212–227. https://doi.org/10.1111/nph.14083.

Cardoso-Silva, C.B., Costa, E.A., Mancini, M.C., Balsalobre, T.W., Canesin, L.E., Pinto, L.R., Carneiro, M.S., Garcia, A.A., de Souza, A.P., Vicentini, R., 2014. De novo assembly and transcriptome analysis of contrasting sugarcane varieties. PLoS One 9 (2), e88462. https://doi.org/10.1371/journal.pone.0088462.

Chen, L., Wang, T., Zhao, M., Tian, Q., Zhang, W.H., October 2013. Identification of aluminum-responsive microRNAs in barley (*Hordeum vulgare*) roots by high-throughput sequencing. Planta 238 (4), 767–779. https://doi.org/10.1007/s00425-013-1939-6.

Chen, X., July 4, 2005. MicroRNA biogenesis and function in plants. FEBS (Fed. Eur. Biochem. Soc.) Lett. 579 (26), 5923–5931. https://doi.org/10.1016/j.febslet.2005.07.071.

Ding, Y., Tao, Y., Zhu, C., 2018. Emerging roles of microRNAs in the mediation of drought stress response in plants. J. Exp. Bot. 69, 3215–3226. https://doi.org/10.1093/jxb/ery160.

Dong, Q., Hu, B., Zhang, C., 2022. microRNAs and their roles in plant development. Front. Plant Sci. 13, 824240. https://doi.org/10.3389/fpls.2022.824240.

Fu, C.X., Sunkar, R., Zhou, C.E., Shen, H., Zhang, J.Y., Matts, J., Wolf, J., Mann, D.G.J., Stewart, C.N., Tang, Y.H., Wang, Z.Y., 2012. Overexpression of miR156 in switchgrass (*Panicum virgatum* L.) results in various morphological alterations and leads to improved biomass production. Plant Biotechnol. J. 10 (4), 443–452. https://doi.org/10.1111/j.1467-7652.2011.00677.x.

Garsmeur, O., Droc, G., Antonise, R., Grimwood, J., Potier, B., Aitken, K., Jenkins, J., Martin, G., Charron, C., Hervouet, C., Costet, L., Yahiaoui, N., Healey, A., Sims, D., Cherukuri, Y., Sreedasyam, A., Kilian, A., Chan, A., D'Hont, A., 2018. A mosaic monoploid reference sequence for the highly complex genome of sugarcane. Nat. Commun. 9 (1), 2638. https://doi.org/10.1038/s41467-018-05051-5.

Gentile, A., Ferreira, T.H., Mattos, R.S., Dias, L.I., Hoshino, A.A., Carneiro, M.S., Souza, G.M., 2015. Effects of Drought on the Microtranscriptome of Field-Grown Sugarcane Plants.

Guo, S., Xu, Y., Liu, H., Mao, Z., Zhang, C., Ma, Y., Zhang, Q., Meng, Z., Chong, K., 2013. The interaction between OsMADS57 and OsTB1 modulates rice tillering via DWARF14. Nat. Commun. 4, 1566. https://doi.org/10.1038/ncomms2542.

Gupta, O.P., Permar, V., Koundal, V., Singh, U.D., Praveen, S., 2017. MicroRNA regulated defense responses in *Triticum aestivum* L. during Puccinia graminis f.sp. tritici infection. Front. Plant Sci. 8, 506.

Hanif, M., Gao, Y., Khan, A., Wang, T., Zhang, F., Zhang, Y., Hu, L., 2021. Genome-wide identification and characterization of microRNAs in wheat (*Triticum aestivum* L.) under drought stress. Genes 12 (2), 286.

Jeyaraj, A., Chandran, V., Gajjeraman, P., October 2014. MicroRNAs in rice: genome-wide identification, functional analysis, and development-related expression profiling. DNA Res. 21 (5), 427–437. https://doi.org/10.1093/dnares/dsu012.

Jiao, Y., Wang, Y., Xue, D., Wang, J., Yan, M., Liu, G., Dong, G., Zeng, D., Lu, Z., Zhu, X., Qian, Q., Li, J., 2010. Regulation of OsSPL14 by OsmiR156 defines ideal plant architecture in rice. Nat. Genet. 42 (6), 541–544. https://doi.org/10.1038/ng.591.

Jones-Rhoades, M.W., Bartel, D.P., 2004. Computational identification of plant microRNAs and their targets, including a stress-induced miRNA. Mol. Cell 14 (6), 787–799. https://doi.org/10.1016/j.molcel.2004.05.027.-.

Kozomara, A., Griffiths-Jones, S., 2014. miRBase: annotating high confidence microRNAs using deep sequencing data. Nucleic Acids Res. 42 (Database issue), D68–D73. https://doi.org/10.1093/nar/gkt1181.

Lakhotia, N., Joshi, G., Bhardwaj, A.R., Katiyar-Agarwal, S., Agarwal, M., 2017. Identification and characterization of miRNAome in root, stem, leaf and tuber developmental stages of potato (*Solanum tuberosum* L.) by high-throughput sequencing. BMC Plant Biol. 17, 93.

Li, D., Wang, L., Liu, X., Cui, D., Chen, T., Zhang, H., Jiang, C., Xu, C., Li, P., Li, S., Zhao, L., Wang, A., 2016. Deep sequencing of maize small RNAs reveals a diverse set of microRNA in dry and imbibed seeds. PLoS One 11 (6) e0156435.

Li, F., Pignatta, D., Bendix, C., Brunkard, J.O., Cohn, M.M., Tung, J., Sun, H., Kumar, P., Baker, B., 2012. MicroRNA regulation of plant innate immune receptors. Proc. Natl. Acad. Sci. U.S.A. 109 (5), 1790–1795. https://doi.org/10.1073/pnas.1118282109.

Li, N., Li, X., Zhao, X., Li, F., Li, H., Li, T., Wang, Y., Li, C., 2019. Wheat microRNA members involved in miRNA399-PHO2-mediated Pi signaling and metabolism regulation. Plant Mol. Biol. 99 (4–5), 459–471.

Li, Y.F., Zheng, Y., Addo-Quaye, C., Zhang, L., Saini, A., Jagadeeswaran, G., Axtell, M.J., Zhang, W., Sunkar, R., July 1, 2010. Transcriptome-wide identification of microRNA targets in maize. Plant J. 62 (1), 108–120. https://doi.org/10.1111/j.1365-313x.2010.04133.x.

Li, Y., Lu, Y.G., Shi, Y., Wu, L., Xu, Y.J., Huang, F., Guo, X., Zhang, Y., Fan, J., Zhao, J.Q., Wang, P.R., Feng, Q., Zhou, B., Peng, Y.L., Zhang, L., Liang, G., Chen, W., Yao, N., Zhang, X., Li, L., 2014. Multiple rice microRNAs are involved in immunity against the blast fungus Magnaporthe oryzae. Plant Physiol. 164, 1077–1092. https://doi.org/10.1104/pp.113.231049.

Li, Y., Zhang, Q., Zhang, J., Wu, L., Qi, Y., Zhou, J.M., Li, X., 2017. Identification of microRNAs involved in pathogen-associated molecular pattern-triggered plant innate immunity. Plant Physiol. 176 (1), 856–872.

Macedo, L.A.F., Oliveira, L.F.V., Reis, R.R., Rodrigues, R.L.M., Tsai, S.M., Camargo, L.E.A., da Silva, F.R., 2015. Genome-wide identification of microRNA targets in the neglected endophytic plant-colonizing fungi, Periconia macrospinosa. BMC Genom. 16, 393.

Raghuram, B., Sheikh, A.H., Sinha, A.K., 2014. Regulation of MAP kinase signaling cascade by microRNAs in Oryza sativa. Plant Signal. Behav. 9 (10). https://doi.org/10.4161/psb.29804 e972130.

Santos Pereira, P., Dantas Dos Santos, C., de Queiroz, C.B., Braga Costa, L., Borges Sampaio, T., Castro Ferreira, W., Silva Girardi, R., Menossi, M., 2019. Identification of miRNAs and their targets in sugarcane infected with sugarcane mosaic virus. Viruses 11, 77. https://doi.org/10.3390/v11010077.

Sharma, D., Tiwari, M., Lakhwani, D., Trivedi, P.K., Tripathi, R.D., 2018. Differential expression of miRNAs in response to salt stress in wheat. Plant Physiol. Biochem. 130, 168–177.

Song, X., Li, Y., Cao, X., Qi, Y., MicroRNA, L., 2019. Biogenesis, functions and insights into plant miRNA biogenesis. Plant J. 100 (4), 650–664.

Takeda, T., Suwa, Y., Suzuki, M., Kitano, H., Ueguchi-Tanaka, M., Ashikari, M., Matsuoka, M., Ueguchi, C., 2003. The OsTB1 gene negatively regulates lateral branching in rice. Plant J.: Cell. Mol. Biol. (Sarreguemines, Fr., Online) 33 (3), 513–520. https://doi.org/10.1046/j.1365-313x.2003.01648.x.

Wang, J., Wang, Q., Yang, Y., Wang, Y., Wu, C., Zhang, Y., Sun, Y., Sun, L., 2018. Identification and characterization of microRNAs involved in early-stage wheat grain development. BMC Plant Biol. 18 (1), 130.

Wang, M., Wang, Q., Zhang, B., 2018. Response of microRNAs to cold treatment in the young spikes of common wheat. BMC Genom. 19, 787. https://doi.org/10.1186/s12864-018-5183-3.

Wheeler, B.M., Heimberg, A.M., Moy, V.N., Sperling, E.A., Holstein, T.W., Heber, S., Peterson, K.J., 2009. The deep evolution of metazoan microRNAs. Evol. Dev. 11 (1), 50–68. https://doi.org/10.1111/j.1525-142X.2008.00302.x.

Xie, K., Chen, J., Wang, Q., Yang, Y., 2018. Direct phytotoxicity and physiological effect of microRNAs in plants. Plant Cell Rep. 37, 1593–1609. https://doi.org/10.1007/s00299-018-2313-3.

Xie, K., Shen, J., Hou, X., Yao, J., Li, X., Xiao, J., Xiong, L., 2012. Gradual increase of miR156 regulates temporal expression changes of numerous genes during leaf development in rice. Plant Physiol. 158 (3), 1382–1394. https://doi.org/10.1104/pp.111.190488.

Xie, K., Wu, C., Xiong, L., 2006. Genomic organization, differential expression, and interaction of squamosa promoter-binding-like transcription factors and microRNA156 in rice. Plant Physiol. 142 (1), 280–293. https://doi.org/10.1104/pp.106.084475.

Xin, M., Wang, Y., Yao, Y., Xie, C., Peng, H., Ni, Z., Sun, Q., 2010. Diverse set of microRNAs are responsive to powdery mildew infection and heat stress in wheat (*Triticum aestivum* L.). BMC Plant Biol. 10, 123. https://doi.org/10.1186/1471-2229-10-123.

Yao, Y., Guo, G., Ni, Z., Sunkar, R., Du, J., Zhu, J.K., Sun, Q., 2007. Cloning and characterization of microRNAs from wheat (*Triticum aestivum* L.). Genome Biol. 8 (6), R96. https://doi.org/10.1186/gb-2007-8-6-r96.

Zamore, P.D., Tuschl, T., Sharp, P.A., Bartel, D.P., March 31, 2000. RNAi: double-stranded RNA directs the ATP-dependent cleavage of mRNA at 21 to 23 nucleotide intervals. Cell 101 (1), 25–33. https://doi.org/10.1016/S0092-8674(00)80620-0.

Zhang, B., 2015. MicroRNA biogenesis: enzyme-driven and argonaute-dependent processing of microRNA precursors in plants. FEBS (Fed. Eur. Biochem. Soc.) Lett. 589 (18), 1–8.

Zhang, B., Pan, X., Anderson, T.A., July 24, 2006. Identification of 188 conserved maize microRNAs and their targets. FEBS (Fed. Eur. Biochem. Soc.) Lett. 580 (15), 3753–3762. https://doi.org/10.1016/j.febslet.2006.05.063.

Zhang, F., Yang, J., Zhang, N., Wu, J., Si, H., 2022. Roles of microRNAs in abiotic stress response and characteristics regulation of plant. Front. Plant Sci. 13, 919243. https://doi.org/10.3389/fpls.2022.919243.

Zhang, J., Luo, D., Xue, S., Chen, X., Jin, Y., Tian, Y., Yang, X., Zhu, L., Huang, W., Wang, R., 2020. Identification of microRNAs involved in regulating the flowering time in sugarcane. Int. J. Mol. Sci. 21, 6610. https://doi.org/10.3390/ijms21186610.

CHAPTER

MicroRNAs in dicot plant like *Arabidopsis*

12

Chanchal Kumari[1], Indu Malik[2], Neha Shekhawat[2], Meenakshi Sharma[3], Riddhima Singh[1], Neha Gochhwal[2]

[1]*Amity Institute of Biotechnology, Amity University, Noida, Uttar Pradesh, India;* [2]*DPG Degree College, Gurugram, Haryana, India;* [3]*Daulat Ram College, University of Delhi, Delhi, India*

1. Introduction

MicroRNAs (miRNAs) are small noncoding RNA molecules in eukaryotes that play important role in synchronization of plants gene expression. They perform numerous biological processes for regulating development, growth, and to combat stress imposed by biotic and environmental factors. miRNAs are involved in modifying the expression of specific target genes at the level of posttranscription by attaching to the mRNA molecules, leading to translation degradation. During plant growth, miRNAs have been revealed to play a crucial role in regulating the organs formation, such as leaves, flowers, and roots. It is also involved shoot branching regulation and transition from vegetative to reproductive growth in plants. These RNA sequences are extremely conserved among species in plant and animal kingdoms. Along with regulation of expression of genes, they have imperative role in synchronization of various processes of developmental through their additional genetic networks (Nag and Jack, 2010; Huijser and Schmid, 2011; Srikanth and Schmid, 2011; Yamaguchi and Abe, 2012). miRNAs have a key part in regulation of developmental progressions such as apical meristem of shoot (SAM) regulation, root and leaf regulation, development of organs and floral changeover of plant. MicroRNAs (miRNAs) are endogenous, single stranded RNA molecule belonging to small, noncoding RNA nucleotide of 19–24 length. The biosynthesis of miRNAs in *Arabidopsis* commences with formation of transcript by using miRNA encoding genes as template, by RNA polymerase II interpreting miRNA transcript with 5′ capped (m7Gppp) and 3′ prime polyadenylated transcripts (Pri-miRNAs). The miRNAs genes are several hundred nucleotides long ranging from miR165a of 319 bp to miR472 of 4975 bp, and located within the intronic region of proteins-coding genes and its expression is controlled through own promoter under regulation by a variety of transcriptional factors. Pri-miRNAs is elongated stem-loop construction transcribed in correspondence to MIR gene and consist of the mature miRNA's precursor sequences. This process undergoes two successive steps to make miRNA/miRNA* duplex in exceedingly specific D-bodies, nuclear dicing bodies. Pri-miRNA is prepared into small predecessor that is pre-mRNA with the help of microprecursors with

MicroRNA Advances and Application in Plant Biology. https://doi.org/10.1016/B978-0-443-21682-4.00011-7

central components of HYPONASTIC LEAVES 1 (HYL1), DICER LIKE 1 (DCL1), RNA-binding double-stranded protein 1 (DRB1), and SERRATE (SE) (Fig. 12.1). Furthermore, Pre-mRNA is processed in the double stranded miRNA duplex by DCL1. To regulate specific expression of gene, minor RNAs and AGO

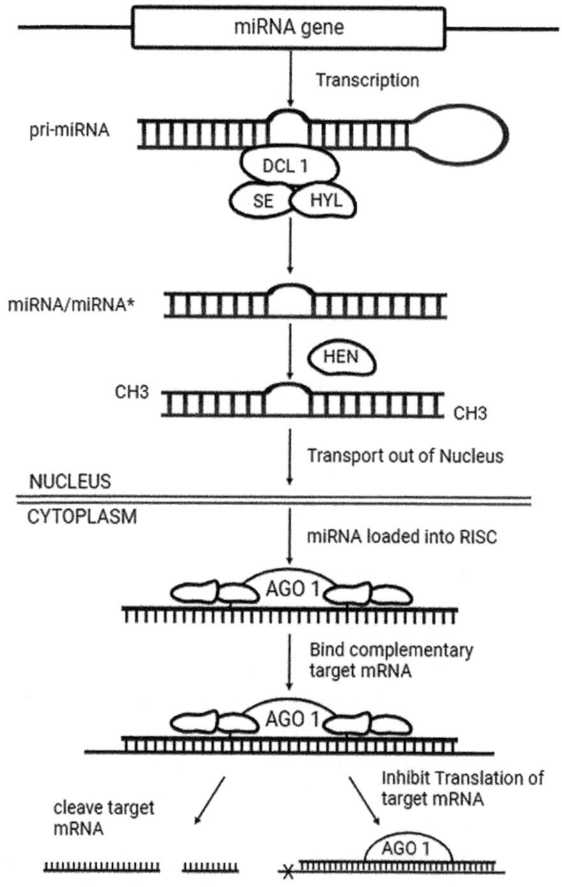

FIGURE 12.1

MiRNA-mediated gene silencing in *Arabidopsis* involves the transcription of pri-miRNAs, which are long stem-loop structures corresponding to MIR genes. The pri-miRNA undergoes processing into a short precursor, or pre-miRNA, facilitated with complex comprising of HYPONASTIC LEAVES 1 (HYL1), DICER LIKE 1 (DCL1), Double-stranded RNA-Binding protein 1 (DRB1), and SE, SERRATE. The methylation of terminal nucleotide, 3′-OH is done by HEN1. The guide RNA strand, in the double-stranded RNA is loaded into Argonaute 1 (AGO1), forming the pre-RISC complex within the nucleus. In the cytoplasm, the guide RNA binds to the target mRNA through complementary binding, leading to the cleavage of the targeted messenger RNA or it inhibits translation.

Image created by Biorender.

proteins accumulate into miRNA-induced silencing (miRISC) complex. The nascent double stranded miRNA duplex contains 2 nucleotide overhang and 2-OH at 3′ ends in each strand. The OH of 3′ terminal nucleotide is proximately methylated by HEN1 and miRNA (guide strand) in dsRNA is loaded into Argonaute 1 (AGO1) and its paralogs to form the pre-RISC complex. During the maturation step, the passenger RNA in duplex is expelled from AGO, and the supplementary strand remains attached to AGO—making RISC—and works as the chaperon to suppress target RNAs and passenger strand is degraded. miRISC assembles inside nucleus, which is accredited to AGO1, a nucleo-cytosolic shutting protein (Fig. 12.1). Furthermore, AGO1 miRNAs are transferred to cytosol through the Exportin-1 (CRM1). Exportin-1 recognizes (by recognizing specific transport signals) few RNAs, such as viral mRNAs, cellular mRNAs, and small nuclear RNAs, and export them to the cytoplasm from the nucleus, through transcription−export complex-2 (TREX-2) and NUP1, nucleoporin protein. AGO1 is altered to reconcile the procedure of assembly of RISC with the help of numerous cofactors such as heat shock protein HSP90 and cyclophilin 40 (CyP40). Additionally, EMA1 (enhanced miRNA activity 1) and TRN1, Transportin-1, belonging to AGO1-interacting importin-β proteins family, affect the miRNA loading into AGO1, EMA1 regulates negatively and TRN1 regulates positively (Sunny et al., 2016).

The miRNAs effect target gene expression either through the suppression of their target miRNAs by transcripts cleavage or by translational repression (Verma et al., 2021). The RISC complex consisting of one strand of a miRNA catalyzes mRNA cleavage by AGO1 protein's PIWI domain or its paralogs. Subsequent cleavage, of 3′ fragments occur, which disintegrated by XRN4 with 5′−3′ exonuclease (Singh et al., 2019). The 5′ fragments are uridylated by HEN1 suppressor 1 (HESO1) and UTP (URT1) at 3′ end both in vivo and in vitro. Moreover, RISC-interacting clearing 3′−5′ exoribonuclease 1 (RICE1) is associated with breakdown of uridylated 5′ fragments. Furthermore, the cytosolic contains SKI2, SKI3, and SKI8 domains, and these controls degradation of the 5′-fragment produced by RISC protein. Gene silencing through translational repression is also achieved. However, this mechanism is less common (Tungare et al., 2022).

An apt amount of evidence has been found from the information obtained from model plant *Arabidopsis* and others such as rice, potato, maize and plants of other species have shown the preservation of these miRNAs and their integration in various regulatory pathways/gene networks (Patani et al., 2023). The role of miRNA is also established in zygotic embryo formation, and this gave clue for its role in somatic embryogenesis, an important route of plant regeneration through tissue culture. Somatic embryogenesis represents a developmental mechanism for generating embryos, which can give rise to whole plants from somatic cells, which are undifferentiated. This process holds significant importance for conservation of germplasm and in vitro clonal propagation. In addition to miRNAs crucial role in plant growth and somatic embryogenesis, it also plays a vital role in responding to biotic and stress imposed by environmental condition (Gangwar et al., 2023).

Biotic stresses are produced by living organisms such as pests, pathogens, and herbivores. On the other hand, abiotic stresses refer to stresses produced by nonliving factors such as drought, salinity, heat, and cold. Both biotic and abiotic stresses can cause substantial impairment to plants, affecting their yield and productivity. Here we will be discussing, miRNAs role in flowering, somatic embryogenesis and stress adaptation in plants.

2. Role of miRNA in various facets of *Arabidopsis* growth and development

Arabidopsis is seed bearing plant and plant growth initiates with the process of embryogenesis from zygote. The embryo apical portion consist of SAM (shoot apical meristem) portion, with vital function in emergence and growth of its aboveground part of plant (shoot) whereas root apical meristem (RAM) give rise to root, underground part of plant (Dong et al., 2022; Venglat et al., 2005). miRNAs play an important role in gene expression tuning during various facets of plant growth, such as root and shoot development (Patani et al., 2024). The primary root (PR) generally is derivative of embryonic radicle, whereas lateral root (LR) and adventitious root (AR) are postembryonically derived (Maddala et al., 2024). miR160 regulates root development by downregulating various Auxin Response Factors (*ARF*s) responsible for orchestrating root development (Meng et al., 2010; Singh et al., 2018). *ARF*s represent a group of factors involved in transcription that facilitate response of auxin by influencing downstream target genes expressions (Woodward and Bartel 2005). miR160 suppresses the *ARF10* and *ARF16* expression, which are responsible for regulating the primary root tip growth and affect the root tip geotropism (Wang et al., 2005). Primary root length and branching is also regulated by repressing *ARF17* (Mallory et al., 2005). In *Arabidopsis* miR169 and its gene targets encoding for Nuclear Factor Y (NF-YA) subunit A is also involved in major regulation of PR length and activity of root apical meristem (RAM) (Rhoades et al., 2002). miR156 is well-preserved family of miRNAs among all plant species and it downregulates the expression of *SPL* genes (Wu et al., 2009). This miRNA and its targeted *SPL* genes are important for controlling the root length. In *Arabidopsis*, miR156−SPL10 module regulates primary root length and development by modulating the auxin and cytokinin phytohormone responses, leading meristematic activity in the root. miR156 increased expression has been found to be associated with small primary length and meristem size (Barrera-Rojas et al., 2019). miRNA also regulated development of lateral roots in *Arabidopsis*.

Recently, one study suggested involvement of miR156/SPL10 regulatory pathway in plant lateral roots regulation by direct activation of AGL79 (Meddya et al., 2023). Furthermore, the gain-of function of AGL79 transgenic plants, showed its involvement in leaf shape regulation of plant, branching of shoot, and time of flowering in *Arabidopsis* (Gao et al., 2018; Parveen et al., 2023). The adventitious roots formation is important for the large-scale propagation of elite genotypes. Adventitious rooting is a heritable quantitative trait influenced by various endogenous and environmental stimuli and miRNA regulating the process of adventitious root formation significantly and precisely. The subtle balance of AUXIN RESPONSE FACTOR (ARF) activator and repressor control formation of adventitious roots.

In *Arabidopsis*, miR160 target, ARF17 and miR167 targets ARF6 as negative regulators and ARF8, as positive regulators of adventitious rooting (Gutierrez et al., 2009). miR160 target ARF16 expression regulation is also important in maintaining standard growth and development of shoots and lateral rooting in *Arabidopsis* (Gutierrez et al., 2009). The shoot apical meristem (SAM) is pivotal in the initiation and growth of aboveground organs (Vora et al., 2023). miRNAs, including miR165/166 and miR394, are involved in the developmental regulation of shoot meristem in *Arabidopsis*. miR394a/b play crucial role in maintenance of SAM by negatively regulating, LEAF CURLING RESPONSIVENESS (LCR) gene belonging to F-box protein family (Kumar et al., 2019). The STM (shoot meristemless)—WUS (Wuschel)—CLV (Clavata) pathway plays important part in the upkeep of meristem activity in SAM. In plants, branching in shoots began by the formation of new meristems in the leaf axil that form secondary growth axes. CUP SHAPED COTYLEDON genes (CUC1, CUC2, and CUC3) in *Arabidopsis* encoding for NAC domain containing transcription factor function redundantly in initiating axillary meristem. Gene expression of CUC1 and CUC2 are regulated by miR164, showing role of miR164 in controlling axillary meristem. miR164 overexpression in the CUC3−2 mutant inhibited meristem formation due to degradation of CUC transcript. Conversely, miR164 mutants showed formation of accessory buds in leaf axils of *Arabidopsis* plants (Raman et al., 2008). miRNA is also important in leaf development and growth, for instance, the miR319 family targets the TEOSINTE BRANCHED1, CYCLOIDEA, and PROLIFERATING CELL NUCLEAR ANTIGEN BINDING FACTOR (TCP) transcription factors, which play important role in leaf development and growth. The upregulation of miR319 results in diminished leaf size and altered leaf shape by influencing the expression of TCP genes In *Arabidopsis*. Conversely, mutations leading to the functional loss in miR319-targeted TCP genes resulted in the development of serrated leaves, affecting expression levels of downstream targets (Baulies et al., 2022). This pathway, mediated by miRNA, has demonstrated role in leaf development regulation across various plants including tomato and rice (Palatnik et al., 2003). The miR396 family targets GROWTH REGULATING FACTORS (GRFs) for root development. In *Arabidopsis*, the overexpression of miR396 leads to decreased root growth, influencing the expression of GRF genes. This miRNA-mediated pathway also plays crucial role in regulating root development in other plant species, including maize and rice (Rodriguez, 2010). The miRNAs playing crucial role in *Arabidopsis* growth and development is listed in Fig. 12.2.

Flowering	miR156	miR159	miR172				
Biotic stress	miR393	miR319	miR159	miR160	miR165/166		
	miR167	miR390	miR398	miR408	miR773		
Somatic embryogenesis	miR156	miR159	miR172				
Abiotic stress	miR169	miR396	miR168	miR167	miR171	miR408	miR157
	miR393	miR165	miR319	miR165	miR319	miR159	miR394
	miR159	miR394	miR159	miR394	miR156	miR158	miR397
	miR172	miR160	miR402	miR164	miR173	miR165ab	miR166
Growth and development	miR167	miR319	miR396	miR160	miR169	miR156	miR165/166
	miR394	miR164	miR319	miR396	miR400		

FIGURE 12.2

miRNA involved in growth, development, and stress in *Arabidopsis*.

3. Role of miRNA in flowering

Knowledge in relation to miRNAs provides us with an auxiliary plant genetic tool for the future of genetic upgradation of crops. The synchronization of flowering time is controlled via flowering time synchronization through the interaction of various biochemical and environmental factors with the association of miRNAs and their targets, impact, and cross-talk with other miRNA-pathways. The crucial transition in a plant's life cycle occurs when it moves through vegetative phase to phase of reproduction. Various complex sets of pathways, controlled by numerous genes, participate to regulate flowering. The flowering process is guided by five major pathways, which include self-regulation, vernalization, photoperiod, aging, and Gibberellin (GA) pathways. Numerous studies have shown the miRNAs' vital role in flowering. miRNAs such as miR156 and miR172 play an essential role in transformation of plants from vegetative to stage of flowering. miR156 attaches to SQUAMOSA PROMOTER BINDING PROTEIN-LIKE (SPL), which influences MADS-box gene expression and causes the vegetative stage to change into the reproductive stage (Xu et al., 2016). In *Arabidopsis*, miR156 high expression promotes juvenile phase, and conversion of flowering sage is achieved by targeting transcription factors SPL3, SPL4, SPL5, SPL9, SPL10, and SPL15. These SPLs' sequences contain miRNA binding sites and response elements (MRE) to which miR156 binds and negatively regulates SPL gene expression. The decrease in expression of miR156/miR157 leads to heightened expression of SBP/SPL proteins as well as this helps in the transformation from the vegetative to the stage of reproduction in plants (He et al., 2018; Fouracre et al., 2021). The transition to flowering in A. thaliana is influenced by the precise negative regulation exerted by miR156 on SPL3. This regulation alters the expression of the *FT* gene in the leaves, leading to a postponement in the flowering process. In relation to flowering, the role of miR156 is evolutionarily conserved and evidence can be found in studies related to overexpression lines of tobacco, maize, *Arabidopsis* where delay in flowering and extension of juvenile phase was found. Similar to miR156, additionally, miR172 controls the time of floral organ development and flowering through degradation as well as target mRNA suppression. miR172 synchronizes the growth of plants from the juvenile to adult stages by acting as an *AP2*-like gene, which includes SM-LIKE 2, SCHNARCHZAPFEN, and TARGET OF EARLY ACTIVATION TAGGED 1/2/3. It is implemented in the regulation of numerous flowering processes such as flowering time, determination of flower organs, morphogenesis, and development of plants through control of transcription factor, AP2. It targets the APETALA2 (AP2) transcription factor, which controls the maturation of flowers. As stated by Nowak et al. (2022), the overexpression of miR172 causes a delay in flowering, whereas its suppression causes an early blooming.

Additional miRNAs, such as miR159 & miR319, which select the transcription factors MYB and TCP (TCP FAMILY TRANSCRIPTION FACTOR), respectively, are also implicated in the development of flowers. Various floral development disorders such as delay in flowering are observed due to overexpression of miR159 and miR319. Research has demonstrated that in A. thaliana, a loss-of-function miR159 causes a severe vast array of abnormalities, impedes growth, and deformities in the petals, anthers, and sepals.

miR156, miR160, miR167, miR172, and miR394 are all upregulated at the time of flower development in Chinese hickory (Carya cathayensis). miR167, miR529, and miR2950 targets the *ARF4* gene, which affects the flower development in Chinese jujube (Ziziphus jujuba) infected by phytoplasm leading to development of leaf-like structures in place of floral organs. Many studies in A. thaliana have suggested that miRNAs help in regulation of development of reproductive morphology; similarly, they are also engaged in the regulation of these organs in other plants too. In tomato, miR156b is an important factor, which controls flowering and fruit morphology through regulation of meristem activity and initial stages of development of fruits. There are a number of miRNAs involved in the flower development and its transition and still a lot of scope is left to find different pathways involved in the same. miRNAs studied during various aspects of development are shown in Fig. 12.2.

4. Role of microRNA in somatic embryogenesis

The most effective method for micropropagation of commercially significant plants is somatic embryogenesis. When expressed in somatic cells, the homeobox transcription factor WUSCHEL (WUS) has been demonstrated to induce dedifferentiation, which is followed by the generation of new stem cells that can improve somatic embryogenesis (Khare et al., 2017). Other transcription factors, such as WUS; BABY BOOM BBM; LEAFY COTYLEDON1, LEC1 and LEC2; AGAMOUS-like 15, AGL15; MYELOBLASTOSIS 118, MYB118; and EMBRYOMAKER, EMK, play a crucial role in the embryogenic transition by controlling pathways related to phytohormones, primarily those related to metabolism, transport, and auxin signaling (Nowak et al., 2022). MicroRNAs regulate SE induction in conjunction with TFs, and a variety of miRNAs exhibit varying degrees of expression, which contribute to the synthesis of somatic embryos in other plants as well as *Arabidopsis*. Numerous studies have suggested the function of miRNAs in various plants in the SE, consisting of miR160 and miR166/165, miR167, miR393, miR156, miR396, and miR528 (Nowak et al., 2022). Recent studies have revealed that the ectopic expression of miR394, in conjunction with the transcription factor WUSCHEL (WUS), stimulates somatic embryogenesis in the *Arabidopsis* Ler accession that is resistant, by downregulating its target gene, *LEAF CURLING RESPONSIVENESS* (*LCR*) (Lu et al., 2023). The other study reported role of miR393 in zygotic embryo (ZE) in *Arabidopsis*, in which it inhibits early embryo maturation of the major regulators of ZE, by repressing inhibition of *LEAFY COTYLEDON* 2 (*LEC*2) and *FUSCA3*. This gave clues for the role of miR393 in somatic embryo induction, and its role was proved in SE induction in *Arabidopsis*. miR393 transcript accumulation in conjunction with downregulation of auxin receptor *TIR1* (*TRANSPORT INHIBITOR 1*) and *AFB1* (*AUXIN F-BOX PROTEIN 1*) has been reported (Wójcik et al., 2016).

5. Role of miRNA in biotic and abiotic stress

Biotic stress pertains to stress caused by living organisms such as pathogens, insects, and further pests. For plant defense against biotic stress, MiRNAs play a vital part. To protect plants from biotic stressors, miRNAs are essential, such as pathogen infections. They specifically target and control the gene expression associated with the immune system of plants, such as genes encoding disease resistance proteins, transcription factors, and enzymes that participate in formation of secondary metabolites. For instance, miR393 regulates the expression of the F-box protein transport inhibitor response 1 (TIR1), which controls the immune response of plants (Navarro et al., 2006). Furthermore, it has been demonstrated that miR160 targets genes related to jasmonic acid (JA) signaling, which is a critical defense against insects in plants (Yang et al., 2013).

With the model plant *Arabidopsis*, the function of miRNA during bacterial contact in plants was first identified. Exposure of *Arabidopsis* to the bacterial PAMP (pathogen-associated molecular pattern) peptide (flg22) was observed to induce the production of miR393, a factor that aids in Pattern Triggered Immunity (PTI). Furthermore, miR393 achieves this by suppressing the auxin receptor gene's expression (Navarro et al., 2006). Research has demonstrated that the bacterial pathogen *Pseudomonas syringae* pv. tomato (*Pst*) DC3000 infected *Arabidopsis* leaves increased the expression of miR393, miR319, miR159, miR160, miR165/166, and miR167, while downregulating the expression of miR390, miR398, and miR408 (Zhang et al., 2011).

Researchers studying the stress nematodes caused in *Arabidopsis* discovered that the miR159abc triple mutant *Arabidopsis* lines were less susceptible to infection when exposed to *Meloidogyne incognita*. This implies that miR159 plays a crucial part in the plant's reaction to the Root-Knot Nematode (RKN) by suppressing its target gene MYB33, whose expression is important in the early stages of infection (Medina et al., 2017). Additionally, an increase in miRNA159 was observed in *Arabidopsis* gall, indicating its involvement in the plant's defense response (Medina et al., 2017). Another study with root knot nematode, *Meloidogyne javanica* showed role of miRNA390 in gall formation in *Arabidopsis* plant, suggesting essential role of miRNAs in infection caused by nematodes (Cabrera et al., 2016).

As in case of stress caused by fungal pathogens, MiRNA773 has been demonstrated to downregulate its target gene, MET2 (methyltransferase 2), hence negatively affecting *Arabidopsis* immunity against hemibiotrophic (*Fusarium oxysporum*) and necrotrophic (*Plectosphaerella cucumerina*) fungal infections (Salvador-Guirao et al., 2018). The above literature suggests the role miRNA plays in various plant pathogens including bacteria, fungi and nematodes.

Abiotic stress encompasses the challenges imposed by nonliving factors such as salt, drought, and extreme temperatures. miRNAs play a crucial role in plant responses to abiotic stresses, including fluctuations in temperature, water scarcity, and salinity. They modulate the expression of genes associated with stress tolerance, including those responsible for ion transport, stress-responsive transcription factors, and enzymes involved in osmo-protection.

The main abiotic factor that has a detrimental effect on crop productivity is drought. Numerous miRNAs have been shown to express more when there is a drought, including miRNA396, miRNA168, miRNA169, miRNA167, miRNA171, miRNA408, miRNA157, and miRNA393. MiRNA169 plays a crucial part in *Arabidopsis*' ability to react to stress of salinity and drought. The downregulation of miRNA169 results in up-regulation of its target gene, the transcription factor Nuclear Factor YA5 (NFYA5). This subsequently causes the expression of genes linked to drought stress response to rise (Li et al., 2008). The miRNA396 targets GRFs (Growth Regulating Factors), and under drought stress conditions, gets upregulated. The miRNA167 combats drought stress by its targeting and regulation of the expression of auxin response factors (ARFs), specifically ARF6 and ARF8. This miRNA gets upregulated, leading to a reduction in the levels of ARF6 and ARF8. Another one, i.e., miRNA168 contributes to drought resistance in *Arabidopsis* by targeting ARGONAUTE1 (AGO1), a main component of the RNA-induced silencing complex (RISC) (Liu et al., 2008). Furthermore, it has been demonstrated that miR399 regulates phosphate homeostasis, an important process for plant maturation and abiotic stress response (Chiou et al., 2006).

Likewise, it has been demonstrated that miRNAs play a part in how various plants react to different abiotic stressors such as salinity and high temperatures. Many miRNAs, including miRNA393, miRNA396, miRNA168, miRNA167, miRNA165, miRNA319, miRNA159, miRNA394, miRNA156, miRNA158, miRNA169, and miRNA171, exhibit elevated expression in *Arabidopsis* under salt stress conditions. Elevated levels of miRNA393 under salt stress conditions result in the upregulation of its target genes, Auxin signaling F-box 2 (AFB2), and Transport inhibitor response 1 (TIR1). TIR1/AFB2 are crucial components of the auxin signaling pathway. Subsequently, Auxin/Indole 3-acetic acid (Aux/IAA) repressors are bound by TIR1/AFB2 proteins, which sets off a series of events that result in ubiquitination and eventual degradation of Aux/IAA through the 26S proteasome pathway (Liu et al., 2008).

In *Arabidopsis* during elevated temperature, miR398 influences expression of genes encoding copper/zinc superoxide dismutase *CSD1* and *CSD2*, playing important role in the scavenging of reactive oxygen species (ROS) generated due to oxidative stress caused by high temperature (Sunkar et al., 2006) (Fig. 12.2). The miR156 increases the maintenance of acquired thermotolerance by targeting transcription factors SQUAMOSA PROMOTER BINDING PROTEIN-LIKE (SPL), which are significantly activated by thermo stress (Stief A et al., 2014). During cold conditions in *Arabidopsis*, the miRNA397, miRNA172, miRNA168 and miRNA169 were discovered to be upregulated (Liu et al., 2008; Zhang et al., 2009; Lv et al., 2010). Transcription factors, or upstream regulatory components in the auxin metabolic pathway, become critical under cold condition. In *Arabidopsis*, it has been demonstrated that miR160 and miR397 react to cold stress. ARF (Auxin Response Factor) is the target of miR160, which negatively controls *ARF* gene expression and increases *Arabidopsis*' resistance from cold shock (Yang et al., 2013). Moreover, it was shown that miRNA402 enhanced *Arabidopsis* resistance to cold stress by encouraging seed germination (Kim et al., 2010).

Heavy metals stress also form important component of abiotic stress and miRNA expression alters in combating this stress. For example, miRNA164 and miRNA173 have been found to be increased and decreased, respectively upon stress imposed by cadmium (Cd) in *Arabidopsis* seedlings (Barciszewska-Pacak et al., 2015). Various miRNAs have also shown significant response to copper (Cu) deficiency stress as well as its excess, which are namely, miRNA165ab, miRNA166, miRNA319, and miRNA400. This implies their overall function in maintaining copper homeostasis in *Arabidopsis* (Barciszewska-Pacak et al., 2015).

6. Conclusion

In plants miRNAs are engaged in a number of physiological processes and are essential in controlling gene expression, including developmental processes, flowering, responses to stresses due to biotic and abiotic factors. It's been proven that miRNAs control the expression of genes related to defense against plant pathogens by targeting pathogenesis-related genes or genes involved in phytohormone signaling pathways. Additionally, by controlling genes involved in osmotic adjustment, detoxification and antioxidant defense, miRNAs can modulate plants response to abiotic stress.

Understanding the role of miRNAs in plant stress is important for developing crops that can provide more resistance to biotic and abiotic factors, which can ultimately lead to increased crop productivity and sustainability. By targeting specific genes with role in stress responses, plant growth and development can be modulated by miRNAs, which can also help plants in adaptation to harsh climatic conditions. miRNAs' roles and mechanisms of action can provide valuable insights into the biology of plants and can support in the development of novel strategies for enhancing plant growth as well as resilience to stress.

References

Barciszewska-Pacak, M., Milanowska, K., Knop, K., Bielewicz, D., Nuc, P., Plewka, P., Pacak, A.M., Vazquez, F., Karlowski, W., Jarmolowski, A., Szweykowska-Kulinska, Z., 2015. Arabidopsis microRNA expression regulation in a wide range of abiotic stress responses. Front. Plant Sci. 6, 410.

Barrera-Rojas, C.H., et al., 2019. miR156-targeted SPL10 controls Arabidopsis root meristem activity and root-derived de novo shoot regeneration via cytokinin responses. J. Exp. Bot. 71 (3), 934–950.

Baulies, J.L., Bresso, E.G., Goldy, C., Palatnik, J.F., Schommer, C., 2022. Potent inhibition of TCP transcription factors by miR319 ensures proper root growth in Arabidopsis. Plant Mol. Biol. 108 (1–2), 93–103.

Cabrera, J., Barcala, M., García, A., Rio-Machín, A., Medina, C., Jaubert-Possamai, S., Favery, B., Maizel, A., Ruiz-Ferrer, V., Fenoll, C., Escobar, C., 2016. Differentially expressed small RNA s in Arabidopsis galls formed by *Meloidogyne javanica*: a functional role for miR390 and its TAS 3-derived tasi RNA s. New Phytol. 209 (4), 1625–1640.

Chiou, T.J., Aung, K., Lin, S.I., Wu, C.C., Chiang, S.F., Su, C.L., 2006. Regulation of phosphate homeostasis by microRNA in Arabidopsis. Plant Cell 18 (2), 412–421.

Dong, Q., Hu, B., Zhang, C., 2022. microRNAs and their roles in plant development. Front. Plant Sci. 13, 824240.

Fouracre, J.P., He, J., Chen, V.J., Sidoli, S., Poethig, R.S., 2021. VAL genes regulate vegetative phase change via miR156-dependent and independent mechanisms. PLoS Genet. 17 (6) e1009626.

Gangwar, P., Johri, P., Datta, R., Singh, S., Trivedi, M., 2023. Overview on 1 phytopharmaceuticals and biotechnology of herbal plants. In: Phytopharmaceuticals and Biotechnology of Herbal Plants. CRC Press, pp. 1–17.

Gao, R., Wang, Y., Gruber, M.Y., Hannoufa, A., 2018. miR156/SPL10 modulates lateral root development, branching and leaf morphology in Arabidopsis by silencing AGAMOUS-LIKE 79. Front. Plant Sci. 8, 2226.

Gutierrez, L., Bussell, J.D., Pacurar, D.I., Schwambach, J., Pacurar, M., Bellini, C., 2009. Phenotypic plasticity of adventitious rooting in Arabidopsis is controlled by complex regulation of AUXIN RESPONSE FACTOR transcripts and microRNA abundance. Plant Cell 21 (10), 3119–3132.

He, J., Xu, M., Willmann, M.R., McCormick, K., Hu, T., Yang, L., Poethig, R.S., 2018. Threshold-dependent repression of SPL gene expression by miR156/miR157 controls vegetative phase change in *Arabidopsis thaliana*. PLoS Genet. 14 (4) e1007337.

Huijser, P., Schmid, M., 2011. The control of developmental phase transitions in plants. Development 138 (19), 4117–4129.

Khare, P., Kishore, K., Sharma, D.K., 2017. A study on the standardization parameters of *Cassia angustifolia*. Asian J. Pharmaceut. Clin. Res. 10 (7), 329–332.

Kim, J.Y., Kwak, K.J., Jung, H.J., Lee, H.J., Kang, H., 2010. MicroRNA402 affects seed germination of *Arabidopsis thaliana* under stress conditions via targeting DEMETER-LIKE Protein3 mRNA. Plant Cell Physiol. 51 (6), 1079–1083.

Kumar, A., Gautam, V., Kumar, P., Mukherjee, S., Verma, S., Sarkar, A.K., 2019. Identification and co-evolution pattern of stem cell regulator miR394s and their targets among diverse plant species. BMC Evol. Biol. 19, 1–16.

Li, W.X., Oono, Y., Zhu, J., He, X.J., Wu, J.M., Iida, K., Lu, X.Y., Cui, X., Jin, H., Zhu, J.K., 2008. The Arabidopsis NFYA5 transcription factor is regulated transcriptionally and post-transcriptionally to promote drought resistance. Plant Cell 20 (8), 2238–2251.

Liu, H.-H., Tian, X., Li, Y.-J., Wu, C.-A., Zheng, C.-C., 2008. Microarray-based analysis of stress-regulated microRNAs in *Arabidopsis thaliana*. RNA 14 (5), 836–843.

Lu, L., Holt, A., Chen, X., Liu, Y., Knauer, S., Tucker, E.J., Laux, T., 2023. miR394 enhances WUSCHEL-induced somatic embryogenesis in *Arabidopsis thaliana*. New Phytol. 238 (3), 1059–1072.

Lv, D.K., Bai, X., Li, Y., Ding, X.D., Ge, Y., Cai, H., Ji, W., Wu, N., Zhu, Y.M., 2010. Profiling of cold-stress-responsive miRNAs in rice by microarrays. Gene 459 (1–2), 39–47.

Maddala, V.K.S., Singh, S., Bandi, J., 2024. Soil properties influencing nutrient signaling in plant soil systems. In: Essential Minerals in Plant-Soil Systems. Elsevier, pp. 373–393.

Mallory, A.C., Bartel, D.P., Bartel, B., 2005. MicroRNA-directed regulation of Arabidopsis AUXIN RESPONSE FACTOR17 is essential for proper development and modulates expression of early auxin response genes. Plant Cell 17 (5), 1360–1375.

Meddya, S., Meshram, S., Sarkar, D., Datta, R., Singh, S., Avinash, G., Thulasinathan, T., 2023. Plant stomata: an unrealized possibility in plant defense against invading pathogens and stress tolerance. Plants 12 (19), 3380.

Medina, C., Da Rocha, M., Magliano, M., Ratpopoulo, A., Revel, B., Marteu, N., Magnone, V., Lebrigand, K., Cabrera, J., Barcala, M., Silva, A.C., 2017. Characterization of microRNAs from Arabidopsis galls highlights a role for miR159 in the plant response to the root-knot nematode *Meloidogyne incognita*. New Phytol. 216 (3), 882–896.

Meng, Y., Ma, X., Chen, D., Wu, P., Chen, M., 2010. MicroRNA-mediated signaling involved in plant root development. Biochem. Biophys. Res. Commun. 393 (3), 345–349.

Nag, A., Jack, T., 2010. Sculpting the flower; the role of microRNAs in flower development. Curr. Top. Dev. Biol. 91, 349–378.

Navarro, L., Dunoyer, P., Jay, F., Arnold, B., Dharmasiri, N., Estelle, M., Voinnet, O., Jones, J.D., 2006. A plant miRNA contributes to antibacterial resistance by repressing auxin signaling. Science 312 (5772), 436–439.

Nowak, K., Morończyk, J., Grzyb, M., Szczygieł-Sommer, A., Gaj, M.D., 2022. miR172 regulates WUS during somatic embryogenesis in Arabidopsis via AP2. Cells 11 (4), 718.

Palatnik, J.F., Allen, E., Wu, X., Schommer, C., Schwab, R., Carrington, J.C., Weigel, D., June 26, 2003. Control of leaf morphogenesis by microRNAs. Nature 425 (6955), 257–263.

Parveen, R., Khare, N., Singh, S., Johri, P., 2023. An in silico outlook for the detection and surveillance of evolving and persistent plant pathogens. In: Detection, Diagnosis and Management of Soil-Borne Phytopathogens. Springer Nature Singapore, Singapore, pp. 17–40.

Patani, A., Patel, A., Prajapati, D., Khare, N., Singh, S., 2023. Various agriculture crop plant-based bioactive compounds and their use in nanomaterial synthesis and applications. In: Nanomaterials from Agricultural and Horticultural Products. Springer Nature Singapore, Singapore, pp. 223–241.

Patani, A., Prajapati, D., Shukla, K., Patel, M., Patani, P., Patel, A., Singh, S., 2024. Environmental stress–induced alterations in the micro-and macronutrients status of plant. In: Essential Minerals in Plant-Soil Systems. Elsevier, pp. 181–195.

Raman, S., Greb, T., Peaucelle, A., Blein, T., Laufs, P., Theres, K., 2008. Interplay of miR164, CUP-SHAPED COTYLEDON genes and LATERAL SUPPRESSOR controls axillary meristem formation in *Arabidopsis thaliana*. Plant J. 55 (1), 65–76.

Rhoades, M.W., Reinhart, B.J., Lim, L.P., Burge, C.B., Bartel, B., Bartel, D.P., 2002. Prediction of plant MicroRNA targets. Cell 110, 513–520.

Rodriguez, R.E., Mecchia, M.A., Debernardi, J.M., Schommer, C., Weigel, D., Palatnik, J.F., July 2010. Control of cell proliferation in *Arabidopsis thaliana* by microRNA miR396. Development 137 (14), 103–112.

Salvador-Guirao, R., Baldrich, P., Weigel, D., Rubio-Somoza, I., San Segundo, B., 2018. The microRNA miR773 is involved in the Arabidopsis immune response to fungal pathogens. Mol. Plant Microbe Interact. 31 (2), 249–259.

Singh, A., et al., 2018. Plant small RNAs: advancement in the understanding of biogenesis and role in plant development. Planta 248 (3), 545–558.

Singh, A., Sengar, R.S., Singh, A.K., Singh, S., Kumar, A., Ahmad, K., 2019. Impact of genetic engineering on agriculture food crops. Biotech Today: Int. J. Biol. Sci. 9 (1), 17–24.

Srikanth, A., Schmid, M., 2011. Regulation of flowering time: all roads lead to Rome. Cell. Mol. Life Sci. 68, 2013–2037.

Stief, A., Altmann, S., Hoffmann, K., Pant, B.D., Scheible, W.R., Bäurle, I., 2014. Arabidopsis miR156 regulates tolerance to recurring environmental stress through SPL transcription factors. Plant Cell 26 (4), 1792–1807.

Sunkar, R., Kapoor, A., Zhu, J.K., 2006. Posttranscriptional induction of two Cu/Zn superoxide dismutase genes in Arabidopsis is mediated by downregulation of miR398 and important for oxidative stress tolerance. Plant Cell 18 (8), 2051–2065.

Sunny, J.S., Snijesh, V.P., Singh, S., 2016. Potential role of mirnas in rheumatoid arthritis. Austin Arthritis 1 (2), 1009.

Tungare, K., Shahu, R., Zambare, V., Agarwal, P., Jobby, R., Nisar, N., Jha, P., 2022. Toxicity mitigation of textile dye reactive blue 4 by hairy roots of Helianthus annuus and testing its effect in in vivo model systems. BioMed Res. Int. 2022.

Venglat, S.P., Xiang, D., Kusshalappa, K., Keller, W., Palmer, D., Selvara, G., Datla, R., 2005. OBPC Symposium: maize 2004 & beyond-developmental and molecular genetics of embryogenesis in plants. Vitro Cell Dev. Biol. Plant 41, 378–387.

Verma, N., Sao, P., Srivastava, A., Singh, S., 2021. Physiological and molecular responses to drought, submergence and excessive watering in plants. Harsh Environ Plant Resilience 305–321.

Vora, Z., Pandya, J., Vaikuntapu, P.R., Singh, S., Asha, S., Purama, R.K., 2023. Genome-wide association of defense hormone crosstalk in plants. In: Hormonal Cross-Talk, Plant Defense and Development. Academic Press, pp. 353–371.

Wang, J.-W., Wang, L.-J., Mao, Y.-B., Cai, W.-J., Xue, H.-W., Chen, X.-Y., 2005. Control of root cap formation by MicroRNA-targeted auxin response factors in Arabidopsis. Plant Cell 17 (8), 2204–2216.

Wójcik, A.M., Gaj, M.D., 2016. miR393 contributes to the embryogenic transition induced in vitro in Arabidopsis via the modification of the tissue sensitivity to auxin treatment. Planta 244, 231–243.

Woodward, A.W., Bartel, B., 2005. Auxin: regulation, action, and interaction. Ann. Bot. 95 (5), 707–735.

Wu, G., Park, M.Y., Conway, S.R., Wang, J.-W., Weigel, D., Poethig, R.S., 2009. The sequential action of miR156 and miR172 regulates developmental timing in Arabidopsis. Cell 138 (4), 750–759.

Xu, M., Hu, T., Zhao, J., Park, M.Y., Earley, K.W., Wu, G., Poethig, R.S., 2016. Developmental functions of miR156-regulated SQUAMOSA PROMOTER BINDING PROTEIN-LIKE (SPL) genes in *Arabidopsis thaliana*. PLoS Genet. 12 (8) e1006263.

Yamaguchi, A., Abe, M., 2012. Regulation of reproductive development by non-coding RNA in Arabidopsis: to flower or not to flower. J. Plant Res. 125, 693–704.

Yang, F., Wang, Q., Schmitz, G., Müller, D., 2013. MicroRNA 160 affects transcript abundance of Arabidopsis auxin response factor 17 during pollen development. Plant J. 74 (6), 947–959.

Zhang, J., Xu, Y., Huan, Q., Chong, K., 2009. Deep sequencing of Brachypodium small RNAs at the global genome level identifies microRNAs involved in cold stress response. BMC Genom. 10, 1–16.

Zhang, W., Gao, S., Zhou, X., Chellappan, P., Chen, Z., Zhang, X., Fromuth, N., Coutino, G., Coffey, M., Jin, H., 2011. Bacteria-responsive microRNAs regulate plant innate immunity by modulating plant hormone networks. Plant Mol. Biol. 75, 93–105.

CHAPTER

13 MicroRNAs in legumes and symbiotic association with rhizobacteria

G. Vamsi Krishna[1], Lellapalli Rithesh[2], P. Kishore Varma[3], V. Prasanna Kumari[1]

[1]*Department of Plant Pathology, Agricultural College, ANGRAU, Guntur, Andhra Pradesh, India;* [2]*Department of Plant Pathology, Kerala Agricultural University, Thiruvanathapuram, Kerala, India;* [3]*Department of Plant Pathology, RARS, ANGRAU, Guntur, Andhra Pradesh, India*

1. Introduction

Pulse crops, also known as grain legumes, are essential food crops belonging to the family Fabaceae. They play a crucial role in global food and fodder production due to their high nutritional value. Some of the prominent pulse crops include *Cicer arietinum* (chickpea), *Arachis hypogaea* (peanut), *Cajanus cajan* (pigeonpea), *Glycine* max (soybean), *Phaseolus mungo* (black gram), *Pisum sativum* (pea), and *Medicago sativa* (alfalfa).These crops are rich sources of protein, oil, fiber, vitamins, and various micronutrients, making them valuable for human and livestock consumption. In India, they have been an integral part of agriculture and play a significant role in the country's economy. One of the main reasons for their agronomic importance is their ability to form symbiotic relationships with soil microbes, particularly nitrogen-fixing bacteria. These beneficial associations enable the legumes to fix atmospheric nitrogen and convert it into a usable form for the plants. As a result, they require less chemical fertilizer compared with other major crop plants, reducing environmental impacts and production costs. Additionally, pulse crops possess deep root systems that allow them to efficiently access limited soil moisture, making them more drought-tolerant and suitable for regions with water scarcity. Their ability to establish symbiotic associations with arbuscular mycorrhiza fungi also contributes to their phosphorus uptake efficiency, aiding in nutrient acquisition under favorable environmental conditions. Overall, pulse crops have remained vital components of sustainable agriculture, promoting soil health, conserving resources, and providing essential nutrients to human and animal diets (Sunny et al., 2016).

Most legumes form a mutualistic relationship between legume plants and nitrogen-fixing rhizobia, as well as their ability to form symbiotic relationships with Arbuscular mycorrhizal (AM) fungi. This symbiosis is indeed crucial for sustainable agricultural output and ecological significance due to the process of biological nitrogen fixation (BNF) (Verma et al., 2021).

Biological Nitrogen Fixation (BNF): Legume plants have the unique ability to form a symbiotic relationship with nitrogen-fixing rhizobia, which results in BNF.

MicroRNA Advances and Application in Plant Biology. https://doi.org/10.1016/B978-0-443-21682-4.00004-X

During BNF, atmospheric nitrogen (N_2) is converted into ammonia (NH_3) in the root nodules of the legume plant. This converted form of nitrogen can be utilized by the plant host as a nutrient.

Symbiosis with Rhizobia: Under nitrogen-starved conditions, legume plants initiate a symbiotic relationship with rhizobial bacteria. These rhizobia live as intracellular symbionts in the root nodules, where they establish a microaerobic environment to protect their oxygen-sensitive nitrogenase enzyme complex, which is responsible for nitrogen fixation. In return, the legume plant provides the rhizobia with sugars and other organic compounds as a source of energy and nutrients.

Symbiosis with AM Fungi: Legume plants can also form symbiotic relationships with Arbuscular mycorrhizal (AM) fungi. In this symbiosis, the AM fungi colonize the roots of the legume plant, forming structures called arbuscules. These arbuscules act as sites of nutrient exchange, particularly facilitating the uptake of phosphorous from the soil to the plant host.

Role in Sustainable Agriculture: The ability of legume plants to engage in both nitrogen-fixing rhizobia symbiosis and AM fungal symbiosis makes them important components of sustainable agricultural practices. By fixing nitrogen and facilitating phosphorous uptake, legume crops can enhance soil fertility and reduce the need for synthetic fertilizers, thus contributing to more sustainable and environmentally friendly agricultural outputs. In summary, legume plants' ability to form mutualistic relationships with nitrogen-fixing rhizobia and AM fungi plays a crucial role in ecological processes and sustainable agriculture by improving soil fertility and nutrient availability (Singh et al., 2019).

Symbiosis refers to a close and long-term interaction between two different biological species. In this case, legume plants and rhizobia bacteria form a symbiotic relationship. During symbiosis, the cells of both organisms undergo cellular reprogramming to facilitate the establishment of a mutually beneficial relationship. Root nodules are specialized structures formed on the roots of legume plants as a result of the symbiotic interaction with rhizobia bacteria. Within these nodules, a process called biological nitrogen fixation (BNF) takes place. Rhizobia bacteria have the ability to convert atmospheric nitrogen into a form that plants can use (ammonia), which helps the plant with its nitrogen requirements (Tungare et al., 2022).

Stages of Nodulation: The successful formation of root nodules involves several stages:

a. **Bacterial Ingress:** Rhizobia bacteria enter the root system of the legume plant.
b. **Movement Across Infection Thread:** The bacteria move through the root through specialized structures called infection threads.
c. **Cortical Division/Dedifferentiation:** The plant's root cells undergo changes to create a suitable environment for the bacteria.
d. **Primordium Development:** The formation of nodule primordia, which are small, undeveloped nodules.
e. **Bacterial Colonization:** The rhizobia bacteria colonize the nodule primordia.

f. Nitrogen Fixation: Inside the nodules, the rhizobia bacteria perform nitrogen fixation to provide nitrogen to the plant.

Flavonoids and Nod Genes: Legume crop species secrete flavonoids and related compounds into their root exudates. These compounds attract specific rhizobial species and stimulate them to express nod genes. The nod-gene products then create nod factors, which are lipochito-oligosaccharide signaling molecules produced by the bacteria. Nod factors play a crucial role in initiating the process of nodulation and establishing the symbiotic relationship.

LysM Receptor-like Kinase: During the interaction between plants and microbes, a receptor-like kinase called LysM receptor-like kinase is involved. This receptor is present in the root epidermis of plants and plays a role in recognizing and responding to the nod factors produced by the rhizobia bacteria. The interaction of nod factors with the LysM receptor-like kinase leads to immediate subcellular changes in the root epidermis and subsequent alterations in the root cortex, facilitating the formation of nodules. The process of nodulation is a fascinating example of how plants and bacteria can interact and collaborate to benefit each other. The plant obtains a vital nutrient (nitrogen), while the bacteria gain a protected environment to thrive and reproduce. This mutualistic relationship has significant implications for agriculture and ecological systems (Patani et al., 2023).

Rhizobia release signaling molecules called Nod factors into the soil. These Nod factors are perceived by the root epidermal cells of the leguminous plant. In response to Nod factors, root epidermal cells undergo depolarization, and this leads to a cytoplasmic calcium ($Ca^{2+)}$ spike in the root hair. The Ca^{2+} spike triggers a signaling pathway that involves several components, including the transcription factor ERN (Ethylene response factor essential for nodulation), the calcium and calmodulin-dependent protein kinase (CaMK), and two members of the GRAS family of proteins, NSP1 and NSP2. The signaling pathway activates responses in the cortical cells of the root. Cortical cells restart their cell cycle and initiate the formation of a nodule meristem, which is a specialized region for nodule development. Selected epidermal cells produce infection threads, which are extracellular matrix conduits lined with plasma membranes. These threads serve as a path for rhizobia to enter the root. By combining cell division and thread matrix tumbling, rhizobia enter the root through the infection thread (Gangwar et al., 2023). In the developing nodule, the infection threads eventually unite with cells. Once the infection threads reach the cells, the bacteria are discharged into the cytoplasm of the host plant. This process resembles endocytosis. Within the cytoplasm of the host plant, the bacteria transform into specialized nitrogen-fixing structures called bacteroids. Bacteroids fix nitrogen from the atmosphere and convert it into a form that the plant can use. In return, the plant provides carbon energy to the bacteroid (Fig. 13.1).

Recent research has revealed that small RNAs, specifically microRNAs (miRNAs), are essential regulators of gene expression and plant growth during symbiotic nitrogen fixation in legumes (Patani et al., 2024). MiRNAs are short, noncoding

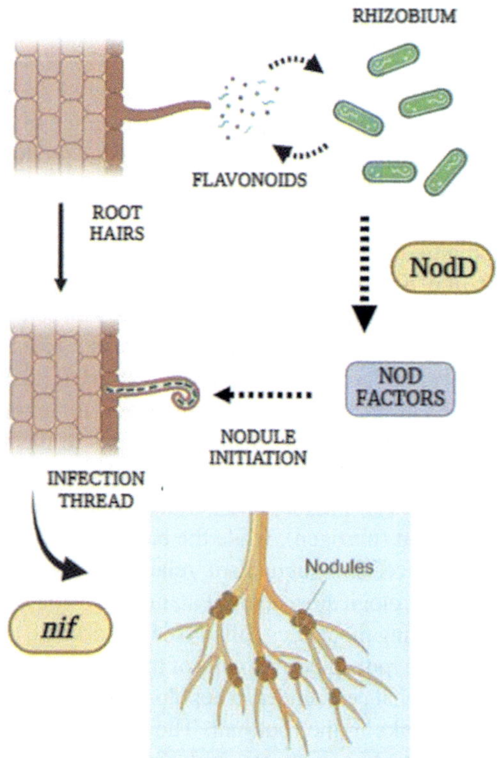

FIGURE 13.1

Schematic overview of the nodulation process and biological nitrogen fixation.

RNA molecules that typically consist of 20–24 nucleotides. They play a crucial role in posttranscriptional gene regulation (PTGS) by negatively controlling the expression of their target messenger RNAs (mRNAs) (Maddala et al., 2024). During the establishment and maintenance of symbiosis, cellular reprogramming occurs, involving the activation or inhibition of specific nodulation genes through miRNAs. High-throughput sequencing of small RNA libraries has allowed researchers to discover miRNAs and their targets in the context of legume-rhizobia symbiosis. These miRNAs are key players in coordinating various processes, such as bacterial colonization, organogenesis, and symbiotic nitrogen fixation (Meddya et al., 2023). The miRNAs achieved target gene regulation by binding to the complementary sequences in the target mRNA through Watson-Crick base pairing. Once bound, miRNAs can inhibit gene expression either by translational repression, preventing the translation of the mRNA into protein, or by mRNA cleavage, leading to the degradation of the mRNA (Treiber et al., 2019). Additionally, some miRNAs can move systemically within the plant through the vascular system, enabling communication between stem/leaf tissues and the roots. This systemic movement of miRNAs contributes to the coordination of various plant processes required for nodule formation and function during legume-rhizobia symbiosis (Parveen et al., 2023).

2. What is a microRNA (miRNA)?

MicroRNAs are short, noncoding RNA molecules with a length of around 22 nucleotides that are crucial for the control of gene activity. They are translated from certain genes known as MIR genes, which come from single-stranded RNA precursors. The miRNA genes are translated into primary miRNA transcripts (pri-miRNAs), which are generally significantly longer, about 70 nucleotides. In the nucleus, Precursor miRNA transcripts (pre-miRNAs), which are around 70 nucleotides long, are produced as a consequence of the processing of the pri-miRNAs by the enzyme Drosha (Vora et al., 2023). Afterwards, with the aid of Exportin-5, the pre-miRNAs are transported from the nucleus into the cytoplasm. After reaching the cytoplasm, the Dicer enzyme continues to process pre-miRNAs and cleaves the pre-miRNAs into mature double-stranded miRNAs, which are 21–24 base pairs in length. The mature miRNA duplex is subsequently loaded with one strand into the RNA-induced silencing complex (RISC), which directs the miRNA to its target mRNA. The miRNA in the RISC complex binds to particular regions in the 3′-untranslated region (UTR) of target mRNAs. Depending on the level of complementarity between both the miRNA and its target mRNA, this interaction might result in either mRNA destruction or translational repression (Khare et al., 2017). MiRNAs may regulate gene expression by binding to target mRNAs, inhibiting translation, or causing mRNA degradation (Fig. 13.2) (Guo et al., 2016).

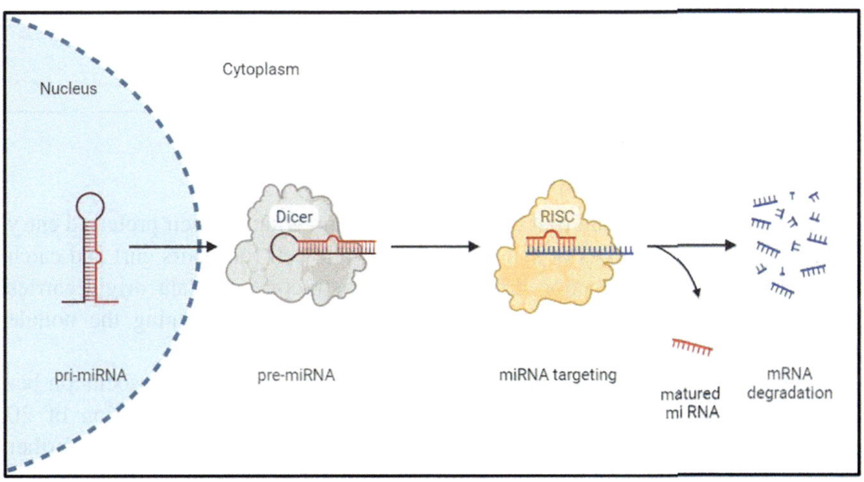

FIGURE 13.2

An overview of the miRNA pathway.

3. Identification of miRNA in legumes

Plant miRNAs were first time detected in *Arabidopsis thaliana* by Park et al. (2002). Since then, miRNAs have been identified in various plant species using different methods. Plant miRNAs have been isolated using genetic screening (Lee et al., 1993; Wightman et al., 1993), direct cloning after small RNA isolation (Fu et al., 2005) and computational prediction strategies (Bartel, 2004; Adai et al., 2005). Direct cloning and sequencing became the preferred method with the advent of Next-Generation Sequencing (NGS) technology since 2005. This approach has been successfully used in many legume species such as *M. truncatula*, common bean, soybean, chickpea, peanut, and lotus (Subramanian et al., 2008; Szittya et al., 2008). Legume species have been found to contain both conserved and novel miRNAs. These miRNAs have the potential to regulate legume-specific cellular processes. Around 2000 miRNA sequences belonging to 285 miRNA families have been identified from legumes and are available in the publicly available miRNA database, miRBase.

The identification of miRNAs that regulate the symbiosis between legumes and rhizobia has increased exponentially over the past 2 decades with the development of NGS technologies. Soybean nodules provided the first miRNA collection that was sequenced. Following this, a number of research used NGS technologies and methods in different legumes, particularly model legumes, to gain understanding of the spatiotemporal regulation of symbiosis by miRNAs. There are two basic methods for identifying miRNA genes and their target mRNAs: Sequencing of sRNA libraries, followed by miRNA-target functional validation, Bioinformatic and computational prediction followed by experimental validation.

4. miRNAs coordinates different phases of legume–rhizobium symbiosis

Rhizobia penetrate the plant root through the root hair, which is their preferred entry route to initiate the process of symbiosis. The single-cell root hairs curl and catch rhizobia there by the infection thread, a tubular structure of plant origin carries the bacteria into the underlying proliferating cortical cells forming the nodule primordium.

Regulation of miRNAs during the early stages of the symbiotic association between soybean roots and *Bradyrhizobium japonicum*. The identification of 20 conserved, 35 unique, and 14 novel miRNAs opens up new avenues for further research into the molecular mechanisms governing this beneficial plant-microbe interaction. To validate the presence and expression of the identified miRNAs, the researchers used a traditional molecular biology technique called northern blot analysis. This technique helps confirm the size and abundance of the miRNAs in the samples (Subramanian et al., 2008).

Hoang et al. in 2020 focused on investigating microRNAs (miRNAs) in soybean root hairs during their interaction with *Bradyrhizobium japonicum*, a nitrogen-fixing bacterium commonly associated with legume plants. They have isolated soybean root hairs at different time points (12, 18, 24, and 48 h postinoculation, hpi) after exposure to B. japonicum. They compared the miRNA expression profiles of these isolated root hairs with the miRNA expression profiles of roots after removing the root hairs (stripped roots) and uninoculated control roots and identified a total of 114 miRNAs from the isolated soybean root hairs, out of which 22 miRNAs were novel, 66 miRNAs showed differential expression between the root hairs and the stripped roots. 48 miRNAs exhibited distinct regulation between infected root hairs (exposed to B. japonicum) and uninoculated root hairs. This indicates that these miRNAs are likely involved in the response of root hairs to the bacterial infection.

5. Role of miRNA involved in symbiosis and nodulation

Leguminous plants and rhizobacteria engage in nitrogen fixation through their symbiotic interactions. This advantageous interaction signifies the development of nitrogen-fixing structures known as root nodules (Yang et al., 2022). The control of this mutual interaction involves several molecular mechanisms. These methods consider the detection of many miRNAs that actively take part in this association and may control it in various ways either favorably or unfavorably (Tiwari et al., 2021a). The underlying molecular process, as well as hundreds of genes involved in atmospheric nitrogen fixation, have been identified (Roy et al., 2020). The presence of miRNA genes in various plants such as *C. arietinum*—Chickpea (Tiwari et al., 2021b), *M. truncatula*—barrel medic (Nandety et al., 2023), *P. vulgaris*—common bean (Ayra et al., 2021), *L. japonicus*—birdsfoot trefoil (Sun et al., 2023), *G. max*—soybean (Yun et al., 2022), and *A. hypogaea*—groundnut (Li et al., 2021) indicate that they are involved in nodulation (Table 13.1).

In light of recent research suggesting the function of miRNAs in nodulation and N_2 fixation-related signaling pathways, According to Chakraborty et al. (2022), regulation of MIR171h targets the nodulation signaling pathway transcription factor *MtNSP2*, which is necessary for nodulation as well as the mycorrhizal communication process during the initial phases of nitrogen-seeking legume root development in *M. truncatula*. Increased expression of MIR396b in the roots of *M. truncatula* inhibited root growth and reduced mycorrhizal colonization by targeting six growth-regulating factor genes (*MtGRF*) and two basic helix-loop-helix transcription factor (*bHLH79*)-like genes, suggesting a crucial role for MIR396b in rhizogenesis and mycorrhizal colonization. *NSP2*, a nonstructural protein transcription factor, is linked to the activation of MIR171c in root nodules, which is important for the formation of nodules. Although MIR397 is only induced in rhizobial infected nodules with active N2 fixation, it targets the laccase copper protein family gene in *L. japonicus* and contributes to copper homeostasis associated with nitrogen fixation

Table 13.1 Role of mi RNAs in controlling nodulation in legume plants with possible molecular mechanisms involved.

Crop	miRNAs	Trait	Target genes	Function	References
G. max	MIR-393J	Nodule formation	*ENOD93*	Targets early Nodulin 93 (*ENOD93*) gene and regulates nodule development	Yan et al. (2015)
G. max	GMA, MIR-2606B, MIR-4416.	Nodule formation	Mannosyl- oligosaccharide 1, 2-alpha-mannosidase, Rhizobium induced peroxidise 1-like gene (*RIP1*)	Regulation of nodulation process	Yan et al. (2016)
G. max	MIR-482, MIR-1512, MIR-1515	Nodule formation	Gm-12g28730, Gm-17g04060, Gm-04g05920, Glyma-09g27690	Regulation of nodulation process	Li et al. (2010)
L. japonicus	MIR-2111	Nodulation	*TOO MUCH LOVE*, A nodulation suppressor	Hypernodulation and aberrant root formation.	Okuma et al. (2020)
M. truncatula	MIR-2111	Symbiosis and nodulation	*TOO MUCH LOVE* 1, *TOO MUCH LOVE* 2	Positively regulates root symbiotic nodulation	Gautrat et al. (2020)
A. hypogaea	AHYMI-399, AHYMIR-159, MIR-3508	Nodule infection	Pectinesterase gene	Controls nodulation process	Figueredo et al. (2020)
M. truncatula	MIR-166	Root and nodule Formation	Class-III HD-ZIP genes	Reduction in number of nodule formation.	Boualem et al. (2008)
M. truncatula	MICRORNA-169	Nodule Development	MtHAP2-1	Controls nodule development	Combier et al. (2006)
Phaseolus vulgaris	MIR-319D	Rhizobium N_2 Fixation	*TCP10*		
M. truncatula	MIR-1507, MIR-2109, MIR-2118	Nodulation and symbiosis	NB-LRR genes	Reduced the activity of NB-LRR genes and nodulation process	

(Xu et al., 2023). After that, Hossain et al. (2019) proved that gma-MIR171q MIRNAs have a negative effect on how soybeans develop nodules. The authors demonstrated that the regulatory expression of two GRAS transcription factor genes, *GmSCL-6* and *GmNSP2*, which are the targets of gmaMIR171o and gma-MIR171q MIRNAs, plays an active role in the expression of Nodule Inception transcription factor (NIN), Early Nodulin ENOD40, and Ethylene Responsive Factor (ERN) genes involved in the nodulation process in soybeans.

According to several studies (Lauressergues et al., 2012; Chakraborty et al., 2022), The growth of nodules and the mycorrhizal communication cascade in *M. truncatula* depend on the downward regulation of miR171h, aiming at *MtNSP2*. Six *MtGRF* and two *bHLH79* genes were specifically targeted by miR396b overexpression in *M. truncatula* roots, which reduced mycorrhizal colonization and inhibited root growth. This suggests that miR396b plays a significant role in both root growth and mycorrhizal colonization. It is necessary to gain further knowledge about the exact regulation of nodulation in soybeans and the intrinsic molecular mechanism of miR172c regulating rhizobial infection (Wang et al., 2014). The authors proposed that *ENOD40* gene transcription in soybeans was inhibited by the dearth of rhizobial nodule number control 1 (*NNC1*). Despite this, when rhizobia were present, the nod factor receptors triggered a signal cascade that resulted in the overexpression of miR172c, which targets the *NNC1* gene. As a result, NNC1 inhibition promotes ENOD40 gene transcription, resulting in nodule organogenesis in soybeans.

Holt et al. (2015) provided evidence in favor of the inductive potential of miR172a in *L. japonicus* roots, which required both active symbiotic bacteria and Nod factor signaling at the early stage of symbiotic infection. It was predicted that miR172a may target the *RAP2-7-like1*, *AP2*, and *AP2-like2* genes during bacterial symbiosis. Further research by Yan et al. (2015) showed that overexpression of miR-393j-3p, a miRNA that targets the Early Nodulin 93 (*ENOD93*) gene, severely inhibited nodule growth in soybean. Turner et al. (2012) demonstrated that the miRNA 169 targets nuclear factor-y alpha-related transcription factor (*Glyma17g05920*), encodes proteins that support nodule formation, exhibit considerable levels of expression.

Prior research has shown that miR169 controls *M. truncatula* nodule formation by specifically focusing the new TF gene *MtHAP2-1* that is unique to symbiotic organisms (Combier et al., 2006). With increased nodule counts at the transgenic level, Li et al. (2010) supported the involvement of miR482, miR1512, and miR1515, hence indicating their function in nodule growth in soybean. Nevertheless, Wang et al. (2015) showed that miR156 overexpression in transgenic plants prevented *L. japonicus* nodule growth. Similar to this, overexpression of miR319, the target teosinte branched 1, cycloidea, pcf1(*TCP10*) TF gene, in common beans promotes the activity of the *LOX2* gene involved in jasmonic acid production and boosts the growth of nodules while inhibiting the rhizobial infection process.

Additionally, to get a better understanding of the role of miRNAs in regulating nodulation and symbiosis, constructed and functionally verified the regulatory mechanism of the nodulation and symbiosis processes in *M. truncatula* by silencing target NB-LRR genes using miR2118, miR2109, and miR1507 miRNAs. Symbiotic bacteria upregulate miR2118, miR2109, and miR1507 miRNAs during nodulation and symbiotic nitrogen fixation at the expense of downregulating NB-LRR genes, impairing the plant's innate immunity. Recent study by Gautrat et al. (2020) have shown that miR2111 targets *TOO MUCH LOVE*, a protein that encodes the nodulation suppressor F-box/kelch-repeat, and that miR2111 may also increase nodulation. Yet, depending on shoot-acting hypernodulation and the aberrant root 1 (*HAR1*) receptor, rhizobial inoculation or infection and nitrate treatment decreased the amount of miR2111s in leaves and roots. Furthermore, Gautrat et al. (2020) hypothesized that the Clavata3/embryo surrounding region 12 (*CLE12*) and *CLE13* signaling peptides synthesized in roots act through *HAR1*/super numeric nodule (SUNN) receptors to inhibit the activity of miR2111 by describing the fine-tuning and autoregulation mechanisms of nodulation (Okuma et al., 2020). A C-terminally encoded peptide (CEP) generated in the root and acting in the shoot through the compact root architecture 2 (*CRA2*) receptor by this miR2111 promotes root symbiotic nodulation in nitrogen-starved situations. Similar to this, Okuma et al. (2020) verified the regulatory function of MIR2111-5 locus-derived *HAR1*-dependent miR2111s. Shoots of *L. japonicus* govern root nodulation, according to functional assessments.

A. hypogaea-specific miRNAs, Ahy-miR3508, Ahy-miR3509, and Ahy-miR3516, have also been discovered; however, it is unknown if they contribute to the nodulation process (Figueredo et al., 2020). A study of the common bean's transcriptome revealed 185 mature miRNAs and 181 potential targets for these miRNAs (Formey et al., 2015). Functional analysis of some miRNAs revealed that miR-Nov153, which targets uridine kinase (Phvul.003 g180800), miR319, which targets a member of the *TCP* TF family (Phvul.011, g156900), and miR-Nov494, which targets aldehyde dehydrogenase (Phvul.004 g162200.1), were upregulated but their corresponding target genes were downregulated (Formey et al., 2015).

These miRNAs also produced a large number of 21-nucleotide phased siRNAs. NB-LRRs were observed in soybean nodules and common bean nodules from PHAS loci corresponding to protein-coding genes (Formey et al., 2015). The establishment of nodules and rhizobial symbiosis in common beans have both been linked to circRNAs (Wu et al., 2020). According to the scientists, they function as ET during the formation of nodules and the nitrogen fixation process, regulating transmembrane transport and positively regulating kinase activity. The interaction of several miRNAs impacting hormone signaling and regulating different regulatory genes during rhizobial infection, nodule organogenesis, and nitrogen fixation has recently been thoroughly described by Tiwari et al. (2021a) and Hoang et al. (2020). Our knowledge of legume symbiosis at the molecular level will be further enhanced by a full understanding of diverse gene networks, their interactions with regulatory ncRNAs, and their exact roles in regulating nodulation and associated events throughout the symbiosis process.

6. miRNAs controlling nutrient uptake and homeostasis in legumes

In order to grow and develop properly, plants need nutrients by engaging a variety of physiological and molecular systems through their roots and soil (Masclaux-Daubresse et al., 2010). Among these systems, ncRNAs have been shown to play a crucial role in controlling the absorption of several macro- and micronutrients (Fan et al., 2017). One of the most significant factors influencing plant growth and development is nitrogen (N), which is a crucial component of the complete nitrogen metabolism and the source of a number of vital amino acids (Uchida, 2000). RNA sequencing, may explain the multitude of nitrate transporter QTLs, genes, and noncoding RNAs (ncRNAs) regulating N use efficiency (NUE) in plants (Simons et al., 2014). Yet, it is still unknown exactly how plants, especially legumes, maintain equilibrium at the molecular level.

There is evidence that miRNAs regulate plant's reactions to nitrogen (Xu et al., 2011; Islam et al., 2022; Jamla et al., 2023). In *Arabidopsis* grown under nitrogen-limited circumstances, pri-miR156 and pri-miR447c were found to be upregulated, whereas pri-miR169 and pri-miR398a were reported to be downregulated. Hsieh et al. (2009) In situations of nitrogen deprivation, many nitrogen-responsive miRNAs, including miR164, miR169, miR172, and miR397 in maize shoots and miR160, miR167, miR168, and miR169 in maize roots, have been described (Xu et al., 2011). Similarly, a number of miRNAs that respond to nitrogen have been discovered in legume crops (Wang et al., 2013). Using short RNA sequencing of low-nitrogen-tolerant and low-nitrogen-sensitive soybean genotypes, Wang et al. (2013) discovered 168 nitrogen-responsive miRNAs. The research found that under short-term low N, gma-miR2606a/b-3p was downregulated in the roots of a low-nitrogen-tolerant genotype and upregulated in the roots of a low-nitrogen-sensitive genotype. Nevertheless, during short-term low N stress, gma-miR396b/c/d/f/g-5p was upregulated in the shoots of low-nitrogen-sensitive genotype and downregulated in the shoots of low-nitrogen-tolerant genotype (Wang et al., 2013).

The second-most critical macronutrient for fundamental biochemical and metabolic functions of plants, including legumes, is phosphorus (P) (Rausch and Bucher, 2002). Inorganic phosphate is typically the type of P that plants absorb (Pi). P shortage therefore restricts the general growth and development of plants. It has been shown that a number of P-responsive ncRNAs are involved in the development of different plant species (Chiou et al., 2006; Pant et al., 2008; Yuan et al., 2016; Wang et al., 2017; Lv et al., 2020). Similarly, previously P-responsive miRNAs have been identified in a variety of plants (Boualem et al., 2008; Valdés-López et al., 2010; Zeng et al., 2010; Gao et al., 2016). Pi homeostasis is regulated by a number of conserved regulatory miRNAs, including miR399 (Huen et al., 2017) and miR156, miR169, and miR2111 (Hsieh et al., 2009). In the roots of *M. sativa* under low Pi stress, Li et al. (2018) validated the inductive roles of miR399 targeting phosphate transporter genes and miR398 targeting copper chaperone for superoxide dismutase (SOD). MiR-156, which targets the *SPL* TF, MiR-159, which targets the *MYB* TF, MiR-160, which targets the auxin response factor TF, MiR-171, which targets the *GRAS* TF, and MiR-2643 were downregulated. In Arabidopsis, a molecular mechanism involving the IPS1 lncRNA acting as an endogenous target mimicry (eTM) transgene for miRNA399 targeting the Pi overaccumulator *pho2* gene and regulating Pi homeostasis has been identified (Franco-Zorrilla et al., 2007). In *M. sativa*, overexpression of the *PHR1* gene and inhibition of the *pho2* gene (encoding a transcript that causes Pi transporter degradation) under low Pi circumstances allowed for significant Pi uptake (Wang et al., 2017). Medtr1g074930, a Pi transporter gene, was able to transport more Pi in low Pi circumstances thanks to the upregulation of phosphate deficiency-induced lncRNAsPDIL2 and PDIL3. Nevertheless, PDIL1 lncRNA acts as a target mimic for miR399 and prevents MtPHO2 transcripts from being degraded, which might block the Pi transport gene and Pi uptake (Wang et al., 2017). Lv et al. (2020) used transcriptome sequencing to identify 120 differentially expressed circRNAs in two distinct P-responsive soybean genotypes grown at various P concentrations to further understand the function of circRNAs that respond to P. The proposed function of the differentially expressed circRNAs is linked to nucleoside binding, organic material catabolic processes, and oxidoreductase activity, according to gene ontology (GO) enrichment study Lv et al. (2020). It may be possible to target low-responsive circRNAs to increase the efficiency with which soybeans utilize phosphorus. Hence, Pi homeostasis in plants is mostly regulated by a complex network of ncRNAs and their matching target gene(s).

7. Conclusion

Legumes are a valuable source of protein, fiber, and minerals for living beings, with high nutritional value. Legume's ecological and agricultural significance is influenced by its capacity to form a symbiotic relationship with organisms that fix nitrogen. Research on symbiotic nitrogen fixation has the long-term objective of gaining sufficient knowledge to enable the transmission of this ability to other, nonleguminous crop plants. But as our knowledge expands, it becomes harder and harder to grasp the biological intricacy of the nodulation mechanism. The miRNAs are one of the key regulators of the nodulation process since they have particular targets at all stages of nodule growth. The discovery of miRNAs has created a fresh opportunity to identify and comprehend the intricate regulatory mechanisms in plants. The miRNAs control hormone homeostasis, nodulation autoregulation, and nodulation start and development. The miRNA movement and involvement in shoot-to-root signaling influence a wide range of processes, including bacterial infiltration, meristem multiplication and differentiation, nitrogen metabolism, and ageing. They also contribute to the spatiotemporal control of phytohormones and their homeostasis. By incorporating root nodule machinery into nonlegumes, it is possible to correct ecological imbalances and achieve agronomic sustainability. A better understanding of the molecular process that stimulates nodulation would either provide candidate genes for the selection of superior cultivars or facilitate the engineering of nitrogen-fixing traits in nonlegume crops with the potential to increase nitrogen assimilation.

References

Adai, A., Johnson, C., Mlotshwa, S., Archer-Evans, S., Manocha, V., Vance, V., Sundaresan, V., 2005. Computational prediction of miRNAs in *Arabidopsis thaliana.* Genome Res. 15, 78–91.

Ayra, L., Reyero-Saavedra, M.D.R., Isidra-Arellano, M.C., Lozano, L., Ramírez, M., Leija, A., Fuentes, S.I., Girard, L., Valdés-López, O., Hernández, G., 2021. Control of the rhizobia nitrogen-fixing symbiosis by common bean MADS-domain/AGL transcription factors. Front. Plant Sci. 12, 679463.

Bartel, D.P., 2004. MicroRNAs: genomics, biogenesis, mechanism, and function. Cell 116 (2), 281–297.

Boualem, A., Laporte, P., Jovanovic, M., Laffont, C., Plet, J., Combier, J.P., Niebel, A., Crespi, M., Frugier, F., 2008. MicroRNA166 controls root and nodule development in Medicago truncatula. Plant J. 54 (5), 876–887.

Chakraborty, S., Valdés-López, O., Stonoha-Arther, C., Ané, J.M., 2022. Transcription factors controlling the rhizobium–legume symbiosis: integrating infection, organogenesis and the abiotic environment. Plant Cell Physiol. 63 (10), 1326–1343.

Chiou, T.J., Aung, K., Lin, S.I., Wu, C.C., Chiang, S.F., Su, C.L., 2006. Regulation of phosphate homeostasis by microRNA in Arabidopsis. Plant Cell 18 (2), 412–421.

Combier, J.P., Frugier, F., De Billy, F., Boualem, A., El-Yahyaoui, F., Moreau, S., Vernié, T., Ott, T., Gamas, P., Crespi, M., Niebel, A., 2006. MtHAP2-1 is a key transcriptional regulator of symbiotic nodule development regulated by microRNA169 in Medicago truncatula. Gene Dev. 20 (22), 3084–3088.

Fan, X., Naz, M., Fan, X., Xuan, W., Miller, A.J., Xu, G., 2017. Plant nitrate transporters: from gene function to application. J. Exp. Bot. 68 (10), 2463–2475.

Figueredo, M.S., Formey, D., Rodríguez, J., Ibáñez, F., Hernández, G., Fabra, A., 2020. Identification of miRNAs linked to peanut nodule functional processes. J. Biosci. 45, 1–7.

Formey, D., Iñiguez, L.P., Peláez, P., Li, Y.F., Sunkar, R., Sánchez, F., Reyes, J.L., Hernández, G., 2015. Genome-wide identification of the Phaseolus vulgaris sRNAome using small RNA and degradome sequencing. BMC Genom. 16 (1), 1–17.

Franco-Zorrilla, J.M., Valli, A., Todesco, M., Mateos, I., Puga, M.I., Rubio-Somoza, I., Leyva, A., Weigel, D., García, J.A., Paz-Ares, J., 2007. Target mimicry provides a new mechanism for regulation of microRNA activity. Nat. Genet. 39 (8), 1033–1037.

Fu, H., Tie, Y., Xu, C., Zhang, Z., Zhu, J., Shi, Y., Jiang, H., Sun, Z., Zheng, X., 2005. Identification of human fetal liver miRNAs by a novel method. FEBS Lett. 579 (17), 3849–3854.

Gangwar, P., Johri, P., Datta, R., Singh, S., Trivedi, M., 2023. Overview on 1 phytopharmaceuticals and biotechnology of herbal plants. In: Phytopharmaceuticals and Biotechnology of Herbal Plants, pp. 1–17. CRC Press.

Gao, R., Austin, R.S., Amyot, L., Hannoufa, A., 2016. Comparative transcriptome investigation of global gene expression changes caused by miR156 overexpression in Medicago sativa. BMC Genom. 17 (1), 1–15.

Gautrat, P., Laffont, C., Frugier, F., 2020. Compact root architecture 2 promotes root competence for nodulation through the miR2111 systemic effector. Curr. Biol. 30 (7), 1339–1345.

Guo, G., Liu, Q., Smith, N.A., Liang, G., Wang, M.B., 2016. RNA silencing in Plants: mechanisms, technologies and applications in horticultural crops. Cur. Genomics 1, 476–489.

Hoang, N.T., Tóth, K., Stacey, G., 2020. The role of microRNAs in the legume–Rhizobium nitrogen-fixing symbiosis. J. Exp. Bot. 71 (5), 1668–1680.

Holt, D.B., Gupta, V., Meyer, D., Abel, N.B., Andersen, S.U., Stougaard, J., Markmann, K., 2015. Micro RNA 172 (miR172) signals epidermal infection and is expressed in cells primed for bacterial invasion in Lotus japonicus roots and nodules. New Phytol. 208 (1), 241–256.

Hossain, M.S., Hoang, N.T., Yan, Z., Tóth, K., Meyers, B.C., Stacey, G., 2019. Characterization of the spatial and temporal expression of two soybean miRNAs identifies SCL6 as a novel regulator of soybean nodulation. Front. Plant Sci. 10, 475.

Hsieh, L.C., Lin, S.I., Shih, A.C.C., Chen, J.W., Lin, W.Y., Tseng, C.Y., Li, W.H., Chiou, T.J., 2009. Uncovering small RNA-mediated responses to phosphate deficiency in Arabidopsis by deep sequencing. Plant Physiol. 151 (4), 2120–2132.

Huen, A.K., Rodriguez-Medina, C., Ho, A.Y.Y., Atkins, C.A., Smith, P.M.C., 2017. Long-distance movement of phosphate starvation-responsive microRNAs in Arabidopsis. Plant Biol. 19 (4), 643–649.

Islam, W., Tauqeer, A., Waheed, A., Zeng, F., 2022. MicroRNA mediated plant responses to nutrient stress. Int. J. Mol. Sci. 23 (5), 2562.

Jamla, M., Joshi, S., Patil, S., Tripathi, B.N., Kumar, V., 2023. MicroRNAs modulating nutrient homeostasis: a sustainable approach for developing biofortified crops. Protoplasma 260 (1), 5–19.

Khare, P., Kishore, K., Sharma, D.K., 2017. A study on the standardization parameters of Cassia angustifolia. Asian J. Pharmaceut. Clin. Res. 10 (7), 329–332.

Lauressergues, D., Delaux, P.M., Formey, D., Lelandais-Brière, C., Fort, S., Cottaz, S., Bécard, G., Niebel, A., Roux, C., Combier, J.P., 2012. The microRNA miR171h modulates arbuscular mycorrhizal colonization of Medicago truncatula by targeting NSP2. Plant J. 72 (3), 512–522.

Lee, R.C., Feinbaum, R.L., Ambros, V., 1993. The *C. elegans* heterochronic gene lin-4 encodes small RNAs with antisense complementarity to lin-14. Cell 75 (5), 843–854.

Li, H., Deng, Y., Wu, T., Subramanian, S., Yu, O., 2010. Misexpression of miR482, miR1512, and miR1515 increases soybean nodulation. Plant Physiol. 153 (4), 1759–1770.

Li, L., Li, Q., Davis, K.E., Patterson, C., Oo, S., Liu, W., Liu, J., Wang, G., Fontana, J.E., Thornburg, T.E., Pratt, I.S., 2021. Response of root growth and development to nitrogen and potassium deficiency as well as microRNA-mediated mechanism in peanut (*Arachis hypogaea* L.). Front. Plant Sci. 12, 695234.

Li, Z., Xu, H., Li, Y., Wan, X., Ma, Z., Cao, J., Li, Z., He, F., Wang, Y., Wan, L., Tong, Z., 2018. Analysis of physiological and miRNA responses to Pi deficiency in alfalfa (*Medicago sativa* L.). Plant Mol. Biol. 96, 473–492.

Lv, L., Yu, K., Lu, H., Zhang, X., Liu, X., Sun, C., Xu, H., Zhang, J., He, X., Zhang, D., 2020. Transcriptome-wide identification of novel circular RNAs in soybean in response to low-phosphorus stress. PLoS One 15 (1) p.e0227243.

Maddala, V.K.S., Singh, S., Bandi, J., 2024. Soil properties influencing nutrient signaling in plant soil systems. In: Essential Minerals in Plant-Soil Systems, pp. 373–393. Elsevier.

Masclaux-Daubresse, C., Daniel-Vedele, F., Dechorgnat, J., Chardon, F., Gaufichon, L., Suzuki, A., 2010. Nitrogen uptake, assimilation and remobilization in plants: challenges for sustainable and productive agriculture. Ann. Bot. 105 (7), 1141−1157.

Meddya, S., Meshram, S., Sarkar, D., Datta, R., Singh, S., Avinash, G., Thulasinathan, T., 2023. Plant stomata: an unrealized possibility in plant defense against invading pathogens and stress tolerance. Plants 12 (19), 3380.

Nandety, R.S., Wen, J., Mysore, K.S., 2023. Medicago truncatula resources to study legume biology and symbiotic nitrogen fixation. Fundam. Res. 3 (2), 219−224.

Okuma, N., Soyano, T., Suzaki, T., Kawaguchi, M., 2020. MIR2111-5 locus and shoot-accumulated mature miR2111 systemically enhance nodulation depending on HAR1 in Lotus japonicus. Nat. Commun. 11 (1), 5192.

Pant, B.D., Buhtz, A., Kehr, J., Scheible, W.R., 2008. MicroRNA399 is a long-distance signal for the regulation of plant phosphate homeostasis. Plant J. 53 (5), 731−738.

Park, W., Li, J., Song, R., Messing, J., Chen, X., 2002. CARPEL FACTORY, a Dicer homolog, and HEN1, a novel protein, act in microRNA metabolism in *Arabidopsis thaliana*. Curr. Biol. 12 (17), 1484−1495.

Parveen, R., Khare, N., Singh, S., Johri, P., 2023. An in silico outlook for the detection and surveillance of evolving and persistent plant pathogens. In: Detection, Diagnosis and Management of Soil-Borne Phytopathogens. Springer Nature Singapore, Singapore, pp. 17−40.

Patani, A., Patel, A., Prajapati, D., Khare, N., Singh, S., 2023. Various agriculture crop plant-based bioactive compounds and their use in nanomaterial synthesis and applications. In: Nanomaterials from Agricultural and Horticultural Products. Springer Nature Singapore, Singapore, pp. 223−241.

Patani, A., Prajapati, D., Shukla, K., Patel, M., Patani, P., Patel, A., Singh, S., 2024. Environmental stress−induced alterations in the micro-and macronutrients status of plant. In: Essential Minerals in Plant-Soil Systems. Elsevier, pp. 181−195.

Rausch, C., Bucher, M., 2002. Molecular mechanisms of phosphate transport in plants. Planta 216, 23−37.

Roy, S., Liu, W., Nandety, R.S., Crook, A., Mysore, K.S., Pislariu, C.I., Frugoli, J., Dickstein, R., Udvardi, M.K., 2020. Celebrating 20 years of genetic discoveries in legume nodulation and symbiotic nitrogen fixation. Plant Cell 32 (1), 15−41.

Simons, M., Saha, R., Guillard, L., Clément, G., Armengaud, P., Cañas, R., Maranas, C.D., Lea, P.J., Hirel, B., 2014. Nitrogen-use efficiency in maize (*Zea mays* L.): from 'omics' studies to metabolic modelling. J. Exp. Bot. 65 (19), 5657−5671.

Singh, A., Sengar, R.S., Singh, A.K., Singh, S., Kumar, A., Ahmad, K., 2019. Impact of genetic engineering on agriculture food crops. Biotech Today: Int. J. Biol. Sci. 9 (1), 17−24.

Subramanian, S., Fu, Y., Sunkar, R., Barbazuk, W.B., Zhu, J.K., Yu, O., 2008. Novel and nodulation-regulated microRNAs in soybean roots. BMC Genom. 9, 1−14.

Sun, Z., Jiang, S., Wang, D., Li, L., Liu, B., Ran, Q., Hu, L., Xiong, J., Tang, Y., Gu, X., Wu, Y., 2023. Single-cell RNA-seq of Lotus japonicus provide insights into identification and function of root cell types of legume. J. Integr. Plant Biol.

Sunny, J.S., Snijesh, V.P., Singh, S., 2016. Potential role of mirnas in rheumatoid arthritis. Austin Arthritis 1 (2), 1009.

Szittya, G., Moxon, S., Santos, D.M., Jing, R., Fevereiro, M.P., Moulton, V., Dalmay, T., 2008. High-throughput sequencing of Medicago truncatula short RNAs identifies eight new miRNA families. BMC Genom. 9, 1−9.

Tiwari, M., Pandey, V., Singh, B., Bhatia, S., 2021a. Dynamics of miRNA mediated regulation of legume symbiosis. Plant Cell Environ. 44 (5), 1279–1291.

Tiwari, M., Singh, B., Yadav, M., Pandey, V., Bhatia, S., 2021b. High throughput identification of miRNAs reveal novel interacting targets regulating chickpea-rhizobia symbiosis. Environ. Exp. Bot. 186, 104469.

Treiber, T., Treiber, N., Meister, G., 2019. Regulation of microRNA biogenesis and its cross talk with other cellular pathways. Nat. Rev. Mol. Cell Biol. 20, 5–20.

Tungare, K., Shahu, R., Zambare, V., Agarwal, P., Jobby, R., Nisar, N., Jha, P., 2022. Toxicity mitigation of textile dye reactive blue 4 by hairy roots of Helianthus annuus and testing its effect in in vivo model systems. BioMed Res. Int. 2022.

Uchida, R., 2000. Essential nutrients for plant growth: nutrient functions and deficiency symptoms. Plant Nutr. Manag. Hawaii's Soils 4, 31–55.

Valdés-López, O., Yang, S.S., Aparicio-Fabre, R., Graham, P.H., Reyes, J.L., Vance, C.P., Hernández, G., 2010. MicroRNA expression profile in common bean (*Phaseolus vulgaris*) under nutrient deficiency stresses and manganese toxicity. New Phytol. 187 (3), 805–818.

Verma, N., Sao, P., Srivastava, A., Singh, S., 2021. Physiological and molecular responses to drought, submergence and excessive watering in plants. In: Harsh Environment and Plant Resilience: Molecular and Functional Aspects, pp. 305–321.

Vora, Z., Pandya, J., Vaikuntapu, P.R., Singh, S., Asha, S., Purama, R.K., 2023. Genome-wide association of defense hormone crosstalk in plants. In: Hormonal Cross-Talk, Plant Defense and Development. Academic Press, pp. 353–371.

Wang, T., Zhao, M., Zhang, X., Liu, M., Yang, C., Chen, Y., Chen, R., Wen, J., Mysore, K.S., Zhang, W.H., 2017. Novel phosphate deficiency-responsive long non-coding RNAs in the legume model plant Medicago truncatula. J. Exp. Bot. 68 (21–22), 5937–5948.

Wang, Y., Wang, L., Zou, Y., Chen, L., Cai, Z., Zhang, S., Zhao, F., Tian, Y., Jiang, Q., Ferguson, B.J., Gresshoff, P.M., 2014. Soybean miR172c targets the repressive AP2 transcription factor NNC1 to activate ENOD40 expression and regulate nodule initiation. Plant Cell 26 (12), 4782–4801.

Wang, Y., Wang, Z., Amyot, L., Tian, L., Xu, Z., Gruber, M.Y., Hannoufa, A., 2015. Ectopic expression of miR156 represses nodulation and causes morphological and developmental changes in Lotus japonicus. Mol. Genet. Genom. 290, 471–484.

Wang, Y., Zhang, C., Hao, Q., Sha, A., Zhou, R., Zhou, X., Yuan, L., 2013. Elucidation of miRNAs-mediated responses to low nitrogen stress by deep sequencing of two soybean genotypes. PLoS One 8 (7), e67423.

Wightman, B., Ha, I., Ruvkun, G., 1993. Posttranscriptional regulation of the heterochronic gene lin-14 by lin-4 mediates temporal pattern formation in *C. elegans*. Cell 75, 855–862.

Wu, Z., Huang, W., Qin, E., Liu, S., Liu, H., Grennan, A.K., Liu, H., Qin, R., 2020. Comprehensive identification and expression profiling of circular RNAs during nodule development in Phaseolus vulgaris. Front. Plant Sci. 11, 587185.

Xu, Y., Cheng, L., Chen, J., Lu, Y., Qin, Y., Yan, Y., Liu, F., Tan, J., 2023. Genome-wide investigation on metal tolerance protein (MTP) genes in leguminous plants: Glycine max, Medicago truncatula, and Lotus japonicus. Acta Physiol. Plant. 45 (1), 18.

Xu, Z., Zhong, S., Li, X., Li, W., Rothstein, S.J., Zhang, S., Bi, Y., Xie, C., 2011. Genome-wide identification of microRNAs in response to low nitrate availability in maize leaves and roots. PLoS One 6 (11), e28009.

Yan, Z., Hossain, M.S., Arikit, S., Valdés-López, O., Zhai, J., Wang, J., Libault, M., Ji, T., Qiu, L., Meyers, B.C., Stacey, G., 2015. Identification of micro RNA s and their mRNA targets during soybean nodule development: functional analysis of the role of miR393j-3p in soybean nodulation. New Phytol. 207 (3), 748–759.

Yan, Z., Hossain, M.S., Valdés-López, O., Hoang, N.T., Zhai, J., Wang, J., Libault, M., Brechenmacher, L., Findley, S., Joshi, T., Qiu, L., 2016. Identification and functional characterization of soybean root hair micro RNA s expressed in response to *Bradyrhizobium japonicum* infection. Plant Biotechnol. J. 14 (1), 332–341.

Yang, J., Lan, L., Jin, Y., Yu, N., Wang, D., Wang, E., 2022. Mechanisms underlying legume–rhizobium symbioses. J. Integr. Plant Biol. 64 (2), 244–267.

Yuan, J., Zhang, Y., Dong, J., Sun, Y., Lim, B.L., Liu, D., Lu, Z.J., 2016. Systematic characterization of novel lncRNAs responding to phosphate starvation in *Arabidopsis thaliana*. BMC Genom. 17 (1), 1–16.

Yun, J., Sun, Z., Jiang, Q., Wang, Y., Wang, C., Luo, Y., Zhang, F., Li, X., 2022. The miR156b-GmSPL9d module modulates nodulation by targeting multiple core nodulation genes in soybean. New Phytol. 233 (4), 1881–1899.

Zeng, H.Q., Zhu, Y.Y., Huang, S.Q., Yang, Z.M., 2010. Analysis of phosphorus-deficient responsive miRNAs and cis-elements from soybean (Glycine max L.). J. Plant Physiol. 167 (15), 1289–1297.

Further readings

Chand Jha, U., Nayyar, H., Mantri, N., Siddique, K.H., 2021. Non-coding RNAs in legumes: their emerging roles in regulating biotic/abiotic stress responses and plant growth and development. Cells 10 (7), 1674.

Han, H., Zhou, Y., 2022. Function and regulation of microRNA171 in plant stem cell homeostasis and developmental programing. Int. J. Mol. Sci. 23 (5), 2544.

Sós-Hegedűs, A., Domonkos, Á., Tóth, T., Gyula, P., Kaló, P., Szittya, G., 2020. Suppression of NB-LRR genes by miRNAs promotes nitrogen-fixing nodule development in Medicago truncatula. Plant Cell Environ. 43 (5), 1117–1129.

Turner, M., Yu, O., Subramanian, S., 2012. Genome organization and characteristics of soybean microRNAs. BMC Genom. 13, 1–16.

CHAPTER

14 Engineered artificial MicroRNA in plant: an overview

Amol D. Gholap[1], Navnath T. Hatvate[2], Deblina Bhowmik[2], Sampada D. Sawant[1], Sonali A. Gholap[3], Kiran M. Londhe[4], Satish Rojekar[5], Md Faiyazuddin[6]

[1]*Department of Pharmaceutics, St. John Institute of Pharmacy and Research, Palghar, Maharashtra, India;* [2]*Institute of Chemical Technology Mumbai, Marathwada Campus, Jalna, Maharashtra, India;* [3]*Department of Data Analytics, CRISIL House, Mumbai, Maharashtra, India;* [4]*Department of Botany, Dr. Babasaheb Ambedkar Marathwada University, Chhatrapati Sambhajinagar, Maharashtra, India;* [5]*Department of Pharmacological Sciences, Icahn School of Medicine at Mount Sinai, New York, NY, United States;* [6]*School of Pharmacy, Al-Karim University, Katihar, Bihar, India*

1. Introduction

The severe acute respiratory syndrome coronavirus II that caused the pandemic sadly reminded the world that RNA is a potent agent of both illness and therapeutics. Research into RNA and its structure will inevitably advance, but sometimes this quick advancement has been accompanied by a loss of common understanding regarding crucial fundamental aspects of RNA structure. The "space" between the bases within a helix is typically depicted by the convenient and helpful ball-and-stick representations of RNA (and DNA) or "ladders" (Vicens and Kieft, 2022).

RNA is similar in structure to DNA but has a single-stranded helical structure, while DNA is double-stranded. RNA comprises nucleotides containing a sugar molecule, a nitrogenous base, and a phosphate group. There are three main types of RNA: messenger RNA (mRNA), transfer RNA (tRNA), and ribosomal RNA (rRNA). mRNA carries genetic information from the DNA in the cell nucleus to the ribosomes, where it is translated into protein. tRNA helps to decode the genetic information in mRNA and brings the appropriate amino acids to the ribosome to form the protein. rRNA makes up the ribosome, which is the site of protein synthesis. RNA is also involved in other biological processes, such as RNA interference (RNAi), which regulates gene expression by targeting specific mRNA molecules for degradation, and catalytic RNA (ribozymes), which can perform chemical reactions like enzymes. RNA plays a crucial role in the central dogma of molecular biology, which states that DNA is transcribed into RNA, which is then translated into protein as presented in Fig. 14.1. This process is essential for the functioning of all living organisms (Cooper et al., 2009; Kukurba and Montgomery, 2015).

MicroRNA Advances and Application in Plant Biology. https://doi.org/10.1016/B978-0-443-21682-4.00013-0

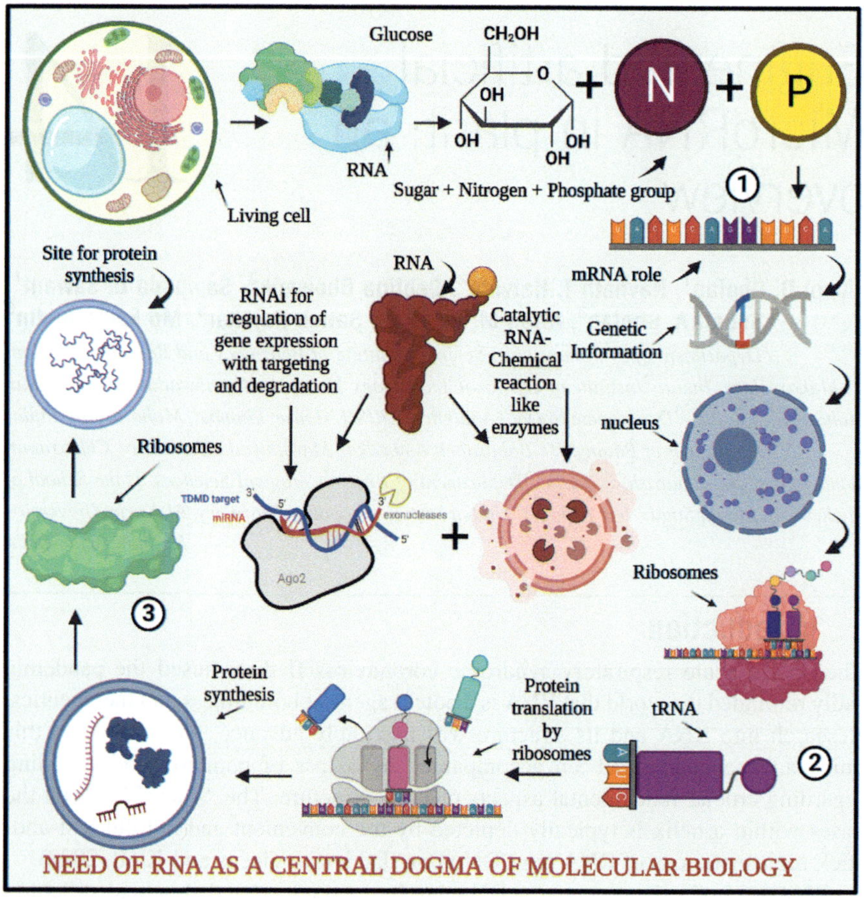

FIGURE 14.1

The RNA is made up of single stranded helical structure containing sugar molecule, nitrogen base along with phosphate group in it. There are three type of the RNA called mRNA, tRNA and rRNA required for central dogma of molecular biology. The mRNA is required for decoding of the genetic information in mRNA and translated into the protein while the mRNA is essential for taking out genetic information from DNA into the nucleus and transfer into that of the ribosomes. The RNA contains ribosomes and they are the site of protein synthesis. The RNAi (interference) is actual suite of the synthesis while catalytic RNA acts as ribozymes required for the chemical reaction like that of the enzymes.

MicroRNAs (miRNAs), which are small noncoding RNA molecules, play crucial regulating roles in the activation of genes. They typically have a length of 21–25 nucleotides. MiRNAs must first be developed by the cellular apparatus after being translated from DNA in order to silence a gene. When mature, miRNAs bind to specific messenger RNAs (mRNAs) and stop those mRNAs from turning into proteins (Kidner and Martienssen, 2005).

MiRNAs are produced by the transcription of DNA, and they then go through several processing steps to become adult miRNAs. The RNA-induced silencing complex (RISC), a protein complex that can recognize and attach to specific mRNA molecules, is then formed using the mature miRNA. After binding to the target mRNA, the miRNA-RISC complex can stop mRNA from being transcribed into protein by either promoting mRNA destruction or stopping mRNA translation. This process, known as gene silencing, may result in a decrease in the production of the specific gene as presented in Fig. 14.2 (Jones-Rhoades et al., 2006).

Numerous types of research have demonstrated that miRNAs play a significant character in controlling metabolic homoeostasis. For the control of *C. elegans*' typical growth, Lee et al. first reported the functions of miRNAs in 1993 (Lee et al., 1993). A few cellular processes, including cell differentiation, growth, and disease, have been shown to depend heavily on miRNAs. The production of miRNAs has been found to be dysregulated in several illnesses, including cancer, cardiovascular disease, and neurological problems. There are ongoing efforts to comprehend the functions of miRNAs and the potential medicinal applications of this area of research (Berezikov et al., 2006).

Recent studies have also suggested that miRNAs, which can influence gene translation by being generated by cells and taken up by other cells, may play a role in intercellular communication (Gholap et al., 2023a, 2023b, 2023c, 2023d, 2024a). Attention has been drawn to the potential of using miRNAs as medicinal and diagnostic goals for a variety of diseases. Overall, research is being done to better comprehend the functions of miRNAs, how they may be used to diagnose and treat diseases, and other vital fields of study.

2. Why amiRNAs were needed, design and advantages?

One of the extremely preserved processes in the plant kingdom is the control of gene expression provided by miRNA. According to studies, the synthesis of miRNAs was not hampered by mutation in the 21 nt mature patterns in the hairpin stem loop (Sablok et al., 2011). These findings created a fresh opportunity that paved the way for the development of amiRNA technology (Schwab et al., 2006). With the aid of all the essential enzymes enlisted in the synthesis of endogenous miRNA, alteration of miRNA/miRNA arrangements in the duplex of hairpin stem loop of endogenous miRNA might result in mature amiRNAs (Sunny et al., 2016).

Comparative modeling-based gene silencing has become a practical method for suppressing gene transcription so that their activities may be investigated. For this goal, many antisense or short interrupting RNA-based gene silencing approaches are employed regularly in plant study (Gholap et al., 2024b; Verma et al., 2021; Ajay Khedkar Dattatray Tupe, 2012; Amol and Rajurkar, 2012a; Singh et al., 2019; Vora et al., 2023). One of these methods that might be used as a device in

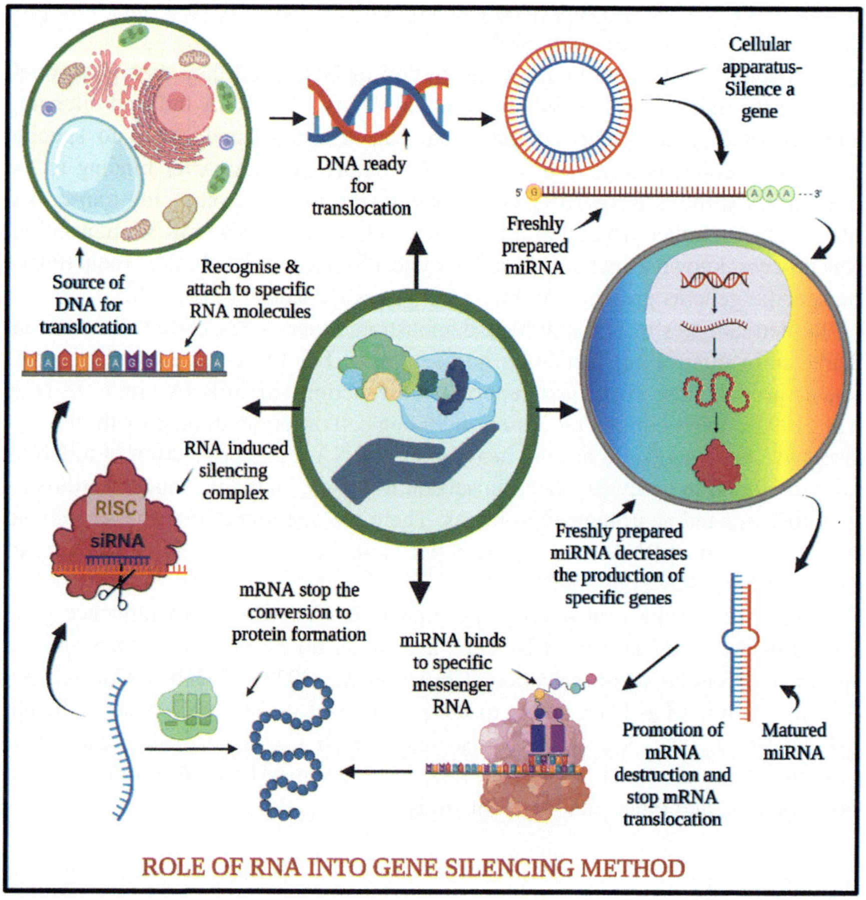

FIGURE 14.2

The DNA of the cells undergoes translocation to give cellular apparatus. The freshly prepared miRNA is induced from this process, which is required for the decrease into the production of specific genes. After the attainment of the maturity the matured miRNA get binded to specific messenger mRNA, which later induced silencing complex and recognizes increase into the attachment to specific RNA molecules. These are used for the promotion mRNA destruction and stopping the mRNA translocation.

functional genomics is gene silencing mediated by artificial microRNAs (amiRNAs). AmiRNAs are single-stranded, 21 nt long, and comparable with microRNAs. They are created by swapping out the adult miRNA duplex segments in pre-miRNAs. These amiRNAs deregulate target expression through the short RNA biogenesis and silencing mechanism.

AmiRNAs are created by replacing the miRNA region with a particular miRNA that is equivalent to the intended target sequence, using an endogenous miRNA precursor as a structural support (Ossowski et al., 2008). If the secondary fold-back structure of the new precursor is preserved, its biogenesis will be identical to that of the endogenous one. As a result, any specific gene, or even many genes, may be targeted by amiRNAs for functional gene research, and an amiRNA sequence with low resemblance to any undesirable plant genes can be chosen to prevent off-target effects (Singh et al., 2019). The silencing action is stable throughout several generations because they are transacting (Molnar et al., 2009). Because of their tiny size, amiRNAs may also be used to create structures that express several unrelated amiRNAs and to target specific alleles or splice variants for a given gene (Patani et al., 2023). Also, it has been asserted that when used in agriculture, amiRNAs are less likely than other methods to compromise biosafety or the atmosphere (Duan et al., 2008). Additional benefits include the potential for synergism with noncleavable targets, tissue expression patterns, and strand-specific silencing as presented in Fig. 14.3 (Liu and Chen, 2010).

AmiRNA technology, which adheres to different improvements, has several benefits over other gene silencing techniques. This is a strong and effective method that might be used to translate findings for enhancing beneficial features in plants as well as to uncover new understanding of metabolic pathways and gene activities across numerous disciplines. This paper outlines the history of short RNAs, advancements in RNAi-based gene silencing, the effects of amiRNA on gene silence, and potential upcoming directions for this technology's application in plant research.

Compared with hpRNAi, the amiRNA technology still has several benefits, most notably a reduction in off-target impacts and efficacy. Additionally, the resilience of this technology is increased by the ease with which amiRNA sequences can be optimized for addressing one or more mRNAs without interfering with the production of other genes simultaneously with a single amiRNA (Schwab et al., 2006). Several studies were also carried out to evaluate the accuracy of amiRNA by means of genome-wide expression assessments, which showed that amiRNAs are a fairly precise technique for suppressing genes and destroy genes without altering the activity of other genes (Khraiwesh et al., 2008).

3. Computational tools and resources for identifying plant miRNAs and their targets

miRDB: The miRDB is the database for the prediction and targeting of the microRNA. The web server application also allows users to provide their own sequences for the identification of miRNA targets, giving them the freedom to research any unique miRNAs or aimed mRNAs of interest. MirTarget, which was developed by analyzing high-throughput profile data derived from CLIP-RNA ligation sequencing investigations, predicts the targets. The functional miRNA annotation is a new component of miRDB. The creators have worked extremely hard to integrate computational studies with literature curation, resulting in the finding of 568

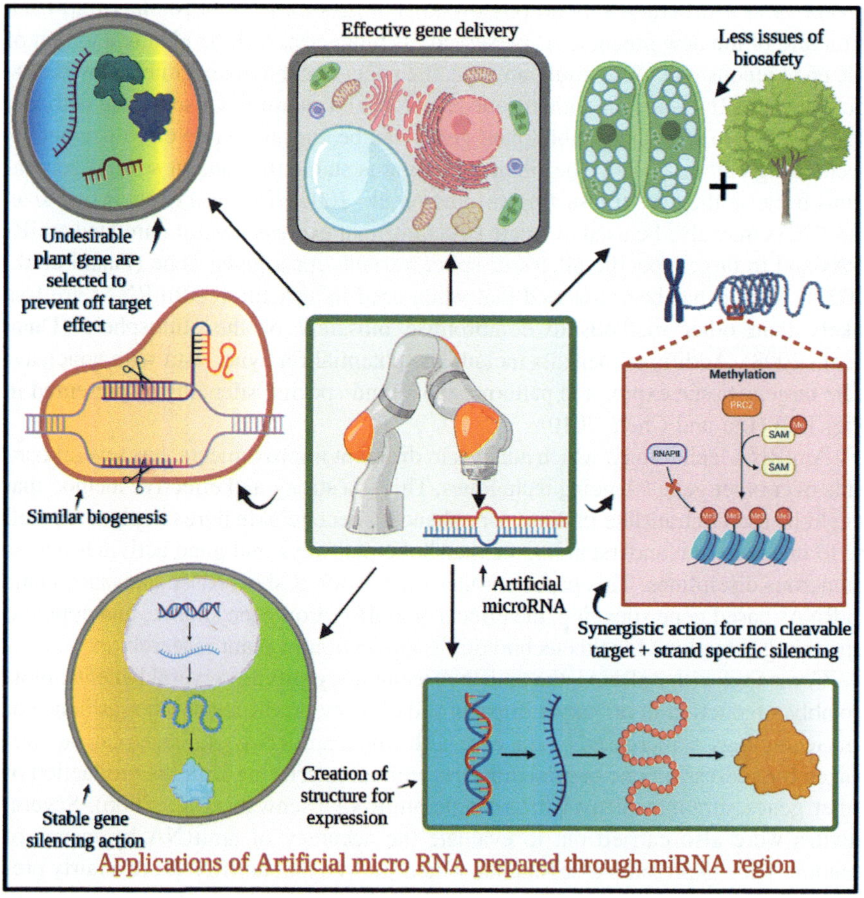

FIGURE 14.3

The AmiRNA is prepared by replacement of the miRNA region by using particular miRNA equivalent target sequence by implementing endogenous miRNA precursors. These are carrying many advantages similar biogenesis, undesirable plan genes selected to prevent off target effect. It also gives less biosafety and atmospheric issues and also gives synergistic actions for noncleavable target and induced tissue expression pattern. These also used gene silencing method and also give stable silencing action for several generations.

functional miRNAs in humans. This program utilizes 30 compensatory pairings, local AU content, free energy, target-site availability, machine learning, evolutionary conservation, and seed sequence complementarity (Wong and Wang, 2015).

Various computational tools are used to determine the miRNAs in the plant. Some of those resources used in plants are:

TAPIR: Plant-specific miRNA targets are predicted and stored in the TAPIR database (Parveen et al., 2023).

PMRD: PMRD is a database of plant miRNAs that enables screening for genomes and comprises the patterns of miRNAs and their target genes, secondary constructions, and expression profiles. Also, it keeps extensive records of plant miRNA data (Cui et al., 2012).

miRbase: Among the most significant public databases containing data on miRNAs, miRBase is an all-inclusive database that offers available miRNA sequence data, annotation, gene target predictions, and other data (Kozomara and Griffiths-Jones, 2014).

TarBase: Over 500,000 verified interactions have been added to the DIANA-TarBase v7.0 database by mechanical curation of the reported and empirically verified miRNA:target interactions. In comparison with other relevant databases, TarBase v8.0 has 9 to 250 times more items as of October 2017 and has 1,077,279 empirically confirmed interactions collected from 1165 papers (Vlachos et al., 2015).

miRExpress: An efficient and cutting-edge technique to identify linear and nonlinear correlations for paired gene expression data in immune cells is made possible by the MIrExpress database (Wang et al., 2009). Three phases make up the procedure by which miRExpress creates the miRNA expression designs. Preprocessing the primary data generated by second-generation sequencing is the initial step. All sequencing reads align with known mature miRNAs in the second stage. The third stage is creating miRNA expression profiles from the alignment's results (Wang et al., 2015).

miRDeep-P: A scoring system and filtering parameters unique to plants were added to the original miRDeep, which is based on a probabilistic model of miRNA synthesis in animals. Users of miRDeep-P can investigate the expression patterns of identified miRNA genes and find new ones. Finding miRNA genes in plant species without extensive annotation is one of miRDeep-most P's useful applications (Yang and Li, 2011). MiRDeep-P has the benefit of producing stem-looped precursor sequences and their placement in reference sequences, which may be used to recognize specific miRNA genes in addition to the arrangements of the putative miRNAs. The tool was created primarily to overcome the difficulties of decoding plant NGS data since plants have more varied precursor lengths and more common big paralogous family sequences (Kuang et al., 2019).

miRDeepFinder: In order to recognize and functionally study plant microRNAs (miRNAs) and their targets using small RNA datasets collected through deep sequencing, the software program miRDeepFinder was created. Preprocessing of raw data, identification of preserved miRNAs, mining and classification of new miRNAs, miRNA expression profiling, prediction of miRNA targets, and evaluation of miRNA-involved gene pathways and gene networks are some of the features offered by miRDeepFinder (Xie et al., 2012a).

Triplet-SVM: Utilizing characteristics from the structure and triplet sequence, triplet-SVM categorizes the genuine and fake microRNA precursors. The miRNA registration database and the pseudomiRNA datasets from the protein coding regions were gathered for the positive training dataset (Sumaira et al., 2016).

PlantMiRNAPred: Another well-known instrument for predicting plant-specific pre-miRNAs is PlantMiRNAPred. When using five knowledge-based energy characters individually or in combination, the training dataset for PlantMiRNAPred, which contains 980 genuine pre-miRNAs and 980 pseudo pre-miRNAs, was chosen. The testing samples used in PlantMiRNAPred were immediately entered into plantMirP for the forecast based on the trained SVM models for plantMirP. The testing datasets are divided into three sections: known pre-miRNAs from eight different classes to test the accuracy of categorizing real pre-miRNAs; updated datasets to test the accuracy of identifying new pre-miRNAs from plants; and negative testing dataset to test the accuracy of recognizing pseudo hairpins (Yao et al., 2016; Khare et al., 2017).

miRPara: For the study of genome-scale patterns, miRPara was created. The most accurate method for identifying novel miRNAs was MiRPara. MiRNA and pre-miRNA sequences can be predicted using miRPara from DNA or RNA segments of any length. MiRPara is mainly made for use in wet-lab research to screen lengthy sequences for potential miRNAs and to pretest potential miRNAs to narrow the search window for bench-based searches (Wu et al., 2011).

miRnovo: Using a species-precisely training set and genome, Mirnovo was used to forecast miRNAs in each subsample and the unsampled dataset. Both hairpins and mature miRNAs can be predicted by Mirnovo, the latter of which also provides count information. However, the initial and improved sequence clustering stages of the Mirnovo workflow may occasionally be inaccurate and as a result, have a minor impact on the expression data that is produced (Vitsios et al., 2017).

The other computational tools for the prediction of miRNAs in animals, humans, and other organisms are:

PicTar: PicTar lists the focus for both individual and groupings of miRNAs. PicTar relies on tissue-precise target estimation scores from four mammalian tissues to increase prediction specificity and efficiency; nevertheless, it only images the 3′ UTR for target hunting. It depends on seed arrangement complementarity, evolutionary preservation, free energy, and target-site abundance. The most recent update to PicTar occurred in March 2007, and it uses hg19 to anticipate human targets (Krek et al., 2005).

TargetScan: A web-based system called TargetScan makes predictions about miRNA targets by looking for preserved and nonconserved locations. Based on seed arrangement similarity, evolutionary management, free energy, target-site availability, target-site abundance, 30 compensatory pairing, GeU wobble, and local AU content, it determines mRNA goals in the 30 UTR and CDS. The most recent 7.1 version additionally makes use of new miRNA families. One of the tools that is updated the most frequently is TargetScan, which was most recently updated in June 2016 (Agarwal et al., 2015).

miRanda: MiRanda was created utilizing a comparison of miRNA complementarity to coding sequence (CDS) regions and 30 UTR, and it uses mirSVR to score the probability of mRNA downregulation. Utilizing complementary seed sequences, evolutionary preservation, free energy, and GeU wobble that is allowed in the seed area, miRanda can identify the target. The mature miRNA sequences utilized in this program came from miRBase v15.0, despite the fact that it has been cited more than 3000 times (John et al., 2004).

RNAhybrid: In large focus RNAs, RNAhybrid identifies many possible miRNA binding sites by looking for the most energetically advantageous hybridization sites. It anticipates targets grounded on seed sequence, target-site abundance, complementarity, free energy, 30 compensatory pairing, GeU wobble, and complementary pairs and the most recent update was in August 2013 (Krek et al., 2005; Rehmsmeier et al., 2004).

ViTa: The ViTa database compiles and links viral information from miRBase, ICTVdB, VirGen, VBRC, and other databases. The information includes known viral miRNAs and related host miRNA target genes identified by miRanda and TargetScan. Moreover, it offers confirmed annotations, such as human miRNA expression and tissues contaminated with viruses (Hsu et al., 2007).

miWalkis: Including validated loci for their target genes, the centralized database MiRWalk contains analytical data for human, mouse, and rat miRNAs.

miRTarBbase: A database called miRTarBase incorporates connections between targets and miRNAs that have undergone experimental verification (Chou et al., 2018).

miRGen: Animal miRNA target genes identified by integrating many popular target gene prediction programs are stored in the database MiRGen, along with positional correlations among animal miRNAs and genomic regions (Georgakilas et al., 2016).

4. miRNA-based biotechnology for improving plant development, fruit, and seed development as well plant biomass yield

Controlling the formation and proliferation of plants focusing miRNAs will significantly improve plant development and further improve plant biomass and output; this is important, especially for products used to make bioenergy (Gholap et al., 2023e; Gholap et al., 2023f; Anand et al.; Bothiraja et al., 2014; Amol and Rajurkar, 2012b; Gholap and Tupe, 2012; Gholap et al., 2023g; Tupe and Gholap, 2012; Gholap and Dighe, 2012). The various factors of the plant, such as root development, floral development, apical meristem, phage development, biomass production, seeds, fruit development, etc. affect the plant's growth and development (Amol and Musmade, 2011; Amol and Rajurkar, 2012a, 2012b, 2012c, 2012d, 2012e; Amol and Tupe, 2012; Gholap et al., 2011; Gholap, 2012; Rajurkar and Gholap, 2012a, 2012b).

Crop and horticultural scientists are very interested in the function of miRNAs in controlling fruit and seeds development. MiRNAs are essential for seed development, according to newly available evidence. The bulk of current research focuses on identifying miRNAs and forecasting their targets during grain filling. When wheat and rice grains fill up, many miRNAs are released effectively (Piriyapongsa and Jordan, 2008).

Plant biomass output is typically impacted by blooming, especially for annual products. Plant biomass output typically increases when flowering and floral growth are postponed. So, by controlling plant flowering and floral growth, we can increase plant biomass output. The ABC model, which includes several critical genes and one gene expression anomaly that results in the aberrant growth of flowers, can be used to describe the complex biological process of floral control (Luo et al., 2013). Recent research demonstrates that miRNAs play a role in this procedure as well. MiRNAs control the development of flower meristems and the beginning of viral changes (Luo et al., 2013). In addition to controlling the creation of floral meristems, the beginning of phage changes, and floral patterning, miRNAs also play a role in controlling the growth of floral organs. Any tactic that alters the floral shape and flowering time can alter plant development and even plant biomass. A growing body of data indicates that changes in miRNA expression have an impact on plant biomass output, either directly or indirectly. Delaying flowering significantly boosted plant biomass production (Fu et al., 2012) (Table 14.1).

5. miRNA-based biotechnology for improving plant tolerance to environmental stresses

miR156: The first miRNA found in plants, miR156, was discovered and demonstrated to be essential for phase shift and plant growth. When plants were exposed to different environmental stressors, miR156 was abnormally developed. MiR156 activity has been verified using transgenic methods. Plants excessively expressing the miR156 gene demonstrated that miR156 was necessary for the recall of heat stress in plants, and these plants displayed increased heat stress endurance. Additionally, miR156 has been used to increase plant growth. In contrast to their control plants, transgenic switchgrass that is extracted excessively with the proper miR156 generated 58%−101% more plant biomass (42) (59). The miR159: miR159 has a significant impact on how plants grow. One research concluded that transgenic rice plants overexpressing miR159 proved more susceptible to heat stress than the wild-type controls, indicating that down-regulating miR159 may help people be more tolerant of heat stress. miR159 reacted to a variety of environmental stressors (60).

miR169: One of the biggest miRNA groups that is present in all plant species is miR169. Both normal plant growth and a plant's ability to respond to environmental duress are significantly influenced by miR169. After 7 days of drought therapy,

Table 14.1 Summary of plant development due to Artificial miRNA.

Sr. no.	Plant	miRNA	Modification	References
1	*Oryza sativa*	miR397	Improve rice yield by increasing grain size and promoting panicle branching	Zhang et al. (2013a)
		miR156	Enhanced biomass	Xie et al. (2012b)
		miR166	Increase uptake of macronutrients	Iwamoto and Tagiri (2016)
		miR408	Enhanced yield of the grains	
2	*Arabidopsis thaliana*	MiR156	Increased biomass	Schwab et al. (2005)
		miR319	Delayed flowering biomass	Schommer et al. (2014)
		miR398	Cu uptake together with assimilation	Sunkar et al. (2006)
		miR399	Phosphorous accession and translocation	Aung et al. (2006), Bari et al. (2006)
		miR408	Increased biomass	
3	Tomato	miR159	Parthenocarpy	Silva et al. (2017)
		miR858	Increase anthocyanin	Jia et al. (2015)
		miR1917	Increase in fruit ripening	Palatnik et al. (2007)
4	Wheat	miR156	Increased biomass	Liu et al. (2018)
		miR402	Increase Pi acquisition and biomass	Bai et al. (2018)
5	Banana	miR397	Increased biomass	Patel et al. (2019)
6	Strawberry	miR399	Enhance fructose and glucose content	Wang et al. (2017a)
7	Barley	miR172	Affect the yield of cereal grains	Houston et al. (2013)

functional overexpression of miR169 in recombinant tomato plants greatly increased the plant's resistance to drought stress. Additionally, miR169 targets genes involved in assimilating nitrogen. MiR169 is a potential candidate as a result for increasing plant resistance to nitrogen deficiency and stress from drought (Zhang et al., 2011).

miR319: Numerous environmental stressors have been linked to miR319, and research have found that miR319 is typically up regulated under stressful circumstances. According to transgenic research, creeping bentgrass plants with constitutive expression of miR319 had substantially higher resistance to salinity and drought stress and different plant growth (62). In this research, recombinant and wild-type saplings were exposed to 12°C for 2 days then exposed to 4°C for 4 days to acclimatize them to frigid temperatures. When correlated to the wild-type standards, which had a survival rate of 13%, mutant rice that overexpressed miR319 had a persistence degree of about 50%. Together, the findings of these studies imply that miR319 overexpression may be used to increase plants' resistance to a variety of external stressors (63).

miR393: Rice and *Arabidopsis* were able to express miR393 owing to the Cauliflower Mosaic Virus (CaMV) 35S promoter. Plants were treated with salt (150 mM NaCl) and alkalinity (75 mM $NaHCO_3$) for 15 days on transgenic and normal plants. The findings of this research demonstrated that, in contrast to wild-type plants, the T_3 generation seedlings of transgenic Arabidopsis that overexpressed miR393 were more susceptible to salinity and alkalinity treatments. The transgenic rice seedlings showed a comparable trait as well. Overexpression of miR393 substantially reduced root and plant growth in both *Arabidopsis* and rice (Gao et al., 2011).

miR394: A new study in Arabidopsis has shown that miR394 expression is greatly increased by high salinity management and is also influenced by iron and sulfate deficit. MiR394 is an evolutionarily conserved gene. Transgenic Arabidopsis plants that overexpressed miR394 limited water loss during leaf transpiration under dry circumstances, eventually enhancing drought stress resistance (Ni et al., 2012).

miR395: Since there is a single nucleotide variation between the mature miRNA sequences of various members of the miR395 family, various substrates may be modulated during these events. As a result, miR395 can contribute to plants' ability to tolerate heavy metals. MiR395-overexpressed transgenic canola displayed increased resistance to cadmium stress. This was demonstrated by the modified plants experiencing less reactive stress from cadmium than their wild-type counterparts (Zhang et al., 2013b).

miR396: A recognized stress-responsive miRNA is miR396. Transgenic Arabidopsis and rice plants that inherently overexpressed miR396 experienced substantially stunted root growth as well as lowered plant growth and development during salinity and alkalinity stress treatments. Interestingly, under dehydration stress, there were no phenotypic changes between wild-type plants and transgenic Arabidopsis seedlings overexpressing miR396. This indicates that miR396 regulates plant responses to salinity and alkalinity stressors in a negative manner (Gao et al., 2010).

miR402: It has been demonstrated that salt, dehydration, and cold stressors cause miR402 to be up-regulated in Arabidopsis. However, transgenic plants that overexpressed miR402 only showed increased plant development under salinity stress and not under dehydration or cold stress circumstances. Overexpression of miR402 boosted seed sprouting in Arabidopsis under all three pressure circumstances (Kim et al., 2010).

miR417: Stresses related to sodium and thirst also changed miR417 expression. Arabidopsis seed germination and survival were adversely affected by constitutive overexpression of miR417 in the presence of ABA and elevated salt stress (Jung and Kang, 2007).

miR828: A wound-induced miRNA is called miR828. Surprisingly, ethylene, hydrogen peroxide (H_2O_2), methyl jasmonate, or nitric oxide did not increase the transcript of miR828. (NO). However, transgenic sweet potatoes overexpressing pre-miR828 showed a rise in the creation of H_2O_2 and lignin, two substances produced to support plant defense systems (Lin et al., 2012).

6. Unique features of plant miRNAs and difficulties in annotation

The distinct characteristics of miRNAs set them apart from other noncoding sRNAs, primarily siRNA in plants and piRNA in mammals.

While siRNAs vary considerably from 20 to 24 nt or even longer, and piRNAs are typically lengthier than 26 nts, miRNAs are primarily 20 to 22 nt long (Axtell et al., 2007). The successive split through these complexes results in the 2-nt overhangs of the duplex of mature miRNA and miRNA, which are derived from the stem-loop structure, which is resultant from a pri- or pre-miRNA and recognized by merging conjugates centered on Drosha or DCL (Bologna et al., 2013). Northern blot or deep sequencing techniques can be used to identify mature miRNAs, which were recognized and protected by the RISC complex (Zhang et al., 2007). A combined amount of 40% of the miRNAs in plants are greater than 100 nts, compared with >95% of the miRNAs in animals, through four plant and three animal model taxa as examples. Additionally, none of the animal miRNAs fall within the 6%–11% spectrum of plant miRNAs that are lengthier than 200 nt.

7. Evolution of plant miRNA annotation methods and criteria

Numerous labelling techniques, as well as research into the biogenesis route and features of plant miRNAs, have been used. To find and identify miRNAs, researchers typically employ one of three methods:

(1) experimental techniques, such as the northern blot, which draw some of their inspiration from the fact that many plant miRNAs have been preserved across species;
(2) sequencing technologies, such as deep arrangement, which are typically used in conjunction with subsequent bioinformatic analyses; and
(3) pure computational techniques for analyzing features such as stem-loop structure, sequence features. The development of deep sequencing, one of these techniques, greatly aided the labelling of plant miRNAs (Kozomara and Griffiths-Jones, 2011).

According to the genetic requirements, 22 nt miRNA should either be found by cloning or discovered by hybridization. The biogenesis requirements comprised the following:

(1) The precursor's variably coiled hairpin framework should have a minimum free energy and include a miRNA duplex despite any major internal loops or bulges;
(2) a miRNA sequence and the secondary structure of the precursor should be phylogenetically preserved; and
(3) the precursor build-up is expected to rise when DCL activity is decreased. The initial version of the criteria emphasized these experimental validations, sequence, and structure preservation, even studies using the DCL mutant, as miRNAs in plants were verified by gel blot or patterns in a cDNA library and their preservation was addressed (Ambros et al., 2003) (Table 14.2).

8. amiRNA studies in model organisms and plant immunity

AmiRNAs have potential applications in plant immunity because they can be used to silence genes that are involved in disease susceptibility, making the plant more resistant to pathogens. For example, researchers have used amiRNAs to target genes that are essential for the replication of certain viruses, leading to a reduction in virus accumulation in infected plants. Another potential application of amiRNAs in plant immunity is the targeting of genes that are involved in the plant's response to fungal or bacterial pathogens. By silencing these genes, researchers have been able to enhance the plant's defense mechanisms and reduce the severity of disease symptoms. The artificial micro RNAs have been tested in many organisms known as model organisms.

8.1 Arabidopsis thaliana

AmiRNAs were initially used in mechanistic experiments in *A. thaliana* to find potential phosphatase elements of the system governing the polar focusing of PIN (pin-formed) auxin transport proteins. Candidates for PIN signaling included the PP2AA1, PP2AA2, and PP2AA3 regulation A components of the protein phosphatase 2A (PP2A) complex (Michniewicz et al., 2007). It was feasible to investigate the triple mutant phenotype, which exhibited more serious growth defects than the single mutants, by creating transgenic lines with two estrogens inducible amiRNAs that concurrently target all three PP2A genes. Genes implicated in maintaining the circadian rhythm in A. Thaliana were studied using amiRNAs by merging the transient expression of amiRNAs with a luciferase-based reporter of circadian clock activity in mesophyll protoplasts, a standardized technique for quick evaluation of gene activity in circadian cycles was developed (Kim and Somers, 2010). It was feasible to reproduce the loss-of-function phenotypes of the circadian cycling-related genes

Table 14.2 Evolution summary of miRNAs.

Sr. No.	Evolution	Details	References
1	Gene inverted duplication	MIR genes resemble those that surround target genes' miRNA-binding domains. These inverted repetitions may create hairpin structures that originally yield many small RNAs. (i.e., siRNAs). Only a crucial section of the hairpin was preserved during evolution, resulting in a MIR gene that encodes just one major type of small RNA, the miRNA.	Allen et al. (2004)
2	Spontaneous evolution	Only 50% of the MIR gene families in Arabidopsis lyrata that can match to protein coding genes produce miRNAs aimed at the identical genes. The target inverted replication theory cannot account for this suggests that MIR genes in plants must have other sources. The genome's hairpin regions, which rarely give birth to miRNA precursors after gaining promoters that allow transcription, may be an origin of miRNAs.	Fahlgren et al. (2010)
3	Miniature inverted-repeat transposable elements (MITE)	MITE RNAs are another potential origin of miRNAs because they have the ability to fold into stem loops that mimic miRNA precursors. It has been found that 38 miRNAs in rice and 10 miRNAs in arabidopsis are generated from MITEs. MITEs, on the other hand, frequently produce numerous short RNAs, also known as siRNAs. Thus, there is ongoing debate over the MITE-origin theory. In support of this hypothesis, TamiR1123, a wheat gene involved in vernalization, was recently discovered to be derived from a MITE region.	Feyissa et al. (2019)
4	Functional divergence of miRNAs	In addition to de novo beginning, repetitions of extant MIR genes result in paralogous genes, extending the MIR gene family and possibly its functional range. Arabidopsis MIR159/319, which came from the same ancestry but split during the common ancestors of land plants and developed to target two different gene families, is one blatant example of functional divergence of paralogous miRNAs. Positively chosen miRNAs and target genes were discovered in harvested rice when studies compared it with wild rice (oryza rufipogon), indicating that miRNAs were engaged in or even drove rice domestication.	Palatnik et al. (2007), Liu et al. (2015)
5	Coevolution of miRNAs and miRNA target genes	Small RNA is amplified in some miRNA-target relationships. Upon miRNA-guided RNA cleavage, some miRNA targets, including PPR, NBS-LRR, MYB, and noncoding TAS loci, produce secondary, phased siRNAs (phasiRNAs). With just one initial miRNA, such small RNA amplification is probably an affordable method to silence a large number of related genes. Throughout the development of plants, this regulatory approach has been widely used.	Fei et al. (2013)

EARLY FLOW- ERING3 (ELF3), ZEITLUPE (ZTL), and GIGANTEA (GI) using transient expression of amiRNAs and luciferase-based reporters. Additionally, it was feasible to investigate the role of the well-preserved regulators of eukaryotic circadian rhythms, the casein kinase II beta subunits (CKB) gene family. RNAi lines were used to demonstrate how the blockage of AGP6 and AGP11 expression by RNAi prevented the development of the pollen tube and hindered pollen discharge (Maier et al., 2009). The decrease in fertilization has been hypothesized to be caused by the abortion of pollen grains during growth rather than the formation of pollen tubes and stamen, using amiRNA scaffolds to silence AGP6 and AGP11 particularly (Coimbra et al., 2009; Eamens et al., 2011).

A customized version of the Arabidopsis MIR390a or TAS1c precursors is found in BsaI/ccdB vectors for amiRNA or syn-tasiRNA cloning and translation. A ccdB cassette bordered by two BsaI sites replaced a portion of the endogenous sequence (Carbonell et al., 2014). On Arabidopsis, 122 amiRNAs aiming at the families of protein kinase, protein phosphatase, and gene transcription factors were evaluated to gauge the promise of an amiRNA-based screen (Hauser et al., 2013). An easier way to replicate and develop amiRNAs in rice and *Arabidopsis* is described in a study. It is built on modified miR319a or miR528 precursors, which have restriction zones to enable a single-step replication of a single PCR product. AmiRNAs can be precisely produced from these altered precursors and operate effectively in plant protoplasts, according to tests using stem-loop reverse-transcriptase quantitative PCR (RT-qPCR) and functional assays (Zhang et al., 2019). A research work of Serrate-Associated Protein 1 (SEAP1), a conserved splicing-related protein, has been studied promoting miRNA in Arabidopsis (Li et al., 2021). In another study, AtSYN4 and AtSCC3 of the *Arabidopsis* cohesin subunit communicate with one another, and AtSCC3 attaches to AtBMI1s. AtSCC3 artificial microRNA silencing or atsyn4 altered plants exhibit decreased amounts of H2Aub1. ChIP-seq experiments show that most AtSYN4 and AtSCC3 binding events are connected to H2Aub1 along the genome in regions where transcription is triggered without the assistance of H3K27me3 (Zhang et al., 2023).

8.2 Chlamydomonas reinhardtii

AmiRNAs effectively suppressed *Chlamydomonas* genes using the endogenous pre-miR-NA1162 as their scaffold. Zhao et al. suppressed the tryptophan synthase b-subunit and rubisco small subunit encoding *Chamydomonas* MAA7 and RBCS1/2 transcripts, correspondingly (Molnar et al., 2009; Zhao et al., 2009). While HydA2's function is still unclear, HydA1 constitutes the majority of hydrogenase activity under anaerobic circumstances, and HYD3 is necessary for normal development because it is necessary for the functioning of cytosolic Fe−S enzymes (Godman et al., 2010). Another study demonstrates the regulation of artificial micro RNA in increasing the hydrogen gas production and the fatty acid production in the green algae *Chlamydomonas* species (Wang et al., 2017b).

8.3 Oryza sativa

A significant portion of the world's human population consumes rice (*Oryza sativa*), which provides a considerable amount of the energy in the average person's diet. The occurrence of diseases brought on by different bacterial, viral, and fungal pathogens is one of the main factors restricting rice output. The discovery of rice miRNAs has advanced significantly in recent years. In rice biotechnology, specific miRNAs have shown promise for controlling agricultural characteristics. MiR393's suppression of auxin signaling brings on bacterial resilience (Navarro et al., 2006). Rice plants overexpressing miR319 had broader leaf blades and improved cold endurance, while miR397 overexpression was reported to increase grain size and promoting panicle branching, improving rice production (Zhang et al., 2013a; Yang and Huang, 2014). A potential new miRNA called Seq119 is linked to a viral sign of rice contaminated with RSV called a poor seed setting rate. Seed setting rates were incredibly low, resembling the impacts of RSV infection, when Seq119 expression was repressed in rice plants by translating a mimic target or through CRISPR/Cas editing.

Rice types Nipponbare (japonica) and IR64 (indica) were effectively modified with amiRNA engineering aimed at three distinct genes: phytoene desaturase (pds, Os03g08570), spotted leaf 11 (Spl11, Os12g38210), and elongated uppermost internode1/CYP714D. (Eui1, Os05g40384). The effects of amiRNA-mediated silencing of these genes were extraordinarily unique and reliable in the progeny, and they resulted in the identical phenotypes as loss-of-function mutants, including an albino phenotype (pds), spontaneous lesion formation in the absence of pathogens (spl11), and lengthening of the foremost internode at heading stage (eui1) (Warthmann et al., 2008). In a recent study, Low seed setting rate, a viral symptom of RSV-infected rice, is correlated with the reduced expression of the short RNA Seq119, a potential new miRNA. Seed setting rates were meager, resembling the impacts of RSV infection, when Seq119 expression was repressed in rice plants by translating a mimic target or through CRISPR/Cas editing. A potential target of Seq119 was also suggested (Yuan et al., 2023). Another study based on the architecture of the rice osa-MIR528 precursor; a dimeric engineered microRNA precursor expression vector that binds the viral MP gene was developed. The MP amiRNAs were effectively expressed in marker-free transgenic plants, which had a high resistance to RSV infection (Zhou et al., 2022).

8.4 Physcomitrella patens

Physcomitrella patens, a moss, has become a model plant for functional gene analysis because of its exceptional capacity to incorporate DNA into the nucleus genome through homologous recombination, enabling the production of mutants with specific genes knocked out (Khraiwesh et al., 2011). Since novel mechanisms of miRNA-mediated regulation were recently discovered with the help of amiRNAs, the demand for miRNA synthesis in *Physcomitrella* has increased significantly.

Surprisingly, decreased levels of PpGNT1 were seen in a mutant form of *Physcomitrella* lacking DCL1 expression after the authors established the amiRNA targeting PpGNT1. Rather than mRNA cleavage or translational repression, the amiRNA could attach to the mRNA in a noncleavable manner and subsequently direct DNA methylation of the responding gene (Khraiwesh et al., 2010) (Table 14.3).

Table 14.3 Artificial microRNAs in plant immunity.

Sr. no.	Plant	miRNA	Modification	References
1	*Oryza sativa*	miR319	Wider leaf blades and enhanced cold tolerance	Yang and Huang (2014)
		miR396	Sensitivity towards alkali and salinity	Gao et al. (2010)
		miR398	Drought salinity sensitivity improvement	
		miR399	Phosphorous accession and translocation	Aung et al. (2006), Bari et al. (2006)
2	*Arabidopsis thaliana*	miR159	Against TYMV virus and delay in flowering	Sunny et al. (2016)
		miR169	Sensitivity towards drought and increase water loss of leaves	Li et al. (2008)
		miR393	Resistance toward disease	Navarro et al. (2006)
		miR394	Sensitivity of salinity	Song et al. (2013)
		miR396	Resistance towards alkali, salinity, and nematode	Chou et al. (2018), Gao et al. (2010)
		miR858	Resistance towards fungus	Camargo-Ramírez et al. (2018)
		miR773	Resistance towards fungus	Salvador-Guirao et al. (2018)
		miR488	Resistance against bacteria and fungus	Lee et al. (2015)
3	Tomato	miR169	Decreased transpiration, reduced water loss from the leaves and drought resistance	Yang et al. (2013)
		miR156	Tolerance towards drought	Feyissa et al. (2019)
		miR482	Resistance towards bacteria	Canto-Pastor et al. (2019)
4	Cotton	miR160	Enhanced heat tolerance	Bai et al. (2018)

9. Patents updates of artificial miRNA in plants

Over the past 10 years, artificial microRNA has been employed in the modification of various plant-related diseases generating plant immunity as well as being effectively used for gene sequencing. They have been demonstrated to regulate several embryonic processes in plants, including the timing of blooming, leaf morphology, organ polarity, floral morphology, and root development. There are many patented inventions present with artificial microRNA in different plant species (Table 14.4).

Table 14.4 Patent update of artificial miRNAs.

Sr. no.	Title of the patent	Application	Patent no.	References
1	MicroRNA compositions and methods for enhancing plant resistance to abiotic stress	A technique for boosting a plant's ability to withstand abiotic stress or for boosting its biomass, vigor, or output.	US 2017/ 0088854 A1	Classification and P
2	Multiple virus resistance in plants	The goal of the present discovery is to use recombinant DNA technologies to posttranscriptionally suppress or inhibit the expression of target coding sequences of plant parasitic viruses in order to have a protective impact on plants.	US 2019/ 0249189 A1	Frizzi and Hagen (2019)
3	Invertebrate microRNAs	This innovation describes a nonnatural transgenic plant cell that expresses at least one invertebrate miRNA for silencing a target gene of an invertebrate pest or of a symbiont linked to the invertebrate pests.	US 10,570,414 B2	Allen et al. (2020)
4	Methods and compositions for enhancing the efficacy and specificity of RNA silencing	This innovation describes a nonnatural transgenic plant cell expressing at least one invertebrate miRNA in planta for suppressing a target gene of an invertebrate pest or a commensal linked to the invertebrate pests.	US 2020/ 0291396 A1	Application and P (2020)

Continued

Table 14.4 Patent update of artificial miRNAs.—*cont'd*

Sr. no.	Title of the patent	Application	Patent no.	References
5	Down-regulation of gene expression using artificial microRNAs	The use of artificial miRNAs and isolated nucleic acid fragments that include precursor miRNAs in down-regulating gene expression are reported.	US 9,546,374 B2	Artificial (2017)
6	Isolated dsRNA molecules and methods of using same for silencing target molecules of interest	SmRNAs formed in the plant and in which the dsRNA molecule also includes a helicase attachment site that is situated to enable the unfolding of the isolated dsRNA molecule's strands to single-stranded RNA (SSRNA) and the capture of an RNA-dependent RNA polymerase to amplify the dsRNA molecule in the plant cell and produce secondary siRNA products from the dsRNA sequence.	US 2019/ 0002878 A1	Meir (2019)
7	Temporal regulation of gene expression by microRNAs	This innovation offers molecular tools and techniques for precisely regulating gene expression in plants, plant diseases, or plant pests. In more detail, this innovation provides plant miRNA genes with unique circadian expression patterns that can be used to create recombinant DNA structures for at least one gene's temporally specific expression.	US 9,976, 152 B2	Wanatha et al. (2013)

Table 14.4 Patent update of artificial miRNAs.—*cont'd*

Sr. no.	Title of the patent	Application	Patent no.	References
8	Plants with useful traits and related methods	The innovation offers strategies for producing plants with beneficial characteristics by momentarily silencing the MSH1 gene in the plants. Techniques for locating genetic loci in plants that generate beneficial traits and plants made from those loci are also given.	US 10,920,286 B2	With and Traits (2021)
9	Compositions and methods to control insect pest	A tiny RNA can make up the silent component. (SRNA). Micro-RNA (miRNA) and short-interfering RNA (siRNA) are both examples of SRNAs.	US 2017/ 0355991 A	Presnail et al. (2017)
10	Gene silencing by using micro-RNA molecules	The innovation also offers a technique for silencing a specific sequence using microRNA.	EP 1 711 613 B1	Tepzz (2013)
11	Engineered miRNA precursors for suppressing target gene expression in plants	The current innovation offers molecular constructs, including constructs with a gene suppression element, and techniques for using them.	US 7,723,568 B2	Lutfiyya et al. (2010)
12	Improved methods controlling gene expression	The innovation offers transgenic expression structures that contain the microRNAs' coding sequences.	CA2604807C	https://patents.google.com/patent/CA2955842A1/en

10. Conclusion

Structures thsat contain the gene silencing into the animal cells are executed mostly with the help of the shRNA rather than amiRNAs. The shRNA are implemented for the silence target gene expression through the RNA interference. The shRNA expression into the cells is completed with the help of plasmids delivery or by using viral vectors. The utilization of bacterial vectors can be done for the same. Different types of constructs are prepared for the expression of amiRNAs, which are utilized efficiently in other animal species from insects to that of the mammals. These types of application are done on basic research category along with therapeutics tests. The intron along with marker gene is used for promoter derivative production, which are further required for the amiRNAs expression. These types of the expressions are can be alone or they are in the form of clusters. The production and design of the amiRNAs are done by using different types of the tools as well as protocols. Such type of evolved molecules are used for the effective targeting of the human genes and currently used in some of the preclinical studies for better exploration in gene silencing sector.

References

Agarwal, V., Bell, G.W., Nam, J., Bartel, D.P., 2015. Predicting effective MicroRNA target sites in mammalian MRNAs. Elife 4, 1–38. https://doi.org/10.7554/eLife.05005.

Ajay Khedkar Dattatray Tupe, A.G., 2012. Cubosomes: emerging drug delivery system. Inven. Rapid NDDS 1–3.

Allen, E., Xie, Z., Gustafson, A.M., Sung, G.-H., Spatafora, J.W., Carrington, J.C., 2004. Evolution of MicroRNA genes by inverted duplication of target gene sequences in arabidopsis thaliana. Nat. Genet. 36, 1282–1290. https://doi.org/10.1038/ng1478.

Allen, I.E., Fallon, O., Us, M.O., Donovan, W.P., Heck, G.R., Lake, C., Us, M.O., Roberts, J.K., Us, M.O., Ursin, V., et al., 2020 United States Patent; 2020; vol. 2; ISBN 2008218813.

Ambros, V., Bartel, B., Bartel, D.P., Burge, C.B., Carrington, J.C., Chen, X., Dreyfuss, G., Eddy, S.R., Griffiths-Jones, S., Marshall, M., et al., 2003. A uniform system for MicroRNA annotation. RNA 9, 277–279. https://doi.org/10.1261/rna.2183803.

Amol, D.G., Musmade, D.S., 2011. Advances in the field of nanooncology: an overview. Inven. Impact NDDS 4, 1–3.

Amol, D.G., Rajurkar, V.G., 2012a. Phonophoresis: ultrasound enhanced transdermal drug delivery. Inven. Rapid NDDS 1–3.

Amol, D.G., Rajurkar, V.G., 2012b. Potential applications of superparmagnetic nanoparticles technology. Inven. Rapid NDDS 1–3.

Amol, D.G., Rajurkar, V.G., 2012c. Potential applications of capillary electromigration methods for the improvement of separation efficiency. Inventi Rapid Pharm Anal. Qual. Assur. 1–3.

Amol, D.G., Rajurkar, V.G., 2012d. New threat of clinical toxicities evoked from nanocarrier systems. Inven. Rapid NDDS 1–4.

Amol, D.G., Rajurkar, V.G., 2012e. Lipid nanocapsules: a novel promising tool as drug delivery system. Inven. Rapid NDDS 1–4.

Amol, D.G., Tupe, D., 2012. Potential applications of botanical extract. Inven. Rapid Ethnopharmacol. 1–4.

Anand M., Mahajan Amol D., Gholap, S.S.B. Recent updates on Q fever: diagnosis and treatment. Pharmacologyonline 254–264.

Application, P. Patent Application Publication (10) Pub . No .: US 2020/0291396 A1. 2020.

Artificial, E.U., 2017. (12) United States Patent. 2.

Aung, K., Lin, S.-I., Wu, C.-C., Huang, Y.-T., Su, C., Chiou, T.-J., 2006. Pho2, a phosphate overaccumulator, is caused by a nonsense mutation in a MicroRNA399 target gene. Plant Physiol. 141, 1000–1011. https://doi.org/10.1104/pp.106.078063.

Axtell, M.J., Snyder, J.A., Bartel, D.P., 2007. Common functions for diverse small RNAs of land plants. Plant Cell 19, 1750–1769. https://doi.org/10.1105/tpc.107.051706.

Bai, Q., Wang, X., Chen, X., Shi, G., Liu, Z., Guo, C., Xiao, K., 2018. Wheat MiRNA Tae-miR408 acts as an essential mediator in plant tolerance to pi deprivation and salt stress via modulating stress-associated physiological processes. Front. Plant Sci. 9, 1–17. https://doi.org/10.3389/fpls.2018.00499.

Bari, R., Datt Pant, B., Stitt, M., Scheible, W.-R., 2006. PHO2, MicroRNA399, and PHR1 define a phosphate-signaling pathway in plants. Plant Physiol. 141, 988–999. https://doi.org/10.1104/pp.106.079707.

Berezikov, E., Cuppen, E., Plasterk, R.H.A., 2006. Approaches to MicroRNA discovery. Nat. Genet. 38, S2–S7. https://doi.org/10.1038/ng1794.

Bologna, N.G., Schapire, A.L., Palatnik, J.F., 2013. Processing of plant MicroRNA precursors. Brief. Funct. Genomics 12, 37–45. https://doi.org/10.1093/bfgp/els050.

Bothiraja, C., Gholap, A.D., Shaikh, K.S., Pawar, A.P., 2014. Investigation of ethyl cellulose microsponge gel for topical delivery of eberconazole nitrate for fungal therapy. Ther. Deliv. 5, 781–794. https://doi.org/10.4155/tde.14.43.

Camargo-Ramírez, R., Val-Torregrosa, B., San Segundo, B., 2018. MiR858-Mediated regulation of flavonoid-specific MYB transcription factor genes controls resistance to pathogen infection in arabidopsis. Plant Cell Physiol. 59, 190–204. https://doi.org/10.1093/pcp/pcx175.

Canto-Pastor, A., Santos, B.A.M.C., Valli, A.A., Summers, W., Schornack, S., Baulcombe, D.C., 2019. Enhanced resistance to bacterial and oomycete pathogens by short tandem target mimic RNAs in tomato. Proc. Natl. Acad. Sci. USA 116, 2755–2760. https://doi.org/10.1073/pnas.1814380116.

Carbonell, A., Takeda, A., Fahlgren, N., Johnson, S.C., Cuperus, J.T., Carrington, J.C., 2014. New generation of artificial MicroRNA and synthetic trans-acting small interfering RNA vectors for efficient gene silencing in arabidopsis. Plant Physiol. 165, 15–29. https://doi.org/10.1104/pp.113.234989.

Chou, C.-H., Shrestha, S., Yang, C.-D., Chang, N.-W., Lin, Y.-L., Liao, K.-W., Huang, W.-C., Sun, T.-H., Tu, S.-J., Lee, W.-H., et al., 2018. MiRTarBase update 2018: a resource for experimentally validated microRNA-target interactions. Nucleic Acids Res. 46, D296–D302. https://doi.org/10.1093/nar/gkx1067.

Classification, P. (12) Patent Application Publication (10) Pub. No.: US 2017/0088854 A1 2017, 1.

Coimbra, S., Costa, M., Jones, B., Mendes, M.A., Pereira, L.G., 2009. Pollen grain development is compromised in arabidopsis Agp6 Agp11 null mutants. J. Exp. Bot. 60, 3133–3142. https://doi.org/10.1093/jxb/erp148.

Cooper, T.A., Wan, L., Dreyfuss, G., 2009. RNA and disease. Cell 136, 777–793. https://doi.org/10.1016/j.cell.2009.02.011.

Cui, X., Wang, Q., Yin, W., Xu, H., Wilson, Z.A., Wei, C., Pan, S., Zhang, D., 2012. PMRD: a curated database for genes and mutants involved in plant male reproduction. BMC Plant Biol. 12, 215. https://doi.org/10.1186/1471-2229-12-215.

Duan, C.-G., Wang, C.-H., Fang, R.-X., Guo, H.-S., 2008. Artificial MicroRNAs highly accessible to targets confer efficient virus resistance in plants. J. Virol. 82, 11084–11095. https://doi.org/10.1128/JVI.01377-08.

Eamens, A.L., Agius, C., Smith, N.A., Waterhouse, P.M., Wang, M.-B., 2011. Efficient silencing of endogenous microRNAs using artificial microRNAs in arabidopsis thaliana. Mol. Plant 4, 157–170. https://doi.org/10.1093/mp/ssq061.

Fahlgren, N., Jogdeo, S., Kasschau, K.D., Sullivan, C.M., Chapman, E.J., Laubinger, S., Smith, L.M., Dasenko, M., Givan, S.A., Weigel, D., et al., 2010. MicroRNA gene evolution in arabidopsis lyrata and arabidopsis thaliana. Plant Cell 22, 1074–1089. https://doi.org/10.1105/tpc.110.073999.

Fei, Q., Xia, R., Meyers, B.C., 2013. Phased, secondary, small interfering RNAs in posttranscriptional regulatory networks. Plant Cell 25, 2400–2415. https://doi.org/10.1105/tpc.113.114652.

Feyissa, B.A., Arshad, M., Gruber, M.Y., Kohalmi, S.E., Hannoufa, A., 2019. The interplay between MiR156/SPL13 and DFR/WD40–1 regulate drought tolerance in alfalfa. BMC Plant Biol. 19, 434. https://doi.org/10.1186/s12870-019-2059-5.

Frizzi, A., Hagen, C., 2019. Pirmont 1.

Fu, C., Sunkar, R., Zhou, C., Shen, H., Zhang, J., Matts, J., Wolf, J., Mann, D.G.J., Stewart, C.N., Tang, Y., et al., 2012. Overexpression of MiR156 in switchgrass (Panicum virgatum L.) results in various morphological alterations and leads to improved biomass production. Plant Biotechnol. J. 10, 443–452. https://doi.org/10.1111/j.1467-7652.2011.00677.x.

Gao, P., Bai, X., Yang, L., Lv, D., Li, Y., Cai, H., Ji, W., Guo, D., Zhu, Y., 2010. Overexpression of osa-MIR396c decreases salt and alkali stress tolerance. Planta 231, 991–1001. https://doi.org/10.1007/s00425-010-1104-2.

Gao, P., Bai, X., Yang, L., Lv, D., Pan, X., Li, Y., Cai, H., Ji, W., Chen, Q., Zhu, Y., 2011. Osa-MIR393: a salinity- and alkaline stress-related MicroRNA gene. Mol. Biol. Rep. 38, 237–242. https://doi.org/10.1007/s11033-010-0100-8.

Georgakilas, G., Vlachos, I.S., Zagganas, K., Vergoulis, T., Paraskevopoulou, M.D., Kanellos, I., Tsanakas, P., Dellis, D., Fevgas, A., Dalamagas, T., et al., 2016. DIANA-MiRGen v3.0: accurate characterization of MicroRNA promoters and their regulators. Nucleic Acids Res. 44, D190–D195. https://doi.org/10.1093/nar/gkv1254.

Gholap, A.D., 2012. Translational nanomedicine: status assessment and opportunities. Int. J. Res. Pharm. Biomed. Sci. 4, 1454–1463.

Gholap, A.D., Dighe, S.B., 2012. Role of sebum in human health and acne. Inven. Rapid Cosmeceuticals 1–3.

Gholap, A.D., Borude, S.S., Mahajan, A.M., Gholap, M.A.D., 2011. Aquasomes: a potential drug delivery carrier. Pharmacologyonline 230–237.

Gholap, A.D., Bhowmik, D.D., Deshmukh, A.Y., Hatvate, N.T., 2023a. Quintessential impact of dendrimer bioconjugates in targeted drug delivery. Compr. Anal. Chem. 103, 257–302.

Gholap, A.D., Said, R.P., Pawar, R.D., Ambore, G.S., Hatvate, N.T., 2023b. Importance of carbohydrate-drug conjugates in vaccine development: a detailed review. Compr. Anal. Chem. 103, 191–256.

Gholap, A.D., Tupe, D., 2012. Artificial viral envelopes: emerging vehicle for gene delivery. Pharmatutor-Art-1439 1–3.

Gholap, A.D., Gupta, J.S., Kamandar, P.A., Banchhod, G.V., Hatvate, N.T., 2023c. Antibody-drug conjugates for cancer therapy: an up-to-date review on the chemistry and pharmacology. Compr. Anal. Chem. 103, 105–190.

Gholap, A.D., Said, R.P., Pawar, R.D., Ambore, G.S., Hatvate, N.T., 2023d. Importance of carbohydrate-drug conjugates in vaccine development: a detailed review. Compr. Anal. Chem. 191–256.

Gholap, A.D., Hatvate, N.T., Khuspe, P.R., Mandhare, T.A., Kashid, P., Gaikwad, V.D., 2023e. Proteomics in oncology: retrospect and prospects. In: Novel Technologies in Biosystems, Biomedical and Drug Delivery. Springer Nature Singapore, Singapore, pp. 243–269.

Gholap, A.D., Sayyad, S.F., Hatvate, N.T., Dhumal, V.V., Pardeshi, S.R., Chavda, V.P., Vora, L.K., 2023f. Drug delivery strategies for avobenzone: a case study of photostabilization. Pharmaceutics 15, 1008. https://doi.org/10.3390/pharmaceutics15031008.

Gholap, A.D., Gupta, J.S., Kamandar, P.A., Banchhod, G.V., Hatvate, N.T., 2023g. Antibody-Drug Conjugates for Cancer Therapy: An Up-To-Date Review on the Chemistry and Pharmacology.

Gholap, A.D., Kapare, H.S., Pagar, S., Kamandar, P., Bhowmik, D., Vishwakarma, N., Raikwar, S., Garkal, A., Mehta, T.A., Rojekar, S., et al., 2024a. Exploring modified chitosan-based gene delivery technologies for therapeutic advancements. Int. J. Biol. Macromol. 260, 129581. https://doi.org/10.1016/j.ijbiomac.2024.129581.

Gholap, A.D., Rojekar, S., Kapare, H.S., Vishwakarma, N., Raikwar, S., Garkal, A., Mehta, T.A., Jadhav, H., Prajapati, M.K., Annapure, U., 2024b. Chitosan scaffolds: expanding horizons in biomedical applications. Carbohydr. Polym. 323, 121394. https://doi.org/10.1016/j.carbpol.2023.121394.

Godman, J.E., Molnár, A., Baulcombe, D.C., Balk, J., 2010. RNA silencing of hydrogenase(-like) genes and investigation of their physiological roles in the green alga chlamydomonas reinhardtii. Biochem. J. 431, 345–352. https://doi.org/10.1042/BJ20100932.

Hauser, F., Chen, W., Deinlein, U., Chang, K., Ossowski, S., Fitz, J., Hannon, G.J., Schroeder, J.I., 2013. A genomic-scale artificial MicroRNA library as a tool to investigate the functionally redundant gene space in arabidopsis. Plant Cell 25, 2848–2863. https://doi.org/10.1105/tpc.113.112805.

Houston, K., McKim, S.M., Comadran, J., Bonar, N., Druka, I., Uzrek, N., Cirillo, E., Guzy-Wrobelska, J., Collins, N.C., Halpin, C., et al., 2013. Variation in the interaction between alleles of HvAPETALA2 and MicroRNA172 determines the density of grains on the barley inflorescence. Proc. Natl. Acad. Sci. USA 110, 16675–16680. https://doi.org/10.1073/pnas.1311681110.

Hsu, P.W.-C., Lin, L.-Z., Hsu, S.-D., Hsu, J.B.-K., Huang, H.-D., 2007. ViTa: prediction of host MicroRNAs targets on viruses. Nucleic Acids Res. 35, D381–D385.

Iwamoto, M., Tagiri, A., 2016. MicroRNA-targeted transcription factor gene RDD1 promotes nutrient ion uptake and accumulation in rice. Plant J. 85, 466–477. https://doi.org/10.1111/tpj.13117.

Jia, X., Shen, J., Liu, H., Li, F., Ding, N., Gao, C., Pattanaik, S., Patra, B., Li, R., Yuan, L., 2015. Small tandem target mimic-mediated blockage of MicroRNA858 induces anthocyanin accumulation in tomato. Planta 242, 283–293. https://doi.org/10.1007/s00425-015-2305-5.

John, B., Enright, A.J., Aravin, A., Tuschl, T., Sander, C., Marks, D.S., 2004. Human microRNA targets. PLoS Biol. 2, e363. https://doi.org/10.1371/journal.pbio.0020363.

Jones-Rhoades, M.W., Bartel, D.P., Bartel, B., 2006. MicroRNAs and their regulatory roles in plants. Annu. Rev. Plant Biol. 57, 19–53. https://doi.org/10.1146/annurev.arplant.57.032905.105218.

Jung, H.J., Kang, H., 2007. Expression and functional analyses of MicroRNA417 in arabidopsis thaliana under stress conditions. Plant Physiol. Biochem. 45, 805–811. https://doi.org/10.1016/j.plaphy.2007.07.015.

Khare, P., Kishore, K., Sharma, D.K., 2017. A study on the standardization parameters of Cassia angustifolia. Asian J. Pharmaceut. Clin. Res. 10 (7), 329–332.

Khraiwesh, B., Ossowski, S., Weigel, D., Reski, R., Frank, W., 2008. Specific gene silencing by artificial MicroRNAs in Physcomitrella patens: an alternative to targeted gene knockouts. Plant Physiol. 148, 684–693. https://doi.org/10.1104/pp.108.128025.

Khraiwesh, B., Arif, M.A., Seumel, G.I., Ossowski, S., Weigel, D., Reski, R., Frank, W., 2010. Transcriptional control of gene expression by MicroRNAs. Cell 140, 111–122. https://doi.org/10.1016/j.cell.2009.12.023.

Khraiwesh, B., Fattash, I., Arif, M.A., Frank, W., 2011. Gene function analysis by artificial MicroRNAs in Physcomitrella patens. In: Kodama, H., Komamine, A. (Eds.), RNAi and Plant Gene Function Analysis, Methods in Molecular Biology, vol 744. Humana Press, Totowa, NJ, pp. 57–79. Methods in Molecular Biology.

Kidner, C.A., Martienssen, R.A., 2005. The developmental role of MicroRNA in plants. Curr. Opin. Plant Biol. 8, 38–44. https://doi.org/10.1016/j.pbi.2004.11.008.

Kim, J., Somers, D.E., 2010. Rapid assessment of gene function in the circadian clock using artificial microRNA in arabidopsis mesophyll protoplasts. Plant Physiol. 154, 611–621. https://doi.org/10.1104/pp.110.162271.

Kim, J.Y., Kwak, K.J., Jung, H.J., Lee, H.J., Kang, H., 2010. MicroRNA402 affects seed germination of arabidopsis thaliana under stress conditions via targeting DEMETER-LIKE Protein3 MRNA. Plant Cell Physiol. 51, 1079–1083. https://doi.org/10.1093/pcp/pcq072.

Kozomara, A., Griffiths-Jones, S., 2011. MiRBase: integrating microRNA annotation and deep-sequencing data. Nucleic Acids Res. 39, D152–D157. https://doi.org/10.1093/nar/gkq1027.

Kozomara, A., Griffiths-Jones, S., 2014. MiRBase: annotating high confidence microRNAs using deep sequencing data. Nucleic Acids Res. 42, D68–D73. https://doi.org/10.1093/nar/gkt1181.

Krek, A., Grün, D., Poy, M.N., Wolf, R., Rosenberg, L., Epstein, E.J., MacMenamin, P., da Piedade, I., Gunsalus, K.C., Stoffel, M., et al., 2005. Combinatorial microRNA target predictions. Nat. Genet. 37, 495–500. https://doi.org/10.1038/ng1536.

Kuang, Z., Wang, Y., Li, L., Yang, X., 2019. MiRDeep-P2: accurate and fast analysis of the microRNA transcriptome in plants. Bioinformatics 35, 2521–2522. https://doi.org/10.1093/bioinformatics/bty972.

Kukurba, K.R., Montgomery, S.B., 2015. RNA sequencing and analysis. Cold Spring Harb. Protoc. 2015. https://doi.org/10.1101/pdb.top084970.

Lee, R.C., Feinbaum, R.L., Ambros, V.T.C., 1993. Elegans heterochronic gene lin-4 encodes small RNAs with antisense complementarity to lin-14. Cell 75, 843–854. https://doi.org/10.1016/0092-8674(93)90529-Y.

Lee, H.J., Park, Y.J., Kwak, K.J., Kim, D., Park, J.H., Lim, J.Y., Shin, C., Yang, K.-Y., Kang, H., 2015. MicroRNA844-Guided downregulation of cytidinephosphate diacylglycerol Synthase3 (CDS3) MRNA affects the response of arabidopsis thaliana to bacteria and fungi. Mol. Plant Microbe Interact. 28, 892–900. https://doi.org/10.1094/MPMI-02-15-0028-R.

Li, W.-X., Oono, Y., Zhu, J., He, X.-J., Wu, J.-M., Iida, K., Lu, X.-Y., Cui, X., Jin, H., Zhu, J.-K., 2008. The arabidopsis NFYA5 transcription factor is regulated transcriptionally and posttranscriptionally to promote drought resistance. Plant Cell 20, 2238–2251. https://doi.org/10.1105/tpc.108.059444.

Li, M., Yu, H., Liu, K., Yang, W., Zhou, B., Gan, L., Li, S., Zhang, C., Yu, B., 2021. Serrate-associated protein 1, a splicing-related protein, promotes MiRNA biogenesis in arabidopsis. New Phytol. 232, 1959–1973. https://doi.org/10.1111/nph.17691.

Lin, J., Lin, C., Lin, H., Chen, Y., Jeng, S., 2012. MicroR828 regulates lignin and H2O2 accumulation in sweet potato on wounding. New Phytol. 196, 427–440. https://doi.org/10.1111/j.1469-8137.2012.04277.x.

Liu, Q., Chen, Y.-Q., 2010. A new mechanism in plant engineering: the potential roles of microRNAs in molecular breeding for crop improvement. Biotechnol. Adv. 28, 301–307. https://doi.org/10.1016/j.biotechadv.2010.01.002.

Liu, Q., Wang, H., Hu, H., Zhang, H., 2015. Genome-wide identification and evolutionary analysis of positively selected MiRNA genes in domesticated rice. Mol. Genet. Genom. 290, 593–602. https://doi.org/10.1007/s00438-014-0943-0.

Liu, Z., Wang, X., Chen, X., Shi, G., Bai, Q., Xiao, K., 2018. TaMIR1139: a wheat MiRNA responsive to pi-starvation, acts a critical mediator in modulating plant tolerance to pi deprivation. Plant Cell Rep. 37, 1293–1309. https://doi.org/10.1007/s00299-018-2313-6.

Luo, Y., Guo, Z., Li, L., 2013. Evolutionary conservation of MicroRNA regulatory programs in plant flower development. Dev. Biol. 380, 133–144. https://doi.org/10.1016/j.ydbio.2013.05.009.

Lutfiyya, I.L., Louis, S., Us, M.O., Luethy, H., Louis, S., Us, M.O., Sara, E., So, O.E.E., 2010. (12) United States Patent, 2.

Maier, B., Wendt, S., Vanselow, J.T., Wallach, T., Reischl, S., Oehmke, S., Schlosser, A., Kramer, A.A., 2009. Large-scale functional RNAi screen reveals a role for CK2 in the mammalian circadian clock. Genes Dev. 23, 708–718. https://doi.org/10.1101/gad.512209.

Meir, O., 2019. Classification, P. Patent Application Publication (10) Pub. No.: US 2019/0002878 A1, p. 1.

Michniewicz, M., Zago, M.K., Abas, L., Weijers, D., Schweighofer, A., Meskiene, I., Heisler, M.G., Ohno, C., Zhang, J., Huang, F., et al., 2007. Antagonistic regulation of PIN phosphorylation by PP2A and PINOID directs auxin flux. Cell 130, 1044–1056. https://doi.org/10.1016/j.cell.2007.07.033.

Molnar, A., Bassett, A., Thuenemann, E., Schwach, F., Karkare, S., Ossowski, S., Weigel, D., Baulcombe, D., 2009. Highly specific gene silencing by artificial MicroRNAs in the unicellular alga chlamydomonas reinhardtii. Plant J. 58, 165–174. https://doi.org/10.1111/j.1365-313X.2008.03767.x.

Navarro, L., Dunoyer, P., Jay, F., Arnold, B., Dharmasiri, N., Estelle, M., Voinnet, O., Jones, J.D.G., 2006. A plant MiRNA contributes to antibacterial resistance by repressing auxin signaling. Science 312, 436–439. https://doi.org/10.1126/science.1126088.

Ni, Z., Hu, Z., Jiang, Q., Zhang, H., 2012. Overexpression of gma-MIR394a confers tolerance to drought in transgenic arabidopsis thaliana. Biochem. Biophys. Res. Commun. 427, 330–335. https://doi.org/10.1016/j.bbrc.2012.09.055.

Ossowski, S., Schwab, R., Weigel, D., 2008. Gene silencing in plants using artificial microRNAs and other small RNAs. Plant J. 53, 674–690. https://doi.org/10.1111/j.1365-313X.2007.03328.x.

Palatnik, J.F., Wollmann, H., Schommer, C., Schwab, R., Boisbouvier, J., Rodriguez, R., Warthmann, N., Allen, E., Dezulian, T., Huson, D., et al., 2007. Sequence and expression differences underlie functional specialization of arabidopsis MicroRNAs MiR159 and MiR319. Dev. Cell 13, 115–125. https://doi.org/10.1016/j.devcel.2007.04.012.

Parveen, R., Khare, N., Singh, S., Johri, P., March 29, 2023. An in silico outlook for the detection and surveillance of evolving and persistent plant pathogens. In: Detection, Diagnosis and Management of Soil-Borne Phytopathogens. Springer Nature Singapore, Singapore, pp. 17–40.

Patani, A., Patel, A., Prajapati, D., Khare, N., Singh, S., August 1, 2023. Various agriculture crop plant-based bioactive compounds and their use in nanomaterial synthesis and applications. In: Nanomaterials from Agricultural and Horticultural Products. Springer Nature Singapore, Singapore, pp. 223–241.

Patel, P., Yadav, K., Srivastava, A.K., Suprasanna, P., Ganapathi, T.R., 2019. Overexpression of native musa-MiR397 enhances plant biomass without compromising abiotic stress tolerance in banana. Sci. Rep. 9, 16434. https://doi.org/10.1038/s41598-019-52858-3.

Piriyapongsa, J., Jordan, I.K., 2008. Dual coding of SiRNAs and MiRNAs by plant transposable elements. RNA 14, 814–821. https://doi.org/10.1261/rna.916708.

Presnail, J., Louis, S.T., Moines, W.D.E.S., 2017. BriAontdRein2G. Vol. 1.

Rajurkar, V.G., Gholap, A.D., 2012a. Recent technologies in nanomedicine for treatment of cancer. Inven. Rapid Pharm. Tech. 4, 1–3.

Rajurkar, V.G., Gholap, A.D., 2012b. Recent technologies in nanomedicine for treatment of cancer. Inven. Rapid Pharm. Tech. 1 (4), 1–4.

Rehmsmeier, M., Steffen, P., Höchsmann, M., Giegerich, R., 2004. Fast and effective prediction of MicroRNA/target duplexes. RNA 10, 1507–1517. https://doi.org/10.1261/rna.5248604.

Sablok, G., Pérez-Quintero, Á.L., Hassan, M., Tatarinova, T.V., López, C., 2011. Artificial MicroRNAs (AmiRNAs) engineering – on how MicroRNA-based silencing methods have affected current plant silencing research. Biochem. Biophys. Res. Commun. 406, 315–319. https://doi.org/10.1016/j.bbrc.2011.02.045.

Salvador-Guirao, R., Baldrich, P., Weigel, D., Rubio-Somoza, I., San Segundo, B., 2018. The MicroRNA MiR773 is involved in the arabidopsis immune response to fungal pathogens. Mol. Plant Microbe Interact. 31, 249–259. https://doi.org/10.1094/MPMI-05-17-0108-R.

Schommer, C., Debernardi, J.M., Bresso, E.G., Rodriguez, R.E., Palatnik, J.F., 2014. Repression of cell proliferation by MiR319-regulated TCP4. Mol. Plant 7, 1533–1544. https://doi.org/10.1093/mp/ssu084.

Schwab, R., Palatnik, J.F., Riester, M., Schommer, C., Schmid, M., Weigel, D., 2005. Specific effects of MicroRNAs on the plant transcriptome. Dev. Cell 8, 517–527. https://doi.org/10.1016/j.devcel.2005.01.018.

Schwab, R., Ossowski, S., Riester, M., Warthmann, N., Weigel, D., 2006. Highly specific gene silencing by artificial MicroRNAs in arabidopsis. Plant Cell 18, 1121–1133. https://doi.org/10.1105/tpc.105.039834.

Silva, E.M., Silva, G.F.F., Bidoia, D.B., Silva Azevedo, M., Jesus, F.A., Pino, L.E., Peres, L.E.P., Carrera, E., López-Díaz, I., Nogueira, F.T.S., 2017. MicroRNA159-Targeted SlGAMYB transcription factors are required for fruit set in tomato. Plant J. 92, 95–109. https://doi.org/10.1111/tpj.13637.

Singh, A., Sengar, R.S., Singh, A.K., Singh, S., Kumar, A., Ahmad, K., 2019. Impact of genetic engineering on agriculture food crops. Biotech Today: Int. J. Biol. Sci. 9 (1), 17–24.

Song, J.B., Gao, S., Sun, D., Li, H., Shu, X.X., Yang, Z.M., 2013. MiR394 and LCR are involved in arabidopsis salt and drought stress responses in an abscisic acid-dependent manner. BMC Plant Biol. 13, 210. https://doi.org/10.1186/1471-2229-13-210.

Sumaira, K.A., Salim, A., Chandra, S.S.V., 2016. In: Das, S., Pal, T., Kar, S., Satapathy, S.C., Mandal, J.K. (Eds.), Proceedings of the 4th International Conference on Frontiers in

Intelligent Computing: Theory and Applications (FICTA) 2015, vol 404. Springer India, New Delhi. Advances in Intelligent Systems and Computing.

Sunkar, R., Kapoor, A., Zhu, J.-K., 2006. Posttranscriptional induction of two Cu/Zn superoxide dismutase genes in arabidopsis is mediated by downregulation of MiR398 and important for oxidative stress tolerance. Plant Cell 18, 2051–2065. https://doi.org/10.1105/tpc.106.041673.

Sunny, J.S., Snijesh, V.P., Singh, S., 2016. Potential role of mirnas in rheumatoid arthritis. Austin Arthritis 1 (2), 1009.

Tepzz, B.T., 2013. Tepzz_7__6_¥b_t (11). 1, 1–69.

Tupe, D., Gholap, A.D., 2012. The threat of photoaging: clinical conditions, evaluation and treatment. Inven. Rapid Cosmeceuticals 1–3.

Verma, N., Sao, P., Srivastava, A., Singh, S., 2021. Physiological and molecular responses to drought, submergence and excessive watering in plants. Harsh Environment and Plant Resilience: Molecular and Functional Aspects, pp. 305–321.

Vicens, Q., Kieft, J.S., 2022. Thoughts on how to think (and talk) about RNA structure. Psroc. Natl. Acad. Sci. USA 119, 1–9. https://doi.org/10.1073/pnas.2112677119.

Vitsios, D.M., Kentepozidou, E., Quintais, L., Benito-Gutiérrez, E., van Dongen, S., Davis, M.P., Enright, A.J., 2017. Mirnovo: genome-free prediction of MicroRNAs from small RNA sequencing data and single-cells using decision forests. Nucleic Acids Res. 45, e177. https://doi.org/10.1093/nar/gkx836.

Vlachos, I.S., Paraskevopoulou, M.D., Karagkouni, D., Georgakilas, G., Vergoulis, T., Kanellos, I., Anastasopoulos, I.-L., Maniou, S., Karathanou, K., Kalfakakou, D., et al., 2015. DIANA-TarBase v7.0: indexing more than half a million experimentally supported MiRNA:MRNA interactions. Nucleic Acids Res. 43, D153–D159. https://doi.org/10.1093/nar/gku1215.

Vora, L.K., Gholap, A.D., Jetha, K., Thakur, R.R.S., Solanki, H.K., Chavda, V.P., 2023. Artificial intelligence in pharmaceutical technology and drug delivery design. Pharmaceutics 15, 1916. https://doi.org/10.3390/pharmaceutics15071916.

Wanatha, H.A.O., Della, M., Matematit, M., 2013. Hao Wanatha Maria Della Maria Matematit. vol. 2.

Wang, W., Lin, F.-M., Chang, W., Lin, K.-Y., Huang, H., Lin, N., 2009. MiRExpress: analyzing high-throughput sequencing data for profiling microRNA expression. BMC Bioinf. 10, 328. https://doi.org/10.1186/1471-2105-10-328.

Wang, L., Mo, Q., Wang, J.M.I.E., 2015. A database for gene coexpression correlation in immune cells based on mutual information and pearson correlation. J. Immunol. Res. 2015, 1–10. https://doi.org/10.1155/2015/140819.

Wang, Y., Zhang, J., Cui, W., Guan, C., Mao, W., Zhang, Z., 2017a. Improvement in fruit quality by overexpressing MiR399a in woodland strawberry. J. Agric. Food Chem. 65, 7361–7370. https://doi.org/10.1021/acs.jafc.7b01687.

Wang, Y., Jiang, X., Hu, C., Sun, T., Zeng, Z., Cai, X., Li, H., Hu, Z., 2017b. Optogenetic regulation of artificial MicroRNA improves H2 production in gsreen alga chlamydomonas reinhardtii. Biotechnol. Biofuels 10, 257. https://doi.org/10.1186/s13068-017-0941-7.

Warthmann, N., Chen, H., Ossowski, S., Weigel, D., Hervé, P., 2008. Highly specific gene silencing by artificial MiRNAs in rice. PLoS One 3, e1829. https://doi.org/10.1371/journal.pone.0001829.

With, P., Traits, U., 2021. Methods, R, vol 2. United States Patent.

Wong, N., Wang, X., 2015. MiRDB: an online resource for microRNA target prediction and functional annotations. Nucleic Acids Res. 43, D146–D152. https://doi.org/10.1093/nar/gku1104.

Wu, Y., Wei, B., Liu, H., Li, T., Rayner, S.M.R.P., 2011. A SVM-based software tool for prediction of most probable MicroRNA coding regions in genome scale sequences. BMC Bioinf. 12, 107. https://doi.org/10.1186/1471-2105-12-107.

Xie, F., Xiao, P., Chen, D., Xu, L., Zhang, B., 2012a. MiRDeepFinder: a MiRNA analysis tool for deep sequencing of plant small RNAs. Plant Mol. Biol. 80, 75–84. https://doi.org/10.1007/s11103-012-9885-2.

Xie, K., Shen, J., Hou, X., Yao, J., Li, X., Xiao, J., Xiong, L., 2012b. Gradual increase of MiR156 regulates temporal expression changes of numerous genes during leaf development in rice. Plant Physiol. 158, 1382–1394. https://doi.org/10.1104/pp.111.190488.

Yang, L., Huang, H., 2014. Roles of small RNAs in plant disease resistance. J. Integr. Plant Biol. 56, 962–970. https://doi.org/10.1111/jipb.12200.

Yang, X., Li, L., 2011. MiRDeep-P: a computational tool for analyzing the MicroRNA transcriptome in plants. Bioinformatics 27, 2614–2615. https://doi.org/10.1093/bioinformatics/btr430.

Yang, C., Li, D., Mao, D., Liu, X., Ji, C., Li, X., Zhao, X., Cheng, Z., Chen, C., Zhu, L., 2013. Overexpression of MicroRNA319 impacts leaf morphogenesis and leads to enhanced cold tolerance in rice (O ryza sativa L.). Plant Cell Environ. 36, 2207–2218. https://doi.org/10.1111/pce.12130.

Yao, Y., Ma, C., Deng, H., Liu, Q., Zhang, J., Yi, M., 2016. PlantMirP: an efficient computational program for the prediction of plant pre-MiRNA by incorporating knowledge-based energy features. Mol. Biosyst. 12, 3124–3131. https://doi.org/10.1039/C6MB00295A.

Yuan, Q., Zhai, Y., Zhou, L., Ai, X., Chen, J., Yan, F., 2023. A novel MiRNA in rice associated with the low seed setting rate symptom of rice stripe virus. Int. J. Mol. Sci. 24, 3675. https://doi.org/10.3390/ijms24043675.

Zhang, B., Wang, Q., Pan, X., 2007. MicroRNAs and their regulatory roles in animals and plants. J. Cell. Physiol. 210, 279–289. https://doi.org/10.1002/jcp.20869.

Zhang, X., Zou, Z., Gong, P., Zhang, J., Ziaf, K., Li, H., Xiao, F., Ye, Z., 2011. Overexpression of microRNA169 confers enhanced drought tolerance to tomato. Biotechnol. Lett. 33, 403–409. https://doi.org/10.1007/s10529-010-0436-0.

Zhang, Y., Yu, Y., Wang, C.-Y., Li, Z.-Y., Liu, Q., Xu, J., Liao, J., Wang, X., Qu, L., Chen, F., et al., 2013a. Overexpression of MicroRNA OsmiR397 improves rice yield by increasing grain size and promoting panicle branching. Nat. Biotechnol. 31, 848–852. https://doi.org/10.1038/nbt.2646.

Zhang, L.W., Song, J.B., Shu, X.X., Zhang, Y., Yang, Z.M., 2013b. MiR395 is involved in detoxification of cadmium in Brassica napus. J. Hazard Mater. 250–251, 204–211. https://doi.org/10.1016/j.jhazmat.2013.01.053.

Zhang, D., Zhang, N., Shen, W., Li, J.-F., 2019. Engineered artificial MicroRNA precursors facilitate cloning and gene silencing in arabidopsis and rice. Int. J. Mol. Sci. 20, 5620. https://doi.org/10.3390/ijms20225620.

Zhang, Y., Ma, M., Liu, M., Sun, A., Zheng, X., Liu, K., Yin, C., Li, C., Jiang, C., Tu, X., et al., 2023. Histone H2A monoubiquitination marks are targeted to specific sites by cohesin subunits in arabidopsis. Nat. Commun. 14, 1209. https://doi.org/10.1038/s41467-023-36788-3.

Zhao, T., Wang, W., Bai, X., Qi, Y., 2009. Gene silencing by artificial MicroRNAs in chlamydomonas. Plant J. 58, 157–164. https://doi.org/10.1111/j.1365-313X.2008.03758.x.

Zhou, L., Yuan, Q., Ai, X., Chen, J., Lu, Y., Yan, F., 2022. Transgenic rice plants expressing artificial MiRNA targeting the rice stripe virus MP gene are hisghly resistant to the virus. Biology 11, 332. https://doi.org/10.3390/biology11020332.

CHAPTER

Role of microRNA in plant defense 15

Pawan Kumar[1], Meenakshi Sharma[2], Kumar Shantanu[3], Sanavar Soham[4], Anita Garg Mangla[5]

[1]*Department of Botany, Shivaji College, University of Delhi, New Delhi, India;* [2]*Department of Botany, Daulat Ram College, University of Delhi, New Delhi, India;* [3]*Department of Botany, Deshbandhu College, University of Delhi, New Delhi, India;* [4]*Department of Botany, Kalindi College, University of Delhi, New Delhi, India;* [5]*Department of Biochemistry, Daulat Ram College, University of Delhi, New Delhi, India*

1. Introduction

Plants in their lifecycle undergo through barrage of perturbations in their environment and are constantly under biotic and abiotic threats (Fig. 15.1). To successfully respond to these threats, plants possess physical barriers, such as the cell wall, cuticle, and periderm, that act as the first line of defense (Jones and Dangl, 2006) and have evolved innate immune system that has the ability to recognize potential pathogens and mount an effective defensive immune response. Immunity is based on the identification of microbe or pathogen or damage-associated molecular patterns (MAMPs or PAMPs or DAMPs) by pattern recognition receptors (PRRs) on

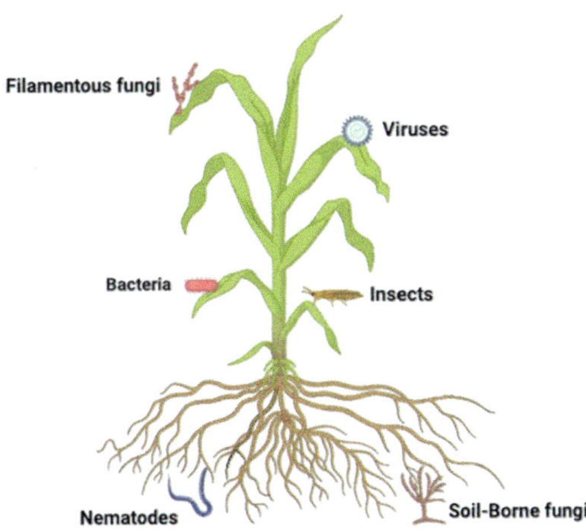

FIGURE 15.1

Schematic diagram showing interaction of a plant with various biotic agents.

Figure is created via BioRender.com.

MicroRNA Advances and Application in Plant Biology. https://doi.org/10.1016/B978-0-443-21682-4.00006-3

plants cell surface and they work with other coreceptors (Gómez-Gómez and Boller, 2000; Zhou and Zhang, 2020). Flagellin sensing 2 (AtFLS2) is one of the first reported PRRs in plants and it recognizes fragments of bacterial flagellin (Gómez-Gómez and Boller, 2000).

A number of MAMPs have been identified, including lipopolysaccharide (LPS), peptidoglycan (PGN), lipopeptides, fungal wall component chitin and chitosan, EF-Tu peptide elf18, peptide flg22 of bacterial flagellin, eicosapolyenoic acid (EP), and β-glucans of oomycete (Versluys et al., 2017) and plants use various lytic enzymes to release these different MAMPs from the invading pathogens. DAMPs are also diverse in nature such as immune-related peptides and proteins, extracellular nucleotides, fragments of plant cell wall-derived polysaccharides, and these are released in response to cell damages (Gust et al., 2017).

PRRs are diverse in nature and consists of RLKs and RLPs. Receptor-like kinases (RLKs) possess an intracellular kinase domain, a transmembrane domain, and an extracellular domain. Receptor-like proteins (RLPs) contain transmembrane as well as extracellular domain, however lacks intracellular kinase domain (Fig. 15.2). These RLKs and RLPs play an essential role as a first line of defense in plants. Interaction of these PRRs with MAMP or PAMP generates immune responses which are referred as PAMP triggered immunity (PTI) in plants. In response to these, pathogens produce effector molecules to inhibit PTI, by secreting various effector molecules/proteins into plant cells thereby interfering with MAMP detection and affecting signal transduction which in turn can lead to disease development. To combat this, plants have evolved another rapid and robust immune system to

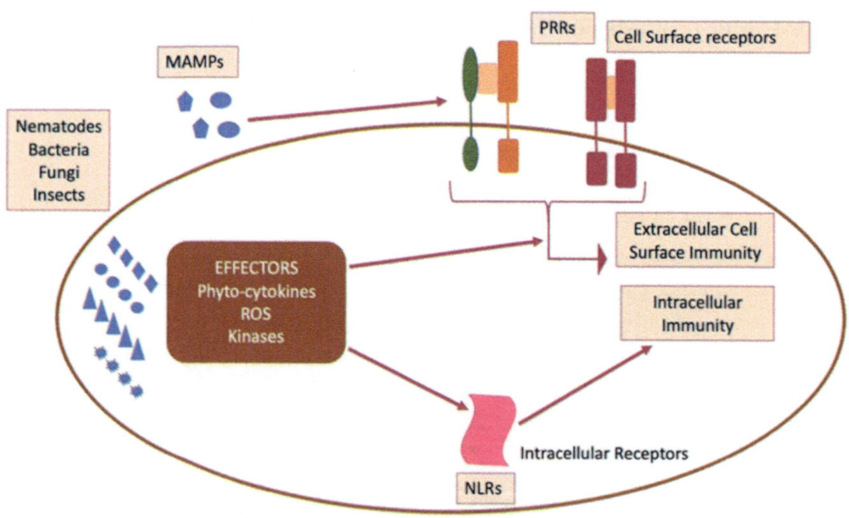

FIGURE 15.2

Schematic diagram illustrating the engagement of two distinct components of plant immunity (PTI and ETI) in combating various microbes and insects.

perceive such effectors through additional receptors namely nucleotide binding leucine rich repeat receptors (NLRs), which is the second layer of plant defense known as effector triggered immunity (ETI). These NLRs have an N-terminal, a central nucleotide binding (NB-ARC) domain, and a C-terminal leucine-rich repeat domain. While PRRs generally function as extracellular immune receptors, NLRs have been found to be exceedingly important for intracellular defense mechanisms. Both PRRs and NLRs cooperate to provide innate defense against the pathogenic assault. Therefore any pathogenic assault is perceived by plants as multiple danger signals by PRRs and NLRs. Activation of these immune receptors (PRRs and NLRs) leads to major downstream signaling events in plants including calcium influx across plasma membrane, release of reactive oxygen species (ROS), activation of series of calcium-dependent kinases (CPKs), leading to downstream mitogen-activated protein kinase (MAPK), G protein activation, production of a number of immunogenic peptides, and a plethora of hormones involved in defense (Tang et al., 2017). Plants actively produce various phytocytokines which are immunogenic peptides and are perceived by PRRs (Hou et al., 2021). These phytocytokines serve as plant peptide hormones and play a crucial role in both plant immunity as well as development and as signals of cellular communication (Luo, 2012) and are secreted/processed when infections occur (Tanaka and Heil, 2021).

In plants, both PTI and ETI together work as a robust surveillance system to identify any pathogenic invasion and mount protective immune response. Pathogens in turn target PRR and downstream complexes so as to suppress plant protective immune responses and lead to development of disease.

In addition to local immunity with PTI and ETI, plants also possess systemic immunity with systemic acquired resistance (SAR) and induced systemic immunity (ISR). Systemic immunity involves spreading of systemic signals from affected tissue to unaffected site in order to enhance resistance to a subsequent secondary stress. An example is seen in *Arabidopsis thaliana* where in low levels of protein PRRs and deficient PTI response is seen in its differentiated outer layer. However, in response to pathogen challenge, damage in neighboring cells leads to the induction of increased expression of PRRs; therefore, nonresponsive cells can now be seen mounting a localized immune response (Zhou et al., 2020).

Recent data have suggested the role of small RNAs (sRNAs) in the development of various diseases by regulating gene expression by posttranscriptional downregulation of protein output from their target mRNAs (Guo et al., 2019). MicroRNAs (miRNAs) represent a class of sRNAs typically 21–23 nucleotides with a role in the regulation of gene expression, and the expression of these growth regulatory miRNAs has been seen to frequently increase during the activation of plant immunity.

In response to pathogens, miRNA plays crucial roles such as:

a. Regulating equilibrium between growth and defense by affecting signaling pathways of various phytohormones such as salicylic acid (SA), jasmonic acid (JA), and ethylene (ET) (Pandey and Baldwin, 2007; Ma and Ma, 2016). SA signaling is associated with defense against biotrophic pathogens, while JA and ET signaling are involved in defense against necrotrophic pathogens and herbivores. Induced systemic resistance (ISR) is a defense mechanism wherein plants develop systemic resistance against a broad range of pathogens following the colonization of roots by beneficial microbes (Pieterse et al., 2014). This mechanism involves the priming of defense responses, resulting in an enhanced ability to counter subsequent pathogen attacks.

b. Modulating the expression of genes involved in plant defense mechanisms, including those encoding pathogen recognition receptors, defense signaling components, and antimicrobial compounds (Navarro et al., 2008). This regulation helps fine tune the defense response and enhances plant resistance against pathogens and pests.

c. Epigenetic modifications of defense-related genes, such as histone modifications and DNA methylation, leading to changes in their expression and hence affecting response to the pathogenic assault (Pikaard and Scheid, 2014).

These recent advancements in understanding the role of miRNAs in plant defense will enable researchers to develop novel strategies for enhancing plant resistance against various stresses.

2. miRNA and bacterial infection in plants

Plants are under constant threats of bacterial pathogens (Nazarov et al., 2020). The defense mechanisms of plants against bacteria entail the activation of elements of PTI and ETI pathways (Yuan et al., 2021). One of best studied example of pathways affected by miRNA is auxin signaling in *A. thaliana* infected with *Pseudomonas syringae*. Bacterial PAMPs/PRRs pair viz. flagellin/FLS2 is very well described in *A. thaliana-P. syringae* system, which is responsible for affecting auxin signaling in plants (Boller and Felix, 2009). Flagellin protein (flg22) of bacterial flagellum having a 22aa binding domain at N-terminus binds with flagellin sensing receptor (FLS2) of plant (Gómez-Gómez and Boller, 2000). An intracellular kinase domain of FLS2 activates MAPK3 cascade. MAPK3 cascade involves phosphorylation by kinases as well as dephosphorylation by phosphatases of their substrates leading to activation and deactivation, respectively (Bigeard and Hirt, 2018). MAPK pathway leads to posttranslational modification (PTM) of target proteins. Downstream phosphorylation of a double-stranded RNA binding proteins (DRB1) in nucleus leads to biogenesis of miR393, miR160, and miR167. These miRNAs affect gene targets that control how auxins are transported, synthesized, and repressed in plants.

Auxin signaling in plants is tightly regulated by a group of three proteins TIR/AFB (for auxin perception), Aux/IAA (a transcription factor (TF) for auxin transport regulation), and ARF (a TF which represses auxin responsive genes) (Powers and Strader, 2016). Aux/IAA and ARF proteins can form a complex with each other.

Presence of enough auxin removes Aux/IAA from ARF. Thereafter, free ARF binds to promotor region and represses transcription of SAUR proteins. SAUR proteins regulate H^+-ATPase activity using phosphatases leading to cell expansion (Powers and Strader, 2016). Eventually, repressing transcription of SUAR proteins can inhibit cell expansion. Simultaneously, free Aux/IAA proteins after their release from (Aux/IAA)-ARF complex can bind to TIR/AFB proteins in the presence of auxins and present them for ubiquitylation and their subsequent degradation (Powers and Strader, 2016). miR393 represses TIR/AFB proteins impairing perception of auxins. Downregulation of auxin receptors results into reduced plant growth leaving plant less usable for pathogen ultimately providing resistance against pathogen (Navarro et al., 2006). Furthermore, miR160 and miR167 can lead to repression of ARF proteins following perception of flg22 during flg22/FLS2 interaction. In the absence of ARF proteins, transcription of SUAR gene can occur and henceforth can cause cell expansion which is a primary function of auxin. Several other miRNAs involved in auxin signaling are enlisted in Table 15.1.

A similar pathway suggesting defensive role of plant-miRNAs during bacterial infection is related to ROS detoxification which is initiated by binding of bacterial PAMP, Ef-Tu with receptors (EFR), triggering similar MAPK cascade differing in terms of miRNA targets and final response that provide defense against bacteria (Boller and Felix, 2009).

It is interesting to note that miRNAs can have multiple gene targets and same miRNA can therefore produce different defense responses. For example, elevated levels of miR393 were also found to downregulate MEMB (SNARE) gene leading to decrease in the production of SNARE proteins which is responsible for membrane fusion. Absence of SNARE protein facilitates exocytosis of a PR1 (pathogenesis related) protein mediated by AGO2 proteins suggesting the role of miR393 in conferring resistance against bacterial pathogen. Additionally, research has shown that blocking pre-miR393 (which is a precursor of miR393) by bacterial effector (AvrPtoB) could hamper AGO2-mediated exocytosis of PR1 protein rendering plants more susceptible to pathogen.

miRNAs can also regulate mRNA of R-genes thereby halting the synthesis of R-proteins. Blocked mRNA transcripts of R-genes are further processed into secondary siRNAs (Luo, 2012). These siRNAs eventually regulate genes involved in nodulation process by blocking mRNA transcripts of their targets during symbiosis. For instance, a significant rise in levels of miR1507 during *Bradyrhizobium* infection in root hairs of *Glycine max* is related to the downregulation of symbiotic process during initial stages of its interaction (Li et al., 2010). miR172 during *Phaseolus vulgaris-Rhizobium etli* symbiosis is known to regulate the process of nodule senescence by targeting APETALA2 (AP2), a key TF involved in flowering (Nova-Franco et al., 2015). Impairing nodulation process can have detrimental effects on overall plant development rendering it unsuitable for associated bacteria and also making it more vulnerable to attack by various other pathogens. Another miR171 which targets nodulation signaling pathway 2 (NSP2), a key nodulation TF of active nodules; and miR397, which is systemically upregulated in nitrogen-fixing nodules is known

Table 15.1 Functional roles of some plant miRNAs in defense against bacteria.

Plant miRNA	Plant	Bacteria	Target	Role	References
miR160	*Arabidopsis thaliana*	*Pseudomonas syringae*	TFs such as ARF	Auxin signaling	Fahlgren et al. (2007)
miR167	*Arabidopsis thaliana*	*Pseudomonas syringae*			Fahlgren et al. (2007)
	Manihot esculenta	*Xanthomonas axonopodis*			Pérez-Quintero et al. (2012)
miR390	*Arabidopsis thaliana*	*Pseudomonas syringae*	ta-siRNA genes		Zhang et al. (2011a)
miR393	*Arabidopsis thaliana*	*Pseudomonas syringae*	TFs such as ARF and TIR		Navarro et al. (2006)
	Manihot esculenta	*Xanthomonas axonopodis*	TFs such as AFB and TIR		Pérez-Quintero et al. (2012)
	Citrus sp.	*Candidatus* Liberibacter asiaticus			Zhao et al. (2013)
miR393b	*Arabidopsis thaliana*	*Pseudomonas syringae*	MEMB12	Exocytosis of a PRR protein	Zhang et al. (2011b)
miR1507	*Glycine max*	*Bradyrhizobium* sp.	NB-LRR	Defense response	Yan et al. (2016)
miR158	*Arabidopsis*	*Pseudomonas syringae*	PPR		Zhang et al. (2011a)
miR472	*Arabidopsis thaliana*	*Pseudomonas syringae*	NBS-LRR		Boccara et al. (2014)
miR482	*Manihot esculenta*	*Xanthomonas axonopodis*			Pérez-Quintero et al. (2012)
miR171	*Lotus corniculatus*	*Mesorhizobium loti*	NSP2	Nodulation signaling pathway	De Luis et al. (2012)
miR172	*Glycine max*	*Bradyrhizobium* sp.	ENOD 40		Wang et al. (2009)
	Beans	*Rhizobium etli*	APE2		Nova-Franco et al. (2015)
miR397	*Zea mays*	*Herbaspirillum seropedicae*	LAC	Copper homeostasis	Thiebaut et al. (2014)

	Lotus corniculatus	*Mesorhizobium loti*			De Luis et al. (2012)
miR398	*Zea mays*	*Herbaspirillum seropedicae*	CSD, LAC		Thiebaut et al. (2014)
miR399	*Citrus* sp.	*Candidatus* Liberibcaters asiaticus	PHO2	Nutrient homeostasis	Zhao et al. (2013)
miR408	*Arabidopsis thaliana*	*Pseudomonas syringae*	Cu protein plantacyanin & Cu binding protein genes	Copper homeostasis	Zhang et al. (2011b)
miR398	*Arabidopsis thaliana*	*Pseudomonas syringae*	Superoxide dismutase	ROS detoxification	Li et al. (2010)
	Manihot esculenta	*Xanthomonas axonopodis*			Pérez-Quintero et al. (2012)
miR773	*Arabidopsis thaliana*	*Pseudomonas syringae*	MET2	Callose deposition	Li et al. (2010)
miR159	*Arabidopsis thaliana*	*Pseudomonas syringae*	TFs such as MYB	ABA signaling	Zhang et al. (2011a)
miR393	*Arabidopsis thaliana*	*Pseudomonas syringae*	AFB2, AFB3		Fahlgren et al. (2007)
miR398	*Arabidopsis thaliana*	*Pseudomonas syringae*	COX5b.1	Regulation of mitochondrial respiration	Jagadeeswaran et al. (2009)
miR398	Beans	*Rhizobium tropici*	CSD1, Nod 19	Facilitate colonization	Naya et al. (2014)
miR400	*Arabidopsis thaliana*	*Pseudomonas syringae*	Pentatricopeptide repeat (PPR)	Regulation of a PRR gene	Fahlgren et al. (2007)
miR825	*Arabidopsis thaliana*	*Pseudomonas syringae*	Remorin, Zn finger homeobox, frataxin-related genes	Disease resistance	Fahlgren et al. (2007)
miR863-3p	*Arabidopsis thaliana*	*Pseudomonas syringae*	ARLPK1/ARLPK2 and SE	Immune signaling; miRNA biogenesis	Niu et al. (2016)

to regulate copper homeostasis by regulating various laccases during nitrogen-fixation in legumes (Huang et al., 2020). Inhibition of laccases in plants can reduce plant's ability to produce certain secretions that are otherwise helpful in wound healing and providing resistance against pathogens. Alternatively, miR398 is reported to regulate Cu-homeostasis in nitrogen-fixing plants by downregulating a key enzyme, Cu-superoxide dismutase (CSD) which is a crucial antioxidant enzyme protecting plant from oxidative stress by converting toxic oxygen radicals into nontoxic molecules of hydrogen peroxide and oxygen molecules. Regulation of Cu-homeostasis by miR398 demonstrates its significance in ROS detoxification (Ding and Zhu, 2009).

Furthermore, miRNAs such as miR398b and miR773 are known to regulate callose deposition in plants infected with bacteria providing disease resistance against pathogen (Li et al., 2010). These miRNAs target glucan synthase like-callose synthases regulating callose deposition during bacterial infection. While miR160a can positively regulate callose deposition, others candidates such as miR398b and miR773 are known to downregulate callose deposition in plants (Li et al., 2010). Callose, a type of glucan, is deposited between the plasma membrane and cell wall at the site of infection, generally around plasmodesmata, limits pathogen entry and its spread into the host cells.

In nutshell, identification of miRNA targets in host plants infected with bacterial pathogen as well as elucidation of their roles in plant defense can help in devising multipronged strategies for pathogen management.

3. miRNA and fungal infection in plants

Plants are subject to a variety of fungal infections which invade their host plants by establishing hyphal connections (Ferreira et al., 2006). miRNAs can regulate plant defense responses against fungi by regulating several target genes. Such a regulation is initiated by various fungal effectors. One such effector is Ecp6 reported in *Cladosporium fulvum*, causing leaf mold disease in *Solanum lycopersicum* (Sánchez-Vallet et al., 2013). Ecp6 binds with chitin elicitor binding protein, (CEBiP), a type of plant-PRR. Ecp-CEBiP interaction is mediated through chitin elicitor receptor kinase (CERK) which leads to a cascade that activates various MIR genes leading to the synthesis of miRNAs (De Jonge et al., 2010). These miRNAs can provide defense against fungi by regulating ROS accumulation, phytoalexins production (an antimicrobial secondary metabolite), production of chitinases and other enzymes, and by recruiting R-proteins (De Jonge et al., 2010).

Among several targets of miRNAs, gene encoding disease resistance proteins (R-proteins) remain central in conferring resistance against fungal pathogens. Fungal effectors usually cause pathogenicity by targeting R-proteins. Several miRNAs in *Populus szechuanica* infected by a fungus *Melampsora larici-populina* were related to regulating genes expression of R-proteins (Chen and Cao, 2015). In another example, of *Sclerotinia sclerotiorum-Brassica napus* system, miR1885a targeting R gene (TIR-NBS-LRR) and miR6030 targeting a At1g12290-like R-protein have been reported from *B. napus*. These miRNAs which are related to ETI response in *B. napus* have been correlated with increase in susceptibility to pathogen with rise in their abundance during fungal inoculation (Chen and Cao, 2015).

Oxidative burst involving the accumulation of ROS is among common defense responses reported during fungal infection which further causes hypersensitive response (HR) responsible for programmed cell death (PCD), eventually stopping the progression of disease. During ROS accumulation, superoxide radicals are produced in chloroplasts (Zurbriggen et al., 2009). The chloroplast enzyme Cu/Zn-SOD (superoxide dismutase) located in stroma converts superoxide radicals (O_2^-) into hydrogen peroxide molecules (H_2O_2) (Asada, 1999). H_2O_2 induces strong HR response by diffusing through cell membrane (Liu et al., 2007). miR393 and miR398 can downregulate Cu/Zn SOD expression, an enzyme solely responsible for neutralizing oxygen radicals in plants. However, miR396 can upregulate targets responsible for enhanced ROS activity rendering host plant either susceptible or resistant to infection, respectively (De Jonge et al., 2010). miR398 not only targets Cu/Zn superoxide dismutase (SOD) expression but also downregulates CCS1, a gene encoding a chaperone essential for stabilizing SOD (Chu et al., 2005). By destabilizing this enzyme, ROS detoxification is severely impaired.

Several reports have highlighted about the defensive roles of miRNAs against fungal infection by targeting TFs. For example, in *Puccinia cucumerina,* a devastating fungus of *A. thaliana*, miR396 can modulate the expression of TFs, viz. GRFs (Soto-Suárez et al., 2017). GRFs regulate diverse processes related to growth and development. Overproduction of miR396 during fungal infection leads to decrease in levels of GRFs eventually producing developmental effects rendering plant susceptible to fungal attack (Soto-Suárez et al., 2017). An artificial transgenic plant that mimics the targets of miR396 (designated as MIM396) was created to overcome such developmental defects and resulting into resistance against the fungal pathogen. As a result, MIM396 plants upon *P. cucumerina* infection showed low levels of miR396 along with high accumulation of ROS, and enhanced callose deposition providing better resistance (Soto-Suárez et al., 2017). In turmeric, *Curcuma longa*, infected with *Pythium aphanidermatum*, reduction in miR164 levels resulted in rise NAC protein (a type of TF implicated in plant growth and development) (Cao et al., 2016). Elevated levels of NAC could prevent HR during fungal infection. HR prevention can lead to susceptibility to fungal infection.

Furthermore, miRNAs can also activate plant defenses by regulating synthesis of key enzymes that are proven to be effective against fungal pathogens. Few examples of such enzymes include nitrate reductase, L-ascorbate oxidase, and thioredoxin reductase. Nitrate reductase is a key enzyme regulating nitric oxide (NO) signaling. NO can trigger the expression of genes conferring resistance to infection, synthesis of secondary metabolites with antifungal properties, and also in generating the HR (Martínez-Medina et al., 2019). Such roles have been demonstrated in *B. napus* infected with *S. sclerotiorum* where rise in miR397a levels has been correlated with decrease in expression of its target nitrate reductase suggesting its involvement in defense mechanisms against fungal infection. Diverse roles of several other miRNAs such as miR408, miR166e, miR5021, and miRNA-PC-14-5p have been related to the regulation of L-ascorbate oxidase (key enzyme involved in the management of oxidative stress), Cu-chaperone for SOD, Cu−Zn SOD, thioredoxin reductase (a key enzyme involved in maintenance of redox potential), respectively (Chen and Cao, 2015).

Additional functional roles of miRNA during fungal infection are related to mediation of phytohormone signaling, regulation of genes involved in root colonization, and other genes related to defense mechanism are summarized in Table 15.2. Although literature contains decent numbers of reports about direct involvement of miRNA in fungal infection, however specific pathways are yet to be elucidated in detail.

4. miRNA and viral infection in plants

Plant viruses infect their host plants either through wounds (e.g., tobacco mosaic virus, TMV in tobacco plants) or are transmitted employing insect as vectors (e.g., tomato yellow leaf curl virus, TYLCV is transmitted through insect *Bemisia tabaci* in papaya plants) (Lefeuvre et al., 2019). Viral infection requires efficient cell-to-cell passage, and its translocation to distant tissues by movement through vascular tissues. Genetic material of plant viruses (DNA/RNA) can be either single (ss) or double stranded (ds) (Lefeuvre et al., 2019). Vast majority of plant viruses contain RNA as genetic material (e.g., TMV) (Lefeuvre et al., 2019). Host plants deal with viral infection by several mechanisms. One such mechanism involves the degradation of viral transcripts by the recruitment of two kinds of proteins, DICER-like (DCL) and ARGONAUTES (AGO) loaded with miRNAs (Chapman and Carrington, 2007). miRNAs can regulate the synthesis of AGO proteins that are involved in degrading viral RNA molecules through a mechanism briefly described here. DCL proteins target viral dsRNA while AGO proteins target viral ssRNA molecules to process them into virus-derived small interfering RNAs (vsiRNAs) molecules (Zhang et al., 2015). Following virus infection, viral dsRNA are cleaved by DCL to produce vsiRNAs. These vsiRNAs are further picked-up by AGO proteins and are transported to complementary RNA or DNA targets (Mlotshwa et al., 2008). vsiRNA loaded on AGO proteins constitute core part of RNA-induced silencing complex (RISC) which can further inhibit the replication of viral DNAs. Some of vsiRNAs are processed further into secondary virus-activated small interfering RNAs (vasiRNAs) with the assistance of plant-derived RNA polymerases viz. (RDR1) (Xie et al., 2001). In order to inhibit viral infection, efficient processing of viral genetic material rely on the upregulation of AGO proteins which is upregulated by miR168 (Várallyay et al., 2010). Activation of AGO genes by miR168 leads to efficient processing of viral dsRNA molecules leading to a strong plant defense response against virus (Várallyay et al., 2010).

Plants defenses against viruses are not limited to the cleavage of viral transcripts at infection site but also involve transcriptional silencing of transposons, repetitive

Table 15.2 Functional roles of some plant miRNAs in defense against fungal infection.

Plant miRNA	Plant	Fungus	Target	Role	References
miR393	*Oncidium*	*Cladosporium fulvum*	AFB and TIR	Auxin signaling	De Jonge et al. (2010)
miR398	*Arabidopsis thaliana*	*Puccinia graminis*	CCS1, a chaperon of Cu/Zn SOD	ROS detoxification	Yin et al. (2019)
miR396	*Arabidopsis thaliana*	*Plectosphaerella cucumerina*	GRF	Immunity	Soto-Suárez et al. (2017)
	Arabidopsis thaliana	*Plectosphaerella cucumerina*	GRF	ROS detoxification; callose deposition	Soto-Suárez et al. (2017)
miR164	*Curcuma longa*	*Pythium aphanidermatum*	SA, NAM, and NAC	Plant growth and development	Chand et al. (2016)
miR397a	*Brassica napus*	*Sclerotinia sclerotiorum*	Nitrate reductase	NO signaling	Cao et al. (2016).
miR408	*Brassica napus*	*Sclerotinia sclerotiorum*	L-Ascorbate Oxidase	Oxidative stress management	Cao et al. (2016).
miR166	*Gossypium hirsutum*	*Verticillium dahliae*	Clp-1	Defense response	Zhang et al. (2016)
miR5021	*Brassica napus*	*Sclerotinia sclerotiorum*	Cu/Zn superoxide dismutase	ROS detoxification	Cao et al. (2016).
miRNA-PC-14-5p	*Brassica napus*	*Sclerotinia sclerotiorum*	Thioredoxin reductase	Maintenance of redox potential	Cao et al. (2016).
miR1885a	*Brassica napus*	*Sclerotinia sclerotiorum*	TIR-NBS-LRR	Disease resistance	Cao et al. (2016).
miR6030	*Brassica napus*	*Sclerotinia sclerotiorum*	R-protein	Disease resistance	Cao et al. (2016).
miR1447	*Populus beijingensis*	*Dothiorella gregaria*	R-protein	Defense response	Chen et al. (2012)
miR1448	*Gossypium hirsutum*	*Verticillium dahliae*	R-protein	Defense response	Yin et al. (2012)

Continued

Table 15.2 Functional roles of some plant miRNAs in defense against fungal infection.—*cont'd*

Plant miRNA	Plant	Fungus	Target	Role	References
miR482	*Gossypium hirsutum*	*Verticillium dahliae*	R-protein	Disease resistance	Yin et al. (2012)
	Solanum tuberosum	*Verticillium dahliae*	NBS-LRR genes	Defense response	Yang et al. (2015)
miR160	*Pinus taeda*	*Cronartium quercuum*	Auxin response factor, Aux/IAA	Auxin signaling	Sun (2012)
	Populus trichocarpa	*Botryosphaeria dothidea*	ARF, Aux/IAA	Auxin signaling	Zhao et al. (2012)
Can-miRn371	*Capsicum annuum*	*Colletotrichum truncatum*	Ethylene response factor (ERFs)	Ehylene signaling	Mishra et al. (2018)
miR1023	*Triticum aestivum*	*Fusarium graminearum*	FGSG_03101	Suppresses fungal invasion	Jiao and Peng (2018)
miR171	*Medicago truncatula*	*Rhizophagus* spp.	LOM1	Root colonization	Couzigou et al. (2017)
miR1448	*Populus trichocarpa*	*Botryosphaeria dothidea*	S-conjugate, ABC transporter, ATP-binding cassette transport protein	Defense response	Lu et al. (2008)
	Populus beijingensis	*Dothiorella gregaria*	Glutathione	Defense response	Zhao et al. (2012)
miR1450	*Populus trichocarpa*	*Botryosphaeria dothidea*	Leucine-rich repeat	Defense response	Zhao et al. (2012)
miR156	*Malus pumila*	*Alternaria alternaria*	TFs such as WRKY	Defense response	Zhang et al. (2017)
miR159	*Gossypium hirsutum*	*Verticillium dahliae*	HiC-15	Defense response	Zhang et al. (2016)
miR5300	*Solanum lycopersicum*	*Fusarium oxysporum*	Solyc05g008650, tm-2	Defense response	Ouyang et al. (2014)

DNA, and viral transgenes (Jones et al., 2001). One such mechanism involves DCL3 protein which can methylate promotor region of transgenes acquired from viruses. Methylation leads to the inactivation of viral transgenes and eventually providing resistance against infection. Plants can also defend viruses by systemic silencing which involves cell-to-cell movement of silencing miRNA molecules along with viral particles through plasmodesmata (Jones et al., 2001). Such a movement of silencing signals can spread to distant tissues through phloem translocation. If the silencing miRNA signal travels ahead of the virus, it can stop the virus from spreading to the distant tissue, but in case virus is leading the movement, infection persists (Kalantidis et al., 2008).

Apart from these, miRNAs also have several endogenous roles that are implicated in defense against viruses. Some of these roles outlined in Table 15.3 include the regulation of auxin signaling and upregulation R-genes through a mechanism as described in previous sections under plant-bacteria/fungus interactions. For instance, rise in miR160 and miR393 levels in soybean plants upon infection by soybean mosaic virus (SMV) leads to repression of auxin signaling TFs (Singh et al., 2023). Targets of such endogenous miRNAs also include genes participating in the growth and development of viral symptoms. For example, during southern rice black-streaked dwarf virus (SRBSDV) infection in rice, miR164 regulates the expression of NAC1 and NAM proteins resulting into symptoms such as the development of rootlets on aerial branches, and branching on nodes of stem (Xu et al., 2014). As a counterdefense mechanism, viruses also produce suppressor molecules such as P1/HC-Pro which can block PTGS pathways in their host plants. Such PTGS suppressors are reported in tobacco etch virus (TEV), potato virus Y (PVY), tomato mosaic virus (ToMV), TMV, and potato virus X (PVX). These viral suppressors can bind to double-stranded siRNA which leads to the inhibition of HEN1 methyltransferase activities necessary for the methylation of regulatory proteins. In the absence of methylation event, AGO proteins cannot sort their vsiRNAs and hence cannot further regulate the degradation of viral RNAs. These viral suppressors can also block siRNA biogenesis by downregulating RDR enzymes as well as can inhibit endogenous plant miRNAs eventually developing a strong viral infection in plants (Kasschau et al., 2003).

5. miRNA in defense against insects

Insects have evolved various feeding strategies to exploit their hosts plants (Jaenike, 1990). While some insects feed by chewing on plant tissues, others use piercing mouthparts to extract sap for nutrients (Bonaventure, 2012). Plants defense responses to insects are either constitutive or induced (Kaplan et al., 2008). miRNA play crucial roles in regulating plant defense against insect herbivory (Bordoloi and Agarwala, 2021). Some of the functions of miRNAs in plant defense against herbivory are outlined in Table 15.4.

Table 15.3 Functional roles of some plant miRNAs in viral defense.

miRNA	Plant	Virus	Target	Role	References
miR156	*Oryza sativus*	Rice black-streaked dwarf virus (RBSDV)	ARF	Auxin signaling	Sun et al. (2015)
miR160	*Glycine max*	Soybean mosaic virus (SMV)	ARF, TIR		Yin et al. (2013)
miR393	Beans	Mungbean yellow mosaic India virus (MYMIV)	ARF, TIR		Kundu et al. (2017)
miR394	*Glycine max*	Soybean mosaic virus (SMV)	ARF, TIR		Yin et al. (2013)
amiR159	*Arabidopsis thaliana*	Turnip yellow mosaic virus (TYMV)	P69, HC-Pro	Inhibits viral replication	Niu et al. (2006)
amiR171	*Nicotiana tabacum*	Cauliflower mosaic virus (CaMV)	2b		Qu et al. (2007)
miR159a	*Nicotiana benthamiana*	Plum poX virus (PPV)	P1/HC-Pro		Simón-Mateo and García (2006)
miR167b	*Nicotiana benthamiana*	Plum poX virus (PPV)	P1/HC-Pro		Simón-Mateo and García (2006)
miR167b	*Nicotiana tabacum*	Potato virus X (PVX)	TGBp1/p25		Ai et al. (2011)
miR171a	*Nicotiana benthamiana*	Plum poX virus (PPV)	P1/HC-Pro		Simón-Mateo and García (2006)
miR171a	*Nicotiana tabacum*	Potato virus X (PVX)	TGBp1/p25		Ai et al. (2011)
miR1885	*Brassica napus*	Turnip mosaic virus (TuMV)	TIR–NBS–LRR	Disease resistance	Wroblewski et al. (2007)
miR2118	*Oryza sativus*	Rice stripe virus (RSV)	NBS-LRR		Lian et al. (2016)
miR399	*Vigna mungo*	Mungbean yellow mosaic India virus (MYMIV)	NBS-LRR		Kundu et al. (2017)
amiR-AV1-1	*Solanum lycopersicum*	Tomato leaf curl New Delhi virus (ToLCNDV)	AV1 and AV2		Van Vu et al. (2013)
miR398	*Oryza sativus*	Rice black-streaked dwarf virus (RBSDV)	CSD1	ROS detoxification	Sun et al. (2015)
miR164	*Oryza sativus*	Rice black-streaked dwarf virus (RBSDV)	SA, NAM and NAC	Plant growth and development	Sun et al. (2015)
miR171	*Glycine max*	Soybean mosaic virus (SMV)	TFs such as Scarecrow-like	Signaling cascades	Chen et al. (2016)
miR396	*Oryza sativus*	Southern rice black-streaked dwarf virus (SRBSDV)	AtGRF	Leaf symptoms	Xu et al. (2014a,b)
miR397	*Vigna mungo*	Mungbean yellow mosaic India virus (MYMIV)	LOX	SA cascades	Kundu et al. (2017)

Table 15.4 Functional roles of some plant miRNAs involved in plant defense against insects.

Plant miRNA	Plant	Insect	Target	Role	References
miR156	*Nicotiana* spp.	*Manduca sexta*	TFs such as SPL	Growth and development	Stief et al. (2014)
miR157	*Triticum aestivum*	*Diuraphis noxia*	TFs such as MYB		Greyling (2012)
miR157	*Solanum lycopersicum*	*Bemisia tabaci*	TFs such as SPL		Stief et al. (2014)
miR172	*Camellia sinensis*	*Ectropis oblique*	CDPKs		Jeyaraj et al. (2017)
miR396	*Oryza sativa*	*Nilaparvata lugens*	OsGRF8		Dai et al. (2019)
miR164	*Cucumis melo*	*Aphis gossypii*	NAC		Sattar et al. (2012)
miR171	*Camellia sinensis*	*Ectropis oblique*	bHLH		Jeyaraj et al. (2017)
miR828	*Camellia sinensis*	*Ectropis oblique*	MYB TFs		Jeyaraj et al. (2017)
miR160	*Cucumis melo*	*Aphis gossypii*	ARF	Auxin signaling	Sattar et al. (2012)
miR167	*Cucumis melo*	*Aphis gossypii*	ARF		Sattar et al. (2012)
miR393	*Cucumis melo*	*Aphis gossypii*	TIR		Sattar et al. (2012)
miR160a	*Chrysanthemum morifolium*	*Macrosiphoniella sanbourni*	ARF		Xia et al. (2015)
miR167	*Nicotiana* spp.	*Manduca sexta*	TIR and ARF		Bozorov et al. (2012)
miR393a	*Chrysanthemum morifolium*	*Macrosiphoniella sanbourni*	TIR1		Xia et al. (2015)
miR319	*Arabidopsis thaliana*	*Macrosteles quadrilineatus*	TCP3	Jasmonic acid signaling	Li (2015)
miR167a-5p	*Oryza sativa*	*Nilaparvata lugens*	Calmodulin-like protein	Calcium signaling	Wu et al. (2017)
miR396d-3p	*Camellia sinensis*	*Ectropis oblique*	CBP, MAPK	Cell Signaling	Jeyaraj et al. (2017)
miR531	*Oryza sativa*	*Nilaparvata lugens*	MAPK cascade family members		Zhang et al. (2017)
miR390	*Solanum lycopersicum*	*Bemisia tabaci*	LRR-RLKs	Disease resistance protein	Wang KeTao et al. (2018)
miR156	*Camellia sinensis*	*Ectropis oblique*	LRR-RLKs		Jeyaraj et al. (2017)
miR172	*Solanum lycopersicum*	*Bemisia tabaci*	TFs such as AP2/ERF	Ethylene signaling	Wang KeTao et al. (2018)
miR159a	*Chrysanthemum morifolium*	*Macrosiphoniella sanbourni*	GAMYB-like 2	Gibberellic acid signaling	Xia et al. (2015)
miR171	*Solanum lycopersicum*	*Bemisia tabaci*	TFs such as GRAS		Wang KeTao et al. (2018)
miR398b	*Oryza sativa*	*Nilaparvata lugens*	Cu/Zn SOD	ROS detoxification	Wu et al. (2017)

Generally, defense responses to insect attack are elicited through various effectors present in oral secretion of larvae (Hogenhout and Bos, 2011). For example, glucose oxidase (GOX) has been extracted from oral secretions of *Helicoverpa armigera* and is known to elicit plant defenses (Yang et al., 2017). These oral elicitors transmit a salicylic acid-induced protein kinase (SIPK) and wound-induced protein kinase (WIPK) cascade (Hogenhout and Bos, 2011). Such a cascade upregulates MIR genes leading to the biogenesis of specific miRNAs. There are several possible routes by which these synthesized miRNAs can generate defense responses against insects. One such route includes the degradation of mRNA transcripts of target genes leading to the development of resistance against insects. This degradation is accomplished only if perfect pairing between miRNA and its target mRNA transcript exists (Yang et al., 2017). Another mechanism includes translational repression of target genes. For example, in *A. thaliana,* miR156 loaded on AGO1 protein (in RISC complex) associates with a SUO protein and binds with 5′ UTR of mRNA of a target gene, SPL3 (squamosa-promoter binding protein-like) and stops translation initiation by obstructing the movement of ribosomes. This translation repression does not follow mRNA decay but can control target gene expression (Yang et al., 2012). SPL3 is needed for the induction of flowering and changes during the vegetative phase of the plant (Cardon et al., 1997). Consequently, if SPL3 is downregulated, it will negatively affect the plant development and eventually render the plant unsuitable to insect feeding.

Furthermore, endogenous production of miRNAs elicited by insect herbivory can lead to changes in the levels of specific genes transcripts such as those encoding TFs *viz.* NAC and bHLH (Bordoloi and Agarwala, 2021). NAC TFs represent one of the abundant groups of TFs in plants having functional roles in plant growth and development (Yuan et al., 2019). NAC proteins contain DNA-binding domain which can bind to CGT[AG] domains of their target genes and thereby felicitating gene function suitable for providing resistance against insects (Olsen et al., 2005). miR164 is reported to positively regulate NAC6 TF leading to the upregulation of their targets in *A. thaliana* (Sattar et al., 2012). Another example of miRNA that targets TFs such as MYB is miR828. MYB is involved in GA synthesis providing resistance against insects by altering plant growth and development (Yu and Wang, 2020).

Furthermore, miRNAs can defend against insects by employing unique mechanisms that can impact the life history of the insects feeding on plants (Sattar et al., 2012; Ortolá and Daròs, 2024). For example, in *Cucumis melo*, miRNAs are deployed in delaying flow of phloem sap which negatively impacts fecundity and overall fitness of aphids' feeding on their host plants (Sattar et al., 2012). In *C. melo* plant containing dominant Vat^{+} gene (a type of R-gene), aphids feeding is deterred due to the presence of R-protein in the phloem sap (Sattar et al., 2012).

Furthermore, ingestion of plant-miRNAs may interact with insect gene expression and modulate their physiological processes. Aphids feeding on *Chrysanthemum* trigger the downregulation of several miRNAs *viz.* miR160a, miR167, and miR393a having auxin response factors (ARFs) and transport inhibitor response 1 (TIR1) as their potential targets (Singh et al., 2023). Downregulation of these miRNAs causes upregulation of ARF and TIR leading to the accumulation of auxin. Auxin promotes plant growth rendering plants more susceptible to insects.

Plant defense responses to phytophagous insects are mediated via unique plant-hormone signaling networks resulting into an outcome that ultimately affects the herbivores (Erb et al., 2012). It is a usual observation that plants subjected to insect herbivory show enhanced production of JA and SA. In *A. thaliana*, miR319 can inhibit the synthesis of TF, TCP3 (Teosinte Branched Cycloidea/Proliferating Cell Factor). TCP3 repression upregulates LOX gene which further leads to enhanced levels of JA (Lan and Qin, 2020). JA signaling can further activate defense responses including the production of plant protease inhibitors (PPIs) and secondary metabolites (Rojo et al., 2003). Protease inhibitors can impair protein digestion in insect gut leading to decreased fitness and increased mortality of insects (Jongsma and Bolter, 1997). In *Nicotiana attenuata*, RdR1, a key protein of miRNA processing machinery, is induced by foliar application of either JA/SA or by larval regurgitant (Pandey and Baldwin, 2007). In silenced rdr1 plants, nicotine production in leaves is significantly lowered, rendering plants susceptible to pests implying role of miRNAs in plant defense (Pandey and Baldwin, 2007). Recent reports have confirmed the transfer of plant-derived miRNA in different tissues upon dietary ingestion by insects. miRNAs, once ingested in dietary tissues, can be transported into cellular compartments of insect gut through transmembrane transporter-like proteins, receptor-mediated endocytosis, and extracellular vesicles (Zhao et al., 2012). Transportation of miRNA in vesicles maintains its bioactivity and ensures proper release at target sites. Ingested miRNAs can impair several physiological processes of insects leading to impairment of insect development. Overall these specific examples of mechanism clearly suggest that miRNAs play diverse functional roles in plant defense against insects comprising a complex network of cross-talks between different phytohormones signaling pathways and other signaling components by producing an outcome that will provide resistance against insects.

6. Conclusion

Functional roles of miRNAs in plant defense against pathogens such as bacteria, fungi, and viruses as well as insects are diverse. Overall, these responses are initiated *via* interaction of specific host plant PRRs with PAMP/effectors leading to the activation of specific MAPK pathways that activates miRNA production. miRNAs thereafter regulate their target gene expression. Most commonly, these targets are the genes involved in growth and development, production of R-proteins, secondary metabolite production, ROS detoxification, nutrient (Cu) homeostasis, regulation of key metabolic enzymes, and TFs. Characterization of specific pathways showing involvement of miRNAs is crucial for understanding the involvement of miRNAs in different metabolic pathways so that better crop improvement strategies be devised.

References

Ai, T., et al., 2011. Highly efficient virus resistance mediated by artificial microRNAs that target the suppressor of PVX and PVY in plants. Plant Biol. 13 (2), 304–316. https://doi.org/10.1111/j.1438-8677.2010.00374.x.

Asada, K., 1999. The water-water cycle in chloroplasts: scavenging of active oxygens and dissipation of excess photons. Ann. Rev. Plant Biol. 50 (1), 601–639.

Bigeard, J., Hirt, H., 2018. Nuclear signaling of plant MAPKs. Front. Plant Sci. 9, 338308.

Boccara, M., et al., 2014. The Arabidopsis miR472-RDR6 silencing pathway modulates PAMP-and effector-triggered immunity through the post-transcriptional control of disease resistance genes. PLoS Pathog. 10 (1).

Boller, T., Felix, G., 2009. A renaissance of elicitors: perception of microbe-associated molecular patterns and danger signals by pattern-recognition receptors', *Annual review of plant biology*. Ann. Rev. 60, 379–406.

Bonaventure, G., 2012. Perception of insect feeding by plants. Plant Biol. 14 (6), 872–880.

Bordoloi, K.S., Agarwala, N., 2021. MicroRNAs in plant insect interaction and insect pest control. Plant Gene 26, 100271.

Bozorov, T.A., Baldwin, I.T., Kim, S.-G., 2012. Identification and profiling of miRNAs during herbivory reveals jasmonate-dependent and-independent patterns of accumulation in Nicotiana attenuata. BMC Plant Biol. 12, 1–13.

Cao, J.-Y., et al., 2016. Tight regulation of the interaction between Brassica napus and Sclerotinia sclerotiorum at the microRNA level. Plant Mol. Biol. 92, 39–55.

Cardon, G.H., et al., 1997. Functional analysis of the *Arabidopsis thaliana* SBP-box gene SPL3: a novel gene involved in the floral transition. Plant J. 12 (2), 367–377.

Chand, S.K., et al., 2016. Identification and characterization of microRNAs in turmeric (*Curcuma longa* L.) responsive to infection with the pathogenic fungus *Pythium aphanidermatum*. Physiol. Mol. Plant Pathol. 93, 119–128.

Chapman, E.J., Carrington, J.C., 2007. Specialization and evolution of endogenous small RNA pathways. Nat. Rev. Genet. 8 (11), 884–896.

Chen, H., et al., 2016. Genome-wide investigation using sRNA-Seq, degradome-Seq and transcriptome-Seq reveals regulatory networks of microRNAs and their target genes in soybean during soybean mosaic virus infection. PLoS One 11 (3).

Chen, L., et al., 2012. Genome-wide profiling of novel and conserved populus microRNAs involved in pathogen stress response by deep sequencing. Planta 235, 873–883.

Chen, M., Cao, Z., 2015. Genome-wide expression profiling of microRNAs in poplar upon infection with the foliar rust fungus melampsora larici-populina. BMC Genom. 16, 1–13.

Chu, C.-C., et al., 2005. A copper chaperone for superoxide dismutase that confers three types of copper/zinc superoxide dismutase activity in Arabidopsis. Plant Physiol. 139 (1), 425–436.

Couzigou, J.-M., et al., 2017. Positive gene regulation by a natural protective miRNA enables arbuscular mycorrhizal symbiosis. Cell Host Microbe 21 (1), 106–112.

Dai, Z., et al., 2019. The OsmiR396–Os GRF 8–OsF3H-flavonoid pathway mediates resistance to the brown planthopper in rice (Oryza sativa). Plant Biotechnol. J. 17 (8), 1657–1669.

De Jonge, R., et al., 2010. Conserved fungal LysM effector Ecp6 prevents chitin-triggered immunity in plants. Science 329 (5994), 953–955.

De Luis, A., et al., 2012. Two microRNAs linked to nodule infection and nitrogen-fixing ability in the legume *Lotus japonicus*. Plant Physiol. 160 (4), 2137–2154.

Ding, Y.-F., Zhu, C., 2009. The role of microRNAs in copper and cadmium homeostasis. Biochem. Biophys. Res. Commun. 386 (1), 6–10.

Erb, M., Meldau, S., Howe, G.A., 2012. Role of phytohormones in insect-specific plant reactions. Trend. Plant Sci. 17 (5), 250–259.

Fahlgren, N., et al., 2007. High-throughput sequencing of Arabidopsis microRNAs: evidence for frequent birth and death of MIRNA genes. PLoS One 2 (2), e219.

Ferreira, R.B., et al., 2006. Fungal pathogens: the battle for plant infection. Crit. Rev. Plant Sci. 25 (6), 505–524.

Gómez-Gómez, L., Boller, T., 2000. FLS2: an LRR receptor–like kinase involved in the perception of the bacterial elicitor flagellin in Arabidopsis. Mol. Cell 5 (6), 1003–1011.

Greyling, S.-M., 2012. Isolation of Russian Wheat Aphid-Induced ncRNA from Wheat. University of Johannesburg, South Africa.

Guo, Z., Li, Y., Ding, S.-W., 2019. Small RNA-based antimicrobial immunity. Nat. Rev. Immunol. 19 (1), 31–44.

Gust, A.A., Pruitt, R., Nürnberger, T., 2017. Sensing danger: key to activating plant immunity. Trend. Plant Sci. 22 (9), 779–791.

Hogenhout, S.A., Bos, J.I.B., 2011. Effector proteins that modulate plant–insect interactions. Curr. Opin. Plant Biol. 14 (4), 422–428.

Hou, S., Liu, D., He, P., 2021. Phytocytokines function as immunological modulators of plant immunity. Stress Biol. 1, 1–14.

Huang, S., et al., 2020. Plant miR397 and its functions. Funct. Plant Biol. 48 (4), 361–370.

Jaenike, J., 1990. Host specialization in phytophagous insects. Ann. Rev. Ecol. Syst. 21 (1), 243–273.

Jagadeeswaran, G., Saini, A., Sunkar, R., 2009. Biotic and abiotic stress down-regulate miR398 expression in Arabidopsis. Planta 229, 1009–1014.

Jeyaraj, A., et al., 2017. Genome-wide identification of microRNAs responsive to *Ectropis oblique* feeding in tea plant (*Camellia sinensis* L.). Sci. Rep. 7 (1), 13634.

Jiao, J., Peng, D., 2018. Wheat microRNA1023 suppresses invasion of *Fusarium graminearum* via targeting and silencing FGSG_03101. J. Plant Interact. 13 (1), 514–521.

Jones, J.D.G., Dangl, J.L., 2006. The plant immune system. Nature 444 (7117), 323–329.

Jones, L., Ratcliff, F., Baulcombe, D.C., 2001. RNA-directed transcriptional gene silencing in plants can be inherited independently of the RNA trigger and requires Met1 for maintenance. Curr. Biol. 11 (10), 747–757.

Jongsma, M.A., Bolter, C., 1997. The adaptation of insects to plant protease inhibitors. J. Insect Physiol. 43 (10), 885–895.

Kalantidis, K., et al., 2008. RNA silencing movement in plants. Biol. Cell. 100 (1), 13–26.

Kaplan, I., et al., 2008. Constitutive and induced defenses to herbivory in above-and below-ground plant tissues. Ecology 89 (2), 392–406.

Kasschau, K.D., et al., 2003. P1/HC-Pro, a viral suppressor of RNA silencing, interferes with Arabidopsis development and miRNA function. Dev. Cell 4 (2), 205–217.

Kundu, A., et al., 2017. High throughput sequencing reveals modulation of microRNAs in *Vigna mungo* upon mungbean yellow mosaic India Virus inoculation highlighting stress regulation. Plant Sci. 257, 96–105.

Lan, J., Qin, G., 2020. The regulation of CIN-like TCP transcription factors. Int. J. Mole. Sci. 21 (12), 4498.

Lefeuvre, P., et al., 2019. Evolution and ecology of plant viruses. Nat. Rev. Microbiol. 17 (10), 632–644.
Li, H., et al., 2010. Misexpression of miR482, miR1512, and miR1515 increases soybean nodulation. Plant Physiol. 153 (4), 1759–1770.
Li, S., 2015. The *Arabidopsis thaliana* TCP transcription factors: a broadening horizon beyond development. Plant Signal. Behav. 10 (7).
Li, Y., et al., 2010. Identification of microRNAs involved in pathogen-associated molecular pattern-triggered plant innate immunity. Plant Physiol. 152 (4), 2222–2231.
Lian, S., et al., 2016. Time-course small RNA profiling reveals rice miRNAs and their target genes in response to rice stripe virus infection. PLoS One 11 (9).
Liu, Y., et al., 2007. Chloroplast-generated reactive oxygen species are involved in hypersensitive response-like cell death mediated by a mitogen-activated protein kinase cascade. Plant J. 51 (6), 941–954.
Lu, S., Sun, Y., Chiang, V.L., 2008. Stress-responsive microRNAs in populus. Plant J. 55 (1), 131–151.
Luo, L., 2012. Plant cytokine or phytocytokine. Plant Signal. Behav. 7 (12), 1513–1514.
Ma, K.-W., Ma, W., 2016. Phytohormone pathways as targets of pathogens to facilitate infection. Plant Mol. Biol. 91, 713–725.
Martínez-Medina, A., et al., 2019. Nitric oxide in plant–fungal interactions. J. Exp. Bot. 70 (17), 4489–4503.
Mishra, R., et al., 2018. Can-miRn37a mediated suppression of ethylene response factors enhances the resistance of chilli against anthracnose pathogen *Colletotrichum truncatum* L. Plant Sci. 267, 135–147.
Mlotshwa, S., Pruss, G.J., Vance, V., 2008. Small RNAs in viral infection and host defense. Trend. Plant Sci. 13 (7), 375–382.
Navarro, L., et al., 2006. A plant miRNA contributes to antibacterial resistance by repressing auxin signaling. Science 312 (5772), 436–439.
Navarro, L., et al., 2008. Suppression of the microRNA pathway by bacterial effector proteins. Science 321 (5891), 964–967.
Naya, L., et al., 2014. Regulation of copper homeostasis and biotic interactions by microRNA 398b in common bean. PloS One 9 (1), e84416.
Nazarov, P.A., et al., 2020. Infectious plant diseases: etiology, current status, problems and prospects in plant protection. Acta Nat. 12 (3), 46.
Niu, D., et al., 2016. miRNA863-3p sequentially targets negative immune regulator ARLPKs and positive regulator SERRATE upon bacterial infection. Nat. Commun. 7 (1), 11324.
Niu, Q.-W., et al., 2006. Expression of artificial microRNAs in transgenic *Arabidopsis thaliana* confers virus resistance. Nat. Biotechnol. 24 (11), 1420–1428.
Nova-Franco, B., et al., 2015. The micro-RNA172c-APETALA2-1 node as a key regulator of the common bean-Rhizobium etli nitrogen fixation symbiosis. Plant Physiol. 168 (1), 273–291.
Olsen, A.N., et al., 2005. DNA-binding specificity and molecular functions of NAC transcription factors. Plant Sci. 169 (4), 785–797.
Ortolá, B., Daròs, J.-A., 2024. RNA interference in insects: from a natural mechanism of gene expression regulation to a biotechnological crop protection promise. Biology 13 (3), 137.
Ouyang, S., et al., 2014. MicroRNAs suppress NB domain genes in tomato that confer resistance to *Fusarium oxysporum*. PLoS Pathog. 10 (10).
Pandey, S.P., Baldwin, I.T., 2007. RNA-directed RNA polymerase 1 (RdR1) mediates the resistance of *Nicotiana attenuata* to herbivore attack in nature. Plant J. 50 (1), 40–53.

Pérez-Quintero, Á.L., et al., 2012. Bioinformatic identification of cassava miRNAs differentially expressed in response to infection by *Xanthomonas axonopodis* pv. manihotis. BMC Plant Biol. 12, 1–11.

Pieterse, C.M.J., et al., 2014. Induced systemic resistance by beneficial microbes. Annual Rev. Phytopathol. 52, 347–375.

Pikaard, C.S., Scheid, O.M., 2014. Epigenetic regulation in plants. Cold Spring Harbor Perspect. Biol. 6 (12).

Powers, S.K., Strader, L.C., 2016. Up in the air: untethered factors of auxin response. F1000Research 5.

Qu, J., Ye, J., Fang, R., 2007. Artificial microRNA-mediated virus resistance in plants. J. virol. Am Soc. Microbiol. 81 (12), 6690–6699.

Rojo, E., Solano, R., Sánchez-Serrano, J.J., 2003. Interactions between signaling compounds involved in plant defense. J. Plant Growth Regul. 22, 82–98.

Sánchez-Vallet, A., et al., 2013. Fungal effector Ecp6 outcompetes host immune receptor for chitin binding through intrachain LysM dimerization. Elife 2, e00790.

Sattar, S., et al., 2012. Cucumis melo microRNA expression profile during aphid herbivory in a resistant and susceptible interaction. Mole. Plant-Micro. Int. Am. Phytopath Soci. 25 (6), 839–848.

Simón-Mateo, C., García, J.A., 2006. MicroRNA-guided processing impairs Plum pox virus replication, but the virus readily evolves to escape this silencing mechanism. J. Virol. Am. Soc. Microbiol. 80 (5), 2429–2436.

Singh, S., Assumi, R., Bhadrecha, P., 2023. Crosstalk of MicroRNAs with phytohormone signalling pathways. In: Plant MicroRNAs and Stress Response. CRC Press, pp. 257–276.

Soto-Suárez, M., et al., 2017. The Arabidopsis miR396 mediates pathogen-associated molecular pattern-triggered immune responses against fungal pathogens. Sci. Rep. 7 (1), 44898.

Stief, A., et al., 2014. Arabidopsis miR156 regulates tolerance to recurring environmental stress through SPL transcription factors. Plant Cell 26 (4), 1792–1807.

Sun, G., 2012. MicroRNAs and their diverse functions in plants. Plant Mol. Biol. 80, 17–36.

Sun, Z., et al., 2015. Genome-wide characterization of rice black streaked dwarf virus-responsive microRNAs in rice leaves and roots by small RNA and degradome sequencing. Plant Cell Physiol. 56 (4), 688–699.

Tanaka, K., Heil, M., 2021. Damage-associated molecular patterns (DAMPs) in plant innate immunity: applying the danger model and evolutionary perspectives. Ann. Rev. Phytopathol. 59, 53–75.

Tang, D., Wang, G., Zhou, J.-M., 2017. Receptor kinases in plant-pathogen interactions: more than pattern recognition. Plant Cell 29 (4), 618–637.

Thiebaut, F., et al., 2014. Genome-wide identification of microRNA and siRNA responsive to endophytic beneficial diazotrophic bacteria in maize. BMC Genom. 15, 1–18.

Várallyay, É., et al., 2010. Plant virus-mediated induction of miR168 is associated with repression of ARGONAUTE1 accumulation. EMBO J. 29 (20), 3507–3519.

Versluys, M., Tarkowski, Ł.P., Van den Ende, W., 2017. Fructans as DAMPs or MAMPs: evolutionary prospects, cross-tolerance, and multistress resistance potential. Front. Plant Sci. 7, 239345.

Van Vu, T., Choudhury, N.R., Mukherjee, S.K., 2013. Transgenic tomato plants expressing artificial microRNAs for silencing the pre-coat and coat proteins of a begomovirus, tomato leaf curl New Delhi virus, show tolerance to virus infection. Virus Res. 172 (1–2), 35–45.

Wang KeTao, W.K., et al., 2018. Identification and Characterization of microRNA during Bemisia Tabaci Infestations in Solanum lycopersicum and Solanum Habrochaites.

Wang, Y., et al., 2009. Identification and expression analysis of miRNAs from nitrogen-fixing soybean nodules. Biochem. Biophys. Res. Commun. 378 (4), 799–803.

Wroblewski, T., et al., 2007. Silencing of the major family of NBS–LRR-encoding genes in lettuce results in the loss of multiple resistance specificities. Plant J. 51 (5), 803–818.

Wu, Y., et al., 2017. Identification and analysis of brown planthopper-responsive microRNAs in resistant and susceptible rice plants. Sci. Rep. 7 (1), 8712.

Xia, X., et al., 2015. MicroRNA expression profile during aphid feeding in chrysanthemum (*Chrysanthemum morifolium*). PLoS One 10 (12).

Xie, Z., et al., 2001. An important role of an inducible RNA-dependent RNA polymerase in plant antiviral defense. Proc. Natl. Acad. Sci. U. S. A. 98 (11), 6516–6521.

Xu, D., et al., 2014a. MicroRNAs responding to southern rice black-streaked dwarf virus infection and their target genes associated with symptom development in rice. Virus Res. 190, 60–68.

Xu, W., Meng, Y., Wise, R.P., 2014. Mla-and Rom1-mediated control of microRNA398 and chloroplast copper/zinc superoxide dismutase regulates cell death in response to the barley powdery mildew fungus. New Phytol. 201 (4), 1396–1412.

Yan, Z., et al., 2016. Identification and functional characterization of soybean root hair micro RNA s expressed in response to B radyrhizobium japonicum infection. Plant Biotechnol. J. 14 (1), 332–341.

Yang, L., et al., 2015. Overexpression of potato miR482e enhanced plant sensitivity to Verticillium dahliae infection. J. Integr. Plant Biol. 57 (12), 1078–1088.

Yang, L., et al., 2017. Expressional divergence of insect GOX genes: from specialist to generalist glucose oxidase. J. Insect Physiol. 100, 21–27.

Yang, L., Wu, G., Poethig, R.S., 2012. Mutations in the GW-repeat protein SUO reveal a developmental function for microRNA-mediated translational repression in Arabidopsis. Proc. Natl. Acad. Sci. U. S. A. 109 (1), 315–320.

Yin, C., et al., 2019. A novel fungal effector from Puccinia graminis suppressing RNA silencing and plant defense responses. New Phytol. 222 (3), 1561–1572.

Yin, X., et al., 2013. Detection and evolutionary analysis of soybean miRNAs responsive to soybean mosaic virus. Planta 237, 1213–1225.

Yin, Z., et al., 2012. Genome-wide profiling of miRNAs and other small non-coding RNAs in the Verticillium dahliae–inoculated cotton roots. PLoS One 7 (4), e35765.

Yu, S., Wang, J.-W., 2020. The crosstalk between microRNAs and gibberellin signaling in plants. Plant Cell Physiol. 61 (11), 1880–1890.

Yuan, M., et al., 2021. PTI-ETI crosstalk: an integrative view of plant immunity. Curr. Opin. Plant Biol. 62, 102030.

Yuan, X., et al., 2019. NAC transcription factors in plant immunity. Phytopathol. Res. 1 (1), 1–13.

Zhang, C., et al., 2015. Biogenesis, function, and applications of virus-derived small RNAs in plants. Front. Microbiol. 6, 163530.

Zhang, Q., et al., 2017. Md-miR156ab and Md-miR395 target WRKY transcription factors to influence apple resistance to leaf spot disease. Front. Plant Sci. 8, 526.

Zhang, T., et al., 2016. Cotton plants export microRNAs to inhibit virulence gene expression in a fungal pathogen. Nat. Plant. 2 (10), 1–6.

Zhang, W., et al., 2011a. Bacteria-responsive microRNAs regulate plant innate immunity by modulating plant hormone networks. Plant Mol. Biol. 75, 93–105.

Zhang, X., et al., 2011b. Arabidopsis Argonaute 2 regulates innate immunity via miRNA393*-mediated silencing of a Golgi-localized SNARE gene, MEMB12. Mol. Cell 42 (3), 356–366.

Zhao, H., et al., 2013. Small RNA profiling reveals phosphorus deficiency as a contributing factor in symptom expression for citrus huanglongbing disease. Mol. Plant 6 (2), 301–310.

Zhao, J.-P., et al., 2012. Involvement of microRNA-mediated Gene Expression Regulation in the Pathological Development of Stem Canker Disease in Populus Trichocarpa. Public Library of Science (San Francisco, USA).

Zhou, F., et al., 2020. Co-incidence of damage and microbial patterns controls localized immune responses in roots. Cell 180 (3), 440–453.

Zhou, J.-M., Zhang, Y., 2020. Plant immunity: danger perception and signaling. Cell 181 (5), 978–989.

Zurbriggen, M.D., et al., 2009. Chloroplast-generated reactive oxygen species play a major role in localized cell death during the non-host interaction between tobacco and *Xanthomonas campestris* pv. vesicatoria. Plant J. 60 (6), 962–973.

CHAPTER

MicroRNA databases and bioinformatics analysis 16

Katarína Ražná[1], Ľubomír Harenčár[2], Simona Čerteková[1], Matúš Kučka[1]

[1]*Slovak University of Agriculture in Nitra, Faculty of Agrobiology and Food Resources, Institute of Plant and Environmental Sciences, Nitra, Slovakia;* [2]*Department of Molecular Biology and Biotechnology, Institute of Plant Genetics and Biotechnology, Plant Science and Biodiversity Centre, Slovak Academy of Sciences, Nitra, Slovakia*

1. MicroRNA-related repositories

Since the first microRNA registry, the *miRBase* (https://www.mirbase.org) was published by its founder Griffiths-Jones in 2004, many other microRNA-related repositories and tools have been created. The *miRBase* is a complex database of known miRNA sequences and annotations. This database has established rules for sequence labeling and provides guidelines for uniform criteria for miRNA annotation (Ambros et al., 2003; Meyers et al., 2008). The database is regularly updated, and a status is represented by 38,589 entries. The search of miRNA sequences within the database is possible based on taxonomy or by miRNA identifier or keyword. When searching it is possible to focus on the precursor sequences or just the mature miRNAs. The single sequence search is supported through *RNAcentral* (the database of noncoding RNA sequences—https://rnacentral.org/). The user page also allows to perform BLASTN and SSEARCH searches of a query sequence against the database of known miRNAs.

Database Commons (https://ngdc.cncb.ac.cn/databasecommons) is an integrated system of biological databases, which was founded in 2015. The databases are classified based on data object (animal, plant, fungi, bacteria, archaea, and virus), data type (deoxyribonucleic acid, ribonucleic acid, and protein), database category (for example, gene, genome, and annotation; phylogeny and homology; modification; structure; interaction; pathway; health and medicine; standard, ontology, and nomenclature; literature; metadata ...).

Under *Database Commons* operates *DIANA-TarBase* (http://diana.imis.athena-innovation.gr/DianaTools), previously *TarBase*, founded in 2006, which is the first database gathering experimentally validated interactions between miRNAs and genes, supported by detailed metadata. *DIANA Tools* (http://diana.imis.athena-innovation.gr/DianaTools/) provides in silico tools, software, and data for interpretation, and archiving data originated from deep sequencing data. The *DIANA Tools* offers algorithms *microT* v4, and *microT-CDS* for various target prediction, databases *TarBase* v7.0, and *LncBase*, which contain miRNA targets for both coding

MicroRNA Advances and Application in Plant Biology. https://doi.org/10.1016/B978-0-443-21682-4.00002-6

and noncoding RNAs and *mirPath* software to identify molecular pathways that are modified by miRNA expression (Sao et al., 2021).

PMRD, Plant MicroRNA Database (http://bioinformatics.cau.edu.cn/PMRD/) founded in 2010, integrates data on miRNAs occurring in plants from public databases, including literature, and data originated from miRNAs expression profiles (Zhang et al., 2010). *PMRD* provides several tools as noncoding RNA keyword search, literature-based search, miRNA-target search, new miRNA prediction toolkit, and others.

PmiRKB, Plant microRNA Knowledge Base (http://bis.zju.edu.cn/pmirkb/), includes miRNA data of Arabidopsis (*Arabidopsis thaliana*) and rice (*Oryza sativa*) (Meng et al., 2010). Of the modules, the main four functional modules are: "SNPs" module—collects data of single nucleotide polymorphism within precursor microRNAs (pre-miRNAs), and duplexes between miRNA and target RNA; "Pri-miR" module—investigates transcriptional activity of precursor and primary microRNAs (pri-miRNAs) in the tissue-specific manner; "MiR-Tar" module—evaluates data between miRNAs and their targets by the parallel analysis of RNA ends (PARE) approach, and "Self-reg" module—investigates the metabolism of miRNA precursors on their host precursors, including PARE data.

PMTED, Plant MiRNA Target Expression Database (http://pmted.agrinome.org), is collecting the data of expression profiles of miRNA targets originated from microarray experiments obtained from Gene Expression Ominibus (GEO). The database also provides basic information for searching miRNA sequences with their targets, gene ontology, and different gene expression profiles (Sun et al., 2013).

MepmiRDB, Medicinal Plant microRNA Database (http://mepmirdb.cn/mepmirdb/index.html), provides data on plant microRNAs incorporated in the regulation of secondary metabolites biosynthesis and accumulation in 29 medicinal plant species (Yu et al., 2019). This database provides miRNA sequences and their expression activity. Interactions between miRNAs and targets are supported by degradome-seq data in eight species that are present in the "Interaction" module of the database. The "Expression" module contains information about the expression profiles of mature miRNAs in certain specific organs or growth conditions (Sao et al., 2022).

TransmiR (http://www.cuilab.cn/transmir) is a database, where can be found regulatory relations between transcription factors and miRNAs (Wang et al., 2010). Last update of the database (Tong et al., 2019) published 3730 TF-miRNA regulations for 19 species. The database is mainly focused on animal and human's virus models but also contains data for some plant species, such as *A. thaliana, Solanum lycopersicum,* and *Glycine max.*

Arabidopsis thaliana microRNA-regulated pathways database (http://ppi.bioinfo.asia.edu.tw/EDMRP) provides the data (microRNA, target genes and protein-protein interaction) from the study of pathogen–host interactions between *A. thaliana* microRNA transcription promoter and a pathogen effector *Xanthomonas campestris* pv *campestris* (Xcc) (Kurubanjerdjit et al., 2014). The accessibility of the database depends on the server and network status.

Plant miRNA Encyclopedia PmiREN is a complex database of functional plant miRNAs (https://pmiren.com/) (Guo et al., 2020). It was updated in 2021 (PmiREN2.0; Guo et al., 2022) and publishes miRNA data from more than 90 plant species, including predicted transcription factors and miRNA targets annotation.

TarDB is a database that contains information on plant miRNAs/targets and miRNAs/siRNAs (http://www.biosequencing.cn/TarDB). The database collects miRNA targets that are supported by degradome/PARE sequencing, that are conserved between species, and phasic small interfering RNA (phasiRNA) loci, triggered by miRNAs (Liu et al., 2021).

sRNAanno is a database of annotated small RNAs in plants (www.plantsRNAs.org). The database contents annotations of small RNA of 143 plant species with sequenced genome (Chen et al., 2021). The annotations include miRNAs, phasiRNAs, and hc-siRNAs classes of small RNA.

According to the current state of content focus of miRNA-focused databases, this ratio is in favor of animals and humans' models. Most of the microRNA-based databases collect the data on human or animal sample providing information about various miRNA expression profiles in different biological conditions. Not all freely available databases are updated on a regular basis, and for some of them, there is a problem with real-time accessibility, which is influenced by the server and network status.

2. Databases for structural studies of microRNAs

A matured miRNA molecule is developed from a pre-miRNA and the latter from a pri-miRNA (Lee et al., 2002). Because they have similar biochemical composition and functions to siRNAs, it was necessary to develop a uniform microRNA annotation system. Several databases are currently available for miRNA structural studies, of which *MirBase* is the largest. Because *MirBase* contains miRNAs that are not reliably supported as miRNA genes or are false positives, new databases have also been created that are often derived from *MirBase* (e.g., *MirGeneDB*, *miRCarta*, etc.). While some databases are general, for miRNAs of all organisms (e.g., *MirBase*), some are specialized only for a specific group of organisms, for example, *PMRD* for miRNAs of plants or *MepmiRDB* for miRNAs of medicinal plants.

There is a problem when storing miRNA structures in databases. MiRNAs are biochemically indistinguishable from siRNAs (duplex RNA molecules with short length) and share the same function (posttranscriptional gene silencing by binding to mRNAs). However, while miRNAs target multiple mRNAs (sometimes more than 100 at a time), siRNAs are highly specific with only one mRNA target. In addition, they also differ subtly in length, with siRNA being composed of 21–23 nucleotides and miRNA of 19–25 nucleotides. In addition, they also differ in the manner in which they silence genes. Because siRNAs are fully complementary to miRNAs, they inhibit gene expression by endonucleolytic cleavage of the mRNA, whereas miRNAs are only partially complementary, mostly targeting only the 3′ UTR of the mRNA, so they silence the gene by translational repression (Lam et al., 2015). To avoid the problem of confusing miRNAs with siRNAs or fragments of other RNAs, a uniform system for microRNA annotation was introduced, which consisted of five criteria. The first two criteria dealt with expression and the other three with biogenesis, and to be confirmed, a miRNA must meet a criterion from both expression and biogenesis:

(1) A distinct transcript of approximately 22 nt in length must be detected by hybridization.
(2) Or a transcript of approximately 22 nt in length can be identified in a cDNA library.
(3) The miRNA precursor with the lowest free energy must be formed with a minimum of 16 bp encompassing the first 22 nt of the miRNA and the second arm of the hairpin, and large asymmetric bulges should not be present.
(4) The mature miRNA must be phylogenetically conserved with its secondary pre-miRNA structure, which need not be the stem-loop structure with the lowest free folding energy.
(5) Presence of higher amounts of accumulated precursors in organisms that exhibit lower Dicer function (Ambros et al., 2003).

These criteria were further supplemented in Meyers et al. (2008):

(1) Duplex miRNA must be formed from two opposite stem arms such that two nucleotide 3′ overhangs are formed.
(2) There are usually four or fewer bases that are not complementary between one and the other arm of the miRNA strand.
(3) There are asymmetric protrusions in the miRNA duplex that are minimal in size and frequency.

Since Lee and Ambros (2001) discovered the first miRNA in *Caenorhabditis elegans*, new miRNA cloning, and sequencing techniques have been gradually developed. Because of this, it was necessary to create a database to serve as a common repository for naming, annotating, and disseminating published miRNAs (Enright and Griffiths-Jones, 2007). For this reason, the first *Rfam* database was created, but it focused on all noncoding RNA sequences. The first release of this database contained 25 families annotating more than 50,000 ncRNA genes (Griffiths-Jones et al., 2003). The goal of *Rfam* was to create a database of noncoding RNAs, and statistical profiling models (covariance models) (Gardner et al., 2011). These covariance models can then be used to annotate RNA families in any sequence of nucleic acids (Kalvari et al., 2018). The database contains mainly those RNAs whose secondary structure is conserved, including cis-regulatory elements of mRNA and RNA genes. The 11th release of *Rfam* was already quite large and contained 89,111,824 sequences (Burge et al., 2013). In 2020, this database was expanded to *Rfam* 14, which contained noncoding RNAs from 3444 families, focusing mainly on microRNAs, and viral and bacterial RNAs (Kalvari et al., 2021). This database is still active today at https://rfam.org/, having been updated to *RFam* version 14.9 in November 2022, and contains a dataset of 4108 ncRNA families.

Subsequently, with the growing interest in miRNAs, the first freely available database that focused only on miRNAs, The *MicroRNA Registry*, was created. This database contained published miRNA sequences that could be searched via the abovementioned web interface. This database aimed to simplify miRNA sequence submission and naming and provide a searchable complex database of all published miRNA sequences. Its second edition was released in July 2003 and contained 506 miRNAs from *Drosophila melanogaster, C. elegans, Caenorhabditis briggsae*, mouse, humans, and *A. thaliana* (Griffiths-Jones, 2004). This database has been progressively expanded as version *The MicroRNA Registry 6.0* enriched with several 1000 new miRNA precursor homologs. For this expansion, it was important that the expression of this homologs was verified in at least one species (Hertel et al., 2006). Since 2006, *The MicroRNA Registry* has been rebranded as *miRBase*. Since its inception, this database has allowed viewing the primary structure of miRNA sequences, annotations, and predicted gene targets. Like the previous database, it also allowed researchers to name new miRNAs (Griffiths-Jones et al., 2006). The *miRBase* has gradually increased in size, with, for example, the 20th release of *miRBase* in 2013, it contained 30,424 mature microRNA products that were derived from 24,521 microRNA loci from 206 species (Kozorama and Griffiths-Jones, 2014). *miRbase* 22nd release contained 8000 more mature miRNAs than release 20 (48,860), which were derived from miRNA sequences from 271 organisms. In addition, the 22nd release collected 1493 files from deep small RNA sequencing (sRNA-seq), and after mapping 5.5 billion reads to microRNA sequences, more than 200 sequences were removed (Kozomara et al., 2019). As the *miRBase* is the largest miRNA database that regularly releases new editions, examining the annotations between editions is challenging. For this reason, *miRBaseMiner* software was developed to examine the miRBase annotations and generate custom annotation files (Zhong et al., 2019).

MiRBase is the largest miRNA database, but as it turns out, a large number of miRNAs from the human and metazoan gene sets in this database are not reliably supported as miRNA genes or are false positives. For this reason, *MirGeneDB* (https://mirgenedb.org/) has been created to consistently name and annotate miRNAs (Fromm et al., 2015). This database has expanded substantially since its inception in 2015, currently containing 75 species comprising 16,670 microRNAs from 1549 families. *MirGeneDB* is part of *RNAcentral* and *Elixir Norway* (Fromm et al., 2022).

The database *miRCarta*, available at https://mircarta.cs.uni-saarland.de/, was created as an extension of *miRBase* when it introduced a database to collect new miRNA candidates. In addition to all miRNAs from *miRBase*, more than 12,000 human miRNA precursors have been added to this database (Backes et al., 2018).

The plant miRNA database (*PMRD*) was created by integrating plant miRNAs from publicly available databases, literature, and new proprietary knowledge. This database contains information of miRNAs' primary and secondary structure, the genes targeted by these miRNAs, expression profiles, and a genome browser. Although this is a smaller database (approximately 8400 miRNAs, from more than 120 plant species), for some species, such as poplar, Arabidopsis, soybean, rice, cotton, bearberry, etc., potential target genes for each miRNA with a predicted interaction site have been added to the database. The database is available at http://bioinformatics.cau.edu.cn/PMRD/ (Zhang et al., 2010).

The database *MepmiRDB* is dedicated to miRNAs of medicinal plants only. It contains information of the structure of mature miRNAs and their precursors from 29 medicinal plant species. It also contains information of expression profiles and interactions with target molecules that have been validated in eight plant species. The database can be found at http://mepmirdb.cn/mepmirdb/index.html (Yu et al., 2019).

3. Databases for functional studies of miRNA

miRNAs are important regulators of genes. They influence many important functions in both plants and animals (e.g., in development, stress responses, hematopoiesis, etc.) (Meyers and Green, 2010). Databases for functional studies of miRNAs mainly focus on the prediction of miRNA targets (e.g., *TarDB*, *PAmiRDB*, *MiRDB*, *miRTarBase*). In addition, some databases address miRNA transcription factors (*CircuitsDB*, *TransmiR*), interactions between miRNAs and other molecules (*DIANA-LncBase*, *StarBase*), miRNA expression profiles (*DIANA-miTED*, *EVmiRNA*), genetic polymorphisms of miRNAs (*PolymiRTS*, *miRdSNP*), information on miRNA-encoding transcripts (*DIANA-MiRGen*), and associations between miRNAs and diseases (*TAM*). As with miRNA structural databases, some functional databases focus only on specific types of organisms, and some are general.

MicroRNAs are short noncoding RNA molecules that serve as regulators of important genes. In recent years, a trend of an increasing number of identified miRNAs has been observed (Lu et al., 2010; Paraskevopoulou et al., 2016). MiRNAs have different functions in plants and animals (Ambros, 2004). In plants, some miRNAs are conserved and regulate plant vital functions, development, and stress responses. Furthermore, due to the presence of fewer conserved miRNAs in each plant species, miRNAs may interfere with every aspect of plant life (Millar, 2020). In addition, miRNAs have been found to have a role in host antiviral defense mechanisms (Satish et al., 2019). miRNAs interfere with various physiological and pathological processes as a result of posttranscriptional regulation of the target mRNA. Thus, their interaction with mRNAs has become one of the main directions of drug development (Huang et al., 2022). Deregulation of miRNA expression causes a transition to a pathological state (Perdikopanis et al., 2021).

Databases for functional studies of miRNAs focus primarily on the prediction of their targets. Some databases predict miRNA targets in general for all organisms, but it has been found that more accurate results can be obtained in predicting miRNA targets when targets are targeted only to a specific group. For example, when predicting miRNA targets of four plant species, it was shown that completely excluding animal data led to more accurate prediction of these targets in plants (Ajila et al., 2023). There are also miRNA databases that deal with the study of miRNA transcription factors, miRNA expression, and the effect of polymorphisms on miRNAs.

The *TarDB* database was developed to predict plant miRNA targets and is available at http://www.biosequencing.cn/TarDB. This database predicts interspecies-conserved miRNA targets, miRNA targets promoted by degradome/PARE sequencing, and miRNA-triggered phasiRNA loci. In addition, it provides information on phasiRNAs that are triggered by miRNAs (Liu et al., 2021). It currently contains 62,888 conserved miRNA targets from 24 dicotyledonous and 12 monocotyledonous plant species, 3 bryophytes, 1 basal plant, 1 gymnosperm, and 2 algal species.

The *miRDB* database (http://mirdb.org) was created for the prediction of miRNA targets and their functional annotations. In 2015, it contained information on 6709 miRNAs regulating 2.1 million predicted gene targets. In addition, this database allows users to predict the targets of their own miRNAs, after submitting miRNA sequences. More than 560 functional miRNAs were identified and deposited in humans and 452 in mice (Wong and Wang, 2015). After an extensive RNA sequencing study where target genes were identified for 25 miRNAs, a new algorithm for *miRDB* site, *MirTarget*, was developed for miRNA target prediction by combining CLIP binding and downregulation target data (Liu and Wang, 2019). Based on the improved miRNA target prediction algorithm, 3.5 million target genes were predicted for 7000 miRNAs in 5 species. In addition, the prediction of cell-specific miRNA targets was added in the update after adding more than 1000 expression profiles of cell lines (Chen and Wang, 2020).

The *miRTarBase* database (https://miRTarBase.cuhk.edu.cn/) started with 3576 interactions between target genes and miRNAs, which were experimentally verified by western blot, reporter assays, or microarray experiments after manual review of the relevant literature. In total, it contained 657 miRNAs derived from 17 species that had 2297 target genes (Hsu et al., 2011). In 2022, after 5 refinements, more than 2,200,449 verified interactions miRNAs/targets were collected in the database. Currently, it contains 4630 experimentally verified miRNAs from 37 species having 27,172 target genes. Most miRNAs are available for human (4,475,477) and mouse (693,784). But in addition, it also contains miRNAs from viruses (e.g., Epstein–Barr virus, Herpes B virus) and plants (tomato and rice) (Huang et al., 2022).

The *TransmiR* database (http://www.cuilab.cn/transmir) is dedicated to transcription factors that regulate miRNAs to better understand their impact on pathologies. In the first release, 243 regulatory relationships between 82 transcription factors and 100 miRNAs were identified and experimentally supported from 86 publications (Wang et al., 2010). In 2019, the database was updated to v2.0, which expanded to 3730 regulatory relationships between transcription factors and miRNAs from 19 species. In addition, a module was created to allow searches for putative regulation of transcription factors and miRNAs, but only in humans (Tong et al., 2019).

4. Bioinformatics analysis for microRNAs secondary structure prediction

The discovery of novel miRNAs and the confirmation of analogous miRNAs in plants is highly dependent on the precise structural characteristics of their precursor molecules. Particularly, accurate prediction of de novo miRNA genes that lack homology with established miRNAs remains a major challenge for researchers (Tempel and Tahi, 2012). Secondary structure determination is a critical component of this process and identifying genomic regions that are likely to fold back while forming characteristic stem-loop structures and exhibit properties akin to known miRNAs is a pivotal step in most miRNA identification procedures (Yousef and Allmer, 2014). The development of rapid and accurate tools for predicting the folding arrangements of miRNA precursors has therefore been a subject of great interest.

The creation of a hairpin structure is essential for the development of mature miRNAs, as it is being cleaved at the precisely specified positions dependent on its structural organization. Even though hairpin loop region often exhibits significant variation, not only between different species but also between homologs within the same species (Bartel, 2004), considerable level of conservation is observed in the stem region of miRNA precursors. Among closely related species, conservation occurs in both primary and secondary structure, whereas for more divergent species, it occurs mainly at the secondary structure level. The conservation within precursor molecules of a specific miRNA family is present not just in the fully processed miRNA duplex but also in the surrounding regions, which is an important fact when searching for miRNA homologs based on their secondary structure (Baldrich et al., 2018).

The existence or lack of loops at various positions, and minimum free energy are the main factors affecting secondary structure prediction. Small symmetrical internal loops with unpaired nucleotides on opposite sides of the loop are more common than the bulges, occurring three times more often in the pri-miRNA secondary structure. There has been a suggestion that the distribution of bulges in pre-miRNA is random, while symmetric internal loops are more prevalent in the primary cleavage site region compared to the region of the second cleavage. In 90% of the analyzed structures, the small symmetric internal loops were present within 1−5 nucleotide distance from the miRNA/miRNA* duplex (Miskiewicz et al., 2017). This duplex is released by two Dicer-mediated cleavages on the pri-miRNA: the first cleavage occurs in the lower region of the stem and is followed by a cleavage localized in the upper loop region (Zhu, 2008). The precise mechanism for selecting cleavage sites in pri-miRNA and pre-miRNA is not well understood; however, several publications suggest that the structural features of the neighboring sequences may could be significant factors in this mechanism (Meyers et al., 2010). The presence of relatively unstructured section in the pri-miRNA, approximately 15 nucleotides below the cleavage site, is important for the initial Dicer-mediated cleavage. Insertion or significant displacement of a bulge in this area can modify the position of the cleavage, thus impacting the entire biogenesis of the functional miRNA (Mateos et al., 2010; Song et al., 2010; Werner et al., 2010).

When a sufficient number of similar annotated miRNA sequences is accessible, comparative approach can be used to determine the secondary structure. However, when identifying new miRNAs, thermodynamics is usually employed to forecast secondary structures from a particular sequence. Stability of a structure is predicted in terms of the changes in its folding free energy. It refers to the process of identifying the structure characterized with the lowest free energy, which is considered the most probable structure within a set of possible structures (Mathews and Turner, 2006). Plant miRNA precursors, the mean of minimum free energy is lower, and simultaneously its distribution is wider than those of animals. Higher variability in the precursor length is likely to be the primary factor responsible for this variance (Thakur et al., 2011). Secondary structure prediction from single sequence using thermodynamics is fairly accurate, especially for shorter sequences, such as miRNA. Additionally, by utilizing a partition function calculation of the probability of base pair occurrences in a structure is allowed (Reuter and Mathews, 2010).

Although structure prediction element is of a high importance in the identification of new miRNAs, this approach alone leads to a high occurrence of false positives since numerous regions within the genome can potentially form hairpin structures. During miRNA analysis, a significant amount of raw small sequence data is produced, requiring the segregation of authentic miRNA precursors from non-miRNA hairpins. It is therefore necessary to validate these putative miRNAs by additional, more complex criteria to differentiate the genomic hairpins that encode functional miRNAs from other non-miRNA hairpin structures (Miskiewicz et al., 2017).

To keep up with the rapidly growing amount of sequencing data, new secondary structure prediction algorithms had to be developed, allowing accurate and efficient prediction coupled with reduced time complexity. Most of these tools are based on calculating the minimum folding energy to determine structural features characteristics of miRNA precursors. In the following section, we present a brief summary of commonly used software tools that are utilized to identify miRNAs by predicting their secondary structure.

miRNAFold is a tool accessible through a web browser and features a user-friendly and intuitive interface. It utilizes an algorithm that approximates the pre-miRNA hairpin structure and then refines it, resulting in a significant reduction of estimated number of operations required to complete a task as well as execution time. The algorithm works by analyzing a sliding window of a specified length to discover potential hairpin structures forming sequences. The initial stage involves searching for exact stems that meet specific criteria and can be considered as possible hairpin anchors. Subsequently, these stems are prolonged to identify the lengthiest possible stems that meet another set of criteria, with each stem serving as an approximate model of a pre-miRNA. The algorithm then constructs every possible hairpin based on the central position of the nonexact stem and selects pre-miRNA candidates that meet additional defined constraints. The selection criteria include factors such as sequence length and composition, proportions of internal and terminal loops, as well as bulges and free energy. The algorithm offers several extra features, for example, performing multiplexed analysis of numerous sequences, sending results to users via email upon request, enabling users to download the output including or excluding plot depicting the secondary structure, and providing information about the execution time (Tav et al., 2016).

Mirinho operates on the similar principle as miRNAFold, initially searching for the principal stem, referred to as the stem-arm, by employing a sliding window approach. The search for the putative stem-arm involves identifying the complementary sequence pair of the hypothetical duplex, spaced by the variable sequence of the terminal loop. After identification of putative stem-arms, the stability of prospected pre-miRNA is evaluated using the Needleman–Wunsh global alignment and nearest-neighbor (NN) model. The total energy increment of each motif, predicted by NN model provides an estimate of the overall change in free energy for the RNA folding process, thus determining the stability of the miRNA precursor (Higashi et al., 2015).

miRcheck analyzes secondary structures of putative miRNA precursors and evaluates them based on rules, which include consistent orientation of base pairing, stem-loop length, maximum number of unpaired nucleotides, extension of base pairing, and requirements for the extended miRNA and miRNA* sequences. These criteria help to ensure that the putative miRNA precursor is likely to be a genuine miRNA precursor and not a false positive (Jones-Rhoades and Bartel, 2004).

mFold is part of an integrated collection of programs called UNAfold (Markham and Zuker, 2008). It employs web-based graphical user interfaces and generates detailed results for the folding of sequences, including energy dot plots, structure plots that include reliability information, and single-stranded frequency plots. Based on the energy dot plot, which displays potential base pairs, those falling within a specific predefined free energy range are utilized in determining the final folding (Zuker, 2003).

RNAfold is a part of the ViennaRNA Package, which provides a complex set of tools for analyzing RNA secondary structures. The output of this tool provides an output containing the secondary structure information, represented as a string, with the minimal folding energy in the dot-bracket notation as well its corresponding folding energy. It uses dynamic programming approach to compute the minimum free energy and then finds the secondary structure with the most optimal features. If the -p option is used, RNAfold computes the partition function, which produces a matrix of the probabilities of base pairs occurrence (Lorenz et al., 2011).

RNAstructure can be accessed through various interfaces, including a Web server, a standalone program, or a collection of command-line utilities. It comprises a range of algorithms for RNA secondary structure prediction, one of which is the computation of base pairing probability mentioned earlier. The algorithms integrated in *RNAstructure* use thermodynamic NN model to determine the stability

Table 16.1 Tools for identifying secondary structure of plant miRNAs.

Software	URL	References
mFold	www.unafold.org	Zuker (2003)
RNAfold	http://www.tbi.univie.ac.at/RNA	Lorenz et al. (2011)
RNAstructure	https://rna.urmc.rochester.edu/RNAstructure.html	Mathews (2014)
Mirinho	https://gitlab.inria.fr/erable/mirinho	Higashi et al. (2015)
miRNAFold	http://EvryRNA.ibisc.univ-evry.fr/miRNAFold	Tav et al. (2016)
SUmirFold	–	Alptekin et al. (2017)
miRcheck	http://bartellab.wi.mit.edu/software.html	Jones-Rhoades and Bartel (2004)

of secondary structures. Additionally, it can comparatively predict secondary structures shared by two unaligned sequences, which is a more accurate method than predicting structures using a single sequence (Mathews, 2014; Reuter and Mathews, 2010).

In addition, the prediction of specific localization of mature miRNA sequence within its precursor is also a critical task that can be accomplished with software such as *MaturePred* (Xuan et al., 2011). A support vector machine (SVM)-based machine learning technique is employed to anticipate the position of mature miRNAs in a pre-miRNA candidate. Thorough feature extraction, selection, and high-quality training sample selection ensured high effectiveness and accuracy of predictions made by the model.

Table 16.1 summarizes the previously mentioned tools utilized for predicting miRNA secondary structures.

5. Bioinformatics analysis for microRNAs gene prediction

Methods for identifying plant miRNAs have evolved from early methods, dependent on experimental procedures with limited discovery potential, to high-throughput bioinformatics predictions based on criteria that have been refined over time. The rapidly increasing amount of genomic and transcriptomic data generated by high-throughput next-generation sequencing (NGS) techniques has resulted in surge of bioinformatics tools, which nowadays play a crucial role in identifying miRNAs. Algorithms predicting miRNAs and their precursors present in the genome and transcriptome of an analyzed plants represent an important initial step in most of the in silico analysis of miRNAs. Subsequent determination of miRNA targets enables the functional annotation and exploration of miRNA-mediated regulatory pathways. Through deepening our knowledge of the functions played by the miRNAs in regulating crucial agronomic traits in plants, innovative approaches can be established to enhance crop yield and quality.

To annotate a genomic sequence as a miRNA coding gene, it must meet a specific set of predetermined criteria that have been revised and refined over the years. Ambros et al. (2003) published the first guidelines for miRNA annotation in both animals and plants, which combined a relatively simple sets of biogenesis and expression criteria. The fast-paced discovery of new miRNAs caused by the progress of deep sequencing methods, and the dissimilarities between plant and animal organisms (particularly the significant abundance of other small RNA types, predominantly siRNAs, found in plant organisms) prompted a reevaluation and tightening of criteria for miRNA annotation specific to plant organisms. This is a constantly evolving process aimed at reducing the occurrence of false positives, while simultaneously increasing the sensitivity for the detection of low-abundance miRNAs (Axtell and Meyers, 2018; Zhao et al., 2021).

Meyers et al. (2008) became the first researchers to establish specific annotation guidelines for plant miRNAs. According to them, the primary characteristic of miRNA annotation, which must be fulfilled, is the ability of miRNA to undergo precise excision, which separates the miRNA/miRNA* duplex from the rest of the hairpin precursor, which must satisfy specific structural requirements. Several ancillary criteria, although insufficient on their own, have been postulated to increase the reliability of miRNA annotation. In addition to conserved miRNAs, these new criteria allowed for the detection of novel miRNAs, either through sequencing with coverage deep enough to analyze the sequence of both miRNA and its corresponding miRNA* or by the additional sequencing of isolated miRNA from multiple independent libraries (Meyers et al., 2008).

Considering the expanding knowledge about plant miRNAs and surging amount of new data, these criteria were revised in 2018, proposing several improvements for annotating miRNAs in plant organism (Axtell and Meyers, 2018). Updated criteria for miRNA annotation according to Axtell and Meyers (2018) include:

- Typically single miRNA/miRNA* duplex with 3' overhangs.
- No secondary stems or extensive loops should be present in the miRNA duplex region.
- Precursor size should not exceed 300 nucleotides.
- Both miRNA and miRNA* should be confirmed by sRNA-seq.
- $\leq$5 mismatched positions are allowed in miRNA/miRNA* duplex including $\leq$3 in asymmetric bulges.
- $\geq$75% of the reads ($\geq$90%, for the high confidence miRNA loci) match the exact sequence of miRNA and its corresponding miRNA* or their respective one-nucleotide positional variants.
- miRNA annotation is supported by at least two independent sRNA-seq libraries.
- Annotations based on homology should be considered tentative and should be verified by fulfilling all the criteria by sRNA-seq.

- RNA molecules that are <20 nucleotides or >24 nucleotides in length should not be classified as miRNAs. Additionally, annotating 23 or 24 nucleotide long miRNAs requires compelling validation (Axtell and Meyers, 2018; Meyers et al., 2008).

The homology-based approach, which presents one of the earliest methods for miRNA prediction, is relatively straightforward and simple method for identifying miRNA homologs. These methods typically rely on sequence alignment to determine the level of sequence similarity and subsequent comparison of secondary structure using predicted energy values (Yu et al., 2020). These methods have a major limitation in that they require prior information and often leads in a higher incidence of false positives. The current state of art methods for examining miRNA sequences rely on bioinformatic analysis of sRNA-seq data, alternatively complemented by supplementary analyses such as degradome sequencing (Ganguli et al., 2020). The use of algorithms that identify miRNAs based on their features and evaluation of annotation criteria or employ machine learning techniques in analyzing sRNA-seq data are being employed tackle issues of homology-based approach (Zhao et al., 2021).

To begin sRNAseq analysis, raw sequencing data are mapped on a reference genome, but adapter sequences must first be removed to obtain pure transcriptome sequence for downstream analyses. Cutadapt (Martin, 2011) and Trimmomatic (Bolger et al., 2014) are in silico tools commonly used for adapter removal. Processed raw reads are then aligned to the reference genome using mapping tools such as BWA (Li and Durbin, 2009) or Bowtie2 (Langmead and Salzberg, 2012). Filtering of different noncoding small RNAs is performed by aligning the sequences to databases such as Rfam (Kalvari et al., 2021) and RepBase (Bao et al., 2015). The filtered data can then be analyzed using miRNA identification tools.

Several tools have been created to detect miRNAs in sRNAseq by examining various aspects of miRNA biogenesis and then employing a statistical scoring system to attribute importance to these traits. Employment of previously mentioned plant-specific miRNA criteria plays a key role in improving the accuracy and sensitivity of miRNA prediction in plants, especially by providing guidelines for differentiating between siRNAs and miRNAs. Ultimately, a threshold score is established to identify probable miRNAs. As the size of plant genomes being analyzed and the depth of sRNA-seq libraries increase, time complexity of this process also grows (Zhao et al., 2021). This approach is often employed by the most widely used tools for miRNA prediction, such as miRDeep-P2-based tools. However, the accuracy of this approach is considerably constrained and reliant upon the current comprehension of miRNA biology.

On the other hand, machine learning-based approaches for miRNA identification are based on the use of algorithms and statistical that learn from a training dataset composed of known miRNAs. Machine learning algorithm uses pattern recognition and makes decisions based on the outcomes of the learning process. The main determinant of quality and accuracy of these tools is the selection of relevant features and the accuracy of the used training data set (Vakilian, 2020). Machine learning algorithms used for miRNA prediction include SVMs (Xue et al., 2005), decision trees (Williams et al., 2012), as well as random forests (RFs) (Jiang et al., 2007), hierarchical hidden Markov model (HHMM) (Kadri et al., 2009), or Naive Bayes (Yousef et al., 2006). A fundamental constraint of this approach is the quality of the training data set, which may be negatively affected by the amount of incorrectly identified miRNAs stored in the databases, which consequently undermines the entire process of miRNA identification utilizing these tools. Machine learning-based algorithms may show lower efficiency for identifying plant miRNAs compared to animal miRNAs (Zhao et al., 2021).

Some of the popular tools used for bioinformatics prediction of miRNA genes include:

MiRDeep-P2 is an algorithm that identifies putative miRNA precursor sequences and evaluates them in terms of the probability of being a genuine miRNA precursor, resulting in a ranked list that includes both known and novel miRNA precursors and mature miRNAs based on their scores. Building on miRDeep-P, miRDeep-P2 identifies and quantifies the distribution of sRNA reads and miRNA precursors, keeping only reads within 19–24 length. The reads within 19–24 length are normalized before processing. Reads that show conservation with annotated miRNAs in miRBase are processed separately. The development of the MiRDeep-P2 package was done using Perl programming language, and all the scripts were linked through a bash script. This tool allows for the adjustment of all parameters. Previously mentioned, Bowtie2 and the Vienna RNA package were incorporated into the package as dependencies (Kuang et al., 2019).

miRkwood utilizes a comprehensive workflow that integrates data on the secondary structure of miRNA precursors and the positions of reads. The workflow includes alignment and filtering, identification of known miRNAs, localization of expression signals, and analysis of the predicted folding of the sequence into a stable hairpin structure. miRkwood provides a quality score for each prediction, based on six criteria (stability of the hairpin precursor, read count, existence of the miRNA, processing precision, occurrence of the miRNA/miRNA* duplex and its stability). It implements RNA fold for secondary structure prediction. The software is available as a web server, standalone version, or Galaxy tool and is a useful tool for researchers with an interest in identifying miRNAs in plant genomes (Guigon et al., 2019).

miRador tool is used for identifying miRNAs in plants and is based on the latest established requirements for accurate identification. It combines target prediction and PARE data to evaluate the precision of miRNA identification (Hammond et al., 2023).

Developing new and improved in silico tools for miRNA prediction is an ongoing process that continues to face the challenge of minimizing false-positive predictions and reducing analysis time (Zhou et al., 2020). Identification of miRNAs is an intricate process that requires a combination of various methods for its optimal effectiveness. Table 16.2 shows the most used tools for identifying plant miRNAs.

Table 16.2 Tools for identifying plant miRNAs.

Software	URL	References
miRador	https://github.com/rkweku/miRador.	Hammond et al. (2023)
miRPlant	http://sourceforge.net/projects/mirplant/.	An et al. (2014)
miRkwood	https://bioinfo.univ-lille.fr/mirkwood/	Guigon et al. (2019)
mirDeep-P2	https://sourceforge.net/projects/mirdp2/.	Kuang et al. (2019)
miRscan	http://hollywood.mit.edu/mirscan/index.html	–
MiRPara	http://www.whiov.ac.cn/bioinformatics/mirpara	Wu et al. (2011)
novoMIR	http://www.biophys.uni-duesseldorf.de/novomir/	Teune and Steger (2010)
ShortStack	https://sites.psu.edu/axtell/software/shortstack/	Axtell (2013)
Triplet-SVM	http://bioinfo.au.tsinghua.edu.cn/mirnasvm/	Xue et al. (2005)
microHARVESTER	http://www.ab.informatik.uni-tuebingen.de/brisbane/tb/index.php	Dezulian et al. (2006)
miReader	https://scbb.ihbt.res.in/2810-12/miReader.php	Jha and Shankar (2013)
HHMMIR	https://www.csb.pitt.edu/Faculty/benos/?page_id=1704	Kadri et al. (2009)
SplamiR	http://www.uni-jena.de/SplamiR.html	Thieme et al. (2011)
PlantMirP2	https://github.com/wuqiansibai/plantMiRP2/releases/tag/v1.0/	Fan et al. (2021)
mirMachine	https://github.com/hbusra/mirMachine#readme	Cagirici et al. (2021)
PmiRDiscVali	https://github.com/unincrna/pmirdv	Yu et al. (2019)
Sumir	–	Alptekin et al. (2017)
CleaveLand	https://github.com/MikeAxtell/CleaveLand4/releases	Addo-Quaye et al. (2009)

6. Bioinformatics analysis for microRNAs target prediction

MicroRNA target site can be localized on promoter, 3′-untranslated region (3′-UTR), coding sequence, or 5′-untranslated region (5′-UTR) (Zhang et al., 2014; Munusamy et al., 2017; Li et al., 2018; Hajieghrari et al., 2022). Among the most used approaches belong an evaluation of Watson–Crick complementarity within whole or seed sequence, information about gaps and mismatches, G:U wobble base pairs, occurrence of adenosine (A) in opposite to 1. nt from 5′ end of miRNA, count of target sites, thermodynamic stability, and free energy or information from evolution biology.

Watson–Crick complementarity of purine and pyrimidine bases lies in binding of guanine (G) with cytosine (C), and adenosine (A) with uracil (U). Those due to the presence of their functional carbonyl and amino groups are able to create amino-imino or keto-enol tautomeric forms, for example, a wobble structure of Guanine–Uracil instead of G–C match (Chandorkar et al., 2021). Gaps and mismatches decrease the probability of pairing and must be taken into consideration (McGeary et al., 2022).

The basic requirement for success paring is a full complementarity of a seed sequence that is created by at least 6 nucleotides (nt 2-nt 7) at the 5′ end of mature microRNA (Chipman and Pasquinelli, 2019). Adenosine within the mRNA sequence localized in opposite to the first nucleotide from 5′ end of miRNA is specifically recognized by protein Argonaut (Sheu-Gruttadauria and MacRae, 2017). The highest possible score is from this point the fully complementary mature (~22 nt long) microRNA including the eight-mer seed sequence in combination with the opposite adenosine at the first position (Fig. 16.1). Several target sites in mutual distance 10–50 nt increase the silencing effect, however target sites very close to each other (<8 nt) have competitive impact (Riolo et al., 2020).

It was observed that efficiency of a gene silencing depends on thermodynamic factors. Thermodynamic stability is positively corelated with melting temperature of duplex seed sequence-target sequence ($Tm_{(2nt\text{-}8nt)}$) and negatively with duplex microRNA–microRNA* ($miTm_{(1nt\text{-}5nt)}$) (Hibio et al., 2012).

For the estimation of the silencing complex efficiency was established the mathematical quantity miScore calculated as $Tm_{(2nt\text{-}8nt)} - 0.5 \times miTm_{(1nt\text{-}5nt)}$ (Tian et al., 2020).

The following factor is thermodynamic accessibility ($\Delta\Delta G$) defined as the difference between the free energy obtained by mRNA–miRNA binding (ΔG) and free energy lost by unpairing of target site nucleotides (ΔG_{open}) (Ghoshal et al., 2015). Although the amount of free energy of the miRNA-mRNA complex is low, a greater amount of energy is required to break its bond, which means that this complex is more stable than others (Akhtar et al., 2019).

MicroRNAs are among the most conserved elements of the genome, and moreover, they can appear de novo (Velandia-Huerto et al., 2022). The precise and complex annotation of microRNA sequences, including mRNAs, within different organisms, increases the success rate of prediction, because there is an assumption

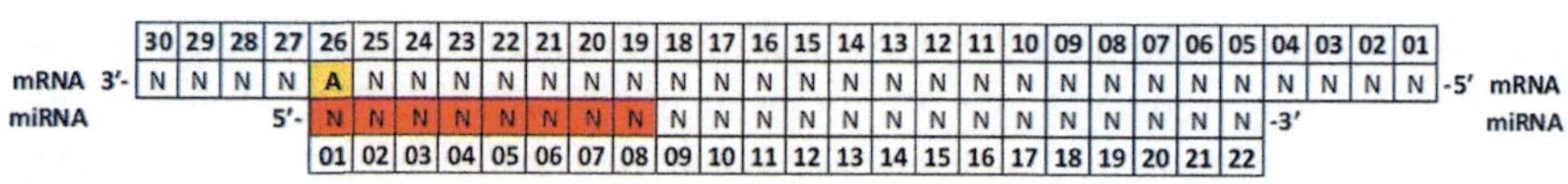

FIGURE 16.1

Fully complementary complex microRNA–mRNA with adenosine in opposite to the first nucleotide from 5′ end of microRNA.

of an occurrence of the same or different form of one sequence. Searching for de novo polymorphisms within mature microRNAs is equally important as searching for polymorphisms within target mRNA sequences. Alignment algorithms such as BLAST can be used for this purpose, which obtain data from various genomic or transcriptomic sources within one or several taxonomic units.

Target prediction should also include indirect effects, such as interactions between microRNA–microRNA*, microRNA–transcription factor, and other forms of negative feed feedback loop (Johnston et al., 2005; Zhang et al., 2015; Li and Yu, 2021; Gou et al., 2022).

By combining all this knowledge, we are able to successfully predict target sites of mature microRNA sequences and subsequently verify their functionality using various biotechnological methods such as overexpression or miRNA silencing (Pan et al., 2018; Othman et al., 2023).

7. Bioinformatics analysis for microRNA related to plant secondary metabolites biosynthesis—model approach

7.1 Identification of microRNAs in the genome of *Linum usitatissimum* L.

MicroRNA sequences were obtained from miRBase Release 22.1 (miRBase, 2023; https://www.mirbase.org/) as compressed gz file including all available mature microRNA sequences in FASTA format with their unique ID (MIMATXXXXXXX). Sequences of hairpins were downloaded automatically based on mature IDs. Dataset of downloaded hairpins was aligned with genome *Linum usitatissimum* v1.0 by blast search on Phytozome 13 (Phytozome 13, 2023, https://phytozome-next.jgi.doe.gov/). The results were sorted according to percentage of positives with a query sequence. For each mature microRNA was assigned the highest found percentage and for further analysis have been selected only miRNAs with percentage equal or higher than 80%.

7.2 Target sequences of SDG key enzymes

Within the metabolic pathway of secoisolariciresinol diglucoside (SDG) were chosen 3 key enzymes (Corbin et al., 2017) and obtained from NCBI (2023)—Dirigent protein 5 and 6 (DIR 5, KM433753.1, DIR 6, KM433752.1), (−)-pinoresinol-(−)-lariciresinol reductase 1 (PLR 1, AJ849359.1) and Uridine glycosyltransferase UGT74S1 (JX011632.1, JN088324.1). Due to a verification of occurrence in the *L. usitatissimum* L., increasing mRNA variability and annotation of contigs were target sequences aligned with a transcriptome *L. usitatissimum* v1.0 (Lus10000001–Lus10043484, Phytozome 13, 2023) using Align Sequences Nucleotide BLAST optimized for somewhat similar sequences (blastn) with default algorithm parameters (NCBI, 2023). The contigs were divided into the separated FASTA files with no more than 10 million characters due to established limit of BLAST server and used as subject sequences. Exported were only FASTAs with a query coverage equal or higher than 50%.

7.3 Final prediction

Selected mature microRNAs (aligning with the genome *L. usitatissimum* v1.0 equal or higher than 80%) and selected target sequences (including aligned parts of contigs coming from transcriptomic data) were uploaded and analyzed using psRNATarget: A Plant Small RNA Target Analysis Server (2017 Update) with default settings of Schema V2 (2017 release) (psRNATarget, 2023, https://www.zhaolab.org/psRNATarget/). From miRBase Release 22.1 were obtained 44,885 mature miRNAs. Aligning of their hairpins proved the occurrence in the genome *L. usitatissimum* v1.0 only for 11,919 mature miRNAs. For further analysis have been selected only 441 mature microRNAs that reach at least 80% percentage of positives within genome *L. usitatissimum* v1.0.

For increasing mRNA variability have been exported aligned parts of contigs coming from the transcriptome *L. usitatissimum* v1.0 (Table 16.3).

From the results (Fig. 16.2) follows the metabolic pathway of SDG in *L. usitatissimum* L. is regulated by 10 microRNA families. Dirigent protein 5 and 6 are probably regulated by family miR160. For (−)-pinoresinol-(−)-lariciresinol reductase 1 (PLR 1) were predicted families miR167, miR171, miR395, miR399, and miR5219. Glucosyltransferase UGT74S1 were identified miR156, miR157, miR159, miR167, miR319, and miR395.

Within this part of the chapter is presented a simple in silico approach for identifying microRNA families involved in the biosynthetic pathway of secondary

Table 16.3 Increasing of mRNA variability by aligned parts of contigs.

Name of enzyme	Accession	Aligned contigs
DIR 5	KM433753.1	Lus10017538; Lus10017539 Lus10024714; Lus10024715 Lus10028749; Lus10032331
DIR 6	KM433752.1	Lus10017538; Lus10017539 Lus10024714; Lus10024715 Lus10028749; Lus10032331
PLR 1	AJ849359.1	Lus10007599; Lus10010403 Lus10012143; Lus10012145 Lus10012147
UGT74S1	JX011632.1	Lus10006351; Lus10006352 Lus10006353; Lus10006721 Lus10008742; Lus10014148 Lus10017825; Lus10024486
	JN088324.1	Lus10006353; Lus10014148 Lus10017825

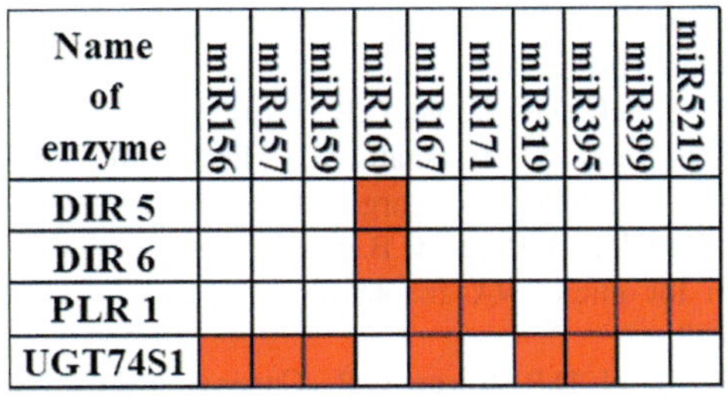

FIGURE 16.2

Matrix of predicted microRNAs.

metabolite SDG. Occurrence of 10 predicted microRNA families in *L. usitatissimum* L. was also proved by Moss (2012), Zhang et al. (2020), and Ali et al. (2021). Within an increasing of mRNA variability were identified 6 contigs for dirigent proteins DIR 5 and DIR 6, 5 contigs for (−)-pinoresinol-(−)-lariciresinol reductase 1 (PLR 1) and 8 for Uridine glycosyltransferase UGT74S1. Only contigs Lus10028749, Lus10017539, Lus10012145, and Lus10017825 were mentioned in the publications Dalisay et al. (2015) and Ahmad et al. (2019). One of our articles pointed to microRNAs involved in the biosynthesis of flax lignans as products of phenylpropanoid pathway (Ražná et al., 2022), however isn't know any publication about microRNAs involved in the translation of enzymes DIR 5, DIR 6, PLR 1, and UGT74S1. This approach can be used for any organism and biosynthetic pathway.

References

Addo-Quaye, C., Miller, W., Axtell, M.J., 2009. CleaveLand: a pipeline for using degradome data to find cleaved small RNA targets. Bioinformatics 25, 130–131. https://doi.org/10.1093/bioinformatics/btn604.

Ahmad, W., Zahir, A., Nadeem, M., Garros, L., Drouet, S., Renouard, S., Abbasi, B.H., 2019. Enhanced production of lignans and neolignans in chitosan-treated flax (*Linum usitatissimum* L.) cell cultures. Process Biochem. 79, 155–165. https://doi.org/10.1016/j.procbio.2018.12.025.

Ajila, V., Colley, L., Ste-Croix, D.T., Nissan, N., Golshani, A., Cober, E.R., Green, J.R., 2023. P-TarPmiR accurately predicts plant-specific miRNA targets. Sci. Rep. 13 (1), 332. https://doi.org/10.1038/s41598-022-27283-8.

Akhtar, M.M., Micolucci, L., Islam, M.S., Olivieri, F., Procopio, A.D., 2019. A practical guide to miRNA target prediction. MicroRNA Target Identification 1–13.

Ali, E., Saand, M.A., Khan, A.R., Shah, J.M., Feng, S., Ming, C., Sun, P., 2021. Genome-wide identification and expression analysis of detoxification efflux carriers (DTX) genes family under abiotic stresses in flax. Physiol. Plantarum 171 (4), 483–501. https://doi.org/10.1111/ppl.13105.

Alptekin, B., Akpinar, B.A., Budak, H., 2017. A comprehensive prescription for plant miRNA identification. Front. Plant Sci. 7. https://doi.org/10.3389/fpls.2016.02058.

Ambros, V., 2004. The functions of animal microRNAs. Nature 431 (7006), 350–355. https://doi.org/10.1038/nature02871.

Ambros, V., Bartel, B., Bartel, D.P., Burge, C.B., Carrington, J.C., Chen, X., Dreyfuss, G., Eddy, S.R., Griffiths-Jones, S., Marshall, M., Matzke, M., Ruvkun, G., Tuschl, T., 2003. RNA 9 (3), 277–279. https://doi.org/10.1261/rna.2183803.

Ambros, V., Bartel, B., Bartel, D.P., Burge, C.B., Carrington, J.C., Chen, X., Tuschl, T., 2003b. A uniform system for microRNA annotation. RNA 9 (3), 277–279. https://doi.org/10.1261/rna.2183803.

Ambros, V., Bartel, B., Bartel, D.P., Burge, C.B., Carrington, J.C., Chen, X., Dreyfuss, G., Eddy, S.R., Griffiths-Jones, S., Marshall, M., Matzke, M., Ruvkun, G., Tuschl, T., 2003. A uniform system for microRNA annotation. RNA 9, 277–279. https://doi.org/10.1261/rna.2183803.

An, J., Lai, J., Sajjanhar, A., Lehman, M.L., Nelson, C.C., 2014. miRPlant: an integrated tool for identification of plant miRNA from RNA sequencing data. BMC Bioinf. 15, 275. https://doi.org/10.1186/1471-2105-15-275.

Asefpour Vakilian, K., 2020. Machine learning improves our knowledge about miRNA functions towards plant abiotic stresses. Sci. Rep. 10, 3041. https://doi.org/10.1038/s41598-020-59981-6.

Axtell, M.J., 2013. ShortStack: comprehensive annotation and quantification of small RNA genes. RNA 19, 740–751. https://doi.org/10.1261/rna.035279.112.

Axtell, M.J., Meyers, B.C., 2018. Revisiting criteria for plant MicroRNA annotation in the era of big data. Plant Cell 30, 272–284. https://doi.org/10.1105/tpc.17.00851.

Backes, C., Fehlmann, T., Kern, F., Kehl, T., Lenhof, H.P., Meese, E., Keller, A., 2018. miRCarta: a central repository for collecting miRNA candidates. Nucleic Acids Res. 46 (1), 160–167. https://doi.org/10.1093/nar/gkx851.

Baldrich, P., Beric, A., Meyers, B.C., 2018. Despacito: the slow evolutionary changes in plant microRNAs. Curr. Opin. Plant Biol. 42, 16–22. https://doi.org/10.1016/j.pbi.2018.01.007.

Bao, W., Kojima, K.K., Kohany, O., 2015. Repbase update, a database of repetitive elements in eukaryotic genomes. Mobile DNA 6, 11. https://doi.org/10.1186/s13100-015-0041-9.

Bartel, D.P., 2004. MicroRNAs: genomics, biogenesis, mechanism, and function. Cell 116, 281–297. https://doi.org/10.1016/S0092-8674(04)00045-5.

Bolger, A.M., Lohse, M., Usadel, B., 2014. Trimmomatic: a flexible trimmer for Illumina sequence data. Bioinformatics 30, 2114–2120. https://doi.org/10.1093/bioinformatics/btu170.

Burge, S.W., Daub, J., Eberhardt, R., Tate, J., Barquist, L., Nawrocki, E.P., Bateman, A., 2013. Rfam 11.0: 10 years of RNA families. Nucleic Acids Res. 41 (1), 226–232. https://doi.org/10.1093/nar/gks1005.

Cagirici, H.B., Sen, T.Z., Budak, H., 2021. mirMachine: a one-stop shop for plant miRNA annotation. JoVE e62430. https://doi.org/10.3791/62430.

Chandorkar, S., Raghunathan, S., Jaganade, T., Priyakumar, U.D., 2021. Multiscale modeling of wobble to Watson–Crick-like Guanine–Uracil tautomerization pathways in RNA. Int. J. Mol. Sci. 22 (11), 5411.

Chen, Y., Wang, X., 2020. miRDB: an online database for prediction of functional microRNA targets. Nucleic Acids Res. 48 (1), 127–131. https://doi.org/10.1093/nar/gkz757.

Chen, C., Li, J., Feng, J., et al., 2021. sRNAanno—a database repository of uniformly annotated small RNAs in plants. Hortic. Res. 8, 45. https://doi.org/10.1038/s41438-021-00480-8.

Chipman, L.B., Pasquinelli, A.E., 2019. miRNA targeting: growing beyond the seed. Trends Genet. 35 (3), 215–222.

Corbin, C., Drouet, S., Mateljak, I., Markulin, L., Decourtil, C., Renouard, S., Lopez, T., Doussot, J., Lamblin, F., Auguin, D., Lainé, E., Fuss, E., Hano, C., 2017. Functional characterization of the pinoresinol-lariciresinol reductase-2 gene reveals its roles in yatein biosynthesis and flax defense response. Planta 246 (3), 405–420. https://doi.org/10.1007/s00425-017-2701-0.

Dalisay, D.S., Kim, K.W., Lee, C., Yang, H., Rübel, O., Bowen, B.P., Lewis, N.G., 2015. Dirigent protein-mediated lignan and cyanogenic glucoside formation in flax seed: integrated omics and MALDI mass spectrometry imaging. J. Nat. Prod. 78 (6), 1231–1242. https://doi.org/10.1021/acs.jnatprod.5b00023.

Dezulian, T., Remmert, M., Palatnik, J.F., Weigel, D., Huson, D.H., 2006. Identification of plant microRNA homologs. Bioinformatics 22, 359–360. https://doi.org/10.1093/bioinformatics/bti802.

Enright, A., Griffiths-Jones, S., 2007. MiRBase: a database of microRNA sequences, targets and nomenclature. In: Altman, S., Ambros (Authors), V., Appasani, K. (Eds.), MicroRNAs: From Basic Science to Disease Biology. Cambridge University Press, Cambridge, pp. 157–171. https://doi.org/10.1017/CBO9780511541766.014.

Fan, D., Yao, Y., Yi, M., 2021. PlantMirP2: an accurate, fast and easy-to-use program for plant pre-miRNA and miRNA prediction. Genes 12, 1280. https://doi.org/10.3390/genes12081280.

Fromm, B., Billipp, T., Peck, L.E., Johansen, M., Tarver, J.E., King, B.L., Peterson, K.J., 2015. A uniform system for the annotation of vertebrate microRNA genes and the evolution of the human microRNAome. Annu. Rev. Genet. 49, 213–242. https://doi.org/10.1146/annurev-genet-120213-092023.

Fromm, B., Høye, E., Domanska, D., Zhong, X., Aparicio-Puerta, E., Ovchinnikov, V., Peterson, K.J., 2022. MirGeneDB 2.1: toward a complete sampling of all major animal phyla. Nucleic Acids Res. 50, 204–210. https://doi.org/10.1093/nar/gkab1101.

Ganguli, S., Singh, P.K., Pal, A., 2020. Chapter 4 - transcriptome-based identification of small RNA in plants: the need for robust prediction algorithms. In: Guleria, P., Kumar, V. (Eds.), Plant Small RNA. Academic Press, pp. 65–97.

Gardner, P.P., Daub, J., Tate, J., Moore, B.L., Osuch, I.H., Griffiths-Jones, S., et al., 2011. Rfam: wikipedia, clans and the "decimal" release. Nucleic Acids Res. 39, 141–145. https://doi.org/10.1093/nar/gkq1129.

Ghoshal, A., Shankar, R., Bagchi, S., Grama, A., Chaterji, S., 2015. MicroRNA target prediction using thermodynamic and sequence curves. BMC Genom. 16, 1–21.

Gou, X., Zhong, C., Zhang, P., Mi, L., Li, Y., Lu, W., Shan, W., 2022. miR398b and AtC2GnT form a negative feedback loop to regulate *Arabidopsis thaliana* resistance against *Phytophthora parasitica*. Plant J. 111 (2), 360–373.

Griffiths-Jones, S., Bateman, A., Marshall, M., Khanna, A., Eddy, S.R., 2003. Rfam: an RNA family database. Nucleic Acids Res. 31 (1), 439–441. https://doi.org/10.1093/nar/gkg006.

Griffiths-Jones, S., Grocock, R.J., Van Dongen, S., Bateman, A., Enright, A.J., 2006. miRBase: microRNA sequences, targets and gene nomenclature. Nucleic Acids Res. 34, 140–144. https://doi.org/10.1093/nar/gkj112.

Griffiths-Jones, S., 2004. The microRNA registry. Nucleic Acids Res. 32, 109–111. https://doi.org/10.1093/nar/gkh023.

Guigon, I., Legrand, S., Berthelot, J.-F., Bini, S., Lanselle, D., Benmounah, M., Touzet, H., 2019. miRkwood: a tool for the reliable identification of microRNAs in plant genomes. BMC Genom. 20, 532. https://doi.org/10.1186/s12864-019-5913-9.

Guo, Z., Zheng, K., Wang, Y., Zhao, Y., Tao, Y., Chen, C., Yang, J., Lu, X., Chen, H., Wang, T., Cao, X., Wei, J., Li, L., Yang, X., 2020. PmiREN: a comprehensive encyclopedia of plant miRNAs. Nucleic Acids Res. 48 (D1), D1114–D1121. https://doi.org/10.1093/nar/gkz894.

Guo, Z., Zheng, K., Zhao, Y., Yang, D., He, H., Wan, M., Tao, Y., Wang, D., Wei, J., Li, L., Yang, X., January 7, 2022. PmiREN2.0: from data annotation to functional exploration of plant microRNAs. Nucleic Acids Res. 50 (D1), D1475–D1482. https://doi.org/10.1093/nar/gkab811.

Hajieghrari, B., Rahmanian-Koshkaki, S., 2022. Plant MicroRNA potential in targeting Sars-CoV-2 genome offering efficient antiviral MiRNA-based therapies. MicroRNA 11 (3), 245–262.

Hammond, R.K., Gupta, P., Patel, P., Meyers, B.C., 2023. miRador: a fast and precise tool for the prediction of plant miRNAs. Plant Physiol. 191, 894–903. https://doi.org/10.1093/plphys/kiac538.

Hertel, J., Lindemeyer, M., Missal, K., Fried, C., Tanzer, A., Flamm, C., Stadler, P.F., 2006. The expansion of the metazoan microRNA repertoire. BMC Genom. 7 (1), 1–15. https://doi.org/10.1186/1471-2164-7-25.

Hibio, N., Hino, K., Shimizu, E., Nagata, Y., Ui-Tei, K., 2012. Stability of miRNA 5′ terminal and seed regions is correlated with experimentally observed miRNA-mediated silencing efficacy. Sci. Rep. 2 (1), 1–10.

Higashi, S., Fournier, C., Gautier, C., Gaspin, C., Sagot, M.-F., 2015. Mirinho: an efficient and general plant and animal pre-miRNA predictor for genomic and deep sequencing data. BMC Bioinf. 16, 179. https://doi.org/10.1186/s12859-015-0594-0.

Hsu, S.D., Lin, F.M., Wu, W.Y., Liang, C., Huang, W.C., Chan, W.L., Huang, H.D., 2011. miRTarBase: a database curates experimentally validated microRNA–target interactions. Nucleic Acids Res. 39 (1), 163–169. https://doi.org/10.1093/nar/gkq1107.

Huang, H.Y., Lin, Y.C.D., Cui, S., Huang, Y., Tang, Y., Xu, J., Huang, H.D., 2022. miRTarBase update 2022: an informative resource for experimentally validated miRNA–target interactions. Nucleic Acids Res. 50 (D1), D222–D230. https://doi.org/10.1093/nar/gkab1079.

Jha, A., Shankar, R., 2013. miReader: discovering novel miRNAs in species without sequenced genome. PLoS One 8, e66857. https://doi.org/10.1371/journal.pone.0066857.

Jiang, P., Wu, H., Wang, W., Ma, W., Sun, X., Lu, Z., 2007. MiPred: classification of real and pseudo microRNA precursors using random forest prediction model with combined features. Nucleic Acids Res. 35, W339–W344. https://doi.org/10.1093/nar/gkm368.

Johnston Jr, R.J., Chang, S., Etchberger, J.F., Ortiz, C.O., Hobert, O., 2005. MicroRNAs acting in a double-negative feedback loop to control a neuronal cell fate decision. Proc. Natl. Acad. Sci. U.S.A. 102 (35), 12449–12454.

Jones-Rhoades, M.W., Bartel, D.P., 2004. Computational identification of plant MicroRNAs and their targets, including a stress-induced miRNA. Mol. Cell 14, 787–799. https://doi.org/10.1016/j.molcel.2004.05.027.

Kadri, S., Hinman, V., Benos, P.V., 2009. HHMMiR: efficient de novo prediction of microRNAs using hierarchical hidden Markov models. BMC Bioinf. 10, S35. https://doi.org/10.1186/1471-2105-10-S1-S35.

Kalvari, I., Nawrocki, E.P., Argasinska, J., Quinones-Olvera, N., Finn, R.D., Bateman, A., Petrov, A.I., 2018. Non-coding RNA analysis using the Rfam database. Curr. Protoc. Bioinformatics 62 (1), e51. https://doi.org/10.1002/cpbi.51.

Kalvari, I., Nawrocki, E.P., Ontiveros-Palacios, N., Argasinska, J., Lamkiewicz, K., Marz, M., Petrov, A.I., 2021. Rfam 14: expanded coverage of metagenomic, viral and microRNA families. Nucleic Acids Res. 49 (D1), 192–200. https://doi.org/10.1093/nar/gkaa1047.

Kozomara, A., Griffiths-Jones, S., 2014. miRBase: annotating high confidence microRNAs using deep sequencing data. Nucleic Acids Res. 42 (1), 68–73. https://doi.org/10.1093/nar/gkt1181.

Kozomara, A., Birgaoanu, M., Griffiths-Jones, S., 2019. miRBase: from microRNA sequences to function. Nucleic Acids Res. 47, D155–D162. https://doi.org/10.1093/nar/gky1141.

Kuang, Z., Wang, Y., Li, L., Yang, X., 2019. miRDeep-P2: accurate and fast analysis of the microRNA transcriptome in plants. Bioinformatics 35, 2521–2522. https://doi.org/10.1093/bioinformatics/bty972.

Kurubanjerdjit, N., Tsai, J.J., Huang, C.H., Ng, K.L., April 2014. Disturbance of *Arabidopsis thaliana* microRNA-regulated pathways by Xcc bacterial effector proteins. Amino Acids 46 (4), 953–961. https://doi.org/10.1007/s00726-013-1646-2.

Lam, J.K., Chow, M.Y., Zhang, Y., Leung, S.W., 2015. siRNA versus miRNA as therapeutics for gene silencing. Mol. Ther. Nucleic Acids 4, e252. https://doi.org/10.1038/mtna.2015.23.

Langmead, B., Salzberg, S.L., 2012. Fast gapped-read alignment with Bowtie 2. Nat. Methods 9, 357–359. https://doi.org/10.1038/nmeth.1923.

Lee, R.C., Ambros, V., 2001. An extensive class of small RNAs in *Caenorhabditis elegans*. Science 294 (5543), 862–864. https://doi.org/10.1126/science.1065329.

Lee, Y., Jeon, K., Lee, J.T., Kim, S., Kim, V.N., 2002. MicroRNA maturation: stepwise processing and subcellular localization. EMBO J 21 (17), 4663–4670. https://doi.org/10.1093/emboj/cdf476.

Li, H., Durbin, R., 2009. Fast and accurate short read alignment with Burrows–Wheeler transform. Bioinformatics 25, 1754–1760. https://doi.org/10.1093/bioinformatics/btp324.

Li, M., Yu, B., 2021. Recent advances in the regulation of plant miRNA biogenesis. RNA Biol. 18 (12), 2087–2096.

Li, Z., Xu, R., Li, N., 2018. MicroRNAs from plants to animals, do they define a new messenger for communication? Nutr. Metabol. 15 (1), 1–21.

Liu, W., Wang, X., 2019. Prediction of functional microRNA targets by integrative modeling of microRNA binding and target expression data. Genome Biol. 20, 1–10. https://doi.org/10.1186/s13059-019-1629-z.

Liu, J., Liu, X., Zhang, S., et al., 2021. TarDB: an online database for plant miRNA targets and miRNA-triggered phased siRNAs. BMC Genom. 22, 348. https://doi.org/10.1186/s12864-021-07680-5.

Lorenz, R., Bernhart, S.H., Höner zu Siederdissen, C., Tafer, H., Flamm, C., Stadler, P.F., Hofacker, I.L., 2011. ViennaRNA package 2.0. Algorithm Mol. Biol. 6, 26. https://doi.org/10.1186/1748-7188-6-26.

Lu, L.F., Boldin, M.P., Chaudhry, A., Lin, L.L., Taganov, K.D., Hanada, T., et al., 2010. Function of miR-146a in controlling Treg cell-mediated regulation of Th1 responses. Cell 142 (6), 914–929.

Markham, N.R., Zuker, M., 2008. UNAFold. In: Keith, J.M. (Ed.), Bioinformatics: Structure, Function and Applications, Methods in Molecular BiologyTM. Humana Press, Totowa, NJ, pp. 3–31. https://doi.org/10.1007/978-1-60327-429-6_1.

Martin, M., 2011. Cutadapt removes adapter sequences from high-throughput sequencing reads. EMBnet J. 17, 10–12. https://doi.org/10.14806/ej.17.1.200.

Mateos, J.L., Bologna, N.G., Chorostecki, U., Palatnik, J.F., 2010. Identification of MicroRNA processing determinants by random mutagenesis of Arabidopsis MIR172a precursor. Curr. Biol. 20, 49–54. https://doi.org/10.1016/j.cub.2009.10.072.

Mathews, D.H., 2014. RNA secondary structure analysis using RNAstructure. CP Bioinformatics 46. https://doi.org/10.1002/0471250953.bi1206s46.

Mathews, D.H., Turner, D.H., 2006. Prediction of RNA secondary structure by free energy minimization. Curr. Opin. Struct. Biol. 16, 270–278. https://doi.org/10.1016/j.sbi.2006.05.010.

McGeary, S.E., Bisaria, N., Pham, T.M., Wang, P.Y., Bartel, D.P., 2022. MicroRNA 3′-compensatory pairing occurs through two binding modes, with affinity shaped by nucleotide identity and position. Elife 11, e69803.

Meng, Y., Gou, L., Chen, D., Mao, C., Jin, Y., Wu, P., Chen, M., 2010. PmiRKB: a plant microRNA knowledge base. Nucleic Acids Res. 39 (Suppl. 1), D181–D187.

Meyers, B.C., Axtell, M.J., Bartel, B., Bartel, D.P., Baulcombe, D., Bowman, J.L., Zhu, J.K., 2008. Criteria for annotation of plant MicroRNAs. Plant Cell 20 (12), 3186–3190. https://doi.org/10.1105/tpc.108.064311.

Meyers, B.C., Green, P.J., 2010. Plant MicroRNAs. Methods and protocols. In: Walker, J.M. (Ed.), Methods in Molecular Biology 592, Series. Springer, Humana Press, ISBN 978-1-60327-004-5.

Meyers, B.C., Simon, S.A., Zhai, J., 2010. MicroRNA processing: battle of the bulge. Curr. Biol. 20, R68–R70. https://doi.org/10.1016/j.cub.2009.12.008.

Millar, A.A., 2020. The function of miRNAs in plants. Plants 9 (2), 198. https://doi.org/10.3390/plants9020198.

miRBase, 2023. miRBase: the microRNA database. https://www.mirbase.org/.

Miskiewicz, J., Tomczyk, K., Mickiewicz, A., Sarzynska, J., Szachniuk, M., 2017. Bioinformatics study of structural patterns in plant MicroRNA precursors. BioMed Res. Int. 2017. https://doi.org/10.1155/2017/6783010, 6783010.

Moss, T., 2012. Characterization of Structural Variants and Associated microRNAs in Flax Fiber and Linseed Genotypes by Bioinformatic Analysis and High-Throughput Sequencing. Case Western Reserve University.

Munusamy, P., Zolotarov, Y., Meteignier, L.V., Moffett, P., Strömvik, M.V., 2017. De novo computational identification of stress-related sequence motifs and microRNA target sites in untranslated regions of a plant translatome. Sci. Rep. 7 (1), 43861.

NCBI, 2023. The National Center for Biotechnology Information. https://ncbi.nlm.nih.gov/.

Othman, S.M.I.S., Mustaffa, A.F., Che-Othman, M.H., Samad, A.F.A., Goh, H.H., Zainal, Z., Ismail, I., 2023. Overview of repressive miRNA regulation by short tandem target mimic (STTM): applications and impact on plant biology. Plants 12 (3), 669.

Pan, J., Huang, D., Guo, Z., Kuang, Z., Zhang, H., Xie, X., Li, L., 2018. Overexpression of microRNA408 enhances photosynthesis, growth, and seed yield in diverse plants. J. Integr. Plant Biol. 60 (4), 323–340.

Paraskevopoulou, M.D., Vlachos, I.S., Karagkouni, D., Georgakilas, G., Kanellos, I., Vergoulis, T., Zagganas, K., Tsanakas, P., Floros, E., Dalamagas, T., Hatzigeorgiou, A.G., 2016. DIANA-LncBase v2: indexing microRNA targets on non-coding transcripts. Nucleic Acids Res. 44 (D1), D231–D238. https://doi.org/10.1093/nar/gkv1270.

Perdikopanis, N., Georgakilas, G., Grigoriadis, D., Pierros, V., Kavakiotis, I., Alexiou, P., Hatzigeorgiou, A., 2021. DIANA-miRGen v4: indexing promoters and regulators for more than 1500 microRNAs. Nucleic Acids Res. 49 (D1), D151–D159. https://doi.org/10.1093/nar/gkaa1060.

Phytozome 13, 2023. The Plant Genomics Resource. https://phytozome-next.jgi.doe.gov/.

psRNATarget, 2023. A Plant Small RNA Target Analysis Server (2017 Update). https://www.zhaolab.org/psRNATarget/.

Ražná, K., Harenčár, Ľ., Kučka, M., 2022. The involvement of microRNAs in plant lignan biosynthesis—current view. Cells 11 (14), 2151. https://doi.org/10.3390/cells11142151.

Reuter, J.S., Mathews, D.H., 2010. RNAstructure: software for RNA secondary structure prediction and analysis. BMC Bioinf. 11, 129. https://doi.org/10.1186/1471-2105-11-129.

Riolo, G., Cantara, S., Marzocchi, C., Ricci, C., 2020. miRNA targets: from prediction tools to experimental validation. Methods Proto. 4 (1), 1.

Sao, P., Chand, Y., Kumar, A., Singh, S., 2021. Potential drug target identification in porphyromonas gingivalis using in-silico subtractive metabolic pathway analysis. Bangladesh J. Med. Sci. 20 (4), 887–896.

Sao, P., Vats, S., Singh, S., 2022. Porphyromonas gingivalis resistance and virulence: an integrated functional network analysis. Gene 839, 146734.

Satish, D., Mukherjee, S.K., Gupta, D., 2019. PAmiRDB: a web resource for plant miRNAs targeting viruses. Sci. Rep. 9, 4627. https://doi.org/10.1038/s41598-019-41027-1.

Sheu-Gruttadauria, J., MacRae, I.J., 2017. Structural foundations of RNA silencing by argonaute. J. Mol. Biol. 429 (17), 2619–2639.

Song, L., Axtell, M.J., Fedoroff, N.V., 2010. RNA secondary structural determinants of miRNA precursor processing in Arabidopsis. Curr. Biol. 20, 37–41. https://doi.org/10.1016/j.cub.2009.10.076.

Sun, X., Dong, B., Yin, L., et al., 2013. PMTED: a plant microRNA target expression database. BMC Bioinf. 14, 174. https://doi.org/10.1186/1471-2105-14-174.

Tav, C., Tempel, S., Poligny, L., Tahi, F., 2016. miRNAFold: a web server for fast miRNA precursor prediction in genomes. Nucleic Acids Res. 44, W181–W184. https://doi.org/10.1093/nar/gkw459.

Tempel, S., Tahi, F., 2012. A fast ab-initio method for predicting miRNA precursors in genomes. Nucleic Acids Res 40, e80. https://doi.org/10.1093/nar/gks146.

Teune, J.-H., Steger, G., 2010. NOVOMIR: de novo prediction of MicroRNA-coding regions in a single plant-genome. J. Nucleic Acids 2010, 495904. https://doi.org/10.4061/2010/495904.

Thakur, V., Wanchana, S., Xu, M., Bruskiewich, R., Quick, W.P., Mosig, A., Zhu, X.-G., 2011. Characterization of statistical features for plant microRNA prediction. BMC Genom. 12, 108. https://doi.org/10.1186/1471-2164-12-108.

Thieme, C.J., Gramzow, L., Lobbes, D., Theißen, G., 2011. SplamiR—prediction of spliced miRNAs in plants. Bioinformatics 27, 1215–1223. https://doi.org/10.1093/bioinformatics/btr132.

Tian, S., Terai, G., Kobayashi, Y., Kimura, Y., Abe, H., Asai, K., Ui-Tei, K., 2020. A robust model for quantitative prediction of the silencing efficacy of wild-type and A-to-I edited miRNAs. RNA Biol. 17 (2), 264–280.

Tong, Z., Cui, Q., Wang, J., Zhou, Y., January 8, 2019. TransmiR v2.0: an updated transcription factor-microRNA regulation database. Nucleic Acids Res. 47 (D1), D253–D258. https://doi.org/10.1093/nar/gky1023.

Velandia-Huerto, C.A., Yazbeck, A.M., Schor, J., Stadler, P.F., 2022. Evolution and phylogeny of microRNAs—protocols, pitfalls, and problems. miRNomics 211–233.

Wang, J., Lu, M., Qiu, C., Cui, Q., January 2010. TransmiR: a transcription factor-microRNA regulation database. Nucleic Acids Res. 38 (Database issue), D119–D122. https://doi.org/10.1093/nar/gkp803.

Werner, S., Wollmann, H., Schneeberger, K., Weigel, D., 2010. Structure determinants for accurate processing of miR172a in *Arabidopsis thaliana*. Curr. Biol. 20, 42–48. https://doi.org/10.1016/j.cub.2009.10.073.

Williams, P.H., Eyles, R., Weiller, G., 2012. Plant MicroRNA prediction by supervised machine learning using C5.0 decision trees. J. Nucleic Acids 2012, 652979. https://doi.org/10.1155/2012/652979.

Wong, N., Wang, X., 2015. miRDB: an online resource for microRNA target prediction and functional annotations. Nucleic Acids Res. 43 (D1), D146–D152. https://doi.org/10.1093/nar/gku1104.

Wu, Y., Wei, B., Liu, H., Li, T., Rayner, S., 2011. MiRPara: a SVM-based software tool for prediction of most probable microRNA coding regions in genome scale sequences. BMC Bioinf. 12, 107. https://doi.org/10.1186/1471-2105-12-107.

Xuan, P., Guo, M., Huang, Y., Li, W., Huang, Y., 2011. MaturePred: efficient identification of microRNAs within novel plant pre-miRNAs. PLoS One 6, e27422. https://doi.org/10.1371/journal.pone.0027422.

Xue, C., Li, F., He, T., Liu, G.-P., Li, Y., Zhang, X., 2005. Classification of real and pseudo microRNA precursors using local structure-sequence features and support vector machine. BMC Bioinf. 6, 310. https://doi.org/10.1186/1471-2105-6-310.

Yousef, M., Allmer, J., 2014. miRNomics. In: MicroRNa Biology and Computational Analysis. Springer, Humana Press, New York, 978-1-62703-747-1.

Yousef, M., Nebozhyn, M., Shatkay, H., Kanterakis, S., Showe, L.C., Showe, M.K., 2006. Combining multi-species genomic data for microRNA identification using a Naive Bayes classifier. Bioinformatics 22, 1325–1334. https://doi.org/10.1093/bioinformatics/btl094.

Yu, D., Lu, J., Shao, W., Ma, X., Xie, T., Ito, H., Wang, T., Xu, M., Wang, H., Meng, Y., January 1, 2019. MepmiRDB: a medicinal plant microRNA database. Database 2019, baz070. https://doi.org/10.1093/database/baz070.

Yu, D., Wan, Y., Ito, H., Ma, X., Xie, T., Wang, T., Shao, C., Meng, Y., 2019. PmiRDiscVali: an integrated pipeline for plant microRNA discovery and validation. BMC Genom. 20, 133. https://doi.org/10.1186/s12864-019-5478-7.

Yu, T., Xu, N., Haque, N., Gao, C., Huang, W., Huang, Z., 2020. Popular computational tools used for miRNA prediction and their future development prospects. Interdiscipl. Sci. Comput. Life Sci. 12, 395–413. https://doi.org/10.1007/s12539-020-00387-3.

Zhang, Z., Yu, J., Li, D., Zhang, Z., Liu, F., Zhou, X., Su, Z., 2010. PMRD: plant microRNA database. Nucleic Acids Res. 38, 806–813. https://doi.org/10.1093/nar/gkp818.

Zhang, Y., Fan, M., Geng, G., Liu, B., Huang, Z., Luo, H., Zhang, H., 2014. A novel HIV-1-encoded microRNA enhances its viral replication by targeting the TATA box region. Retrovirology 11 (1), 1–15.

Zhang, J., Yang, Y., Yang, T., Yuan, S., Wang, R., Pan, Z., Zhou, W., 2015. Double-negative feedback loop between microRNA-422a and forkhead box (FOX) G1/Q1/E1 regulates hepatocellular carcinoma tumor growth and metastasis. Hepatology 61 (2), 561–573.

Zhang, T., Li, Z., Song, X., Han, L., Wang, L., Zhang, J., Pei, X., 2020. Identification and characterization of micrornas in the developing seed of linseed flax (*Linum usitatissimum* L.). Int. J. Mol. Sci. 21 (8), 2708. https://doi.org/10.3390/ijms21082708.

Zhao, Y., Kuang, Z., Wang, Y., Li, L., Yang, X., 2021. MicroRNA annotation in plants: current status and challenges. Briefings Bioinf. 22. https://doi.org/10.1093/bib/bbab075 bbab075.

Zhong, X., Heinicke, F., Rayner, S., 2019. miRBaseMiner, a tool for investigating miRBase content. RNA Biol. 16 (11), 1534–1546. https://doi.org/10.1080/15476286.2019.1637680.

Zhou, X., Khare, T., Kumar, V., 2020. Recent trends and advances in identification and functional characterization of plant miRNAs. Acta Physiol. Plant. 42, 25. https://doi.org/10.1007/s11738-020-3013-8.

Zhu, J.-K., 2008. Reconstituting plant miRNA biogenesis. Proc. Natl. Acad. Sci. U.S.A. 105, 9851–9852. https://doi.org/10.1073/pnas.0805207105.

Zuker, M., 2003. Mfold web server for nucleic acid folding and hybridization prediction. Nucleic Acids Res. 31, 3406–3415. https://doi.org/10.1093/nar/gkg595.

Index

Note: 'Page numbers followed by *f* indicate figures, *t* indicate tables, and *b* indicate boxes.'

K

L

M

R

S

T

U

V

W

9780443216824